THE ALEXANDRIAN EPITOMES OF GALEN
VOLUME 1

MIDDLE EASTERN TEXTS INITIATIVE

Islamic Translation Series
Al-Ḥikma

NEAL A. MAXWELL INSTITUTE
FOR RELIGIOUS SCHOLARSHIP

BRIGHAM YOUNG UNIVERSITY

The Alexandrian Epitomes of Galen

VOLUME 1

On the Medical Sects for Beginners
The Small Art of Medicine
On the Elements According to the Opinion of Hippocrates

An edition and parallel English translation of three Arabic texts, with notes and introduction, by

John Walbridge

Brigham Young University Press • Provo, Utah • 2014

Library of Congress Cataloging-in-Publication Data

Galen, author.
[Works. Selections. English]
The Alexandrian epitomes of Galen. Volume 1, On the medical sects for beginners ; The small art of medicine ; On the elements according to the opinion of Hippocrates / an edition and parallel English translation of three Arabic texts, with notes and introduction by John Walbridge. — First edition.
p. ; cm. — (Graeco-Arabic sciences and philosophy)
On the medical sects for beginners
Small art of medicine
On the elements according to the opinion of Hippocrates
Includes bibliographical references and index.
Summary: "The second-century physician and philosopher Galen is not known for brevity. About fourteen hundred years ago, one or possibly several professors put together a series of epitomes of Galen's work. This edition presents the Arabic and English versions side by side, with a translation by scholar John Walbridge."—Provided by publisher.
Parallel English and Arabic translation of Greek text.
ISBN 978-0-8425-2840-5 (hardback : alk. paper)
I. Walbridge, John, translator, writer of added commentary. II. Title. III. Title: On the medical sects for beginners. IV. Title: Small art of medicine. V. Title: On the elements according to the opinion of Hippocrates. VI. Series: Islamic translation series
[DNLM: 1. History of Medicine. 2. History, Ancient. WZ 290]

R138
610.938—dc23

2014007461

Printed in the United States of America.

First Edition

To four women who cared for my wife Linda in her last illness:

Mary Lou Mayer, her doctor;

Mary Vandeveld, her sister-in-law;

Mary Edwards and April Sievert, her friends.

Contents

Foreword to the Series xiii

Preface xv

Introduction xix
- The Alexandrian medical curriculum xix
  - The medical school in Alexandria xix
  - The Alexandrian medical syllabus xxiii
  - Alexandria and the Islamicate medical curriculum xxv
- The Alexandrian epitomes xxxii
  - Genre, form, and title xxxii
  - Style and content of the epitomes xxxiii
  - History and authorship of the epitomes xxxiii
  - Possible authors of the epitomes xxxviii
  - Plausibility of the Arabic accounts and dating the epitomes xliii
  - The Arabic translation xliii
  - Galen's three texts and their Alexandrian epitomes xlv
- The edition and translation liii
  - Previous versions, editions, and translations liii
  - Descriptions of manuscripts liv
  - Textual history and editing methods lxiii
  - Other editorial policies lxvi
  - Divisions of the texts lxvii
  - Glosses and scholia lxvii
  - Translation lxix
  - Annotation lxx

Abbreviations and Conventions lxxi

◆ ◆ ◆

Preliminary Glosses 1

Manuscript table of contents 1
The eight heads 3
A gloss on the art of medicine 5

◆ ◆ ◆

The Alexandrian Epitome of Galen's Book *On the Medical Sects* 7
The parts of medicine 7
The sects of medicine 10
Commentary on chapter 1 11
The definition of medicine 11
Commentary on chapter 2 14
Medical experience 14
Commentary on chapter 3 18
The necessary causes 18
The differences between the Empiricists and the Rationalists 23
Commentary on chapter 4 30
The Rationalists' criticism of the Empiricists 30
Commentary on chapter 5 32
The Empiricists' criticism of the Rationalists 32
Commentary on chapter 6 35
The opinions of the Methodists 35
Commentary on chapter 7 37
The differences among the sects 37
Commentary on chapter 8 40
Galen's criticism of the Methodists 40
Commentary on chapter 9 41
The Empiricists' criticism of the Methodists 41
Commentary on chapter 10 44
The Rationalists' criticism of the Methodists 44

◆ ◆ ◆

The Alexandrian Epitome of Galen's Book Known as *The Small Art of Medicine* 50
Introduction 50
Methods of instruction 50
Chapter 1 57
The definition of medicine 57

Chapter 2 66
Bodies 66
Chapter 3 69
Signs 69
Chapter 4 71
The best states of health 71
Chapter 5 77
The genera of the organs 77
Chapter 6 78
The diagnosis of the brain 78
Chapter 7 83
The moderate temperament of the brain 83
Chapter 8 84
Immoderate temperaments of the brain 84
Chapter 9 85
The temperament of the eye 85
The structure of the eye 88
Chapter 10 90
The temperament of the heart 90
Chapter 11 92
Compound temperaments of the heart 92
Chapter 12 93
The temperament of the liver 93
Chapter 13 95
The temperament of the testicles 95
Chapter 14 97
The temperament of the entire body—that is, the flesh 97
Chapter 15 98
Its simple temperaments 98
Chapter 16 99
Its compound temperaments 99
Chapter 17 101
The temperament of the stomach 101
Chapter 18 102
The temperament of the lungs 102
Chapter 19 104
Disorders 104
Chapter 20 106
Diagnosis of diseased states 106

Chapters 21 and 22 110
Signs 110
Chapter 23 111
Causes 111
Chapter 24 113
The causes of health 113
Chapter 25 114
The cure of diseases 114
Chapter 26 118
Classes of organic diseases and their treatment 118
Chapter 27 125
Dissolution of continuity 125
Chapter 28 126
Treatment, prophylaxis, and convalescence 126

◆ ◆ ◆

The Alexandrian Epitome of Galen's Book *On the Elements According to the Opinion of Hippocrates* 130
The eight headings to the epitome of Galen's *On the Elements* 132
Chapter 1 135
The genera of the elements 135
Chapter 2 137
Their disagreement about the elements 137
The difference between the element and the principle 139
The principles of things 139
Chapter 3 141
Whether the elements sense and suffer 141
The occurrence of pain 142
Compounds 142
The tools of inference 144
The refutation of the others' arguments 145
Their views on the elements 146
Their disagreement about the temperament 147
The genera of qualities 148
The compound by contiguity 149
Absurdities 149
A syllogism, premises, and conclusion 151
Chapter 4 153
That the element is not numerically one 153

Chapter 5 156
That human bodies do not come to be from a single humor 156
Chapter 6 159
Hot and cold 159
Chapter 7 161
The states of bodies 161
Alteration of quality and quantity 161
Instruction 162
Chapter 8 164
Compounds 164
Chapter 9 166
The qualities 166
Their disagreement about titles of books on the elements 170
Their disagreement about the temperament 171
Two bodies 173
Chapter 10 174
What is said to be potential 174
Chapter 11 175
Their disagreement about the humors 175
The rest of the drugs 180
The division of the elements 181
Chapters 12 and 13 183
Their disagreement about drugs 183
Chapter 14 185
The location of the humors 185

◆ ◆ ◆

Appendix 1: Greek and Islamicate Physicians 187

Appendix 2: The Three Schools of Medicine 203

Appendix 3: The Structure and Terminology of the Eye in the Epitome of *The Small Art* 209

Arabic-Greek-English Glossary 213

Bibliography 257

Index 269
English 269
Greek 289

Foreword to the Series

Brigham Young University and its Middle Eastern Texts Initiative are pleased to sponsor and publish the Islamic Translation Series (ITS). Islamic civilization represents nearly fourteen centuries of intense intellectual activity, and believers in Islam comprise approximately one quarter of the world's population. The texts that appear in ITS are among the treasures of this great culture. But they are more than that. They are properly the inheritance of all the peoples of the world.

As an institution of The Church of Jesus Christ of Latter-day Saints, Brigham Young University is honored to assist in making these texts available to many for the first time. In doing so, we hope to serve our fellow human beings of all creeds and cultures. We also follow the admonition of our own tradition, to "seek . . . out of the best books words of wisdom," believing, indeed, that "the glory of God is intelligence."

—Daniel C. Peterson
—D. Morgan Davis

Preface

This project has bittersweet associations for me. It began as an outgrowth of research on Stoic influences in Islamic philosophy, a project I have pursued off and on for about fifteen years. In 2001, the American Research Institute in Turkey and the National Endowment for the Humanities were kind enough to fund my research on such issues in Istanbul for a summer, a time I remember with the greatest pleasure. When I looked at the manuscripts of the Alexandrian epitomes, it seemed clear to me that they ought to be published and that at least the five I plan to publish had philosophical interest. I began editing the epitome of *On the Elements According to the Opinion of Hippocrates* in my evenings in the Istanbul hostel of the American Research Institute and continued to work on the project in my idle moments during a Rockefeller Fellowship in the Department of the History of Science at the University of Oklahoma the following fall. That winter, my wife, Linda Strickland Walbridge, was diagnosed with a recurrence of breast cancer. I decided to focus on this project since text editing lent itself to hospitals and doctors' waiting rooms. I also thought it would be a good transitional project that could be finished quickly—a piece of naïveté that I doubtless share with many others who have innocently undertaken to edit a text. Linda, *raḥimahā Allāh*, died the following winter, by which time I had collated most of the manuscripts for two of the three texts presented here. (I had also found out that Galen's three schools of physicians are still very much with us.) Over the next five years, the project progressed slowly as I shouldered heavy departmental responsibilities and gradually put my life back together. At any rate, the first three texts are now finished, and I hope to finish two more—the epitomes of *On the Temperament* and *On the Natural Faculties*—before too long.

◆

I am grateful to acknowledge the following organizations for funding and other support while I was working on this project: the American Research Institute in Turkey, the National Endowment for the Humanities, the Turkish Fulbright Commission, the Rockefeller Foundation, the Department of the History of Science at the University of Oklahoma, Indiana University, the İslâm Araştıralırı Merkezi (İSAM) in Istanbul, and the Guggenheim Foundation.

The publication of this book is funded in part by the Sorenson Legacy Foundation, the creation of the biotechnology pioneer James LeVoy Sorenson and his wife, education philanthropist Beverley Taylor Sorenson. I hope that this book, a testimony to the ancient human enterprise of educating those whose profession it is to fight disease and preserve health, will be worthy of their efforts in medicine and education.

The following libraries supplied photographs of manuscripts used in this edition: the British Library in London, the Süleymaniye Kütüphanesı in Istanbul, and the Manisa İl Halk Kütüphanesı in the lovely town of Manisa, near Izmir. I also used the library facilities of the Süleymaniye, the British Library, the University of Oklahoma, Princeton University, İSAM, and Indiana University, and I am grateful to the generosity and kindness of the librarians at those institutions. Only someone who has worked in the Turkish manuscript libraries—and particularly the Süleymaniye, the greatest Islamic manuscript library in the world—can appreciate the achievement of Turkish librarians over the centuries. The manuscripts that I work with are not beautiful, being doctors' and professors' books, usually in bad handwriting with notes in the margins and on the flyleaves and entirely without elegant illustrations; but learned princes and ministers recognized their value, brought them back as trophies to Constantinople, and saw to it that they would be cared for across the centuries. One of the manuscripts used in this project was copied by a Christian doctor in Acre when it was ruled by the Crusaders. It bears the elegant inscription of the royal librarian who checked it into the library of the Aya Sofya mosque three hundred years ago. It was carefully repaired, probably at that time, and then repaired again sometime in the twentieth century. And it was waiting in the same collection, ready to hand as the royal donor and the long-dead librarian intended, when I called for it at the beginning of the twenty-first century. One stands humbled by the persistence of learning and professionalism represented by the survival and cherishing of these books.

I would like to thank the two graduate students who assisted me on this project, Murat Yılmaz and Naser el-Hujelan, the latter of whom has done the basic collations for the second volume of this project. Their care and enthusiasm have been a great help to me. I would also like to thank a number of individuals who have helped me in various ways, particularly during my stays in Istanbul. The staff of the American Research Institute in Turkey, particularly Tony Greenwood, assisted in a variety of practical ways when I was staying in Istanbul in the summers of 2001 and 2005, as did Nüri Tınaz, my host at İSAM in 2007–2008. I would particularly like to mention the hospitality of Hıdır Metin of the Manisa Library, who, with his assistant Ali Arı, stood on the front steps of his library for a day and a half, patiently turning the pages of manuscripts while I photographed them. Jamil and Sally Ragep were generous hosts at the University of Oklahoma. I would also like to acknowledge the staff at the Middle East Texts Initiative at Brigham Young University: Daniel Peterson and Glen Cooper, who originally commissioned the project; Morgan Davis, the directing editor; Elizabeth Watkins, who meticulously copyedited the English text; Muḥammad Eissa, who carefully reviewed the Arabic; and Andrew Heiss and Jonathan Saltzman for their skillful typesetting.

Mostly, I would like to thank my family for their patience and forbearance as I worked on this project—particularly Linda, who did not live to see the project completed, and Frances Trix, my Turkish interpreter and new bride.

Introduction

About fourteen hundred years ago, one or several professors of medicine in Egypt prepared epitomes in Greek to accompany the sixteen works of Galen that constituted the larger part of the standard syllabus of medicine in the medical schools of Alexandria. These epitomes were study guides, similar to the CliffsNotes of modern American students. In contrast to the rambling and argumentative style of Galen's original works, the epitomes are full of the lists, tables, and systematic categorizations of concepts, symptoms, diseases, and organs that medical students have always had to memorize. Besides the Alexandrian students, others must have found them useful, for they were translated into Arabic—supposedly by the famous Christian translator Ḥunayn ibn Isḥāq—and then into Hebrew. Arabic manuscripts of these epitomes are about as common as manuscripts of the Arabic versions of the corresponding original works of Galen. Citations of them, or signs of their influence, are not difficult to find in Arabic medical literature. Thus they are historically important and worthy of closer study than they have received so far. They are also an admirably clear, if sometimes tedious, survey of Galenism as it was understood at the very end of antiquity.

The Alexandrian medical curriculum

The medical school in Alexandria

Greek medical education probably took place most often in the context of a master-apprentice relationship, commonly within a family. In the Hippocratic oath, the physician vows to pass on the art only to his sons, the sons of his own master, and sworn apprentices. Nevertheless, the oath itself is evidence that this system was changing, at least to the extent that those wishing to acquire a medical education could do so by paying a teacher and attending his classes. In all likelihood, the most common pattern was for aspiring doctors to study locally, either in their family or with local teachers, and then—if their ambition, talent, and financial resources allowed—go on to attend lectures of internationally known teachers at one or more of the famous centers of medicine. This was how Galen himself acquired his education. He was the son of what today we might call a real estate developer, who was able to pay for his son to become a doctor. Galen studied with local teachers in Pergamon and Smyrna before going off for several years of advanced study in Alexandria.[1]

From the time of the Ptolemies up to the fall of the city to the Arabs, Alexandria had a reputation as the best place for advanced medical education. Since we do not know the institutional details of medical instruction, it is wise to resist reifying the "Medical School of Alexandria," as has been done with the "University of Gondeshapur." Certainly, instruction was offered there under something like official auspices, with one or more salaried professors; but beyond that we know little. When Galen was there in the middle of the second century, Alexandria was still living off its reputation for instruction in anatomy, acquired four centuries earlier.[2] We know still less about Alexandrian medical instruction in the century or two before the Arab conquest, when even the names of the teachers are unknown, garbled, or unhelpfully common. The medical textbooks do not say much about actual medical practice, so the social role of doctors and medical students has to be pieced together from fragmentary data—notably from the smug accounts of Christian saints whose miraculous cures humiliated proud pagan doctors.[3]

By the fifth and sixth centuries, the triumph of Galenism in the Roman world was complete.[4] Galen had successfully tilted with his rivals: Empiricists, Methodists, Pneumatists, and other Rationalists.

1. On the role of formal medical instruction, see Duffy, "Byzantine Medicine."

2. Galen, *On Anatomical Procedures*, trans. Singer, 3; cited in Iskandar, "Attempted Reconstruction," 235.

3. Nutton, "From Galen to Alexander," 5–7.

4. On medicine in late antiquity and Byzantium, see Nutton, *Ancient Medicine*; Nutton, "From Galen to Alexander"; Prioreschi, *History of Medicine*, vols. 3–4. Documentation for the period is fragmentary. On the philosophical relevance of Galenic medicine, see Pellegrin, "Ancient Medicine." On the increasing

The professors of medicine in Alexandria had now settled down to expounding a common system based mainly on the works of Galen, with his characteristic polemics largely removed and inconsistencies ironed out by drawing together passages from across his writings. This instruction seems to have had a decidedly bookish quality, leading to the professors being called "iatrosophists." The best evidence for instructional practices in late antique Alexandria comes from Arabic sources, which in turn allow us to make sense of the fragmentary instructional literature in Greek. Ḥunayn ibn Isḥāq reports:

> They read only these books [the sixteen books of the Alexandrian curriculum] in the place of medical instruction in Alexandria. They read them in the order that I have mentioned them, assembling each day to hear one of the authoritative texts read and explained, just as our Christian brethren do to this day in the places of instruction called *schole*, where they read from the book of one of the authorities, either one of the Ancients or one of the other books. [The Alexandrians] would read the specialized books individually after study of these books that we have mentioned, just as our brethren today read commentaries on the books of the Ancients. Galen, however, did not intend for his books to be read in this order. Instead, he specified that after his book *The Medical Sects*, his books on anatomy should be read.[5]

In a chapter on judging the qualifications of physicians, the ninth-century physician Isḥāq ibn ᶜAlī al-Ruhāwī writes:

> Galen undertook to determine each one of these natural principles, without which there can be no real knowledge of the states of the human body. Having distinguished them, he wrote a book about each one of these principles, attributing it to that principle and naming the book by it, seeing that it contained that principle and its branches. He continued in this manner until he had covered all the principles of medicine. Now, the Alexandrians were those most knowledgeable about this art, and when they met and gathered the students of the art of medicine, they realized that most of the young men of their day lacked the enthusiasm to read all of these books, especially those that Galen had composed. Wishing to make the art of medicine accessible

---

tendency for philosophers and medical professors to be the same people, see Westerink, "Philosophy and Medicine."

5. Ḥunayn ibn Isḥāq, "Risālah," 18–19.

> to the students, they selected sixteen of Galen's books, also making epitomes of most of them, seeking greater concision thereby. They introduced them into the *schole*—that is, the place that they had for teaching. Thus it is that nowadays anyone who claims to know the nature of the human body and that he is able to preserve health and treat diseases must know these books well in their order and must have read them with a knowledgeable teacher. Anyone who claims to know them ought to be questioned about the first of them, which is Galen's *On the Medical Sects [for Beginners]*.[6]

The thirteenth-century medical historian Ibn Abī Uṣaybiᶜah says much the same:

> These Alexandrians confined themselves to reading sixteen books of Galen in the medical school in Alexandria (*fī mawḍiᶜ taᶜlīm al-ṭibb bi-al-Iskandarīya*). They read them systematically (*ᶜalá al-tartīb*) and met every day to read and understand a part of them. Then they turned to abridgments and epitomes (*al-jumal wa-al-jawāmiᶜ*) to make it easier to memorize and understand them, each one of them being assigned to comment on the sixteen books.[7]

The medical autodidact ᶜAlī ibn Riḍwān (d. 453/1061) condemned the popularity of the resulting compendia and commentaries for having made them substitutes for the original texts of Hippocrates and Galen:

> In the time of Oribasius, when kingdoms had become dominated by Christianity, Oribasius thought of reviving the art [of medicine] and compiled his popular *Compendium* [*Kunnāsh*] for the laity, thus familiarizing the Christian kings with medicine. Paul [of Aegina] followed his path, and when their successors saw these two compendia, they continued to compile their own up to the present day. Even Abū Bakr al-Rāzī ordered each physician to compile a compendium for himself! Accordingly, medical books became abundant, and each doctor acquired a compendium for his own use. . . .
>
> Galen wrote his commentaries in order to bring the medical works of Hippocrates to perfection. His abstracts and commentaries have left nothing out. Consequently, later books are superfluous, and

6. MS Edirne, Selimiye 1658, 97a–b, reprinted in facsimile in al-Ruhāwī, *Conduct of the Physician*, 193–94; al-Ruhāwī, *Adab al-ṭabīb*, 255–56, trans. M. Levey in al-Ruhāwī, *Medical Ethics of Medieval Islam*, 84; and Gutas, *Greek Thought, Arabic Culture*, 93.

7. Ibn Abī Uṣaybiᶜah, 1:103–4.

> to transcribe or reflect on their contents would hinder students from studying medicine. Studied closely, later compendia and similar works are found to represent the doctrine of the Methodists, whose art was rejected by Galen, for he informed us of their harmful influence on medicine. According to their tradition, the Methodists described each disease, saying that it ought to be treated with a special [group of] drugs. This is precisely what is done by the compilers of compendia, which are, therefore, as harmful to medicine as the doctrine of the Methodists. Summaries and commentaries of Galen's books are not self-sufficient and should not replace his books. Summaries fail to encompass all Galen's ideas, while commentaries increase the length of the art and distract [students] from studying, since, of necessity, these would have to be read for verification together with their [original] medical works.[8]

Ibn Riḍwān belonged to the Great Books school of medicine, and his criticisms are scarcely fair. Students struggling to master established medical theory and busy doctors trying to follow the standards of sound practice might admire classical and innovative works of medical theory, but they needed clear, reliable, and well-organized manuals. These the compendia provided, however unstimulating they might be as reading.

The Alexandrian medical syllabus

The Alexandrian medical curriculum, as it is preserved for us by ᶜAlī ibn Riḍwān, was as follows:

> (1) Liberal arts: intermediate knowledge of grammar, mathematics, astrology, and compounding of drugs
>
> (2) Works of Aristotle on logic: *Categories*; *On Interpretation* (propositions); *Prior Analytics* (syllogism); and *Posterior Analytics* (demonstration)[9]

---

8. ᶜAlī ibn Riḍwān, *al-Kitāb al-nāfiᶜ*, 66, 90, trans. Iskandar in "Attempted Reconstruction," 241–42, slightly adapted. Iskandar cites al-Rāzī to show that what he actually meant was that physicians should keep notes on cases and treatments. A similar account is found in Ibn Jumayᶜ, *Treatise to Ṣalāḥ ad-Dīn*, 18–19. A more sympathetic assessment of the compendia is given by Nutton, *Ancient Medicine*, 295–96.

9. Roueché, "Did Medical Students Study Philosophy?" 153–69, says that apparently they did. Logic was still required for nursing students in my own Indiana University as late as the 1970s.

(3) Works of Hippocrates: *Aphorisms; Prognostics; Regimen in Acute Diseases*; and *Airs, Waters, Places*

(4) Galen's works:

First grade: *On the Medical Sects for Beginners; The Small Art of Medicine; On the Pulse for Teuthras*; and *Therapeutics for Glaucon* (two books)

Second grade: *On the Elements According to the Opinion of Hippocrates; On the Temperament* (three books); *On the Natural Faculties* (three books); and *On Anatomy for Beginners* (five books)

Third grade: *On Diseases and Symptoms* (six books, compiled from Galen's *On the Differentiae of Diseases; On the Causes of Diseases; On the Differentiae of Symptoms*; and *On the Causes of Symptoms*)

Fourth grade: *The Diagnosis of the Diseases of the Internal Organs* (six books) and *The Large Pulse* (sixteen books)

Fifth grade: *The Kinds of Fevers* (two books) and *On Crises* (three books)

Sixth grade: *On the Method of Healing* (fourteen books)

Seventh grade: *The Regimen of the Healthy* (six books)[10]

This yields a coherent curriculum:

(1–2) Premedical: liberal arts and basic sciences; practical pharmacy techniques

(3–4.1) Introductory survey of medicine: short medical classics (four works of Hippocrates); introductory works on differing

10. ʿAlī ibn Riḍwān, *al-Nāfiʿ*, 107–13. Iskandar, "Attempted Reconstruction," 248–52. For more on the sixteen books, see Garofalo, "La traduzione araba"; Garofalo, "I Sommari degli Alessandrini"; Lieber, "Galen in Hebrew, 167–83; Savage-Smith, "Galen's Lost Ophthalmology," 121–38; *GAS* 3:80–98, 140–50; Steinschneider, *Arabischen*, 329–31; Steinschneider, *Hebraeischen*, 654–56; Ullman, *Medizin*, 65–67, 343.

A number of Arabic manuscripts of Galen's works and commentaries on them reflect this list of sixteen works. The numbers in parentheses are the numbers of these works in table 1, in the order that these works appear in each manuscript.

Galen's works: Aya Sofya 3593 (5–7, 3), Aya Sofya 3701 (1–5), Istanbul University A4712 (1–4), Sarai Ahmet III 2110 (7–9), Paris 2860 (1–4), Princeton Garrett 1075 (12–15), Florence Laurent. 235 (1, 2, 8–12?), Tehran Majlis 521 (1–6), Majlis 3974

scientific methodologies in medicine; introductions to the clinical application of humoral theory, diagnosis by the pulse, and therapeutics

(4.2) Scientific foundations of medicine: elements; temperaments; normal functions of the body; anatomy

(4.3) Diseases and their diagnosis

(4.4) Advanced diagnosis of internal diseases; advanced diagnosis by the pulse

(4.5) Fevers; expected course of diseases

(4.6) Advanced therapeutics

(4.7) Advanced prophylactic care

This curriculum is corroborated from other sources—particularly Arabic—and is thoroughly Galenic in spirit, as attested by the stress on the importance of a sound grounding in philosophy, logic, and science; the presence of Hippocratic works favored by Galen; and, naturally, the predominance of Galen's own works. Moreover, most of the same books, along with other similar works, are found in a similar order in the suggestions for further reading at the end of *The Small Art*.[11] From our point of view, what is most important is that the works of Galen are those found in the Alexandrian epitomes, usually in that order or close to it.

Alexandria and the Islamicate medical curriculum

The Alexandrian curriculum exercised a strong but not stultifying effect on the medical curriculum in the medieval Islamic world. As with the Alexandrians, we know surprisingly little about actual medical instruction; again, what we know best are the textbooks and reference works (see table 1). Here, as in philosophy, Ibn Sīnā played a key role as systematizer. The enduring popularity of his *Canon of Medicine* no doubt

(9, 10, 3, 11, 14, 12, 13), Majlis 6400 (14, 12, 13), Tehran University Med. Fac. 291 (8, 9), Escurial 797 (14, 13, 12), Escurial 799 (5, 6), Escurial 848 (6, 7), Madrid Bib. Nac. 130 (5, 6). Commentaries: Yaḥyā al-Naḥwī, *Ikhtiṣār* British Library Or. 17 (1–14, 16, 15). Ibn al-Ṭayyib, *Tafsīr* Manisa 1772 (1–4).

See also the schematic tables in MS Vienna med. gr. 16 containing Greek tabular summaries of works 1, 2, 4, 3, 9; Hunger and Kresten, *Katalog der griechischen Handschriften*, 2:60–62; and Gundert, "*Tabula Vindobonenses*."

11. K 1:407–12; G[a] 181–94.

Table 1. The sixteen books and other ancient texts studied in Alexandria

	English Title	Greek Title	Latin Title	Arabic Title	Editions or Manuscripts
	Works of Galen in the Alexandrian Curriculum				
1	*On the Medical Sects for Beginners* (hereafter *The Medical Sects*)	Περὶ αἱρέσεων τοῖς εἰσαγομένοις	*De sectis ad eos qui introducuntur*	*Firaq al-ṭibb li-l-mutaʿallimīn*	K 1:64–105; ed. Helmreich; trans. Frede
2	*The [Small] Art of Medicine* (hereafter *The Small Art*)	Τέχνη ἰατρική	*Ars medica; Ars parva; Microtechne; Tegni*	*al-Ṣināʿah al-ṣaghīrah fī al-ṭibb*	K 1:305–412; ed., trans. Boudon-Millot; trans. Singer
3	*On the Pulse for Teuthras*	Περὶ σφυγμῶν τοῖς εἰσαγομένοις	*De pulsibus ad tirones*	*Fī al-nabḍ al-ṣaghīr ilá Ṭuthrūn*	K 8:453–92
4	*Therapeutics for Glaucon*	Πρὸς Γλαύκωνα θεραπευτικῶν	*Ad Glauconem de methodo medendi*	*[Fī mudāwāt al-amrāḍ] ilá Ighlīqūn*	K 11:1–146
5	*On the Elements According to the Opinion of Hippocrates* (hereafter *On the Elements*)	Περὶ τῶν καθ' Ἱπποκράτην στοιχείων	*De elementis ex Hippocrate, De elementis secondum Hippocratem*	*Fī al-isṭiqsāt ʿalá raʾy Ibbuqrāṭ*	K 1:413–508; ed., trans. De Lacy
6	*On the Temperament*	Περὶ κράσεων	*De temperamentis*	*Fī al-mizāj*	K 1:509–65; trans. Singer
7	*On the Natural Faculties*	Περὶ φυσικῶν δυνάμεων	*De facultatibus naturalibus*	*Fī al-quwá al-ṭabīʿīyah*	K 2:1–124; ed., trans. Brock
8	*On Anatomy for Beginners*, comprising the following five books:			*Fī al-tashrīḥ li-l-mutaʿallimīn*	

	English Title	Greek Title	Latin Title	Arabic Title	Editions or Manuscripts
a	*Anatomy of the Bones*	Περὶ ὀστῶν τοῖς εἰσαγομένοις	*De ossibus ad tirones*	*Tashrīḥ al-*^{c}*iẓām*	K 2:732–78
b	*Anatomy of the Muscles*	Περὶ μυῶν ἀνατομῆς	*De musculorum dissectione*	*Tashrīḥ al-*^{c}*aḍal*	K 18B:926–1026
c	*Anatomy of the Nerves*	Περὶ νεύρων ἀνατομῆς	*De nervorum dissectione*	*Tashrīḥ al-*^{c}*aṣab*	K 2:831–56; ed., trans. Al-Dubayan
d	*Anatomy of the Veins*	Περὶ φλεβῶν καὶ ἀρτηριῶν ἀνατομῆς	*De venarum arteriarumque dissectione*	*Tashrīḥ al-*^{c}*urūq al-ghayr al-ḍawārib*	K 2:779–830
e	*Anatomy of the Arteries*			*Tashrīḥ al-*^{c}*urūq al-ḍawārib*	
9	*On Diseases and Symptoms*, comprising the following four books:			*Fī al-*^{c}*ilal wa-al-a*^{c}*rāḍ*	
a	*On the Differentiae of Diseases*	Περὶ διαφορᾶς νοσημάτων	*De differentiis morborum*	*Fī aṣnāf al-amrāḍ*	K 6:836–80; trans. Johnston
b	*On the Causes of Diseases*	Περὶ τῶν ἐν τοῖς νοσήμασιν αἰτίων	*De causis morborum*	*Fī asbāb al-amrāḍ*	K 7:1–41; trans. Johnston
c	*On the Differentiae of Symptoms*	Περὶ τῆς τῶν συμπτωμάτων διαφορᾶς	*De symptomatum differentiis*	*Fī aṣnāf al-a*^{c}*rāḍ*	K 7:42–84; trans. Johnston
d	*On the Causes of Symptoms*	Περὶ αἰτίων συμπτωμάτων	*De symptomatum causis*	*Fī asbāb al-a*^{c}*rāḍ*	K 7:85–272; trans. Johnston

	English Title	Greek Title	Latin Title	Arabic Title	Editions or Manuscripts
10	*The Diagnosis of Diseases of the Internal Organs*, also known as *On Affected Parts*	Περὶ τῶν πεπονθότων τόπων	*De locis affectis*	*Fī taʿarruf ʿilal al-aʿḍāʾ al-ālimah (al-Mawāḍiʿ al-ālimah, ʿIlal al-aʿḍāʾ al-bāṭinah)*	K 8:1–452; trans. Siegel
11	*The Large Pulse*, comprising the following books:	Περὶ τῶν σφυγμῶν πραγματεία	*De pulsibus*	*al-Nabḍ al-kabīr (Fī al-nabḍ)*	
a	*The Kinds of Pulse*	Περὶ διαφορᾶς σφυγμῶν	*De differentia pulsuum*	*Aṣnāf al-nabḍ*	K 8:493–765
b	*Diagnosis by the Pulse*	Περὶ διαγνώσεως σφυγμῶν	*De dignoscendis pulsibus*	*Taʿarruf al-nabḍ*	K 8:766–961
c	*Causes of the Pulse*	Περὶ τῶν ἐν τοῖς σφυγμοῖς αἰτίων	*De causis pulsuum*	*Asbāb al-nabḍ*	K 9:1–204
d	*Prognosis by the Pulse*	Περὶ προγνώσεως σφυγμῶν	*De praesagitione ex pulsibus*	*Taqdimat al-maʿrifah min al-nabḍ (Sābiq al-ʿilm bi-mā yadillu ʿalayhi al-nabḍ)*	K 9:205–430
12	*On Crises*	Περὶ κρίσεων	*De crisibus*	*al-Buḥrān*	K 9:550–768
13	*Days of Crisis*	Περὶ κρίσεμων ἡμερῶν	*De diebus decretoriis*	*Ayyām al-buḥrān*	K 9:769–941
14	*The Kinds of Fevers*	Περὶ διαφορᾶς πυρετῶν	*De typis febrium*	*Aṣnāf al-ḥummayāt*	K 7:273–405
15	*The Method of Healing*	Θεραπευτικὴ μέθοδος	*De methodo medendi*	*Ḥīlat al-burʾ*	K 10:1–1021; ed., trans. Johnston and Horsley

	English Title	Greek Title	Latin Title	Arabic Title	Editions or Manuscripts
16	*The Regimen of the Healthy*	Ὑγιεινῶν λόγοι	*De sanitate tuenda*	*Tadbīr al-aṣiḥḥāʾ*, *Fī al-ḥīlah li-ḥifẓ al-ṣiḥḥah*	K 6:1–452
			Other Works of Galen		
	On the Constitution of the Art of Medicine [?]	Πρὸς Πατρόφιλον περὶ συστάσεως ἰατρικῆς	*De constitutione artis medicae ad Patrophilum*	*Fī ithbāt al-ṭibb*	K 1:224–304 [?]; not positively identified; see p. 56, n. 25, below
	On Antecedent Causes	Περὶ τῶν προκαταρκτικῶν αἰτίων	*De causis procatarcticis*	*Fī al-asbāb al-muttaṣilah fī al-maraḍ*	ed. Kalbfleisch, trans. Lyons
	On the Humors	Περὶ χυμῶν?	*De humoribus*	*Fī al-akhlāṭ*	Unidentified; cf. *GAS* 3.130
	On the Doctrines of Hippocrates and Plato	Περὶ τῶν Ἱπποκράτους καὶ Πλάτωνος δογμάτων	*De placitis Hippocratis et Platonis*	*Fī ārāʾ Buqrāṭ wa-Falāṭūn*	K 5:181–805; ed., trans. De Lacy
	On Medical Experience	Περὶ τῆς ἰατρικῆς ἐμπειρίας	*De experientia medica*	*Fī al-tajribah al-ṭibbīyah*	ed. Walzer, trans. Frede
	An Outline of Empiricism	Ὑποτυώσεις ἐμπειρικαί	*De subfiguratio empirica*	*Fī jumal al-tajribah*	ed. Deichgräber, trans. Frede
	On the Uses of the Parts of the Human Body	Περὶ χρείας τῶν ἐν ἀνθρώπου σώματι μορίων	*De usu partium corporus humani*	*Manāfiʿ al-aʿḍāʾ*	K 3–4, ed. Helmreich, trans. Tallmadge

English Title	Greek Title	Latin Title	Arabic Title	Editions or Manuscripts
		Works of Hippocrates		
Aphorisms	Ἀφορισμοί	*Aphorismi*	*al-Fuṣūl*	Littré 4:458–609
On the Nature of Man	Περὶ φύσιος ἀνθρώπου	*De natura hominis*	*Ṭabīʿat al-insān*	Littré 6:32–68
Prognostics	Προγνωστικόν	*Prognosticon*	*Taqdimat al-maʿrifah*	Littré 2:140–90
Regimen in Acute Diseases	Περὶ διαίτης ὀξέων	*Regimen acutorum*	*Tadbīr al-amrāḍ al-ḥāddah*	Littré 2:224–376
Airs, Waters, Places	Περὶ ἀέρων, ὑδάτων, τόπων	*De aëre aquis et locis*	*al-Ahwiyah wa-al-azminah wa-al-miyāh wa-al-buldān*	Littré 2:12–39
		Works of Aristotle		
Categories	Κατηγορίαι	*Categoriae*	*al-Maqūlāt*	Bekker 1–15
On Interpretation	Περὶ ἑρμηνείας	*De interpretatione*	*al-ʿIbārah*	Bekker 16–24
Prior Analytics	Ἀναλυτικὰ πρότερα	*Analytica priora*	*al-Qiyās*	Bekker 24–70
Posterior Analytics	Ἀναλυτικὰ ὕστερα	*Analytica posteriora*	*al-Burhān*	Bekker 71–100
Physics	Φυσική	*Physica*	*al-Ṭabīʿah*	Bekker 184–224
On the Heavens	Περὶ οὐρανοῦ	*De caelo*	*al-Samāʾ*	Bekker 268–313
On Genesis and Corruption	Περὶ γενέσεως καὶ φθορᾶς	*De generatione et corruptione*	*al-Kawn wa-al-fasād*	Bekker 314–38

owes as much to its clarity and excellent organization as to its reliability. There are five books: (1) medical theory; (2) simple drugs, arranged alphabetically with an introduction on their preparation and use; (3) diseases of particular organs, arranged from head to extremities; (4) systemic diseases, such as fevers; and (5) compound drugs. The first book, entitled *General Matters (al-Umūr al-kullīyah)*, is what corresponds to our curriculum. Ibn Sīnā begins with the definition and subject matter of medicine, which correspond to the beginning of *The Small Art* and, to a lesser degree, the beginning of *The Medical Sects*. He then moves immediately to the elements, the temperaments, the humors, basic anatomy, and bodily faculties, corresponding to *On the Elements, On the Temperament, On Anatomy for Beginners*, and *On the Natural Faculties*, which form section 4.2 of the curriculum and the fifth through eighth books of the Alexandrian epitomes. The chapter on anatomy is arranged in the same order as the epitomes. The next major subsection deals with diseases, causes, and symptoms, corresponding to *On Diseases and Symptoms* and *The Diagnosis of Diseases of the Internal Organs*; the pulse, corresponding to *The Large Pulse*; and diagnosis by urine and feces, which has no counterpart in the Alexandrian curriculum but which is important in Galenic medicine and is mentioned among the subjects for additional reading in *The Small Art*. Together, these form sections 4.3 and 4.4 of the Alexandrian curriculum. The next section deals with the care of healthy patients, corresponding to *The Regimen of the Healthy*, which is section 4.7. The chapter closes with general therapeutics, corresponding either to the general parts of *On the Method of Healing*, which is section 4.6, or perhaps *Therapeutics for Glaucon*, which was covered in 4.1. At any rate, the treatment of particular diseases is covered in detail in the third and fourth books of the *Canon*. Ibn Rushd's *General Matters* follows the same outline, though it omits the introductory sections on elements, temperaments, and humors.[12] A similar outline prevails in other Islamic medical textbooks.

12. Ibn Rushd, *al-Kullīyāt fī al-ṭibb*; Ibn Rushd, *Kitāb al-kullīyāt*; Ibn Rushd, *Medical Manuscripts*. He also wrote epitomes of seven of the Galenic works included in the Alexandrian epitomes—*On the Elements, On the Temperament, On the Natural Faculties, The Kinds of Fevers, On Diseases and Symptoms, The Regimen of the Healthy,* and *Method of Healing*—along with epitomes of one or two other medical works of Galen; see Ibn Rushd, *Rasāʾil*.

The Alexandrian epitomes

We know the Alexandrian epitomes mostly through their manuscripts, which, unfortunately, do not tell us much about their origins, purposes, authorship, or histories.

Genre, form, and title

Jawāmi^{c} is a common term for various sorts of abridgments, epitomes, short commentaries, introductions, and study guides based on some other text read by students. The root meaning of the Arabic term is "to collect," and the term is used as a collective plural. Whatever the precise meaning of *jawāmi*^{c} might be, it was commonly used for short expositions of famous philosophical and medical texts. Some sixty *jawāmi*^{c} of Greek medical texts are known by mention or manuscript in Arabic. How they differ from other, similar kinds of short works such as *ikhtiṣār, mukhtaṣar, talkhīṣ* (all of which mean "abridgment"), and *thimār* (literally, "fruits") is not clear. The term *jawāmi*^{c} only tells us that these are comparatively concise works intended to aid in reading the books they comment on.

Some of the manuscript title pages of the Alexandrian epitomes add ^{c}*alá al-sharḥ wa-al-talkhīṣ* (by means of commentary and abridgment). This expression clearly had a technical sense and does seem to be a good description, since—as the notes to the translations will show—the epitomes are divided into short sections that typically either elaborate on some point that Galen mentions briefly or in passing or else provide a clarifying summary of Galen's original text. This explains how an "epitome" can be only slightly shorter than the original text and thus "short" only in comparison to a full-scale commentary.

I have used the term "epitome" to render *jawāmiʿ* following the rendering used for Ibn Rushd's *jawāmiʿ* of Aristotle's works. Our epitomes have been most commonly known to scholarship as the *Summaria Alexandrinorum* (Alexandrian summaries). I have avoided this, both in deference to the venerable tradition of the Latin Averroes translations and because this book is already burdened with three languages—Arabic, English, and the presumed Greek—and there seemed no particular reason to give a Latin title to a book that never existed in Latin. In any case, "summary" seemed a particularly bad term for what is going on in the epitomes. Fuat Sezgin speculates that the Greek title was Συνόψεις τῶν Ἀλεξανδρεῶν (Alexandrian synopses), which could well be right, though there is no direct evidence.[13]

Style and content of the epitomes

As a rule, the epitomes edited and translated here do not make continuous arguments. Instead, they list and categorize topics that Galen had dealt with in a less organized fashion, or they restate and explain particular points in Galen's text. Reading the epitomes side by side with Galen's texts makes it perfectly obvious why they were written. Galen meanders, interrupts himself to expand on some side issue, partially develops alternative categorizations, interrupts himself again to savage at length some opponent over a long-forgotten theoretical dispute, and in the end convinces the reader that Galen's vast literary output was achieved by not wasting time on revision. *"Musawwadatuhu mubayyaḍatuhu"* (His rough draft was his fair copy), as a biographer remarked about a particularly long-winded Islamic scientist. The epitomes give the student reader the background he needs to understand the text in a clear and comprehensive manner.

History and authorship of the epitomes

Internal evidence. Each text starts with some variant of "The Alexandrians' epitome of Galen's book x, translated by Ḥunayn ibn Isḥāq," and there is usually a similar colophon. Both vary considerably, so presumably they were not part of the original text. There are glosses by Ibn al-Tilmīdh and others (of which more is said below). One manuscript has introductory glosses to the collection as a whole and to most of the individual books that it contains; but these glosses do not say anything of bibliographical interest, confining themselves to such worthy but unhelpful topics as the order in which Galen's books should be read by the student.

---

13. *GAS* 3:249. Though there are medical works from late antiquity with this title, I have not identified a case where a Greek work entitled σύνοψις is translated as *jawāmiʿ*, but I know of one such work whose title is translated as *jumlah* and another as *jumal*; see Ḥunayn ibn Isḥāq, "Risāla," item 66, *"Jumlat kitābihi al-kabīr fī al-nabḍ,"* and item 111, *"Fī jumal al-tajribah."*

Greek literary evidence and parallels. So far as I know, not one of these texts, in full or in part, exists in Greek or in Latin translation; nor do we have clear references to them or to all the individuals who are mentioned as their authors in Arabic sources. What we do have are examples of similar texts, usually in fragmentary form, with ambiguities of authorship, date, authenticity, and relationship to other similar texts. The nonspecialist attempting to make use of this literature is quickly disabused of any illusion that classicists have done everything there is to do in their field. For earlier periods, generally speaking, only classic works have survived, while the messy notebooks, lectures, and half-finished drafts have perished. For the late Alexandrian period, though, a substantial body of miscellaneous medical literature survives, much of it in the form of commentaries on works of Hippocrates and Galen taught in Alexandria, including several on *The Medical Sects.* These commentaries show varying degrees of completeness and polish, ranging from fragmentary notes on lectures to carefully composed expositions. Because they discuss the same texts for the same classes, they tend to overlap so much that it is not always clear whether they are variations of the same work or different works drawing on the same sources. There are usually questions of authorship, with manuscripts giving ambiguous identifications. In addition, these works survive in Greek, Latin, and Arabic in manuscripts copied—and, occasionally, printed—over a thousand-year period, not always by people who knew what they really were. Finally, many of these works, including almost all those in Arabic, have yet to be edited, studied, or translated into a modern language. This means that even in the occasional case where there happens to be a modern edition and study, it is not yet possible to be sure how that particular work relates to the rest of the surviving literature.

Nevertheless, it is clear that the Alexandrian epitomes are typical of the late antique study guides composed in the form of commentaries, questions and answers, tables, and diagrams for the use of medical students. Similar works also exist for other disciplines, notably philosophy. Such works continued to be written in the Islamic world until works such as Ibn Sīnā's *Canon* supplanted the actual works of Hippocrates and Galen for student use.

Arabic literary evidence. Arabic sources are much more helpful than Greek sources in understanding Alexandrian medical instruction, providing descriptions of methods of instruction and lists of books studied.

In particular, they stress the role of epitomes, short commentaries, and compendia as teaching tools. The most specific account of the origin of the epitomes comes from the thirteenth-century medical historian Ibn al-Qifṭī in a biography of Anqīlāʾus:

> Anqīlāʾus was a learned physician and natural philosopher, an Egyptian who lived in Alexandria. He was one of the Alexandrians who undertook to epitomize Galen's teaching and abridge his books. The excellence of their abridgments, composed as questions and answers, indicates their knowledge of the whole of Galen's teaching (*jawāmiʿ al-kalām*) and their mastery of the art of medicine. This Anqīlāʾus was their chief. He was the one who collected from Galen's scattered remarks thirteen books on the secrets of movements, which he compiled for him who has sexual intercourse and in which he mentioned when it is indicated [reading *yadillu* with Ṣāʿid] and how to prevent its harm.[14] This Anqīlāʾus compiled the books and wrote most of them. For this reason, most people attributed the epitomes to him. Ḥunayn ibn Isḥāq mentioned this in his translation of them from Greek into Syriac.
>
> The Alexandrians were those who administered the school (*dār al-ʿilm*) and the classes in Alexandria. They read Galen's books and assigned what was to be read each day. They prepared commentaries and epitomes on them, summarizing their contents and making it easier for the reader to memorize them and to summarize[15] them in notebooks. According to what Isḥāq ibn Ḥunayn wrote, the first of them was Stephanus of Alexandria, then Gesius, Anqīlāʾus, and

---

14. In both Ibn al-Qifṭī and Ṣāʿid al-Andalusī, this sentence reads: "Wa-kāna Anqīlāʾus hādhā raʾisahum wa-huwa ʾlladhī jamaʿa min manthūr Jālīnūs thalāth ʿashar maqālah fī asrār al-ḥarakāt allafahā fī man jāmaʿa wa-bihi ʿilla muzminah wa-dhakarah mā yūlidu ["yadillu," in Ṣāʿid] ʿalayhi dhālika wa-mā yudfaʿu bihi ḍararuhu." Given the recurrence of the root j-m-ʿ, it seems more likely that there is a problem with the text caused by additional biographical tidbits being inserted and that the meaning is something like "This Anqīlāʾus was their leader. (He was the one who epitomized thirteen books on the secrets of movements from throughout Galen's writings.) He collected them from those who had made epitomes. He had a chronic illness and discussed what caused it and what prevented its harm." For one thing, thirteen books would be almost a thousand pages in Kühn, nearly as long as *The Method of Healing*, which seems like rather a lot for this particular topic, important though it may be for men of a certain age.

15. Reading *jaml* for *ḥaml*.

> Marinus. These four were the chiefs of the Alexandrian physicians, and they were the ones who prepared the epitomes and commentaries. Anqīlāʾus was the one who put them in order in their final form, as was explained before.[16]

Ibn al-Qifṭī has little more to say about these individuals other than that Yaḥyā al-Naḥwī was associated with them, that Stephanus of Alexandria wrote a commentary on the categories, and that Anqīlāʾus's name was sometimes given as Niqulāʾus.[17] A similar account is given in Ibn Juljul's *Classes of Physicians and Philosophers:*

> When the sovereignty of Jesus Christ was manifested and his call spread throughout the lands, victorious everywhere, a group of expert philosophers appeared in Alexandria. They examined carefully the contents of the ancient books that they had found. They abridged all of Galen's works, laying out their contents in summaries and epitomes (*al-jumal wa-al-jawāmiʿ*) in order to make it easier for them to memorize and understand them but without altering their content. The translator Ḥunayn found these books both in the original and in the form of epitomes (*ʿalā al-aṣl wa-al-jawāmiʿ*), in which form they exist up to this day. The chief of the Alexandrians was Anqīlāʾus of Alexandria.[18]

The *Fihrist* of Ibn al-Nadīm alludes to the epitomes but without its usual bibliographic detail:

> Isḥāq ibn Ḥunayn said, "There were 815 years between the death of Galen and the year 290 of the Hijra [903 CE], the year in which the discussion between Ibn Firās and Ibn Shimʿūn took place.[19] The physicians who have been mentioned from the days of Galen up to this year were Stephanus the Alexandrian, Gesius the Alexandrian, Anqīlāʾus the Alexandrian, and Marinus the Alexandrian. These four Alexandrians were among those who commented on the books of Galen and epitomized

---

16. Al-Qifṭī, *Tārīkh al-ḥukamāʾ*; al-Andalusī, *Kitāb ṭabaqāt al-umam*, ed. Cheikho, 40; al-Andalusī, *Kitāb ṭabaqāt al-umam*, ed. ʿAlwān, 109–110; al-Andalusī, *Kitāb ṭabaqāt al-umam*, ed. Muʾnis, 54–55; al-Andalusī, *Science in the Medieval World*, 37. The thirteen books on movement are not known to survive, either under Anqīlāʾus's name or under Galen's. See also the passages cited in connection with Alexandrian medical education on pp. xxi–xxiii above.

17. Al-Qifṭī, *Tārīkh al-ḥukamāʾ*, 35, 356; cf. *GAS* 3:160–61.

18. Ibn Juljul, *Ṭabaqāt al-aṭibbāʾ*, 51.

19. If solar years are meant, this would put Galen's date of death in 88 CE, and in about 113 if lunar years are meant. Both are about a century too early.

and abridged them, making their texts concise, especially the sixteen books."[20]

A more thorough list is given by the eleventh-century Baghdad Christian physician Ibn Buṭlān in his controversy with ᶜAlī ibn Riḍwān:

> That would be more useful and rewarding for [Ibn Riḍwān] than criticizing in front of the young people that which was written by the Alexandrians in their commentaries and epitomes to the sixteen books. [The Alexandrians] were Stephanus, Marinus, Gesius, Arkīlāᵓus, Anqīlāᵓus, Palladius, and John the Grammarian [Philoponus]. Perhaps he may be excused for not knowing their names correctly in Arabic. These were the commentators on the books of the art of medicine. I fail to understand how he can blame them for producing epitomes of the books in which they commented on the original texts and explained their content.[21]

Ibn Buṭlān is also quoted in a passage by Ibn Abī Uṣaybiᶜah:

> Al-Mukhtār ibn Ḥasan Ibn Buṭlān said, "There were seven Alexandrians who epitomized the sixteen books of Galen and commented on them. They were Stephen, Gessius, Theodosius, Akīlāᵓus, Anqīlāᵓus, Palladius, and John the Grammarian. They were all Christians. It is said that Anqīlāᵓus of Alexandria was the first among the Alexandrians and that he was the one who compiled the sixteen books of Galen."[22]

Finally, a gloss in MS Aya Sofya 3588, f.2a, mentions eight "commentators": the seven mentioned by Ibn Abī Uṣaybiᶜah's quotation from Ibn Buṭlān, with the addition of Abū al-Faraj ibn al-Ṭayyib, a commentator of the Islamic period.

These accounts differ somewhat in detail, but the sources generally portray a process leading to the compilation of the epitomes rather than a single event—a literary trend to produce such works culminating in the Alexandrian epitomes that we now have in Arabic.

---

20. Isḥāq ibn Ḥunayn, *Tārīkh al-aṭibbāᵓ*, 69, 79. The passage is paraphrased by Ibn al-Nadīm, *al-Fihrist*, ed. Sayyid, 2.1.283; Ibn al-Nadīm, *al-Fihrist*, ed. Tajaddud, 351; Ibn al-Nadīm, *The Fihrist*, trans. Dodge, 2:689. The passage is quoted without citation by Ibn Abī Uṣaybiᶜah, 1:103.

21. Schacht and Meyerhof, *Medico-Philosophical Controversy*, 93–94 (Eng.), 59–60 (Arab.); my translation.

22. Ibn Abī Uṣaybiᶜah, 1:103.

Possible authors of the epitomes

There are nine individuals named in one source or another as associated with the production of the Alexandrian epitomes:

Table 2. Supposed authors of the Alexandrian epitomes in Arabic sources

Name	**Date**	Ḥunayn ibn Isḥāq, cited by Ibn al-Qifṭī, 71	Isḥāq ibn Ḥunayn, 69, 79; Ibn al-Nadīm, *al-Fihrist*, 2:283; Ibn al-Qifṭī, 71, 356; Ibn Abī Uṣaybiʿah, 1:103	Ibn Buṭlān, *Medico-Philosophical*, English 93–94, Arabic 59–60	Ibn Buṭlān, cited by Ibn Abī Uṣaybiʿah, 1:103	S^h: Aya Sofya 3588 2a Gloss "commentators" (*mufassirū*)	Ibn Juljul, 51; Ibn Ṣāʿid, 40
Gesius [of Petra] (Jāsiyūs)	d. ca. 520		◆	◆	◆	◆	
Anqilāʾus		◆*	◆	◆	◆*	◆	◆*
Akilāʾus					◆		
Arkīlāʾus				◆		◆	
Palladius [of Alexandria] (Fallādiyus)	ca. 550–600			◆	◆	◆	
John Philoponus (Yaḥyā al-Naḥwī)	ca. 490–570			◆	◆	◆	
Stephanus [of Athens or Alexandria] (Isṭafan)	6th–7th century		◆	◆	◆	◆	
Theodosius (Thāwdhusiūs)					◆	◆	
Marinus (Mārīnūs)			◆	◆			
Abū al-Faraj ibn al-Ṭayyib	d. 1043					◆	

* Identified as leader of the group

As can be seen from the table and the passages cited earlier, there are three accounts of the authorship. The first, apparently deriving from a remark by Ḥunayn ibn Isḥāq in his translation of the epitomes into Syriac, mentions only Anqīlāʾus, identifying him as the one who put them into their final form. The second, deriving from the *History of the Physicians*, by Ḥunayn's son Isḥāq, and quoted by Ibn al-Nadīm, Ibn al-Qifṭī, and Ibn Abī Uṣaybiʿah, mentions four individuals, one of whom—Marinus of Alexandria—is not mentioned elsewhere. The third version, deriving from Ibn Buṭlān, mentions seven, three of whom are also in Isḥāq's list, though Ibn Abī Uṣaybiʿah's quotation of Ibn Buṭlān substitutes Theodosius for Marinus.

Gesius (or Gessius) of Petra, who died about 520, was a well-known professor of medicine in Alexandria. He studied philosophy under Ammonius of Alexandria (ca. 440–520), the Neoplatonic commentator on Aristotle who also taught Philoponus. Gesius was more interested in medicine than philosophy and became wealthy and famous.[23] Ibn Abī Uṣaybiʿah mentions that "the best of [the Alexandrian commentaries and epitomes] is the commentary of Gesius on the sixteen books, for he shows learning and discernment in it," a judgment confirmed by the *Suda*, a tenth-century Byzantine encyclopedia.[24] None of his works are extant, with the possible exception of the Latin commentary on *The Medical Sects* otherwise attributed to Agnellus, which is attributed to Gesius in one manuscript.[25]

Anqīlāʾus, whose name is also given as Nīqulāʾus, is mentioned by all the sources that give names of the compilers of the epitomes, some specifically naming him as the leader of the group. Vivian Nutton alluded to the possibility that he might be the Archelaos or Arch[. . .]des who was the author of an extant Greek commentary on *The Medical*

23. Bayard Dodge identifies "Jāsīyūs" as Cassius Felix (or Iatrosophista) and dates the latter to the early fifth century. He was the obscure author of a small work on difficult medical questions and is usually dated between the first and third centuries. I do not find this suggestion convincing. Ibn al-Nadīm, *The Fihrist*, trans. Dodge, 2:689, 976.

24. Ibn Abī Uṣaybiʿah, 1:104. Adler, *Suidae lexicon*, 1:520–21, s.v. "Gesios"; http://www.stoa.org/sol-entries/gamma/207.

25. Nutton, "John of Alexandria Again," 510.

Sects.[26] Second, Nutton also considered the possibility that he is the Angeleuas mentioned by Stephanus as an early commentator on Galen.[27] He could be Agnellus of Ravenna, a sixth-century iatrosophist to whom a Latin commentary on *The Medical Sects* is attributed in one of its two manuscripts.[28] Finally, the name could be a misreading of Asclepius, but this seems unlikely because Asclepius was a familiar name and thus unlikely to be completely distorted. Moreover, the form Anqīlāʾus appears in what seem to be independent sources.[29] Thus, barring the discovery of a new Greek source, any attempt to identify Anqīlāʾus is mere speculation.

Arkīlāʾus occurs twice—in Ibn Buṭlān's controversy with Ibn Riḍwān and in a gloss in manuscript **S**. It appears in the form "Akīlāʾus" in the quotation from Ibn Buṭlān in Ibn Abū Uṣaybiʿah; but since we have a direct text from Ibn Buṭlān with the spelling Arkīlāʾus, corroborated by the gloss in **S**, speculations based on the form Akīlāʾus can be dismissed. [30] Thus, the Greek name must be Archelaus, as Meyerhof and Schacht long ago suggested.[31]

Yaḥyā al-Naḥwī, John the Grammarian, is a notorious problem. He is conventionally identified with John Philoponus, otherwise known as John the Grammarian or John of Alexandria, a major philosopher and scientist of the sixth century (ca. 490–570).[32] Ibn Abī Uṣaybiʿah, citing

---

26. Nutton, "John of Alexandria Again." For the editions of the papyrus and MS containing this commentary, see Nachmanson, "Ein neuplatonischer"; Baffioni, "Inediti di Archelao." A survey of the issues relating to the identification of Anqilāʾus is found in Irvine and Temkin, "Akīlāōs."

27. Dickson, *Stephanus the Philosopher*, 35.

28. Cf. Nutton, "John of Alexandria Again," 511. *BNP* 1:345. The text is published as Agnellus of Ravenna, *Lectures*.

29. Wolska-Conus, "Commentaires de Stéphanos"; Irvine and Temkin, "Akīlāōs," 15–17; cf. the possible reading of lsqblāws given by Rosenthal in his edition of Isḥāq ibn Ḥunayn's *Tārīkh al-aṭibbāʾ*.

30. These speculations are that it is a double of Anqīlāʾus—as Max Meyerhof argues in "Von Alexandrien nach Bagdad," 397, n. 4—or that it derives somehow from the name of the city of Aquileia in northern Italy, as Irvine and Temkin, "Akīlāōs," 19–24, have suggested.

31. Schacht and Meyerhof, *Medico-Philosophical Controversy*, 93, n. 9.

32. Sorabji, *Philoponus and the Rejection of Aristotelian Science*. See also appendix 1, s.v. "Philoponus." Translations of a number of his works have recently been published in the Ancient Commentators on Aristotle series; none have to do with medicine.

several earlier Arabic sources and "Christian histories," reports that John was a sailor who became interested in scholarship and studied with the scholars who happened to travel on his ship. Having come late to scholarship, he was discouraged at his poor prospects; but inspired by the example of a persistent ant, he applied himself and soon excelled, writing a large number of books on medicine and philosophy.[33] If we are to believe Ibn Abī Uṣaybiᶜah, who places him both at the Council of Chalcedon in 451 and the conquest of Egypt by ᶜAmr ibn al-Āṣ in 641, he would have been over 230 when he died, which would explain how he overcame his late start in scholarship. Modern scholars are more inclined to assume that as many as three lesser-known individuals have been conflated with the famous philosopher John Philoponus.[34] If so, our John is presumably the John of Alexandria (fl. first half of the seventh century?) to whom a commentary on *The Medical Sects* is tentatively attributed.[35] A set of Arabic commentaries on the works of the Alexandrian curriculum attributed to him is very much in the style of the Alexandrian epitomes, as is a commentary to *On the Uses of the Parts*.[36] For our purposes, the exact identification is not critical; it is clear that there was a John writing medical works in the general style of our epitomes in sixth- or seventh-century Alexandria.

Marinus cannot be identified with certainty and is perhaps unknown in Greek sources. [37] There was a prominent physician named Marinus about two generations before Galen, but he must be excluded on chronological grounds. It is possible that the name is a corruption of Magnus of Nisibis, a prominent lecturer on medical theory in Alexandria late in

---

33. Ibn Abī Uṣaybiᶜah, 1:154–56; al-Qifṭī, *Tārīkh al-ḥukamāʾ*, 354–57.

34. Savage-Smith, "Ophthalmology," 127. *GAS* 3:157–60 gives references to the works attributed to him in Arabic. Numerous other references can be traced through the indices to the volumes of *GAS* under "Johannes Alexandrinus," "Johannes Grammatikos," and "Yaḥyā al-Naḥwī." See also Meyerhof, "Johannes Grammatikos."

35. John of Alexandria, *Commentaria*.

36. British Library Arund. Or. 17; cf. Temkin, "Late Alexandrian Medicine," 414. They have not been studied seriously.

37. His name is given as Bārsīyūs (Persius) in the manuscript of Ibn Buṭlān's *Medico-Philosophical Controversy*, but this was corrected—rightly, given the forms found in Ibn Isḥāq ibn Ḥunayn and the sources dependent on him—by Schacht and Meyerhof. Schacht and Meyerhof, *Medico-Philosophical Controversy*, 93, n. 9; cf. Irvine and Temkin, "Akīlāōs," 19.

the fourth century. Even less likely is that he was Marinus of Neapolis, Proclus's successor and biographer at the Academy in Athens.

Palladius of Alexandria is presumably the iatrosophist who wrote several commentaries on works of Hippocrates and Galen, including an extant fragment of a commentary on *The Medical Sects* rather similar to the Alexandrian epitomes. He is placed, without strong evidence, in sixth-century Alexandria. (An older reference book unhelpfully dates him between the third and ninth centuries, since he quotes Galen and is quoted by al-Rāzī.)[38]

Stephanus of Athens. There are three possible individuals named Stephanus: Stephanus of Athens, a medical writer; Stephanus of Alexandria, the last significant philosopher in Alexandria before the Persian and Arab conquests; and a Stephanus to whom an alchemical work is attributed. It is possible, though not certain, that they are the same person. At any rate, three surviving medical works are attributed to Stephanus of Athens—two commentaries on Hippocrates and one on Galen's *Therapeutics for Glaucon*, a work in the Alexandrian curriculum.[39] If we accept the identification of the Athenian and Alexandrian Stephanus, we can add a work on urine and a commentary on Aristotle's *On Interpretation*.

Theodosius remains unidentified.

Abū al-Faraj ibn al-Ṭayyib occurs in a list of commentators on the Alexandrian canon but does not concern us here, he being an author of the Islamic period.[40]

Though only one or two of these individuals can be identified with certainty—and some are perhaps relics of the difficulties of transmitting foreign names accurately in Arabic script—two points are striking. First, all of those who can be both tentatively identified and dated are in roughly the same period of the fifth to seventh centuries: (1) Gesius, fifth to sixth century; (2) John of Alexandria, mid-sixth to mid-seventh century; (3) Palladius, sixth century; (4) Stephanus, sixth to seventh century; and (5) Archelaus, sixth century. Second, all of those whose writings are either extant or attested wrote commentaries on works of the Alexandrian curriculum: Gesius and John of Alexandria commented on the entire corpus, Palladius and Archelaus commented on *The Medical Sects*, and Stephanus commented on the *Therapeutics for Glaucon*.

---

38. Smith, *Greek and Roman Antiquities*, 3:94; Baffioni, "Scolii inediti."

39. Dickson, *Stephanus the Philosopher*; Wolska-Conus, "Commentaires de Stéphanos"; Wolska-Conus, "Stéphanos d'Athènes et d'Alexandrie."

40. See appendix 1, s.v. "Ibn al-Ṭayyib."

Plausibility of the Arabic accounts and dating the epitomes

The above survey of individual authors associated with the composition of the epitomes leads us to two conclusions. First, the general account given in Arabic sources of the production of the epitomes is plausible. At least some of the people listed as associated with the project were in Alexandria at about the right time and were engaged in writing commentaries of various sorts on works of Galen, as well as Hippocrates. Other similar works also seem to have emerged from Alexandria around this time. Second, we cannot assume that these four to seven individuals formed a sort of committee charged with drafting medical textbooks; their ages are not close enough. Gesius, for example, was evidently already a person of standing during the reign of Zeno, which seems to make him considerably older than any of the others. The most plausible overall explanation is that during the sixth century, it was customary for iatrosophists—the professors of medicine in Alexandria—to write such works, of which we have a number of whole or fragmentary examples in Greek, Latin, and Arabic. Anqīlāʾus, probably toward the end of the period, compiled the works we know as the Alexandrian epitomes, drawing from similar works written by some or all of the other individuals whose names are associated with the project. These works should thus be taken as representing the tradition of sixth-century Alexandria as it was codified sometime around the year 600.

The Arabic translation

So far as we know now, the translator of the epitomes was the famous Ḥunayn ibn Isḥāq (809–873). Most of the manuscripts for the three texts in this volume say so explicitly, and none gives another translator. The manuscripts that I have examined of the other epitomes are similar. The clearest indication of his authorship of the translations is an offhand remark in the biography of Anqīlāʾus: "Ḥunayn mentioned [that Anqīlāʾus had compiled the epitomes] in his translation of them from Greek into Syriac."[41] Ibn Juljul confirms this somewhat obliquely: "The translator Ḥunayn found these books both in the original and in the form of epitomes (*ʿalā al-aṣl wa-al-jawāmiʿ*), in which form

41. Al-Qifṭī, *Tārīkh al-ḥukamāʾ*, 71.

they exist up to this day."[42] On the other hand, Ḥunayn does not mention them in the list of his Galen translations—but, of course, these are not Galen's books as such. The other early sources, notably the *Fihrist*, do not attribute translations of the epitomes to Ḥunayn. Moreover, translations of Greek material into Arabic are invariably ascribed to the famous Ḥunayn unless they are very clearly stated to be by someone else. It does not help that at least two other major translators—Ḥunayn's nephew Ḥubaysh and his son Isḥāq ibn Ḥunayn—had names that could easily be corrupted into the more familiar *Ḥunayn*. There also seem to be discrepancies between renderings of Greek terms in the epitomes and those in the translations of the underlying works. Nevertheless, it is plausible enough that Ḥunayn translated these works through a Syriac intermediary, as he did with many other medical works. The other difficulties are easy enough to account for. If these books happened not to circulate widely and were not in Ḥunayn's autobibliography, they could easily have escaped the attention of bibliographers like al-Nadīm. As for inconsistencies between the translations of the original texts and the epitomes, anyone who has done technical translations knows how easily such inconsistencies may occur. Given the sheer volume of translations that passed through Ḥunayn's office, inconsistencies are not surprising. Moreover, according to Ḥunayn's own list of the translations he had done of Galen's works, by the time he was about forty he had translated Galen's *The Medical Sects, The Small Art*, and *On the Elements*,[43] so the epitomes could easily have been translated much earlier or much later, allowing time for Ḥunayn's choice of renderings to change.

A further complication is the existence of three different recensions.[44] One version is unquestionably late and can thus be disregarded for purposes of determining authorship, but one might speculate that the two earlier recensions might be an example of when Ḥunayn went back to revise a translation he had done earlier, as we know he sometimes did.

As for the style of the translation, these books are written in the tone of wooden pedantry first adopted by authors of textbooks at the dawn of civilization and faithfully maintained down to our own day. Their virtue is clarity and information, not elegance. There are many

---

42. Ibn Juljul, *Ṭabaqāt al-aṭibbāʾ*, 51.
43. Ḥunayn ibn Isḥāq, "Risālah," items 4–6, 9–10; Meyerhof, "New Light."
44. See pp. lxiv–lxvi below.

deviations from what we would now consider standard classical Arabic. Some could be charitably excused as Middle Arabic, though the copyists bear some of the blame, as in the case of one scribe who spells *ḥāḍir* as *ḥāẓir*. Even conceding that most dialects of spoken Arabic reduce both *ḍ* and *ẓ* to *z*, it is a startling solecism for anyone literate in Arabic. Another curiosity is the habit in one group of manuscripts of using *ākhar* (other) in place of *thānī* (second) in lists.[45] Other errors are best explained as grammatical and spelling errors resulting from a text being written from dictation. Yet others can scarcely be explained as anything other than grammatical error: problems of gender agreement with numbers, for example. Nevertheless, deviations from the norms of classical Arabic are too systematic to be solely the fault of Christian or Jewish copyists and point to an author who either was not a native speaker of Arabic or did not come from a literary community that stressed classical grammar.

Galen's three texts and their Alexandrian epitomes

The three texts edited and translated here were chosen for their general interest and, in particular, for their philosophical significance. They include a work on scientific method as applied to medicine, a survey of medicine that includes a discussion of the epistemology of diagnosis and choice of treatment, and a discussion of theories of the elements as applied to basic physiology. Since the epitomes make most sense in the context of the three works of Galen that they are based on, I will discuss each of Galen's works in turn along with its epitome.

On the Medical Sects for Beginners (Περὶ αἱρέσεων τοῖς εἰσαγομένοις; De sectis ad eos qui introducuntur; Firaq al-ṭibb li-l-mutaʿallimīn).[46] The first text is an essay on the methodological issues separating three schools of Greek medicine: the Dogmatists or Rationalists, the Empiricists, and the Methodists. Galen lived at the end of a period of theoretical ferment in

45. This is probably a calque from the original Greek, for Galen commonly uses ἕτερος in lists to mean "second"; for example, see *Small Art* in K 1:367, 369, 385.

46. K 1:64–105. The most recent edition is still Galen, *Scripta minora*, 3:1–32. For an English translation with an excellent introduction on Galen and ancient philosophy of science, see Galen, *Three Treatises*, 3–20. Ḥunayn ibn Isḥāq's Arabic translation is in Galen, *Firaq al-ṭibb*. For published ancient Greek and Latin commentaries and adaptations, see p. xlviii, n. 55.

medicine, with groups of medical theoreticians linked with the various Hellenistic philosophical schools in ways that are not always clear.[47] Galen wrote *The Medical Sects* as an introduction to the theoretical issues involved in the disputes among the medical schools. This little book, particularly when considered with two works on empiricism, is thus one of the more important texts on philosophy of science from ancient times. Ḥunayn remarked:

> He wrote it for beginners with the intention of describing what each class of the three differing sects had to say in defense of their claims and how they refuted the views and criticisms of those who disagreed with them. I would correct this by saying that they differ in genus, since each one of these three sects contained other sects differing in species. The one who is beginning in medicine learns to distinguish the views of one from another so that at the end, after careful study, he knows what the worth is of each group of them and how each distinguishes truth from error. Galen wrote this when he was young, about the time he went to Rome for the first time.[48]

The Rationalists[49] were, in Galen's view, Hippocrates and his followers, ancient and modern, though the earlier among them would certainly not have known themselves as such. The Rationalists used reason to infer the inner states of the body and from them were able to infer the correct treatment. The Rationalists therefore needed a thorough scientific understanding of the functioning of the body. In that sense, Galen was a classic Rationalist; and it is mainly from the Rationalist physicians that late antique and medieval medical theory came.

The Empiricists were the direct rivals of the Rationalists. Nominally, they rejected all theoretical speculation about the nature of the body and its inner states in favor of long and careful observation of its behavior under various circumstances. An Empiricist physician, having seen a patient recover after a particular treatment, would try the same treatment on another patient with similar symptoms. Galen interpreted the

---

47. Michael Frede, in Galen, *Three Treatises*, ix–xxxiv.

48. Ḥunayn ibn Isḥāq, "Risālah," item 3. Galen confirms Ḥunayn's dating of this work; see K 19:11–12 (*Περί τῶν ἰδίων βιβλίων, On My Own Books, De libris propriis* 1.1–3); Galen, *Scripta minora*, 2:93–94; Galen, *Galien*, 1:136–37; Galen, *Selected Works*, 4–5.

49. The more common modern term for this school was "Dogmatists;" but though the term did not then bear its modern pejorative sense, "Rationalist" seems to me a better rendering.

Empiricists' acquisition of experience as an exercise in memory. In practice, it was more complex than that, since there were techniques for discovering new treatments by analogy—known as "transfer"—and for using written sources to supplement the personal experience of the physician. There was, in other words, a systematic quality to the Empiricists' acquisition of medical experience. Since transfer is really a form of analogy, the whole Empiricist method seems to have been induction guided by analogy. Despite his deeply theoretical bent, Galen was surprisingly respectful of the Empiricists and wrote two more advanced works on their theories: *The Outline of Empiricism* and *On Medical Experience*.[50]

The Methodists held, generally speaking, that all disease was caused by the improper flow of atoms through the microscopic pores of the body. The physician could examine the patient and see by external signs whether he was in a state of constriction or dilation, thus eliminating the need to make inferences about the internal states of the patient. He needed only to decide whether the patient was in a state of dilation or flux—something that should be obvious from cursory examination—and treat accordingly. The physician did not need to make inferences about the internal states of the patient, as the Rationalists thought necessary, or possess extensive experience of patients in various conditions, as the Empiricists claimed.[51] As the epitome observes, "The Methodists profess to reject experience and the use of syllogism; but, in fact, they

50. *Περὶ τῆς ἰατρικῆς ἐμπειρίας* (*Fī al-tajribah al-ṭibbīyah, On Medical Experience*); and *Ὑποτυώσεις ἐμπειρικαί* (*Fī jumal al-tajribah, An Outline of Empiricism*). They are translated in Galen, *Three Treatises*, 23–106. The last survives mainly in Arabic.

51. On the Methodists and the philosophical implications of their theory, see Michael Frede in Galen, *Three Treatises*, xxix–xxxii. A more detailed discussion appears in Frede, "Method of the So-Called Methodical School," 261–78. This collection contains several other articles relevant to our themes. A thorough discussion of the Methodist founders Themison and Thessalus is Pigeaud, "L'introduction du Méthodisme à Rome," 565–99. ANRW 2.37.1–4 contains many detailed articles on medicine in the Roman period relevant to the texts in this volume. For surviving evidence on the Methodists, see Tecusan, *Fragments of the Methodists*, vol. 1. The major complete surviving Methodist text is Soranus, *Γυναικείων*, ed. and French trans., *Maladies des femmes* (4 vols.; Paris: Les Belles Lettres, 1988–2000); *Soranus' Gynecology*, trans. Owsei Temkin (Baltimore: Johns Hopkins University Press, 1991).

employ both of them."[52] This seems a fair methodological criticism—if we are to believe Galen's hostile account of them. Galen, however, was more critical of their failure to take other medically relevant factors into account; he accused them of malpractice rather than bad theory. They claimed, he indignantly reported, that one and the same treatment was appropriate regardless of the patient's age, the climate, the season, or any of the other factors affecting the patient's constitution. Whereas Hippocrates had said that life is brief, but the art of medicine is long, the Methodists said that medicine needs only six months.[53]

Experienced readers of Galen will not be surprised to find that his ideal Empiricist and Methodist, particularly the latter, smell strongly of straw; and elsewhere Galen cited individual Methodists with more respect than his theoretical discussion would indicate. But whatever may be the defects of *The Medical Sects* as medical history, it is deeply interesting as an exercise in the philosophy of science. Galen had recommended it as an introductory work. Arabic sources listed it as the first of Galen's works to be studied.[54] Late antique epitomes and commentaries of various sorts on this book survive, attributed to John of Alexandria, Agnellus of Ravenna, Palladius, Archelaus, and Yaḥyā al-Naḥwī. The long popularity of the work is intriguing. Clearly, teachers continued to find the book useful, despite the extinction of the schools it criticized, using it as a way to introduce students to medical epistemology.[55] It seems to have finally dropped out of the curriculum with Ibn Sīnā, whose *Canon* contained no counterpart to it. Although it continued to be copied, both by itself and in the form of the Alexandrian epitome, the last independent works on the topic that I know of are Abū al-Faraj ibn al-Ṭayyib and ʿAlī ibn Riḍwān, both in the eleventh century.[56]

The Alexandrian epitome of *The Medical Sects* follows the order of Galen's book closely and tends to either elaborate or clarify Galen's less

52. See p. 12 below.

53. K 1:14–15.

54. See pp. xxi–xxiv above; cf. *On My Own Books* 1.1–3; K 19:11–12; ed. Marquardt et al., 2.93–94; ed. Boudon-Millot, 1.136–37; trans. Singer, 4–5; and *On the Order of My Own Books*, *Περὶ τῆς ταξέως τῶν ἰδίων βιβλίων* 2.4, K 19:54; ed. Marquardt et al., 2.84; ed. and trans. Boudon-Millot 1.92; trans. Singer, 25.

55. Epitomes and commentaries on *The Medical Sects* and its place in the Alexandrian curriculum are discussed in Pormann, "Jean le Grammarien," 233–63; Pormann, "Alexandrian Summary"; and Temkin, "Alexandrian Medicine," 405–30.

organized presentation. It does not quote the text being explained, though once in a while the glosses in some of the manuscripts do. Thus, the epitome begins with an account of the parts of medicine, giving alternative divisions into two or five parts. This is followed by a historical introduction listing the prominent members of the three schools—knowledge that Galen could presume in his audience. Galen's first chapter dealt with three things: a definition of the aim of medicine, the problem of the relative importance of theory and experience in medicine, and the various names by which the Rationalist and Empiricist schools were known. The epitome gives two additional definitions of medicine; categorizes the causes of health and disease, a topic mentioned in Galen's definition of medicine; glosses Galen's account of the disagreement between the Rationalists and Empiricists; and gives a more systematic account of the various names by which the sects are known and the principles by which such names are derived.

This is typical of the organization of the three epitomes edited here. The basic unit is a paragraph elaborating on something mentioned by Galen, though sometimes it is not clear exactly what part of Galen's text is being referred to. The text is thus not a continuous exposition or a summary of Galen's text, but a series of notes, often without any transition or even connection between one paragraph and the next. The information presented is not drawn solely from the text being epitomized but comes also from other sources, mostly from Galen's other works. The epitomist is particularly fond of lists and categorizations, something that we know to have been characteristic of Alexandrian medical writing.[57] There is no reference to clinical experience, except when Galen himself has mentioned a case. The personality of the author is totally absent, and there is no attempt at any literary effects. The epitomes are roughly the same length as the original text.

---

56. *GAS* 3:80 lists eight surviving MSS, the earliest a copy that had supposedly belonged to Ibn Sīnā in the fifth/eleventh century (though there is suspicion that the owner's mark is a forgery), and the latest an exquisite undated Ottoman manuscript, Aya Sofya 3557; but this last was produced by a professional scribe for an imperial library rather than by a scholar for practical use. For the *Thimār* of Abū al-Faraj ibn al-Ṭayyib (d. 435/1043), see *GAS* 3:82, 146; it is preserved in one manuscript, Manisa 1772/1, ff. 1b–37a; cf. Dietrich, *Medicinalia Arabica*, 22. For the commentary of ʿAlī ibn Riḍwān, see Ibn Abī Uṣaybiʿah, 2:103; cf. *GAL* G1.484.

57. Savage-Smith, "Ophthalmology," 121–38; Gundert, "*Tabula Vindobonenses*," 91–144.

The [Small] Art of Medicine (Τέχνη ἰατρική; Ars medica, Ars parva, Microtechne, Tegni; al-Ṣināʿah al-ṣaghīrah fī al-ṭibb).[58] The second work of Galen, whose epitome is presented in this volume, is an admirably clear and organized introduction to medicine that retained its popularity into early modern times. Ḥunayn remarks:

> Galen did not intend this book for beginners, since the benefit in reading it is not confined to beginners to the exclusion of those more advanced. That is because Galen's purpose in it was to describe concisely all the general principles of medicine, which is useful both to beginners and to the advanced. It allows beginners to conceive the whole of medicine in their minds by way of description and then return to each part of it later and learn the commentary, abridgment, and demonstrations of it from the books that explain it in detail. For those who are more advanced, it serves as a review of all that they have read and learned through detailed exposition. The professors who taught medicine in the past in Alexandria placed this book after *The [Medical] Sects*; after it, they put *On the Pulse for Beginners* [*for Teuthras*], then, after that, *Therapeutics for Glaucon*, treating them as a single work of five books [*Therapeutics for Glaucon* having two books], to which they gave the title "For Beginners."[59]

Galen began with a review of the possible approaches to teaching a subject, explaining that in this book he was using the method of dialysis of the definition, in which the subject is expounded in accordance with the parts of its definition, a method that makes it easy for the student to understand and remember what he is learning. Galen then went on to analyze the basic concepts of medicine: health, disease, and what is neither healthy nor diseased; the various classes of healthy, diseased, and indifferent causes and signs; and the concept of the temperament and its balance. He then discussed the signs of the simple and composite temperamental imbalances in the brain, eyes, heart, liver, and testicles; in the body as a whole; and in the stomach and lungs. He then moved on to organic defects—diseases in which parts of the body were missing,

58. K 1:305–412. The most modern Greek edition of Galen's *The [Small] Art of Medicine*, with thorough introduction and commentary, is Galen, *Galien*, ed. and trans. Boudon-Millot, vol. 2, *Exhortation*. The one modern English translation is "The Art of Medicine," in Galen, *Selected Works*, trans. Singer, 345–96. Ḥunayn ibn Isḥāq's Arabic translation, G[a] in the present work, is Galen, *al-Ṣināʿah al-ṣaghīrah*, ed. Sālim.

59. Ḥunayn ibn Isḥāq, "Risālah," item 4.

excessive, or deficient in size or number, or deformed. He then analyzed signs and causes in reference to prognosis and diagnosis of acute and chronic diseases. He next discussed surgery, the general principles of therapy, and prophylaxis. He concluded, in proper textbook fashion, with recommendations for additional reading. This being Galen, they were all his own works and comprised most of the books in the Alexandrian curriculum. *The Small Art* constituted an admirable introduction to Galenic medicine and was still being reprinted, translated, adapted, and commented on for the use of practicing physicians and medical theoreticians at least as late at the seventeenth century. It was, as well, held in honor by the inhabitants of Sir Thomas More's Utopia.[60]

Since *The Small Art* is well organized to begin with, the epitome is able to follow it closely, using the Alexandrians' characteristic method of elaborating on points that Galen alludes to, filling in details, and summarizing Galen's arguments. It has an elaborate logical introduction on the methods of instruction. Its most striking feature is the use of tables or numbered lists (the manuscript families disagree on how this information is presented) giving all the possible combinations of various sets of categories. The first, for example, gives the nine possible combinations of body, sign, and cause with healthy, diseased, and neither. These sometimes produce results that border on the incomprehensible, as in "a sign prognosticating a state that is neither healthy nor diseased in the extreme." This is the nonclinical abstractness for which the iatrosophists are criticized, though we might also fairly interpret such oddities as classroom exercises. Certainly, the stress on the many possible relations of signs and symptoms to underlying diseases would not be out of place in a modern medical classroom.

On the Elements According to the Opinion of Hippocrates (Περὶ τῶν καθ' Ἱπποκράτην στοιχείων; De elementis secondum Hippocratem; Fī al-isṭiqsāt ʿalá raʾy Ibbuqrāṭ),[61] the third work of Galen whose epitome appears in this

---

60. See, for example, Galen, *Galen's Art of Physick*, trans. Culpepper. More's reference to it is in chapter 6 of *Utopia*. Ottosson, *Scholastic Medicine and Philosophy*, examines Renaissance medical theory through the commentaries on *The Small Art* written by European professors of medicine.

61. There are three modern editions of *On the Elements:* K 1:413–508; Galen, *Galeni De elementis*, ed. Helmreich, 1–69; and Galen, *On the Elements*, ed. De Lacy. De Lacy's edition is a model of classical philology whose only serious defects are

volume, is a polemical defense of the views on the elements and humors expressed in Hippocrates's *On the Nature of Man* 1–7. Galen explained that he referred to Hippocrates's work as "On the Elements" because contemporary authors tended to use that title for works of this sort, though earlier authors had used some variant of the title "On Nature."[62] Galen conceived of this work as one of a series, to be followed by *On the Temperament, The Method of Healing*, and other works. It is not dated but is generally thought to have been written during Galen's second residence in Rome, around 169. There is also reason to suppose that it was written in two parts, breaking at the end of chapter 9. (The Greek tradition, though not the Arabic, divided it into two books.) In it Galen launched spirited attacks on various other medical theorists, with the most bile reserved for Asclepiades and Athenaeus. Topics arose as the rhetoric of the argument happened to demand. It thus differs in character from its epitome, which is a passionless catalogue of theories, methods, and medical entities that came up in Galen's discussion, with little attempt at overall expository or argumentative structure. Though the epitome is less spirited than Galen's text, it makes a far better textbook. Ḥunayn summarized the content of Galen's book as follows:

> His purpose in this book is to explain that all bodies subject to genesis and corruption—the bodies of animals, plants, and the materials that are generated in the bowels of the earth—are compounded from four principles, which are earth, water, air, and fire. These, he explains, are the remote first principles of the human body. The proximate secondary principles from which are constituted the bodies of human beings and every animal having blood are the four humors—that is, blood, phlegm, and the two biles. This is one of the books that absolutely must be read before *The Method of Healing*.[63]

Galen began by defining "element" (στοιχεῖον). He then criticized the Atomists, arguing that there must be more than one element and that Atomism could not explain how a living body was affected and able to feel. He made similar arguments against those who held that the

---

the lack of a subject index and an emphasis on philological and textual issues at the expense of the scientific and philosophical. The notes contain extensive lists of parallel passages. The Arabic translation adds "the Opinion of" to the Greek title, which I retain when translating the Arabic title.

62. Galen, *On the Elements*, 9.25–30; K 1:487–88; G^a 110–112.

63. Ḥunayn ibn Isḥāq, "Risālah," item 11.

body was composed of only one of the four humors. He criticized the Ionian physicists for their accounts of the transmutations of elements, each of which involved the compression or rarefying of a single element. Galen insisted that, contrary to what some interpreters had claimed, the views of Hippocrates and Aristotle on the four elements and their transmutation are in harmony. Moving to a discussion of qualities, Galen criticized Athenaeus for his theory that the elements are visible and argued that he had confused qualities with elements. He then defended Hippocrates against the charge that he had done the same thing. He next discussed the humors, alluding to how tissues and organs were formed from them. The humors, he claimed, like many of the tissues and organs, were common to all animals having blood—as opposed to those, such as worms, that did not have blood. He then addressed the question of whether the body was created from blood only or from all the humors, as Hippocrates had held. He pointed out that blood is a mixture, as can be seen from its variation, and that it has aspects of the other humors. He criticized Asclepiades for his account of the way that drugs worked to purge humors.

The epitome seems particularly disjointed—often more a series of glosses than a coherent and organized text. It roughly follows the order of Galen's text, though it is sometimes difficult to be sure, since it is not always obvious what passage the epitome is referring to. Sometimes it presents a summary or clarification of Galen's argument, sometimes an explication of something that happens to be mentioned in the text, sometimes background that Galen assumed but that students might not have. Not surprisingly, there is a fair amount of repetition. Some sympathy for the epitomist is in order, however; Galen's text defies easy summary.

The edition and translation

Previous versions, editions, and translations

The works known as the Alexandrian epitomes apparently do not survive in Greek, nor do they seem to have been translated into Latin. The Arabic version of the Alexandrian epitomes was translated into Hebrew as *Kibbutzei Galenos* by Shimshon ben Shlomo in 1322.[64] Sezgin

64. Steinschneider, *Hebraeischen*, 654–55; Lieber, "Galen in Hebrew," 167–83, which contains a good discussion of the history of the Alexandrian epitomes. Ullman, *Medizin*, 67.

lists an *ikhtiṣār* (abridgment) by Yaḥyā al-Naḥwī (John of Alexandria) and a *thimār* (selection) by Abū al-Faraj ibn al-Ṭayyib, but it is not yet clear whether these are works of a similar sort based on Galen's underlying texts or abridgments of the Alexandrian epitomes as such.[65]

The entire Arabic text of the epitomes has been published in the form of a facsimile of Fatih 3588–89—my manuscript **F**—with Tehran Majlis 6037 supplying the three works missing from the Fatih MSS: *Days of Crisis, The Method of Healing,* and *The Regimen of the Healthy.*[66] Muḥammad Salīm Sālim quotes many short passages from the three texts presented here as glosses to his editions of the Arabic translations of *The Medical Sects, The Small Art, On the Pulse for Teuthras,* and *On the Elements.* These passages were edited from manuscripts Aya Sofya 3588 and British Library Add. 23407—herein designated **S** and **D**, respectively. In the case of *On the Pulse for Teuthras,* **S** and Istanbul University A6158—my manuscript **U**—were used. Sālim identifies these passages, oddly and incorrectly, as from *Sharḥ Ḥunayn* (Ḥunayn's commentary). The section on the anatomy of the nerves has been published with translation and commentary by Ahmad M. Al-Dubayan.[67]

Descriptions of manuscripts

There are twenty-two manuscripts known or thought to contain one or more of the Alexandrian epitomes. Ten are known or said to contain one or more of the works edited here. The present edition is based on six manuscripts that contain both *The Medical Sects* and *The Small Art* and eight that contain *On the Elements,* including all of those copied prior to the sixteenth century. I will discuss them in roughly chronological order and grouped by family—at least as the families appear in the first two epitomes (see table 3). The comments on the relationships among the manuscripts apply mainly to those of the epitomes on *The Medical Sects* and *The Small Art;* the manuscript relations for *On the Elements* will be discussed separately.[68]

---

65. *GAS* 3:146–49 passim, 159; cf. Pormann, "Jean le Grammarien," where Yaḥyā al-Naḥwī's version of *The Medical Sects* is discussed and edited. The manuscripts in which these two texts are found are respectively British Library Arund. Or. 17 and Köprülü, Fazel Ahmed Paşa 961.

66. Sezgin, *Alexandrian Compendium.*

67. Galen, "Jawāmiʿ tashrīḥ al-ʿaṣab."

68. I have found a number of problems with the cataloging and identification of the manuscripts of the Alexandrian epitomes and related works. I plan to discuss the bibliography of these texts in a separate article.

Table 3: Arabic manuscripts of the Alexandrian epitomes

			MSS used in this edition			
			A: Aya Sofya 3609*	**D**: Br. Lib. Add 23407*	**F**: Fatih 3538*	**M**: Manisa 1759*
Date (est.) CE		**No. of Books**	1242	(17th c.)	13th c.	1240
1. *Firaq al-ṭibb*	*The Medical Sects*	1	◆	◆	◆	◆
2. *al-Ṣināʿah al-ṣaghīrah*	*The Small Art*	1	◆	◆	◆	◆
3. *Fī al-nabḍ al-ṣaghīr ilá Ṭuthrūn*	*On the Pulse for Teuthras*	1	◆	◆	◆	◆
4. *Ilā Ighlūqun*	*Therapeutics for Glaucon*	2	◆	◆	◆	◆
5. *al-Isṭiqsāt ʿalá raʾy Ibbuqrāṭ, al-ʿAnāṣir*	*On the Elements*	1	◆	◆	◆	◆
6. *Fī al-mizāj*	*On the Temperament*	3	◆	◆	◆	◆
7. *Fī al-quwā al-ṭabīʿiya*	*On the Natural Faculties*	3	◆	◆	◆	◆
8. *Fī al-tashrīḥ*	*On Anatomy*	5	◆	◆	◆	◆
9. *Fī al-ʿilal wa-al-aʿrāḍ*	*On Diseases and Symptoms*	6	†			◆
10. *Taʿarruf ʿilal al-aʿḍāʾ al-bāṭina*	*The Diagnosis of Diseases of the Internal Organs*	6				
11. *al-Nabḍ al-kabīr*	*The Large Pulse*	16				
12. *al-Buḥrān*	*On Crises*	3				
13. *Ayyām al-Buḥrān*	*Days of Crisis*	3				
14. *Aṣnāf al-Ḥummayāt*	*The Kinds of Fevers*	2				
15. *Ḥīlat al-burʾ*	*Method of Healing*	14				
16. *Tadbīr al-aṣiḥḥāʾ*	*Regimen of the Healthy*	6				

* Manuscript seen personally.

† In place of the Alexandrian epitomes of *Diseases and Symptoms*, **A** has two similar abridgments entitled *Jumal maʿānī al-ʿilal waʾl-aʿrāḍ* and *Jumal al-ʿilal waʾl-aʿrāḍ*.

	MSS used in this edition				Other MSS		
	R: Br. Lib. Or. 9202*	**S**: Aya Sofya 3588*	**U**: Ist. Univ. Lib. A6158*	**Y**: Yeni Cami 1179*	Azhar Ṭibb 79	Berlin Staats. Or. Oct. 1122	Chester Beatty 4001
Date (est.) CE	(1100)	(13th c.)	1760	1507, (12th c.)	(17th c)	(11–13th c.)	1329
1. *Firaq al-ṭibb*		◆		◆			
2. *al-Ṣināʿah al-ṣaghīrah*		◆		◆			
3. *Fī al-nabḍ al-ṣaghīr ilá Ṭuthrūn*		◆		◆			
4. *Ilā Ighlūqun*		◆	◆	◆			
5. *al-Isṭiqsāt ʿalá raʾy Ibbuqrāṭ, al-ʿAnāṣir*	◆	◆	◆	◆			
6. *Fī al-mizāj*	◆	◆	◆	◆			
7. *Fī al-quwā al-ṭabīʿīya*	◆	◆	◆	◆		◆†	
8. *Fī al-tashrīḥ*	◆	◆		◆		◆	
9. *Fī al-ʿilal wa-al-aʿrāḍ*		◆	◆	◆		◆	
10. *Taʿarruf ʿilal al-aʿḍāʾ al-bāṭina*			◆	◆		◆	
11. *al-Nabḍ al-kabīr*			◆	◆			
12. *al-Buḥrān*				◆			
13. *Ayyām al-Buḥrān*							
14. *Aṣnāf al-Ḥummayāt*							
15. *Ḥīlat al-burʾ*			◆				◆
16. *Tadbīr al-aṣiḥḥāʾ*					◆		

* Manuscript seen personally.
† Incomplete at beginning.

	Other MSS						
	Escurial 849	F: Fatih 3539*	Princeton Garrett 1G	Haidarabad Āṣaf. ṭibb 44	Haidarabad Āṣaf. Falsafa 371	Saray Ahmet III 2043*	Tehran Majlis 3999
Date (est.) CE	1295	1175	1176	(18th c.)	1613	(1300)	
1. *Firaq al-ṭibb*				◆			
2. *al-Ṣināʿah al-ṣaghīrah*				◆			
3. *Fī al-nabḍ al-ṣaghīr ilá Ṭuthrūn*				◆			
4. *Ilā Ighlūqun*				◆	◆		
5. *al-Isṭiqsāt ʿalá raʾy Ibbuqrāṭ, al-ʿAnāṣir*				◆			
6. *Fī al-mizāj*				◆			
7. *Fī al-quwā al-ṭabīʿīya*		◆		◆			
8. *Fī al-tashrīḥ*		◆		◆			
9. *Fī al-ʿilal wa-al-aʿrāḍ*		◆		◆			
10. *Taʿarruf ʿilal al-aʿḍāʾ al-bāṭina*	◆†	◆		◆			◆
11. *al-Nabḍ al-kabīr*		◆		◆			◆
12. *al-Buḥrān*		◆	◆	◆			◆
13. *Ayyām al-Buḥrān*			◆				◆
14. *Aṣnāf al-Ḥummayāt*		◆	◆	◆			◆
15. *Ḥīlat al-burʾ*			◆			◆‡	
16. *Tadbīr al-aṣiḥḥāʾ*			◆				

* Manuscript seen personally.
† A *jawāmiʿ* but not confirmed to be the Alexandrian epitome.
‡ An abridgment (*ikhtiṣār*) by "one of his companions," but it is not clear that it is the Alexandrian epitome.

	Other MSS							
	Princeton NS 1532	Tehran Majlis 6036	Tehran Sanā 3190	Tehran Univ., Med. Fac. 167	Tehran Univ. 4914	Tehran Univ. 5217	Ist. Univ. Lib. A3559*	Wellcome Hist. Or. 62
Date (est.) CE	1651	1657	1599		(16–17th c.)	(18th c.)	1754	(16th c.)
1. *Firaq al-ṭibb*					◆			
2. *al-Ṣināʿah al-ṣaghīrah*					◆			
3. *Fī al-nabḍ al-ṣaghīr ilá Ṭuthrūn*				◆	◆			
4. *Ilā Ighlūqun*				◆	◆		◆	◆
5. *al-Isṭiqsāt ʿalá raʾy Ibbuqrāṭ, al-ʿAnāṣir*					◆			
6. *Fī al-mizāj*					◆			
7. *Fī al-quwā al-ṭabīʿīya*					◆			
8. *Fī al-tashrīḥ*					◆			
9. *Fī al-ʿilal wa-al-aʿrāḍ*	◆				◆			
10. *Taʿarruf ʿilal al-aʿḍāʾ al-bāṭina*	◆	◆	◆		◆			
11. *al-Nabḍ al-kabīr*	◆	◆						
12. *al-Buḥrān*	◆	◆			◆			
13. *Ayyām al-Buḥrān*	◆	◆			◆			
14. *Aṣnāf al-Ḥummayāt*	◆	◆			◆			
15. *Ḥīlat al-burʾ*		◆				◆		
16. *Tadbīr al-aṣiḥḥāʾ*		◆						

* Manuscript seen personally.

R: *British Library Or. 9202* (131 ff., 17.5 × 25.5 cm., 21 ll., *naskh*) is the oldest-known manuscript of any of our epitomes, covering only the fifth through eighth treatises. It is written in a clear, plain scholar's hand with many marginal corrections in several hands. The manuscript is undated but can be placed at approximately the beginning of the sixth/twelfth century through a series of owners' and collation notes. Folio 1a reports in several hands that it belonged to ᶜAbd al-Wāḥid ibn Muḥammad al-Ṭabīb, then that it passed into the hands of Hibat Allāh ibn [Hay]kal al-Muṭabbib, then ᶜAbd Allāh ibn al-Ḥusayn al-Mutaṭabbib, then his son Ḥasan.[69] The last acquired the book on 14 Ṣafar 547/21 May 1152. A final note places it in the hands of the late Nāṣir al-Dīn Muḥammad ibn ᶜAlī ibn Muḥammad al-Balīnī [?] al-Shāfiᶜī al-Azharī in 984/1576–77. Moreover, a note on the last page, 131b, reports that it was corrected against a copy read for correction to Ibn al-Tilmīdh, who died at a very advanced age in 560/1165. It is thus the oldest of our manuscripts by about a century. Unfortunately, it does not contain either *The Medical Sects* or *The Small Art*. Another volume must once have existed, since ours is labeled as volume two. It was acquired by the British Library in 1923.[70]

F: *Fatih 3538* (291 ff., 23.5 × 16 cm., *naskh*, undated) contains the first eight Alexandrian epitomes, including the earliest-known copy of the first two treatises edited here and an early copy of the third. Fuat Sezgin and the catalog of Turkish medical manuscripts assume that this was copied as part of a set with Fatih 3539, which was completed at the end of Rabīᶜ I 571/October 1175 by ᶜUthmān ibn ᶜAlī ibn Muḥammad al-Samarqandī. This is not the case, however; Helmut Ritter and Richard Walzer—and, following them, Ahmad Al-Dubayan—date it to the seventh/thirteenth century. [71] The manuscript of the first volume, containing all three of the treatises presented here, was in the hands of three successive physicians, natives of Asqalan. The first received it in the 610s/1210s. The remaining two were in Damascus, "Dār al-Maḥrūsa." The owner's notes are, as usual, not very legible. The oldest, partially

69. Unless specifically noted, I have been unable to identify the scribes and owners mentioned in connection with these manuscripts.

70. Hamarneh, *Catalogue*, 222.

71. *GAS* 3:146–48; İhsanoğlu, *Fihris makhṭūṭāt al-ṭibb*, 170–73; Sezgin, *Alexandrian Compendium*, preface to vol. 3; Ritter and Walzer, "Arabsiche Übersetzungen," 44; Al-Dubayan, *Galen*, 33–34.

effaced, seems to read, "Passed into the possession of the servant in need of God . . . ibn al-Shaykh al-Makkī al-Mutaṭabbib . . . on the sixth of the month of Dhū al-Qaᶜda 610 [or 810/19 March 1214 or 3 April 1408]. The second note is from Manṣūr [?] ibn Muḥammad ibn al-Zakī al-ᶜAsqalānī al-Ṭabīb "now in Damascus." The third is from another ᶜAsqalānī "now in Damascus." The non-Christian population of ᶜAsqalān was expelled in 548/1153 after its surrender to the Crusaders. The town was again occupied by Muslims in 583–87/1187–91, reoccupied by Richard Lion-Heart in 1192, and occupied once again by Muslims in 645/1247. In 668/1270 it was destroyed and its harbor blocked. The city was not rebuilt until the twentieth century. Therefore, Muslims who identified themselves as "ᶜAsqalānī, now in Damascus," were most likely part of the population expelled in 548/1153 and probably lived before the middle of the seventh/thirteenth century, thus indicating an early date for our manuscript. Eventually it was endowed by Sultan Maḥmūd I to the Fatih mosque library and has been reproduced in facsimile.[72] The text of this manuscript is most closely linked to **S** but is sloppily copied and contains many readings not supported by other manuscripts.

S: *Aya Sofya 3588* (260 ff., 24.5 × 16 cm., 22 ll., *naskh*) is a single volume containing the first nine epitomes. It is undated, but Ritter and Walzer plausibly date it to the seventh/thirteenth century.[73] It is clearly and carefully written and contains a great many glosses, including some of those that likely came from Ibn al-Tilmīdh, but also many unique to itself. These include introductions to most of the books. The table of contents lists the sixteen books, so there must once have been a second volume. The owners' notes, to the extent that they can be read, point to Jewish owners: "al-ᶜAbd al-Faqīr" Jār Allāh al-Abrahāmī and Ibrāhīm ibn Sulaymān ibn Ḥakīm al-Ruhāwī. The text tends to follow that of **F**, though not always.

M: *Manisa 1759* (342 ff., 25.5 × 17.7 cm, 21 ll., *naskh*) is a single manuscript containing the first nine epitomes. It was copied by one Sallām ibn Ṣāliḥ ibn Khiḍr ibn Ibrāhīm, "known as the Teacher *(muᶜallim)* from Shafarᶜām," a village east of Acre. Its parts are dated between 26 April 6748 and 27 May 6749 according to the Byzantine era of the world,

---

72. Sezgin, *Alexandrian Compendium*, vol. 1.

73. Ritter and Walzer, "Arabsiche Übersetzungen," 39; Al-Dubayan, *Galen*, 37–39.

corresponding to 1240–41. It is carefully written, and, for *The Medical Sects* and *The Small Art*, its text is almost identical to that of **A** (detailed below), which was copied a year later, except that here the glosses are added in the margin in a different hand. It also has the proper Alexandrian epitomes of *Diseases and Symptoms*, rather than the two sets of *jumal* found in **A**. A note at the end of *On the Pulse for Teuthras* notes that it—probably meaning the epitomes of the first three books, *The Medical Sects, The Small Art*, and *On the Pulse for Teuthras*—was copied from a manuscript copied from a manuscript in the hand of Abū al-Khayr Sahl ibn ᶜAbd Allāh ibn Ṭūmā and that both were read for correction to Ibn al-Tilmīdh.[74] It would seem that the first few epitomes in **A** and **M** were copied from the same manuscript written by Ibn Ṭūmā, and that Jirjis ibn Tādrus, the scribe of **A**, borrowed the original to copy when Sallām ibn Ṣāliḥ was finished with it. There is an owner's mark of one Anṭūn al-Ḥakīm al-Yāfūtī, "Anthony, the physician of Jaffa." There is no indication of how or when the manuscript came to be in the Manisa library. The fate of the presumed second volume is also unknown.[75]

A: *Aya Sofya 3609* (299 ff., 25 × 17 cm., 21–24 ll., *naskh*) is the first volume of what was perhaps a complete collection of the epitomes in two volumes, copied by a Christian physician in Crusader Acre. The manuscript was copied for his own use by Jirjis ibn Tādrus in Shawwāl 639/April 1242.[76] Like **M**, it bears the note that it was copied from a manuscript copied from a manuscript written by Ibn Ṭūmā and read for correction to Ibn al-Tilmīdh. This doubtless explains why Ibn al-Tilmīdh's glosses are written with the main text flowing around them. The copying is very accurate, in a clear but not ornate *naskh*. The manuscript contains various more or less illegible owner's notes, culminating in a *waqf* notice of Sultan Maḥmūd I, who donated the books that initially comprised the Aya Sofya library. It is now stored with the rest of the Aya Sofya collection in the Süleymaniye Library. The manuscript contains the epitomes of the standard works through *On Anatomy for Beginners*; then it gives two *jumal* of *Diseases and Symptoms*, neither of which is the Alexandrian epitome proper. These were copied from a

74. See f. 70a.

75. A very detailed description of **M** is in Dietrich, *Medicinalia Arabica*, 32–42; see also Al-Dubayan, *Galen*, 34–35.

76. See ff. 254a, 299b, the former dated 23 Ramaḍān 639/27 March 1242.

manuscript in the hand of the distinguished fourth/tenth century Jacobite physician of Damascus, Abū al-Faraj Jirjis ibn Yūḥannā ibn Sahl al-Yabrūdī, whose manuscripts of Galen's works and whose commentaries on them were highly esteemed.[77]

The text of this manuscript for the first two epitomes, as well as for the glosses, is almost identical with that of **M.**[78]

Y: *Yeni Cami 1179* (398 ff. total, ff. 114–243, 19 × 14 cm., 19 ll., 913/1507, *naskh*) is a composite volume consisting of four works of Aristotle and Ibn Rushd,[79] a tenth/sixteenth-century copy of the first eight epitomes, and a much older copy of the ninth, tenth, and twelfth epitomes, that is, *Diseases and Symptoms, The Large Pulse*, and *Crisis*. 398 ff. The younger portion of the text, which is what is used in this edition, was copied by Junayd ibn Kūnj ibn Junayd [?] in the vicinity of Konya *(ṣaḥrā-yi Qūnya)* on Friday, at the beginning of Rabīᶜ I, 913, which corresponds to 16 July 1507. It is likely that the younger portion of the text was copied to go with the much older existing manuscript. The copying is excellent—careful and clear—but the text is late and eclectic, drawing from both earlier manuscript groups with many additional minor variations. The older portion of the text (16.5 × 12 cm., 16 ll.) is written in oxidized brown ink and is undated, though there is an owner's mark dated 4 Muḥarram 679/6 May 1280.[80]

D: *British Library Or. Add. 23407* (291 ff., 13 × 24.5 cm., 18 ll., *naskh*) is a late manuscript containing the first eight epitomes—that is, through the *Small Anatomy.* Hamarneh dates it to the eleventh/seventeenth century. There is no information about the origins of the manuscript, and it does not show many signs of use. The text for the epitomes edited here is that of the late recension represented by **Y.** Moreover, **D** contains the same eight epitomes as the newer part of **Y.** There are enough independent differences between **D** and **Y** to indicate that neither was copied directly from the other and that the text thus comes from an earlier manuscript.[81]

---

77. Ibn Abī Uṣaybiᶜah, 2:140–43. Since Ibn al-Tilmīdh was another bibliophilic physician who lived about a century later, the account of the manuscripts makes sense. I have not identified Jirjis ibn Tādrus or Abu'l-Khayr Sahl.

78. İhsanoğlu, *Fihris makhṭūṭāt al-ṭibb*, 170–74; Ritter and Walzer, "Arabsiche Übersetzungen," 40; Al-Dubayan, *Galen*, 35–36.

79. Walzer, "Arabische Aristoteles-Übersetzungen," 277–80.

80. Ritter and Walzer, "Arabsiche Übersezungen," 46–47.

81. Hamarneh, *Catalogue*, 19–21, which mostly discusses the epitomes and has no information about the MS beyond what I have reproduced here.

U: *Istanbul University Library A6158* (301 ff., 13.8 × 21.3 cm., 19 ll., *naskh*, 1174/1760), a handsomely written but useless manuscript, containing the fourth through the seventh, the ninth through eleventh, and the fifteenth epitomes. It is dated at folio 61a. Though the manuscript is in excellent condition, it was clearly written by a professional scribe who knew nothing about medicine and was working from a badly damaged original. There are errors of every sort, misreadings, words that the scribe obviously could not read and so drew in their approximate shape, and empty spaces corresponding to gaps or illegible words in the exemplar. The text is eclectic and therefore not of much use in constructing a stemma. In this edition, I have ignored its many unique readings. If it has value, it is for the epitomes of *The Large Pulse, On Affected Parts*, and *The Method of Healing*, for which manuscripts are much less common.

Other manuscripts: I also know of the following manuscripts that I have not used in this edition:

- *Tehran, Majlis 6036*, an almost-complete manuscript dated 1068/1657, missing only *The Medical Sects, On Diseases and Symptoms*, and most of *Anatomy for Beginners*.[82]
- *Tehran University 4914*, a seventeenth-century manuscript containing all the epitomes except for *The Large Pulse, The Method of Healing*, and *The Regimen of the Healthy*.[83]
- *Haidarabad, Āṣafiya, Ṭibb 44*, an eighteenth-century manuscript containing all the treatises except *Days of Crisis, The Method of Healing*, and *The Regimen of the Healthy*.[84]

For manuscripts of other epitomes not edited here, see table 3 above.

Textual history and editing methods

The textual tradition is too contaminated to draw a stemma. Not only is there clear internal and external evidence of cross-contamination of the manuscript traditions, but the formulaic nature of the texts often allows scribes to correct errors. Moreover, the textual history of the epitome of *On the Elements* is different from and somewhat more complex

82. Ḥāʾirī, *Fihrist*, 20–29.
83. Dānish-Pashūh, *Fihrist*, 14:4016–4023.
84. Al-Kantūrī, *Fihrist*, vol. 4.

than that of the other two works. However, some aspects of the texts do allow plausible judgments about their textual histories.

The epitomes of The Medical Sects *and* The Small Art. None of the six manuscripts that I have used to edit these two works (nor either of the two additional manuscripts of the epitome of *On the Elements*) seems to be a copy of another, so all six need to be considered. The six manuscripts divide into three fairly consistent families: **FS**, **AM**, and **DY**. Moreover, the three groups do not vary by accumulation of random copying errors but are clearly distinct recensions.

DY, the third recension, can be dealt with most easily. These two manuscripts, of the eleventh/seventeenth and ninth/fifteenth centuries respectively, almost always agree with each other; and when they disagree, it is almost always by random copying errors and in no particular pattern. These two works are about equally likely to side with either of the two other groups against the other, so in that sense they are not of great textual value; their source must have been a manuscript consulting manuscripts from both the **FS** and **AM** lines.

AM, the second recension, is a second close family, both having been copied—almost certainly from the same manuscript of the first three epitomes—in the Acre area between 1240 and 1242. Thus, for *The Medical Sects* and *The Small Art*, **A** and **M** agree closely with each other. **A** is the least eccentric of the manuscripts, being carefully written with very few unique readings, but **M** is also quite careful. For once, we can identify the source of the recension. It represents the text of the unidentified Ibn Ṭūmā as corrected by Ibn al-Tilmīdh, probably in the second quarter of the twelfth century, approximately 275 years after they were translated by Ḥunayn. This means that while **AM** is a clear witness to a form of the text at least half a century older than any of the other manuscripts, it is almost certainly not the original form of the text.

FS, the first recension, is more problematic. **F** has the most unique readings—almost three hundred in *The Small Art*—followed by **S**, with over 150. While they most often agree with each other, they also often diverge, agreeing either with one of the other manuscript groups or disagreeing with all. On the other hand, they do not have the evidence of systematic revision found in **AM** and **DY**. Moreover, they both—and particularly **F**—have an archaic feel, with many deviations from standard Arabic grammar and usage. Finally, **F** is simply sloppy. **F** and **S** nevertheless seem to represent the oldest form of the text. It is possible that the four hundred or so unique readings in **F** represent the original form of the text, with the readings of **S** and **DSY** representing one attempt at revision and **AM** another. If so, it is tempting to speculate that **F** represents Ḥunayn's original version and that **S** or **AM** represents a revision of the sort he is known to have made to his translations. However, it is also possible that **F** is simply the product of a careless scribe who did not spell very well, and that **S**, or even **AM**, represents the oldest form of the text. In any case, we have no evidence for how Ḥunayn handled this particular text. Based on these facts, I have made the following assumptions in editing the first two texts:

- **FS** represents the oldest form of the text.
- **AM** and **DY** represent two later recensions.
- **F** is a very imperfect witness to its family and thus cannot be used unless it is supported by **S** or by **AM**.

Thus, I have produced an eclectic text—a text-critical sin, to be sure, but the epitomes are neither Homer nor the Bible. I almost always follow the majority of **AM**, **F**, and **S** (or **DSY**). I have mostly omitted variants that cannot plausibly provide evidence against the readings I have chosen—particularly readings found only in one manuscript or only in the **DY** group—though I have kept occasional unique readings from the older manuscripts when they seem particularly interesting. In all, I have omitted about two-thirds of the variants I originally recorded.[85]

The epitome of On the Elements. In the epitome of *On the Elements*, **A** has broken company with **M** to ally itself with **R**, the oldest manuscript available for any of the epitomes. The textual history of this treatise is thus distinctly different from that of the previous two.

- **DY** again is a distinct group. Here it has a clear relationship to **S**, forming a group **DSY**, which is more faintly discernible in the previous two treatises.
- **MS** forms a loose group, with **S** more and **M** less associated with **DY**.
- **AFRU** forms a distinct group with **F** more and **A** less closely associated with **R**. The execrable **U** usually supports this group.

---

85. Readers needing all the variants will be able to find PDF files of my complete collation, which contains all the variants, through Indiana University Scholarworks, http://hdl.handle.net/2022/14416, though the text is not quite the same as my final text.

Since **AFRU** contains two of the three oldest manuscripts and three of the five older ones, I follow it in this edition. Where there is disagreement, I follow **R** if it has support from other old manuscripts. Thus, this text is somewhat less eclectic than those of the other two works, since I have four clear witnesses to what seems the oldest form of the text. Again, I omit almost all variants that cannot plausibly be used as evidence against my chosen readings.

Other editorial policies

Arabic grammar. The Arabic of these translations is not especially good. Moreover, it is clear that part of the confusion in the textual history is the result of scribes trying to correct or polish the text. In the case of the **DY** group, this was obviously done centuries after the composition of the text, so its improvements can be disregarded; but the situation with **FS** and **AM** is less clear. In editing the text, I have tried to walk a middle line, covering the translator's nakedness when possible—as when gender agreement can be handled by changing dots—but otherwise letting the "errors" stand, especially when they have good manuscript support. It is, perhaps, a warning to view the grammatical exactitude of modern Arabic text editions with some suspicion. Nevertheless, I have not been totally consistent in handling this problem, for which I ask the reader's indulgence.

Orthography. I have generally not recorded variants that are simply matters of orthography (for example, *idh*^{an} and *idhan*); nor have I usually recorded variants consisting solely of variant placement or omission of dots (notably the very frequent variation of *yā* and *tā* in imperfect verbs, signalizing a gender or active/passive distinction). In some cases, though, I have done so where the word is significantly different and one or more manuscripts make a clear distinction, as in *yubayyinūna* and *yuthbitūna*.

Greek names. In many cases, the scribe is uncertain as to how Greek names should be spelled. More often than not, dots are omitted and even the basic shape of the word is formed ambiguously. I have reconstructed the form as best I can, having the advantage of knowing, in most cases, what the Greek name actually should be. Versions of the name in my manuscripts incompatible with the form I have chosen have been listed separately, but I have not tried to reproduce all the versions, which are useless even for reconstructing relationships among manuscripts.

Divisions of the texts

I have supplied the numbered paragraph divisions in the three texts.

The chapter divisions of *The Medical Sects* are apparently ancient and are found in the Arabic translation and in the Alexandrian epitome. These chapter divisions differ slightly from those of modern editions of the Greek text but agree in the two Arabic texts. The divergences are noted in my text.

I have added chapter numbers to *The Small Art* and *On the Elements* and numbered them according to the chapter numbers of the modern editions. In the former, I have supplied most of the titles myself. In the latter, they are mostly supplied by a later reader (**R²**) of an early manuscript. Thus, in both cases they have no textual authority and are added simply as an aid to the reader. *On the Elements* is noticeably disorganized, so while the epitomist has more or less followed Galen's order, he has also often pulled together points from various places in the book. Thus, my assignment of chapter numbers has been somewhat arbitrary.

The Arabic textual tradition considers *On the Elements* to be one book in opposition to part of the Greek tradition, which considers it to be two.[86]

Glosses and scholia

I have included the glosses and scholia in the apparatus and have translated almost all of them. They give an idea of the relationship between the readers and the text and are often helpful. Moreover, they convey something of the experience of reading these texts from a manuscript.[87] I have not usually tried to distinguish hands, and all such additions are marked with a superscript *h*—for example, **A**^{h}—after the manuscript sigla. While some are random readers' notes, two groups deserve special mention:

The glosses of Ibn al-Tilmīdh. Throughout the epitomes of *The Medical Sects* and *The Small Art*, occasional glosses are found in manuscripts **A**, **M**, **S**, and sometimes **Y**. These are not just casual readers' notes, since

86. De Lacy's introduction in Galen, *Galeni De elementis*, 20, 44.

87. For an example of a translation that has tried to convey the experience of reading the author's copy by including the marginalia, see the Guyer and Wood translation of Kant, *Critique of Pure Reason*.

they are usually found in several manuscripts. In **A**, one of the older manuscripts, they are actually embedded in the text block and so were obviously planned as part of the content of the manuscript. The first such gloss is introduced as *ḥāshiya li-sanadinā Amīn al-Dawla* (a gloss by our authority, Amīn al-Dawla), and subsequent glosses are introduced as *ḥāshiya lahu* (a gloss by him). This is Abū al-Ḥasan Hibat Allāh ibn Ṣāᶜid ibn al-Tilmīdh, an eminent Christian physician, medical educator, and bibliophile of Baghdad who died at a great age in 560/1165. What is relevant for us is that he produced glosses on various works of Galen and Hippocrates and that he wrote a completion of the Alexandrian epitome of *The Method of Healing*. The glosses here are not comprehensive enough to be considered a separate work, but they could very well have been copied from the margins of a manuscript of the epitomes that had belonged to Ibn al-Tilmīdh.[88] Most of the larger and more significant glosses seem to come from this source. Moreover, as mentioned above, at least the first three epitomes in **A** and **M** were copied from a manuscript connected with him.

The introductions to the books. **S** contains introductions to two of the three books edited here, as well as a general introduction. They contain the familiar "eight heads," a standard medieval Islamic format for introducing books. They are written in a small, unpleasant, and not very legible hand, but I have transcribed and translated them as best I can. These are interesting for what they show about the views of Muslim scholars as to the place of these books and their subjects in medical literature.

Galen's underlying texts. There are occasional quotations in the glosses from the original text *(faṣṣ)* of Galen's works and, in one place (**A**, f. 11b, **M**, f. 14b to paragraph 40 of the epitome of *The Medical Sects*), from Abū al-Faraj ibn al-Ṭayyib's commentary on *The Medical Sects*.

Thersites. One gloss occurs in all the manuscripts of *The Small Art*, an explanation of a reference to Thersites, a badly deformed soldier in the Greek army at Troy. The gloss accurately explaining this obscure reference must have been in the margins of Ḥunayn's translation and, as likely as not, came from a gloss on the Greek manuscript.

88. See appendix 1, s.v. "Ibn al-Tilmīdh."

Translation

The translation is as literal as is compatible with clarity and precision. I have had little success in giving it a grace that the Arabic—and presumably also the Greek—never had. A chronic problem is when to translate medical terms literally and when to use modern medical terminology. As a rule, medieval Arabic uses ordinary words for most medical terms, so there seems no particular reason to translate *baṭn* as "abdomen" when it is the usual Arabic word for "belly." As a result, I usually use ordinary English terms when they are available rather than Latinate medical terminology, which in any case generally derives from the everyday Greek or Latin words that underlie their Arabic renderings.

Some terms posed special problems, which in many cases I discuss in notes. *Ghayr majrā al-ṭabʿ* might reasonably be translated as "abnormal," the term a modern doctor would use; but "contrary to nature," the term that English-speaking Galenists used several centuries ago, preserves the connection to the philosophical usage of "nature." On the other hand, the terms *al-aʿḍāʾ al-mutashābihat al-ajzāʾ* and *al-aʿḍāʾ al-murakkaba* correspond almost exactly to the modern terms "tissues" and "organs." I considered keeping the literal renderings "homoeomerous organs" and "compound organs," since that would have preserved a distinction important in Galenic medicine; but the temptations of concision and elegance won out. Another problem has to do with the scope of terms. "Causes" for *asbāb* fits both traditional and modern usage when used for the causes of disease; but the text also uses it for treatments, for which a modern doctor might prefer a term like "agent." Another problem word is *mizāj*, which can mean "mixture," "constitution," or "temperament." I have almost always used "temperament," though the result is sometimes odd. I have been consistent most of the time, and my usual renderings of significant terms are given in the glossary. Nevertheless, it is not possible always to render Arabic technical terms with the same English word, nor is the translator infallible.

As for translation at the level of sentences and above, the meaning is almost always perfectly clear. When it is not, the general sense is usually obvious enough that a demure obscurity of translation is able to bridge the difficult passage.

In the epitome of *The Medical Sects*, which includes philosophical terminology, I have usually followed the renderings of technical terms in Michael Frede's translation of *The Medical Sects*.[89]

Annotation

For the most part, I have not annotated the translations beyond citing the text that the epitome is commenting on and summarizing what Galen was saying at that point. Given that these epitomes were written at the very end of late antiquity, attempting to cite parallels systematically would have involved dealing with the whole literature of ancient medicine, which I did not think was either necessary or feasible. I have also identified individuals mentioned in the text. Occasionally, I gloss a passage or explain technical terms when clarity is at stake or a reader might have difficulty. In two cases, I have provided appendices. The first gives biographical and source information on individuals and schools mentioned in the introduction and text, notably the putative or possible authors of the epitomes and the members of the ancient medical sects mentioned in the epitome of *The Medical Sects*. Second, in the case of the epitome of *The Small Art*, there is a section on the eye sufficiently complex to require a short appendix explaining the terminology for the anatomy of the eye. In general, however, the texts are clear in themselves; they were, after all, written to explain the difficulties of Galen's texts to students new to them.

I translate almost all the glosses and meaningful variants in the notes.

89. Galen, *Three Treatises*, 3–20. Frede's introduction is particularly valuable for indicating the philosophical issues implicit in these works of medical methodology.

Abbreviations and Conventions

For complete information on works listed here, consult the bibliography.

A	Manuscript Aya Sofya 3609
ANRW	*Aufstieg und Niedergang der römischen Welt*
BNP	*Brill's New Pauly*
Br. Lib. Or.	British Library Oriental Manuscript
CDSB	*Complete Dictionary of Scientific Biography*
D	Manuscript British Library Or. Add. 23,407
DK	Diels, *Die Fragmente*
DPA	*Dictionnaire des philosophes antiques*
EI	*Encyclopedia of Islam*, 2nd edition.
F	Manuscript Fatih 3538
G[a]	Arabic translation of Galen's original text, ed. Muḥammad Salīm Sālim
GAL	Carl Brockelman, *Geschichte der arabischen Literatur*
GAS	Fuat Sezgin, *Geschichte des arabischen Schriftums*
H	Galen, *Scripta minora*, ed. Helmreich
IGAIW	Institut für Geschichte der Arabisch-Islamischen Wissenschaften
IHAIS	Institute for the History of Arabic-Islamic Science, Johann Wolfgang Goethe University
Is. Med.	Fuat Sezgin, ed., Islamic Medicine Reprint Series
Is. Phil.	Fuat Sezgin, ed., Islamic Philosophy Reprint Series
IT	Ibn al-Tilmīdh
K	Galen, *Opera omnia*, ed. Kühn

KR	Kirk and Raven, *Presocratic Philosophers*
M	Manuscript Manisa 1759
MS, MSS	manuscript, manuscripts
PIHAIS	Publications of the Institute for the History of Arabic-Islamic Science
R	Manuscript British Library Or. 9202
RE	Pauly and Wissowa, *Real-Encyclopädie*
S	Manuscript Aya Sofya 3588
U	Manuscript Istanbul University A6158
Vienna ed. gr.	Österreichischen Nationalbibliothek Greek manuscript, medical collection
Y	Manuscript Yeni Câmi 1179

For other abbreviations used in the critical apparatus, see the section "Description of Manuscripts," pp. liv, lix–lxiii, above.

In the English portions of this work, transliterated words follow the Romanization tables established by the American Library Association and the Library of Congress (ALA-LC Romanization Tables: Transliteration Schemes for Non-Roman Scripts, compiled and edited by Randall K. Barry [Washington, DC: Library of Congress, 1997]: available online at http://www.loc.gov/catdir/cpso/roman.html).

THE ALEXANDRIAN EPITOMES OF GALEN
VOLUME 1

[Manuscript table of contents][*]

In the Name of God, the Merciful, the Compassionate

(1) The epitomes of the treatises of Galen
that were studied under the teachers in Alexandria.

There are sixteen books.[1]

First,	*[On] the Medical Sects*	one book
Second,	*The Small Art [of Medicine]*	one book
Third,	*The Small Pulse [for Teuthras]*	one book
Fourth,	his treatise *[Therapeutics] for Glaucon*[2]	two books

* The table of contents of the Alexandrian Epitomes found at the beginning of MS **S**, in slightly different form in **A** and **M**, and in incomplete form in **Y**.

1. Other MSS read: "The index of the epitomes of the sixteen books of Galen that were read in Alexandria." Another MS mentions that they are "commentary and abridgment." One MS has the gloss: "The first volume of the epitomes of Galen's books on medicine, which are among what was translated by Ḥunayn [ibn Isḥāq] the physician, beginning with *[On] the Medical Sects* and ending with *[On] the Causes of Symptoms* [the last part of *On Diseases and Symptoms*, which is the last treatise contained in that MS]. There were eight commentators: Gesius, Anqīlāʾus, Archelaus, Palladius, John the Grammarian, Stephanus, Theodosius, and Abū al-Faraj ibn al-Ṭayyib." On the lists of compilers of the epitomes, see pp. xxxv–xliii above.

2. One MS adds the gloss: "Two books. The first is on the word 'nature' and what is connected with it. After he has explained the word 'nature,' he begins to explain fevers and their states. The second book is on the curing of diseases."

[محتويات النّسخ]

بسم الله الرّحمن الرّحيم[1]

جوامع الكتب التي كانت تقرأ على المعلّمين بالإسكندريّة من كتب جالينوس

وهي ستّ عشر كتاباً[2]

الأوّل منها	كتاب فرق الطّبّ	مقالة
والثّاني	كتاب الصّناعة الصّغيرة	مقالة
والثّالث	كتاب النّبض الصّغير	مقالة
والرّابع	كتابه إلى اغلوقن[3]	مقالتين

١ M: + ربِّ اختمِ بخير الحاضر | ٢ AM: فهرست جوامع الكتب السّتّة عشريّة لجالينوس التي كانت تقرأ بالإسكندريّة ـــهـ؛ Y: الأوّل من جوامع الإسكندرانيّين بكتب جالينوس السّتّة عشر على الشّرح والتّلخيص، وهي الكتب التي كانت تقرأ في الإسكندريّة؛ Sh: مجلّد أوّل من جوامع كتب جالينوس ممّا نقله جنين المتطبّب مبتدئاً بفرق الطّبّ ومختتماً بأسباب الأعراض في الطّبّ مفسّروا [؟] ثمانية جاسيوس وأنقيلاوس وأريكلاوس وفلاديوس ويحيى النّحويّ واسطفن وثاوذسيوس وأبوالفرج بن الطّيّب | ٣ Y: في اسم الطّبيعة وما يتعلّق بها وبعد ما تبيّن اسم الطّبيعة شرع إلى بيان أحوال الحمّى؛ Yh: وهو مقالتان الأولى في اسم الطّبيعة وما يتعلّق بها وفي الحمى وأحوالها والثّانية في شفاء الأمراض

Fifth,	*The Elements [According to the Opinion of Hippocrates]*[3]	one book
Sixth,	*[On] the Temperaments*[4]	three books
Seventh,	*[On] the Natural Faculties*	three books
Eighth,	*[On] Anatomy*[5]	five books
Ninth,	*[On] Diseases and Symptoms*	six books
Tenth,	*The Large Pulse*	sixteen books
Eleventh,	*The Diagnosis of Diseases of the Internal Organs*[6]	six books
Twelfth,	*On the Crises*	three books
Thirteenth,	*Days of Crisis*	three books
Fourteenth,	*Fevers*[7]	two books
Fifteenth,	*The Method of Healing*	fourteen books
Sixteenth,	*The Regimen of the Healthy*	six books

In all, there are sixteen treatises in seventy-three books.

3. One MS adds: "and the Compounds of the Elements."
4. Two MSS read: "*The Temperament*."
5. One MS adds: "*for Beginners*," which is part of the usual title for this work.
6. Two MSS reverse the order of this and the previous work.
7. Some MSS read: "The Kinds of Fevers."

والخامس	كتاب الإسطقسات[٤]	مقالة
والسّادس	كتاب الأمزجة[٥]	ثلث مقالات
والسّابع	كتاب القوى الطّبيعيّة	ثلث مقالات
والثّامن	كتاب التّشريح[٦]	خمس مقالات
والتّاسع	كتاب العلل ولأعراض	ستّة مقالات
والعاشر[٧]	كتاب النّبض الكبير	ستّة عشر مقالات
والحادي عشر[٨]	كتاب تعرّف علل الأعضاء الباطنة	ستّة مقالات
والثّاني عشر	كتاب البحران	ثلث مقالات
والثّالث عشر	كتاب أيّام البحران	ثلث مقالات
والرّابع عشر	كتاب الحمّيات[٩]	وهو مقالتين [كذا]
والخامس عشر	كتاب حيلة البرء	أربعة عشر مقالة
والسّادس عشر	كتاب تدبير الأصحّاء	ستّة مقالات

جملة هذه السّتّة عشر كتابًا، ثلث وسبعون مقالة[١٠]

---

٤ Y^h: ومركّبات العناصر | ٥ AM: المزاج | ٦ Y: + للمتعلّمين | ٧ AM: الحادي عشر | ٨ AM: العاشر | ٩ AM: أصناف الحمّيات | ١٠ AM: جملة. . . مقالة > فجملة المقالات ثلث وسبعون مقالة والكتب ستة عشر كتابًا

[The eight heads][*]

In the name of God, the Merciful, the Compassionate!

(2) The Sheikh,[8] may God have mercy upon him, said:

Each art has a subject peculiar to it and an end. The art of medicine has a subject and an end. Its subject is the human body, and its end is the preservation of health existing in the body or the restoration and acquisition of the health that is absent from the body. Because the human body possesses both a natural and an unnatural state, necessity compels the physician to investigate each of them in theory and in practice. The practice in the case of each state involves either something natural that is to be preserved, or something unnatural that is to be restored to its natural state. Since this is the case, medicine is divided into theory and practice. Theory is divided into the theory of natural things and the theory of unnatural things. The theory of natural things is divided into

1. the elements, which Galen discusses in his treatise *On the Elements [According to the Opinion of Hippocrates]*. Because the humors arise from

2. the mixture of the elements, he talks about them in his book *On the Temperament*; and because the organs are generated from the humors, he talks about these in his treatise *On the Humors*.[9]

---

* The following gloss is found at the beginning of **S**. Similar glosses are found before most of the treatises in this MS.

8. Not positively identified, but possibly Ibn al-Tilmīdh, whose glosses are found elsewhere.

9. There is no book of this title by Galen. Since this gloss is otherwise an explanation for the choice of the sixteen books of the Alexandrian curriculum and this book would make seventeen, the scholiast is probably still referring to *On the*

S^h: [الرّؤوس الثّمانية]

بسم الله الرّحمن الرّحيم

قال الشّيخ، رحمه الله

كلّ صناعة لها موضوع يخصّها وغاية، وصناعة الطّبّ صناعة لها موضوع وغاية، وموضوعها بدن الإنسان، وغايتها حفظ صحّة موجودة فيه أو ردّ واجتلاب صحّة قد فقدت منه، ولأنّ بدن الإنسان توجد له حالتانِ طبيعيّة وغير طبيعيّة تؤدّ الطّبيب الضّرورة إلى النّظر فيهما جميعاً وعلمهما وعملهما، والعمل فيهما جميعاً إمّا الأمر الطّبيعيّ فبأن يحفظ، فإمّا الخارج عن الطّبيعة فبأن يردّ إلى الحال الطّبيعيّة، وإذا كان الأمر على هذا انقسم إلى العلم وإلى العمل، والعلم ينقسم إلى علم الأشياء الطّبيعيّة وعلم الأشياء الخارجة عن الطّبيعة، وعلم الأشياء الطّبيعيّة ينقسم إلى

آ. علم الإسطقسات، وجالينوس يتكلّم فيها في كتابه «في الإسطقسات».

ب. ولأنّ الأخلاط تحدث بامتزاج الإسطقسات فهو يتكلّم فيها في كتابه «المزاج» ولأنّ الأخلاط تكوّنت عنها الأعضاء، فهو يتكلّم فيها في مقالته «في الأخلاط».

3. From the tissues are compounded the organs, so he talks about these in his treatise *On Anatomy [for Beginners].*

4. These organs have faculties, so he talks about those faculties that are natural in his treatise *[On] the Natural Faculties*, about the vital faculties in the treatise *On the Pulse*, and about the psychic faculties in the treatise *[On the Doctrines of] Hippocrates and Plato.*

5. Our organs have functions, so he discusses these in his treatise *On the Uses of the Parts.*

6. The theory of unnatural things is divided into diseases, the causes of diseases, and the symptoms attendant upon diseases. He teaches us about symptoms in his treatise *[On] Diseases and Symptoms.*

7. Since some diseases are manifest and can be observed, while others are hidden and have indications, he speaks about the latter in the treatise *On the Affected Parts*. This is *The Internal Diseases.*

8. Because some common diseases are fevers, which have so many classes, it is therefore not excessive for him to devote a treatise to enumerating *The Species of Fevers*,

9. or because one of the indications is the pulse, he teaches us about that in the treatise *The [Large] Pulse.*

10. Because crisis is a correlate of diseases, he teaches us about that in *[On] Crises.*

11. Because crises occur on particular days, he teaches us about that in his treatise *Days of Crisis.*

12. Instruction is also divided in the first instance into the preservation of natural things, which he discusses in the treatise *The Regimen for the Healthy*,

13. and into the reversal of unnatural things, which he discusses in the treatise *The Method of Healing.*

14. As for his treatise *To Glaucon*, it is derived from the treatise *The Method of Healing.*[10]

15. His treatise *The Small Pulse [for Teuthras]* is derived from *The Large Pulse.*

Elements, a work that is considered in some Greek manuscripts to consist of two books, the second of which does deal with the humors. Since the Arabic tradition always considers it as a single book, it is possible that this gloss depends on a Greek source.

10. The scholiast accidentally added a number in the middle of this entry, which I have corrected in the text and translation.

ج. و من الأعضاء المتشابهة تتركّب الآليّة، فهو يتكلّم فيها في كتابه «في التّشريح».

د. ولهذه الأعضاء قوى فهو يتكلّم في الطّبيعيّة منها في «كتاب القوى الطّبيعيّة» وفي الحيوانيّة في «كتاب النّبض» وفي النّفسانيّة في «كتاب بقراط» وأفلاطون.

هـ. ولآلاتنا أفعال، فهو يتكلّم فيها في كتابه «في منافع الأعضاء».

و. فأمّا علم الأشياء الخارجة عن الطّبيعة فينقسم إلى الأمراض وأسباب الأمراض والأعراض التّابعة لها للأمراض، فهو يعلّمنا من الأعراض في «كتاب العلل والأعراض».

ز. ولمّا كانت الأمراض منها ظاهرة وهذه مشاهدة ومنها خفيّة وهذه لها دلائل، فهو يتكلّم فيها في «كتاب مواضع» [كذا] «الآلمة»، وهو الأمراض الباطنة

ح. ولأنّ من جملة الأمراض الحميات وأقسامها كثيرة، فلهذا ما أفرط لها كتاباً عدّد منه أنواع الحميات.

ط. أو لأنّ أحد الاستدلالات النّبض، فهو يعلّمنا عنها في «كتاب النّبض».

ي. ولأن يلزمها البحران فهو تعلّمنا عنه في «البحران».

يا. ولأنّ البحران له أيّام معروفة فهو يعلّمنا عنها في كتابه «في أيّام البحران».

يب. فأمّا التّعلّم فينقسم أيضاً إلى علم حفظ الأشياء الطّبيعيّة وهو يتكلّم فيها في «كتاب تدبير الأصحّاء».

يج. وإلى ردّ الأشياء الخارجة عن الطّبيعة وهو يتكلّم فيها في «كتاب حيلة البرء».

يد. فأما كتابه «إلى اغلوقن»، فإنّه فرع على «كتاب حيلة البرء»

[يه.] وكتابه «في النّبض الصّغير» فرع على كتابه «في النّبض الكبير»

16. His treatise *The [Small] Art [of Medicine]* is, as it were, a complete summary of all of his books.

17. *[On] the [Medical] Sects* is an introductory book that should be read before any other theoretical or practical book, so that a person will know the path that should be taken in discovering what drugs are appropriate.

This is the order followed by Galen's books on medicine.

[A gloss on the art of medicine]

(3) Some arts are theoretical, like the art of geometry; some purely practical, like the arts of the carpenter and blacksmith; and some both theoretical and practical, like the art of music. The musician is the one who sings and performs the variations by which the melody is formed into the melody. Our art is both theoretical and practical, for the physician needs knowledge of the disease, the cause of the disease, and its indications. It is his place to administer drugs and perform surgery. It is not one of the arts that pursues an end unrestrictedly, but, rather, is one of those that is for the sake of some good. What is it that would make you forget the path that the rule requires and will require? It may be that you will attain the goal, and it may be that you will not.

[يو.] وكتابه في «الصّناعة» يجري مجرى الجملة لكتبه بأسرها.

[يز.] و«الفرق» هو كتاب مقدّمة تجب قراءته قبل الكتب بأسرها العلميّة والعمليّة ليتبيّن منها الإنسان الطّريق التي يجب أن يسلكها في استنباط ما يجب من الأدوية.

فعلى هذا يجري ترتيب كتب جالينوس في الطّبّ. هـ

[صناعة الطّبّ]

S^h: حـ[اشية]

لمّا كانت الصّنائع منها علميّة كصناعة الهندسة ومنها عمليّة حسب كصناعة النّجار والحدّادين ومنها علميّة وعمليّة كصناعة الموسيقاويّة فإنّ الموسيـ[قي] وهو الذي ينغم ويرقي العلل التي من أجلهما صارت هذه النّغمة تركيب هذه النّغمة، فصناعتنا علميّة وعمليّة، فإنّ الطّبيب مفتقر إلى معرفة المرض وسبب المرض ودلائله، وشأنه أن يعطي أدوية ويستعمل الحديد، وليست من الصّنائع التي يتبع غرضاً لا محالة لكن هي ممّا لجودة، فما هو أن ينساك[؟] الطّريق التي أوجبها ويوجبها القانون، وقد يجوز أن تصيب الغرض ويجوز ألا تصيب.

THE ALEXANDRIAN EPITOMES OF GALEN

VOLUME 1

The Alexandrian Epitome of Galen's Book *[On] the Medical Sects*

In the Name of God, the Merciful, the Compassionate![1]

The Alexandrian Epitome of Galen's Book

[On] the Medical Sects

<Entitled *Hairesis*>[2]

Using the method of commentary and abridgment

[The parts of medicine][3]

(1) Some people divide medicine into two parts, and some divide it into five parts.

(2) Those who divide it into two parts say that some of it is theory[4] and some is practice.[5] The theory is divided into three parts: first, the theory of natures; second, the theory of causes; and, third, the theory of

1. Five manuscripts add formulas, such as "My Lord, make it easy by Your mercy," that are not part of the original text.

In a late MS, a reader has copied the following sensible bit of doggerel: "Among the sayings of al-Shaykh al-Raʾīs Abū ʿAlī Ibn Sīnā [Avicenna] is:

Medicine all in two verses is found;	That brevity ever gives beauty is sure.
If little you eat, then from eating refrain,	Good health in digestion will always be yours.
The heaviest burden to be laid upon souls	Is food thrust straight down upon food."

2. Replacing the following phrase in some MSS.

3. Paragraphs 1–4 are introductory material not epitomizing a specific section of Galen's book.

4. This word could also be translated "knowledge" or "science."

5. The outline of medicine given here is the one used in Ibn Sīnā's *Canon*. It

بسم الله الرحمن الرحيم[1]

جوامع الإسكندرانيّين لكتاب جالينوس[2]

في فرق الطّبّ

على الشّرح والتّلخيص[3]

‹المسمّى إراسيس[4]›

[أجزاء الطّبّ]

(١) إنّ بعض النّاس قسّم الطّبّ قسمينِ وبعضهم قسّمه[5] خمسة أقسام.

(٢) والذين قسّموه قسمينِ قالوا إنّ منه علم ومنه عمل[6]، والعلم ينقسم ثلثة أقسام، أحدها علم الطّبائع، والثّاني علم الأسباب، والثّالث علم العلامات والدّلائل، وعلم

---

١ Y^h: + من أقوال الشّيخ الرّئيس أبي عليّ بن سينا

جميع الطّبّ في البيتينِ درج ... وحسن القول في قصر الكلام
تقلّل إن أكلت وبعد أكل ... تجتنب فالشّفاء في الانهضام
وليس على النّفوس أشدّ حالاً ... من إدخال الطّعام على الطّعام

D: ربِّ يسّر وتمّم لي الحجّة؛ FM: + ربِّ يسّر برحمتك؛ S: + ربِّ وفِّق؛ Y: ربِّ أنعمتَ فزده | ٢ F: جوامع كتاب جالينوس | ٣ ADMY: – على. . . والتّلخيص | ٤ FS: – المسمّى أراسيس؛ F: + للإسكندرانين | ٥ ADMY: – قسّمه | ٦ ADMY: علماً ومنه عملاً

signs and symptoms. The theory of natures comprises six things: first, the theory relating to the elements; second, the theory relating to the temperaments; third, the theory relating to the humors; fourth, the theory relating to the organs; fifth, the theory relating to the faculties; and, sixth, the theory relating to actions.[6] The theory relating to the causes comprises the theory of antecedent causes, the theory of preceding causes, and the theory of cohesive causes.[7] One part of the theory of signs and indications is the knowledge of what is presently the case. Such signs are specifically called "indications." Another part is the knowledge of what will be, which is called "prognosis[8] of what will be." The last part is the knowledge of what formerly was,[9] which is called "mnemonics." Practice is divided into two parts, one of which is hygiene and the other therapy. Hygiene is divided into three parts: first, hygiene itself; second, prophylaxis; and, third, nutrition and recuperation by regimen. This regimen may be a regimen for the bodies of old men, a regimen for the bodies of children,[10] or a regimen for those convalescing

---

has parallels in Galen's *On the Parts of Medicine* and fits with the curriculum implied by the sixteen books of Galen studied in the Alexandrian curriculum. See Galen, *On the Parts of Medicine*, 24–49. This particular work is lost in Greek but survives in Arabic and Latin translation. It has, incidentally, a good deal to say about the views of the Empiricists.

6. Three manuscripts contain the following marginal gloss: "A gloss by our authority Amīn al-Dawla [Ibn al-Tilmīdh]: There is no need to mention spirits, since they are included under the humors, the spirits being their vapors." This apparently refers to the science of the faculties, corresponding to Galen's *On the Natural Faculties*, also part of the Alexandrian medical curriculum and of which the Alexandrians made an epitome. This is the first of the glosses apparently by the eminent Baghdad physician Ibn al-Tilmīdh; see pp. lxvii–lxviii above and appendix 1, s.v "Ibn al-Tilmīḍh."

7. In Greek, καταρκτικόν, προηγούμενον, and συνεκτικόν, respectively. They refer respectively to the sensible external conditions leading to disease or health, such as exposure to cold; the internal, not directly sensible conditions of the body that lead to disease, such as an imbalance in the humors; and the sufficient cause for a state of disease or health. The last is also translated in Arabic as *sabab māsik* and in English as "containing cause." For the historical issues connected with these terms, see Galen, *On Antecedent Causes*, esp. 31–37 and 81–125.

8. Some MSS read: "knowledge."

9. Some MSS add: "and is no more."

10. One MS has the marginal gloss: "It might be asked how the regimen of healthy people may be divided into regimens for children and old men, without mentioning adolescents and young adults. The reply is that only the two extremes are mentioned, and the other two are contained within them."

الطبائع يحوي ستّة أشياء، أحدها العلم بأمر الأسطقسات، والثاني العلم بأمر الأمزاج، والثّالث العلم بأمر الأخلاط، والرّابع العلم بأمر الأعضاء، والخامس العلم بأمر القوى، والسّادس العلم بأمر الأفعال[٧]. فأمّا[٨] العلم بأمر الأسباب، فمنه العلم بأمر الأسباب البادية، ومنه العلم بالأسباب[٩] السّابقة، ومنه العلم بالأسباب[١٠] الواصلة، وأمّا علم العلامات والدّلائل، فمنه العلم بما هو حاضر ويقال لهذه العلامات خاصّةً دلائل، ومنه العلم بما سيكون، ويقال لهذا سابق النّظر[١١] بما يكون[١٢]، ومنه العلم بما قد[١٣] سلف[١٤]، ويقال لها مذكّرة بما قد سلف ومضى، وأمّا العمل، فينقسم قسمين أحدهما حفظ الصّحّة، والآخر اجتلاب[١٥] الصّحّة، وحفظ الصّحّة ينقسم ثلثة أقسام، أحدها الحفظ المطلق، والآخر التّقدّم بالحفظ، والثّالث التّغذية والإنعاش بالتّدبير، وهذا التّدبير منه تدبير أبدان الشّيوخ ومنه تدبير أبدان الصّبيان[١٦]، ومنه تدبير أبدان النّاقهين من الأمراض، وأمّا اجتلاب[١٧] الصّحّة، فمنه ما يكون بالتّدبير ومنه ما

---

٧ $A^hM^hS^h$: حاشية لسندنا أمين الدّولة [S^h: ح‹اشية› له]: إنّما ألغى ذكر الأرواح لدخولها في الأخلاط لأنّها بخارها؛ Y: الأفعال > الأسباب البادية | ٨ ADMY: وأمّا | ٩ ADMY: بأمر الأسباب | ١٠ ADMY: بأمر الأسباب | ١١ ADMY: العلم | ١٢ F: بما يكون > وتسمّى منذرة؛ DY: سيكون | ١٣ AM: − قد | ١٤ ADMY: + ومضى | ١٥ FS^2: اختلاف | ١٦ S^h: قد يقال كيف قسّم تدبير الأصحّاء إلى تدبير الصّبيان والمشايخ وألغى ذكر الغلمان والشّباب. الجواب أنّه ذكر الطّرفين والوسطان محصوران. | ١٧ F: فأمّا اختلاف الصّحّة

from diseases. Therapy may be accomplished by regimen, by surgery,[11] or by the use of drugs. Surgery may be to the flesh (such as incision,[12] stitching, cutting, or cauterizing), or it may involve the bones (setting a broken bone or reducing a dislocated bone).

(3) Those who divide medicine into five parts say that its parts are the theory of natural things, which are the six that we mentioned; the theory of causes, which are the three previously mentioned; the theory of symptoms, which are the three we described before; hygiene; and therapy. They say that hygiene is divided into three parts: first, the part that preserves the healthy temperament as it is, which is said to be hygiene strictly speaking; second, the part that removes the causes from which diseases arise[13] while the body is healthy, which is called prophylaxis; and, third, nutrition and recuperation by regimen. The kinds of regimen are those three that we mentioned.[14] They say that therapy is also divided into three parts: first, the regimen of food and drink, exercise and rest—both of the soul and of the body—and sleep and wakefulness; second, surgery and its varieties, which are those we mentioned before; and, third, the use of drugs, both those that cause change and those that purge.

---

11. Literally, "treatment by the hand," translating the Greek χειρουργία, "working by hand," or surgery.

12. A late MS adds the gloss: "That is, to be cut."

13. One MS has the gloss: "In a general sense, disease occurs by the introduction of that which is not natural and which is such as to cause actual harm in the first instance."

14. A gloss in several MSS reads: "That is, the regimen of the old, the young, and the middle-aged."

يكون بعلاج اليد ومنه ما يكون باستعمال الأدوية، والعلاج باليد منه ما هو في اللّحم مثل البط[١٨] والخياطة والقطع والكيّ، ومنه ما هو في العظام مثل جبر العُظم المكسور وإصلاح العظم المخلوع.

(٣) وأمّا الذين قسّموا الطّبّ خمسة أقسام، فقالوا إنّ أقسامه العلم بالأشياء الطّبيعيّة، وهي تلك السّتّة التي ذكرناها، والعلم بالأسباب، وهي تلك الثّلثة التي تقدّم ذكرها، والعلم بالدّلائل، وهي تلك الثّلثة التي وصفناها قبلُ، وحفظ الصّحّة واجتلاب الصّحّة، وقالوا إنّ حفظ الصّحّة ينقسم ثلثة أقسام، أحدها الجزء الذي يحفظ المزاج الصّحيّ على ما هو عليه، ويقال له حفظ الصّحّة المطلق[١٩]، والآخر الجزء الذي يقطع كون الأسباب التي يتوقّع منها إحداث الأمراض[٢٠] ما دام البدن صحيحًا، ويقال له التّقدّم بالحفظ، والثّالث التّغذية والإنعاش بالتّدبير، وأصناف التّدبير هي تلك الثّلثة[٢١] التي ذكرناها، وأمّا[٢٢] اجتلاب[٢٣] الصّحّة، فقالوا إنّه ينقسم أيضًا ثلثة أقسام، أحدها التّدبير بالمطعم والمشرب والحركة والسكون ما كان من ذلك للنّفس وما كان منه للبدن والنّوم واليقظة، والآخر علاج اليد وأصنافه هي تلك التي ذكرناها قبلُ، والثّالث استعمال الأدوية ما كان منها يغيّر وما كان منها يستفرغ.

---

١٨ Y^h: أي اشتقّ | ١٩ ADMY: مطلقًا؛ S: بالمطلق | ٢٠ S^h: المرض يحدث على الإطلاق بإدخال ما خارجة عن الطّبيعة شأنها أن يضرّ بالفعل إضرارًا أوّليًا | ٢١ $A^hM^hS^h$: له أي تدبير الشّيوخ والصّبيان والنّاقهين | ٢٢ ADMY: فأمّا | ٢٣ F: اختلاف

[The sects of medicine][15]

(4) There are three sects of medicine: first, the sect of Empiricists,[16] whose adherents employ experience alone; second, the sect of Rationalists,[17] whose adherents employ both experience and inference; and, third, the sect of the Methodists,[18] who employ neither experience nor inference. The prominent adherents of the Empiricist sect were Acron of Agrigentum,[19] Philinus of Cos,[20] Serapion of Alexandria, Sextus [Empiricus], and Apollonius [Empiricus]. The prominent adherents of the Rationalist sect were Hippocrates, Diocles [of Carystus], Praxagoras, Philotimus, Erasistratus, and Asclepiades. Those who arose to support the Methodist sect were Themison of Laodicea, Thessalus [of Tralles], Mnaseas, Menemachus, and Soranus.[21]

---

15. The Arabic word *firqah*—like the Greek αἵρεσις, which it translates—does not necessarily carry all the connotations of the English "sect" and may mean no more than a shared trend of thought. It does tend to be a term used to refer to those one disagrees with.

16. Rendering ἐμπειρικοί, "experienced."

17. Rendering λογικοί (Rationalists), or δογματικοί (Dogmatists). The Arabic name is literally "the companions of *qiyās*," which can mean syllogism, analogy, or inference.

18. *Aṣḥāb al-ḥiyal* (the people of tricks), a slightly pejorative rendering of the Greek μεθοδικοί.

19. Appendixes 1 and 2 below contain a discussion of these lists of prominent members of the three medical sects, as well as information on the various ancient and medieval physicians and schools mentioned here and elsewhere in this volume.

20. The MSS read this name as the more familiar Philip. I have corrected it, but since there is no MS support for the correction, the error could have been present in the Greek.

21. MSS **A** and **M** probably read: "Themison, Thessalus, Menedotus, Menemachus, Mnesitheus, and Mnaseas."

[فرق الطّبّ]

(٤) فرق الطّبّ ثلثة[٢٤]، الواحدة فرقة أصحاب التّجارب، وأهلها يستعملون التّجربة وحدها، والأخرى[٢٥] فرقة أصحاب القياس، وأهلها يستعملون التّجربة والقياس معاً، والثّالثة فرقة أصحاب الحيل، وأهلها ليس يستعملون لا التّجربة ولا القياس، والذين قاموا بتثبيت فرقة أصحاب التّجارب أقرن الأقراغنيطيّ وفيلنوس القزاقيّ[٢٦] وسارافيون الإسكندرانيّ وسخطس[٢٧] وأبولونيوس[٢٨]، والذين قاموا بتثبيت فرقة أصحاب القياس إبّقراط[٢٩] وديوقليس[٣٠] وفركساغورس وفولوطيموس وأراسسطراطيس[٣١] وأسقليبياذس، والذين قاموا بتثبيت فرقة أصحاب الحيل ثاميسن[٣٢] الأوذيقيّ وثاسلس[٣٣] ومناماخوس ومناساس وسورانوس[٣٤].

٢٤ ADMY: ثلث | ٢٥ ADMY: والثّانية | ٢٦ A: وفيلفوس القوافيّ؛ D: وفيلقوس العواميّ؛ F: وفيلبس القوافيّ؛ M: وفيلفوس القزاقيّ؛ S: وفيليبسس القواقيّ؛ Y: وفيلقوس الفواقيّ | ٢٧ AM: سخيطس؛ D: سنحطس | ٢٨ D: وابولوفيوس؛ F: أبولونموس؛ S: وأبولونمسوس؛ Y: وأبولوفيوس | ٢٩ AM: بقراط | ٣٠ F: ديوقلس | ٣١ F: وأسطراطس؛ S: إراسطراطس | ٣٢ AM: ثامينس؛ F: قاميس؛ S: ثاميسس | ٣٣ AM: + مناتاذس | ٣٤ AM: + ومنيسثاوس ومناساوس

Commentary on chapter 1 of Galen's book *On the Medical Sects*[22]

[The definition of medicine]

(5) Soranus defined medicine in this way: "Medicine is the knowledge of matters related to health, matters related to disease, and matters that relate neither to health nor to disease." Herophilus said, "Medicine is the knowledge of matters related to health, which are healthy bodies, causes preserving or effecting health, and signs indicating health; matters relating to disease, which are diseased bodies, causes effecting disease, and signs indicating disease; and matters related neither to health nor disease, which are the body that has that state, the cause effecting it, and the signs indicating it."[23]

(6) Causes are of two kinds, some healthful and some causing disease.[24] There are two species of healthful causes: some preserving existing health, and some restoring and bringing health after it has been destroyed. There are also two species of causes leading to disease: those that preserve existing disease, and those that attract a disease that had not been there before. The causes of health that preserve existing health are called the regimen of the healthy and are based on food,

22. K 1:64–65; H 3:1–2; G[a] 11–15. In this chapter, Galen states that the aim of the art of medicine is health and that an ancient definition says that medicine is the science of the healthy and the unhealthy. Empiricists rely on experience alone to gain this knowledge, while the Rationalists also use reason. He finishes with the various names of the two schools.

23. Though Galen does not mention the third category of the neutral in *The Medical Sects*, this threefold account of the subject of medicine is mentioned or used elsewhere in Galen's works, including *The Small Art*. Its attribution to Herophilus is well known; see Von Staden, *Herophilus*, 89–98, T42–T49. I have not traced the version of the definition attributed to Soranus—which is, in fact, the definition of Herophilus—but at the beginning of his *Gynecology*, his major surviving work, Soranus mentions that "some divide [gynecology] into physiology, pathology, and therapy," which is apparently what Herophilus meant; see Soranus of Ephesus, *Maladies des femmes*, 1.5; and Soranus of Ephesus, *Gynecology*, 3. Most likely, the quotation comes from a lost work of Soranus.

24. A gloss in three MSS reads: "When he says, 'Causes are of two kinds,' he is indicating that the state that is neither health nor sickness is something

شرح الباب الأوّل
من كتاب جالينوس في الفرق

[حد الطّبّ]

(٥) قال سورانوس في حدّ الطّبّ إنّ الطّبّ معرفة الأمور الصّحيّة والأمور المرضيّة والأمور التي ليست بصحيّة ولا مرضيّة[٣٥]، وقال إيروفيلس إنّ الطّبّ معرفة الأمور الصّحيّة، وهي الأبدان الصّحيحة والأسباب الحافظة والفاعلة للصّحّة، والعلامات الدّالّة على الصّحّة، والأمور المرضيّة، وهي الأبدان المريضة والأسباب الفاعلة للمرض والعلامات الدّالّة على المرض، والأمور التي ليست بصحيّة ولا مرضيّة، وهي البدن الذي حاله هذا[٣٦] الحال، والسّبب الفاعل لذلك والعلامات[٣٧] الدّالّة عليه.

(٦) الأسباب صنفانِ[٣٨] منها صحيّة ومنها مرضيّة، والصّحيّة[٣٩] نوعانِ، منها ما يحفظ الصّحّة الموجودة ومنها ما يردّ ويجلب الصّحّة بعد فسادها، والمرضيّة أيضاً نوعانِ، منها ما يحفظ المرض الموجود، ومنها ما يجذب[٤٠] مرضاً لم يكن، وما كان أيضاً[٤١] من أسباب الصّحّة حافظاً للصّحّة الموجودة، فهو يسمّى تدبير الأصحّاء ويكون

---

٣٥ DSY: – والأمور التي . . . مرضيّة | ٣٦ DSY: هذه | ٣٧ A^2؛ S: والعلامة | ٣٨ $A^hM^hS^h$: حاشية له [S^h: – حاشية له] وقوله [A^h: فقوله؛ S^h: في قوله] الأسباب صنفان تنبيه على أنّ الحال التي [M^h: الذي] ليست بصحّة ولا مرض قول [A^h: يقول] مطلق [S^h: – ل مطلق] داخله في قسمي الصّحّة والمرض الذين جعل الأسباب أسباباً لهما فقط. [S^h: – باباً لهما فقط]. | ٣٩ F: مريضة والصّحّة؛ SM: الصّحيّة | ٤٠ D: يحذف؛ M: يحدث | ٤١ AM: – أيضاً

drink, moderation in exercise, and bathing. The causes that bring about health that had not previously existed are called treatment. Some of these causes evacuate what ought to be removed from the body, such as letting blood from a vein and purging by means of a drug; and some change a state that needs to be changed—either from outside, such as by a poultice, or from inside, such as drinking cold water.

(7) All physicians are in agreement about the general purpose of medicine and share a common view, for all of them seek to acquire health for the body. However, they disagree concerning the existence of the things[25] by which health may be gained and how they may be known. This is because the Empiricists claim that these things may become known only through experience. The Rationalists claim that they do not become known by experience alone but, rather, by experience that is combined with an inference indicating it. The Methodists profess to reject experience and the use of syllogism, but in fact they employ both of them.[26]

(8) The first two sects are known by various names. Since the Empiricist sect seeks to know the things by which health may be acquired by experience only, they are called the empirical sect or the sect that relies on memory.[27] The Rationalist sect, which employs deduction, is called rationalist, dogmatic, and the ones who ascend from matters apparent to sensation to matters apparent to the mind.[28]

---

referred to only in a general way. It actually falls into the two divisions of health and disease, for which reason he makes the causes causes for these two classes only." There is a grammatical problem with the gloss, but the point is clear—to explain why the epitomist does not mention causes of states related neither to health nor disease.

25. One MS glosses this as "That is to say, drugs."

26. Galen himself does not mention the Methodists until chapter 6.

27. K 1:65; H 2; G[a] 15: τηρητικήν, μνημονευτικήν.

28. K 1:65; H 2; G[a] 15: λογικήν, δογματικήν.

بالمطعم والمشرب والقصد في الرّياضة والاستحمام، وما كان منها يحدث صحّة ليست بموجودة، فهو يسمّى مداواة وبعض هذه الأسباب يستفرغ من البدن ما يحتاج إلى استفراغه بمنزلة فصد العرق والإسهال بالدواء وبعضها يغيّر هيئة ما يحتاج إلى تغييره[٤٢] إمّا من خارج بمنزلة الضّماد وإمّا من داخل بمنزلة شرب الماء البارد.

(٧) وجميع[٤٣] الأطبّاء متّفقون على تمام غرض[٤٤] الطّبّ[٤٥] مشتركون فيه إذ كان جميعهم إنّما يطلبون إفادة البدن الصّحّة، إلاّ أنّهم يختلفون في وجود الأشياء[٤٦] التي تستفاد بها الصّحّة واستخراجها، وذلك لأنّ[٤٧] أصحاب التّجارب يزعمون أنّ هذه الأشياء تستخرج بالتجربة وحدها، وأصحاب القياس يزعمون أنها لا يستخرج بالتجربة وحدها لكنّ بالتّجربة التي يكون معها قياس يستدلّ به، وأمّا أصحاب الحيل، فإنّهم يلتمسون إرذال[٤٨] التّجربة واستعمال القياس بالكلام[٤٩]، فأمّا بالفعل، فهم يعيدون منهما جميعاً.

(٨) كلّ واحدة[٥٠] من الفرقتين الأوّلتين تسمّى بأسماء شتّى، أمّا فرقة أصحاب التّجارب، وهي التي تستخرج الأشياء التي تستفاد بها الصّحّة بالتّجربة وحدها، فتقال لها المجرّبة والحافظة والمتذكّرة، وأمّا فرقة أصحاب القياس التي تستعمل الاستدلال[٥١]، فتقال لها القياسيّة وذات الرّأي وذات الارتقاء من الأمور الظّاهرة للحسّ إلى

---

٤٢ DMY: تغيّره | ٤٣ ADMY: جميع | ٤٤ F: تمام غرض > غاية؛ S: – غرض؛ A^h: أي غايته؛ M^h: الغاية | ٤٥ S^h: غايات الصّنائع إمّا جيّدة أو رديّة إمّا ح . . . فمثل غرض صناعة الطّبّ التي غايتها ان . . . الصّحّة أولصناعة البنا، فإنّهما غاي. . . وأمّا الصّناعات التي غاياتها . . . كصناعة اللّصوصيّة ل. . . . ثياب النّاس وما اشته . . . Partly cut off in rebinding | ٤٦ S: + أي الأدوية | ٤٧ AMY: أنّ؛ D: – لأن | ٤٨ AM: رفض؛ DY: رفض العقل وإرذال؛ S: رفض وارذال | ٤٩ F: بالقول؛ DY: + والقوّة | ٥٠ DMY: واحد | ٥١ A: الاستدلالات؛ M: بالاستدلالات؛ S: بالاستدلال

Those who belong to each of these two sects are called by names derived from the names of their sects. Those who employ experience are called Empiricists and practitioners of memory. Those who employ inference are called Rationalists, Dogmatists, and ascenders from the thing that is apparent to sensation to the thing that is understood by the mind. Derived names require three things: first, that the derived name have something in common with the name from which it is derived; second, that the meaning of the derived name have something in common with the meaning of the name from which it is derived; and, third, that the final syllable of the derived name be different from the final syllable of the name from which it is derived.

الأمور الظّاهرة للعقل، وأهل كلّ فرقة من هاتيْنِ الفرقتين يسمّون بأسماء مشتقّة من أسماء فرقتهم، فأمّا أصحاب التّجارب، فيقال لهم المجرّبون والحافظون والمتذكّرون، وأمّا أصحاب القياس، فيقال لهم القياسيّون وذوي[٥٢] الرّأي وذوي[٥٣] الارتقاء من الشّيء الظّاهر للحسّ[٥٤] إلى الشّيء الذي يعرف بالعقل. يحتاج في الأسماء المشتقّة إلى ثلثة أشياء، أحدها أن يكون الاسم المشتقّ مشاركًا للاسم الذي منه اشتِقّ، والآخر أن يكون معناه مشاركًا بمعنى[٥٥] ذلك، والثّالث أن يكون آخر مقطع الاسم المشتقّ مخالفًا لآخر مقطع الاسم الذي اشتقّ منه ــهـ

٥٢ ASY: ذوو | ٥٣ M: ذو؛ AY: ذوو | ٥٤ F: الشّيء الذي بالحسّ؛ M: إلى الحسّ | ٥٥ DSY: لمعنى

Commentary on chapter 2 of Galen's book *On the Medical Sects*[29]

[Medical experience]

(9) Experience has five parts:[30] first, natural, such as a nosebleed, sweating, releasing the bowels, and vomiting; second, accidental, such as drinking cold water and wine and other such things; third, voluntary, those whose occurrence[31] is from a dream, omen, or soothsayer; fourth, imitation, which is when the physician imitates, either by nature, by accident, or by incidence;[32] and, fifth, transition from the thing to what is similar to it, either from an organ to an organ, as in transferring from the upper arm[33] to the thigh, or from a disease to a disease, as from the disease known as erysipelas to the disease known as herpes,[34] or from a drug to a drug, as in transferring from quince to medlar.

---

29. K 1:66–69; H 2–4; G[a] 16–22. In this chapter, Galen summarizes the Empiricist theory of the acquisition of medical knowledge through experience gained from observation and experiment. Some MSS consistently add "Of Galen's Book *On Medical Sects*" to each chapter head.

30. K 1:66–67; H 2–3, G[a]17–18, 21: (1) φυσικόν, (2) τυχικόν, (3) αὐτοσκεδιόν, (4) μιμητικόν, (5) τοῦ ὁμοίου μετάβασις, respectively.

31. Some MSS read: "explanation."

32. K 1:66; H 2; G[a] 17: Περίπτωσις (incidence), a neologism in Greek, is an Empiricist technical term for an observed natural or chance medical phenomenon. The earliest MS reads "voluntary experience."

33. Some MSS read: "muscle."

34. Erysipelas, ἐρυσίπελας—literally "red skin" in Greek—is an inflammation of the skin. Ibn Sīnā defines herpes or shingles, ἕρπης, *namlah*, in the *Canon*, 3.117, as "one or more small pustules breaking out and forming a harmful swelling that may ulcerate or become infected, . . . yellowish in color."

شرح الباب الثاني من كتاب جالينوس في فرق الطّبّ[٥٦]

[التّجربة الطّبيّة]

(٩) أجزاء التّجربة خمسة، أحدها الطّبيعيّ بمنزلة الرّعاف والعرق واستطلاق البطن والقيء، والثّاني العرضيّ بمنزلة شرب الماء البارد والشّراب وغير ذلك ممّا أشبهه والثّالث الإراديّ الذي كونه وثباته[٥٧] إمّا من المنام[٥٨] وإمّا من الزّجر، وإمّا من المتكهّن[٥٩]، والرّابع التّشبيه[٦٠]، وهو أن يشبّه الطّبيب إمّا بالطّبع وإمّا بالعرض وإمّا بالاتّفاق[٦١]، والخامس نقل الشّيء إلى ما هو شبيه به إمّا من عضو إلى عضو بمنزلة النّقلة من العضد[٦٢] إلى الفخذ، وإمّا من علّة إلى علّة بمنزلة النقلة من العلّة المعروفة بالحمرة إلى العلّة المعروفة بالنّملة وإمّا من دواء إلى دواء بمنزلة النّقلة من السّفرجل إلى الزّعرور.

---

٥٦ ADMY: – من كتاب جالينوس في فرق الطّب | ٥٧ D: وبيانه؛ FY: كونه وثباته > وبيانه | ٥٨ F: + وإمّا من المتكهّن | ٥٩ S: وإمّا من التّكهين وإمّا من الزّجر؛ DY: التّكهين | ٦٠ S^h: قال الشّيخ أبوا. . . القسمة يجب أن. . . [؟] آخر التّجربة تن. . .

/ \

المشبّهة والنّقلة

بالطّبيعة بالإرادة بالاتّفاق من دواء إلى دواء

فأمّا القسمة والنّقلة فيقو[مان] مقام الآلة المستخرج بها ال. . . المفيدة للصّحّة وأمّا التّشـ[بيه] بالطّبيعة والإرادة والاتّفـ[اق] فيجري مجرى المادّة التي . . . | ٦١ F: + الإراديّ | ٦٢ FM: العضل

(10) This division may be made in another way, for it is said that experience has four parts: first, incidence, whether with respect to nature or with respect to accident; second, the volitional; third, the imitative; and, fourth, the transition from the thing to what is similar to it.[35] Two of the four parts into which experience is divided serve as the matter from which it is drawn; these are natural and accidental incidence and the voluntary. Two serve as the instrument for the deduction of the things by which health is acquired; these are imitation and transition from the thing to that which is similar to it.

(11) There are three sorts of transition from the thing to that which is similar to it.[36] The first is when drugs are transferred from a disease to another disease resembling it—as when cooling drugs are transferred from the swelling known as erysipelas to the disease known as herpes, since these are two diseases that resemble each other in heat and red color. The second is when drugs are transferred from an organ to another organ similar to it—as when something is transferred from the upper arm to the thigh, since each of the two organs resembles the other in nature and form. The third is when treatment is transferred from one drug to another—as when medlar is used in place of quince in treating diarrhea, due to these two drugs resembling each other in causing constipation.

(12) Imitation is also of three sorts.[37] This is because, in what he does, the physician follows an example either of nature, of accident, or of volition. An example of his following the example of nature is when he sees that someone afflicted by a fever of the blood has a nosebleed and that his nosebleed benefits him. Therefore, when he encounters someone else with a fever of the blood, he lets blood from a vein. An

35. K 1:66; H 2; G[a] 17-18. One MS reads: "nomenclature."
36. K 1:68; H 4; G[a] 20; this passage follows Galen's order and examples exactly.
37. K 1:67; H 3; G[a] 18.

(١٠) وقد تقسم هذه القسمة[٦٣] بضرب آخر ، فيقال إنّ أجزاء التّجربة أربعة، أحدها الاتّفاق إمّا من قِبَل الطّبع وإمّا من قبل العرض، والآخر الإراديّ، والثّالث المشبّه، والرّابع النّاقل من الشّيء إلى شبهه، ومن هذه الأجزاء الأربعة[٦٤] التي تنقسم التّجربة عليها اثنانِ يقومانِ مقام المادّة التي تستمدّ منها، وهما الاتّفاق الطّبيعيّ[٦٥] والعرضيّ[٦٦] والإراديّ، واثنانِ يقومان مقام الآلة التي بها يستخرج الأشياء التي تستفاد بها الصّحّة، وهما التّشبيه ونقل الشّيء إلى ما هو شبيه به.

(١١) ونقل الشّيء إلى ما هو شبيه به يكون على ثلثة أوجه، أحدها أن ينقل الأدوية من علّة إلى علّة تشبهها بمنزلة ما ينقل[٦٧] الأدوية المبرّدة من الورم المعروف بالحمرة إلى العلّة المعروفة بالنّملة لأنّ هاتينِ علّتانِ متشابهتانِ[٦٨] في الحرارة وحمرة اللّون، والثّاني أن ينقل الأدوية من عضو إلى عضو شبيه به بمنزلة ما ينقل الشّيء من العضد إلى الفخذ لمشابهة كلّ واحد من هذينِ العضوينِ للآخر[٦٩] في الطّبع وفي الهيئة، والثّالث أن ينقل العلاج من دواء إلى دواء بمنزلة ما يستعمل في الاستطلاق[٧٠] مكان السّفرجل الزّعرور[٧١] لمشابهة كلّ واحد من هذينِ الدّوائينِ الآخر[٧٢] في القبض.

(١٢) والتّشبيه أيضاً يكون على ثلثة أضرب، وذلك أنّ الطّبيب يتمثّل[٧٣] فيما يفعله إمّا بالطّبع وإمّا بالعرض وإمّا بالإرادة، وامتثاله[٧٤] الطّبع يكون[٧٥] بمنزلة ما إذا هو رأى أنّ صاحب حمّى الدّم لمّا رعف انتفع برعافه استعمل في غيره من أصحاب هذه

---

٦٣ F: التّسمية | ٦٤ F: الأربعة أقسام؛ S: الأربعة الأجزاء | ٦٥ $A^h M^h$: حاشية له: الاتّفاق الطّبيعيّ والعرض والإراديّ كأنّه قال اثنانِ يقومانِ مقام [A: مغاير] المادّة التي تستمدّ منها، وهما الاتّفاق بقسميه، أعني ما اتّفقّ من الطّبيعة وما اتّفق بالعرض والإراديّ | ٦٦ F: والعرض؛ | ٦٧ ADSY: ما ينقل > نقل؛ M: > ينقل | ٦٨ F: علتيّن متشابهتين؛ S: العلّتيّن متشابهتان | ٦٩ DMSY: الآخر | ٧٠ F: استطلاق البطن | ٧١ AM: زعرور | ٧٢ AS: للآخر | ٧٣ A^2؛ ADS: يمتثل | ٧٤ F^2: وامتياله | ٧٥ ADMY: – يكون

example of his following the example of accident is when he sees that someone who has a fever of the blood happens, for some reason, to bleed from a cut that he has received in a part of his body and that he is helped by it. The physician then cuts the vein of another such person and draws blood from him. He follows the example of what occurs by volition in the case when he sees a man in a dream, is led by an omen in his soul, or is told by a soothsayer to have blood let. He thus feels an urge to do so and submits to bloodletting by his own volition. He then benefits from the bloodletting, which the physician thus employs on others who have diseases similar to his.

(13) According to the Empiricists, there are two ways in which things[38] are apprehended and understood: by vision, which is called autopsy, and hearing, which is called history.[39]

(14) There are five species of experiences, which are the parts of experience: first, natural; second, accidental; third, voluntary; fourth, imitative; and, fifth, transition from something to its like. If these five parts are combined as genera, they become four: incidence,[40] volition, imitation, and transition. There are seven if they are distinguished as species: natural; accidental; volitional; imitative; and transitional, this last including three species—from disease to disease, from place to place, and from drug to drug. This division can also be made in another way, for it is said that experience is established by two things, one of them being the matter from which the person acquires experience, and

38. A gloss in one MS reads: "He means those things employed by the Empiricists."

39. K 1:67; H 3; G[a] 20: αὐτοψία, *mushāhadah* in G[a] but *mubāsharah* here; and ἱστορία, *khabar* in G[a] but *riwāyah* here. These are technical terms of the Empiricists referring respectively to one's own accumulated medical observations and to one's knowledge of the experience of others.

40. Three MSS have the gloss: "The natural and the accidental have been combined in incidence so that they become four, in the same way that he says concerning the natural, 'By chance there was an incidence of nosebleed . . . ,' and concerning the accidental, 'There was an incidence in which he was cut and blood flowed from it.'"

الحمّى فصد العرق، وأمّا امتثاله العرض[٧٦] فبمنزلة ما إذا هو رأى أنّ صاحب حمّى الدّم عندما اتّفق أنّ موضعاً من بدنه انخرق بسبب من الأسباب فسال منه دم، فانتفع بذلك استعمل في غيره فجر العرق وإخراج الدّم منه، وأمّا امتثاله الإرادة، فبمنزلة ما إذا رأى إنسان في المنام أو وقع في نفسه بالزّجر أو من التّكهّن[٧٧] أن يفتصد، فمالت نفسه إلى ذلك وافتصد بإرادته وانتفع بفصده استعمل الطّبيب في غيره ممّن به مثل تلك العلّة فصد العرق.

(١٣) إدراك علم الأشياء[٧٨] ومعرفتها تكون على ضربيْن عند أصحاب التّجارب[٧٩]، إمّا بالبصر ويقال له المباشرة، وإمّا بالسّمع، وتقال له الرّواية.

(١٤) التّجارب خمسة أنواع، وهي أجزاء التّجربة[٨٠]، أحدها الطّبيعيّ والثّاني[٨١] العرضيّ والثّالث الإراديّ والرّابع المشبّه والخامس النّاقل من الشّيء إلى شبهه[٨٢]، وهذه الخمسة الأجزاء إذا حصلت أجناساً كانت أربعة، وهي الاتّفاق والإرادة والتّشبيه والنّقل، وإذا فصلت أنواعاً كانت سبعة، وهي الطّبيعيّ والعرضيّ والإراديّ والمشبّه والنّاقل، وهو ثلثة أنواع، إمّا من علّة إلى علّة، وإمّا من موضع إلى موضع، وإمّا من دواء إلى دواء، وهذه القسمة تقسم[٨٣] على وجه آخر، فيقال إنّ التّجربة تثبت بشيئيْن، أحدهما المادّة[٨٤] التي تستمدّ منها صاحب التّجربة والآخر النّوع[٨٥] الذي

---

٧٦ F: امتياله بالعرض | ٧٧ AM: أمر بالتّكهّن؛ $A^{2}M^{2}$: من أمر التّكهّن؛ S: من التّكهين | ٧٨ S: إدراك علم الأشياء > إذ قال علم الأمور؛ S^{h}: يريد الأمور التي يستعملها أصحاب التّجارب | ٧٩ DY^{1}: عند أصحاب التّجارب تكون على ضربيْن؛ ADM: + عند أصحاب التّجارب | ٨٠ F^{h}: وهي أجزاء التّجربة ويقال له صحّ الدّواء | ٨١ FS: والآخر | ٨٢ $A^{h}M^{h}Y^{h}$: حاشية له: قد جمع الطّبيعيّ والعرضيّ من الخمسة في الاتّفاق حتّى صارت أربعة [M: تابعة] كأنّه قال في الطّبيعيّ اتّفق أنّ رعف وفي العرضيّ اتّفق أن شجّ فسال منه دم | ٨٣ S: تنقسم؛ M: تقتسم | ٨٤ $A^{2}M^{2}$: + وتكون الحركة إليها أمّا | ٨٥ $A^{h}M^{h}$: ح ظ أي من صاجب [M: واحب] الإرادة وإن كان كذلك فالجيد أن يقال ويكون من فاعليها الحركة إليها إمّا

the other being the form that he uses. Its matter includes the things that occur by nature, the things that occur by incidence, and the things that are done by volition, whether motivated[41] by a dream, an omen, or soothsaying. Its form includes imitation and the transfer of something to what is like it. Imitation occurs when the physician imitates either what is by nature, what is by volition, or what is by accident. That which is from volition follows either a dream, an omen, or soothsaying. The transfer of something to what is similar to it is either from disease to disease, from organ to organ, or from drug to drug. If nature is mentioned to you here, understand it to mean that faculty governing the body of an animal. That is because this word—that is, "nature"—is employed in three senses: first, the substance and existence of each thing; second, the governing faculty of the body of an animal; and, third, the temperament and usual state of the body.

41. Two MSS have the gloss: "That is, the one exercising volition. If that is the case, then it would be better to say, 'Among the efficient causes of the motion to it is . . . '"

يستعمله، ومادّتها هي الأشياء التي تكون بالطّبع والأشياء التي تكون بالاتّفاق، والأشياء التي تفعل بالإرادة وتكون[٨٦] الحركة[٨٧] إليها إمّا من المنام وإمّا من الزّجر وإمّا من التّكهّن، ونوعها التّشبيه ونقل الشّيء إلى شبهه، والتّشبيه هو أن يشبّه الطّبيب إمّا بما يكون من الطّبع وإمّا بما يكون من الإرادة وإمّا بما يكون من العرض، والذي يكون من الإرادة إمّا أن يتبع المنام وإمّا أن يتبع الزّجر وإمّا أن يتبع التّكهّن، ونقل الشيء إلى شبهه يكون إمّا من علّة إلى علّة وإمّا من عضو إلى عضو وإمّا من دواء إلى ودواء، وإذا قيل لك هاهنا طبع أو طبيعة، فافهم أنّ معنى ذلك هو[٨٨] القوّة المدبّرة لبدن الحيوان، وذلك أنّ هذا الاسم، أعني طبعاً أو طبيعةً تتصرّف على ثلثة وجوه، أحدها جوهر[٨٩] كلّ واحدٍ من الأشياء ووجوده، الثّاني القوّة المدبّرة لبدن الحيوان[٩٠]، والثّالث مزاج البدن وعادته ـه

٨٦ ADSY: + منه | ٨٧ DY: + أي من صاحب الأرادة | ٨٨ ADMY: - هو | ٨٩ Sʰ: أي شكل | ٩٠ ADMY: للبدن

Commentary on chapter 3 of Galen's book *On the Medical Sects*[42]

[The necessary causes]

(15) Some of the causes that alter the body alter it necessarily. These are six: first, the air surrounding it; second, exercise and rest; third, the things that are eaten and drunk; fourth, sleep and wakefulness; fifth, evacuation and retention; and, sixth, the disturbances of the soul, such as grief, worry, fear, joy, and anger.[43] There are other causes that do not alter it necessarily, such as the sword, beasts of prey, stones, arrows, and fire.

(16) The temperament of air may be exactly moderate, in the way that it is in spring; for the air in spring is moderate with respect to heat, coldness, moisture, and dryness. It may be immoderate overall, in the way that it is in summer and winter; for the air in summer is excessively hot and dry and in the winter is excessively cold and moist. It may be moderate in some respects and immoderate in others, as is the case in autumn; for the air in autumn is moderate with respect to heat and coldness but immoderate with respect to moisture and dryness, since it is inclined to be dry. In another respect, it is immoderate with respect to heat and coldness because its temper in the day is

42. K 1:69–74; H 4–9; G^a 23–31. This chapter, like G^a, combines the third and fourth chapters of the modern Greek editions. In this chapter, Galen says that the Rationalists require the physician to know the nature of the body, its inner states and functioning, and all the causes that affect its health and illness, as well as the suitability of particular treatments after consideration of factors such as climate, season, and the patient's age and occupation. In general, Rationalists and Empiricists agree about appropriate treatment, though they disagree on how it is to be discovered. The epitome is a systematic discussion of various things mentioned in the chapter rather than an explanation of Galen's argument itself.

43. K 1:69; H 4–5; G^a 23–24. This chapter is an adaptation of *The Small Art*, B 23; K 1:365–70, G^a 111–19. The causes that alter the body necessarily are the famous "six non-naturals" of later Galenic thought: the things that are neither natural in the sense of being part of the normal physiology nor contrary to nature, like disease, but that impact the balance of the body and can thus be used to treat or prevent disease. They are "necessary" in the sense that we cannot

شرح الباب الثّالث
من كتاب جالينوس في فرق الطّبّ[٩١]

[الأسباب الضّروريّة]

(١٥) الأسباب المغيّرة للبدن منها ما لا بدّ من أن يغيّره ضرورةً[٩٢]، وهي ستّة، أحدها الهواء المحيط به، والثّاني الحركة والسّكون، والثّالث[٩٣] الأشياء التي تؤكل وتشرب، والرّابع النّوم واليقظة، والخامس الاستفراغ والاحتباس[٩٤]، والسّادس عوارض النّفس مثل الغمّ والهمّ والفزع والفرح والغضب، ومنها ما ليس تغييرها أيّاه ضرورةً بمنزلة السّيف والسّبع والحجر والسّهم والنّار.

(١٦) ومزاج الهواء إمّا أن يكون على غاية الاعتدال بمنزلة ما يكون كذلك في الرّبيع، فإنّ الهواء في الرّبيع[٩٥] معتدل[٩٦] في الحرّ والبرد والرّطوبة واليبس[٩٧]، وإمّا أن يكون على خلاف الاعتدال جملةً بمنزلة ما يكون كذلك في الصّيف أو في[٩٨] الشّتاء، فإنّ الهواء في الصّيف يفرط عليه الحرّ واليبس وفي الشّتاء يفرط عليه البرد والرّطوبة، وإمّا إن يكون في بعض الحالات معتدلاً وفي بعضها غير معتدل بمنزلة ما يكون كذلك في الخريف لأنّ هواء الخريف معتدل في الحرارة والبرودة غير معتدل في الرّطوبة واليبوسة، وذلك أنّه إلى اليبس أميل مع أنّه من جهة أخرى أيضاً غير معتدل في الحرارة والبرودة لأنّ مزاجه في النّهار كلّه لا يستوي، وذلك أنّه بالغدوات يكون

---

٩١ F²S؛ ADMY: – من كتاب جالينوس في فرق الطّبّ | ٩٢ F: يغيّرو ضرورة | ٩٣ M: Note in Hebrew | ٩٤ ADMY: والاحتقان | ٩٥ DSY: – فإنّ . . . الرّبيع | ٩٦ DSY: معتدلاً | ٩٧ ADMY: واليبوسة | ٩٨ ADMY: وفي

uneven, since in the early morning it is colder, while in the middle of the day it is hotter. The temper of the air can be divided in another way, for it is said that the temper of the air can be moderate in the way that it is in spring; for even though people think that it is hot and moist, it is not actually so but is actually moderate. It may be hot and dry in the way that it is in summer, or cold and moist in the way that it is in winter. Finally, it may be moderate with respect to heat and coldness but immoderate with respect to moisture and dryness in the way that it is in autumn, when people imagine that it is cold though it actually is not.

(17) The temper of the air must either be natural—which is the temper appropriate to the time of year, as we have described above—or unnatural. If it is unnatural, then it has changed. Its change is either because its substance has changed—which gives rise to epidemics—or because its quality has changed. The change of its quality can occur in one of two ways: either because its natural quality has increased, as can occur in summer when the heat and dryness are excessive, or because it has changed and been converted into an opposite quality, in the way that summer can become moist if there is excessive rain. The change of quality in each of these two ways can occur either in the four seasons of the year, or in three of them, or in two, or in one. The natural temper of the air occurs in accordance with the current time of year and in accordance with the country. That is because some lands are northern and thus cold and dry, some southern and thus hot and moist, some eastern and thus moderate in temper, and some western and thus moderate in temper. Change of air may also occur in accordance with the orientation of the place, for it may face north, south, east, or west.

---

avoid being exposed to them, unlike swords and wild beasts. Galen mentions most of these in his third chapter but not systematically. Since Galen nowhere lists the six in the form that became known later, the Alexandrians may be their source. On the six non-naturals, see Rather, "Six Things Non-Natural"; Jarcho, "Galen's Six Non-Naturals"; and Fitzpatrick, "Galen's Six Non-Naturals."

أبرد وفي إنصاف النّهار أحرّ، وقد يقسم مزاج الهواء قسمةً أخرى ، فيقال إنّ مزاج الهواء إمّا معتدل بمنزلة ما يكون كذلك في الرّبيع الذي ظنّ قوم أنّه حارّ رطب وليس هو كذلك بل هو معتدل، وإمّا حارّ يابس بمنزلة ما يكون كذلك في الصّيف، وإمّا بارد رطب بمنزلة ما يكون كذلك في الشّتاء وإمّا معتدل في الحرّ والبرد غير معتدل في الرّطوبة واليبس بمنزلة ما يكون كذلك في الخريف الذي ظنّ قومٌ أنّه بارد وليس هو ببارد.

(١٧) ومزاج الهواء لا يخلوا من أن يكون إمّا طبيعيّاً، وهو المزاج الذي يكون بحسب الوقت الحاضر من أوقات السّنة كما وصفنا قبل، وإمّا خارجاً عن الطّبيعة ، وإذا كان كذلك فقد تغيّر، وتغيّره يكون إمّا لأنّ جوهره يتغيّر فيحدث عن ذلك وباء، وإمّا لأنّ كيفيّته يتغيّر، وتغيّر كيفيّته يكون على أحد وجهيْنِ، إمّا لأنّ كيفيّته الطّبيعيّة تزيد بمنزلة ما يعرض للصّيف إن يكون مفرط الحرارة مفرط اليبس، وإمّا لأنّه يتغيّر وينقلب إلى كيفيّة مضادّة لكيفيّته[٩٩] بمنزلة ما يعرض للصّيف إن يكون كثير المطر رطباً، وتغيّر الكيفيّة في كلّ واحد من هذيْنِ الوجهيْنِ يكون إمّا في أربعة أوقات السّنة وإمّا في ثلثة منها وإمّا في اثنيْنِ وإمّا في واحد، ومزاج الهواء الذي هو له طبيعيّ يكون بحسب الوقت الحاضر من أوقات السّنة وبحسب البلد، وذلك لأنّ البلدان منها شماليّة، وهي باردة يابسة، ومنها جنوبيّة ، وهي حارّة رطبة، ومنها شرقيّة، وهي معتدلة المزاج، ومنها غربيّة، وهي[١٠٠] معتدلة المزاج، وتغيّر الهواء أيضاً يكون من قبل وضع الموضع[١٠١] بأن يكون يستقبل إمّا الشّمال وإمّا الجنوب وإمّا الشّرق وإمّا الغرب.

---

٩٩ F: يتغيّر. . . لكيفيّته > ينتقل إلى خلافتها | ١٠٠ DFY: + غير؛ F^2: باردة صحّ | ١٠١ A^2M^2؛ AM: الوضع

(18) The effect of exercise[44] and rest varies in that exercise has two effects, for if it is moderate it tends to warm,[45] while if it is excessive it tends to cool.[46] Rest has a single effect since it is always followed by coldness; and moisture always follows coldness since there is no heat to destroy the moisture. The effects of motion vary in three ways: first, the quality of the motion; second, its quantity; and, third, the quantity of the rest that is mingled with it. If its variation with respect to quality is because it is extremely strong and harsh, it heats, dries, and becomes harder; but if it is because it is weak and is not harsh to the one who does it, in this case its effect is much less, as we have explained. It varies with respect to quantity in that if it is frequent it has the same effect as strong exercise, whereas if it is infrequent it has the same effect as light exercise. It also varies with respect to the quantity of the rest that is mingled with it, for if the mixture is rapid and continuous, it will have the same effect as strong exercise; but if it is slow and intermittent, it will have the same effect as weak exercise. Another kind of variation can be associated with motion: the variation with respect to the matter used in the profession in which the person is engaged. In the case of the man who is a bath attendant, his profession heats and moistens, whereas the profession of the fisherman cools and moistens, as does that of the seafarer. The profession of the hunter heats and desiccates, as does the profession of blacksmiths and goldsmiths; but the art of the plowman cools and desiccates.

44. Literally, "motion."
45. A late MS adds: "and moisten."
46. A late MS adds: "and dry."

(١٨) الحركة والسّكون فعلهما[١٠٢] مختلف[١٠٣]، وذلك أنّ الحركة تفعل فعليْنِ لأنّها إذا[١٠٤] كانت معتدلة فمن شأنها أن تسخّن، وإن أفرطت فمن شأنها أن تبرّد، وأمّا السّكون فإنّه يفعل فعلاً واحداً لأنه في كلّ وقت إنّما تتبعه البرودة وتتبع البرودة رطوبة لفقد الحرارة التي تفنى الرّطوبة. أفعال الحركة تختلف من ثلثة وجوه[١٠٥]، أحدها كيفيّة الحركة، والآخر مقدارها، والثّالث مقدار ما يخالطها من السّكون، واختلافها من قِبَل الكيفيّة يكون لأنّها إمّا أن تكون قويّة شديدة عنيفة فتسخّن وتجفّف وتصلّب أكثر، وإمّا إن تكون ضعيفة لا تعنف بصاحبها، فيكون فعلها لما وصفنا أقلّ، وأمّا اختلافها من قبل مقدارها، فهو أنّها إمّا أن تكون كثيرة، فتفعل ما تفعله الحركة القويّة، وإمّا إن تكون يسيرة فتفعل ما تفعله[١٠٦] الضّعيفة، وأمّا اختلافها من قِبَل مقدار ما يخالطها من السّكون وهو[١٠٧] أنّها إمّا أن تكون سريعة متواترة فتفعل ما تفعله القويّة، وإمّا أن تكون بطئة متفاوتة، فتفعل ما تفعله الضّعيفة، وقد يتبع الحركة أيضاً اختلاف آخر من قِبَل اختلاف المادّة التي تستعملها أصحاب الصّناعات، وهي أن يكون الإنسان قيّم حمام، فإنّ هذه الصّناعة تسخّن وترطب، أو[١٠٨] يكون صيّاد السّمك لأنّ هذه الصّناعة تبرّد وترطّب، وكذلك الملاحة، أو[١٠٩] يكون قناصاً[١١٠] لأنّ هذه الصّناعة تسخن وتجفّف، وكذلك صناعة الحدّادين والصّناعة أو أن[١١١] يكون حرّاثاً لأنّ هذه الصّناعة تبرّد وتجفّف[١١٢].

١٠٢ F: + في البدن | ١٠٣ DY: تختلف؛ F^2: مختلفة | ١٠٤ A^2M^2: إن؛ F: إن | ١٠٥ DSY: أوجه | ١٠٦ AF^2M: + الحركة | ١٠٧ F: فهو؛ S: هو | ١٠٨ S: أو إن | ١٠٩ ADY: أو إن | ١١٠ D: قنّاساً؛ S: – لأنّ هذه الصّناعة . . . قناصاً؛ S^2: تبرّد وترطّب أو أن يكون قناصاً و[أ]نّ هذه الصّناعة | ١١١ DSY: وأن | ١١٢ AM: + في أصناف المياه

(19) Water is used necessarily[47] in some circumstances, as when it is used for drink, and in some places not necessarily, as in bathing. There are various kinds of water. Consider stagnant water, which does not flow and is coarse and foul. There is also spring water, which is excellent. The best is from flowing springs, for it heats and cools quickly and is light in weight; and if you look at it, you will see that it is clear and pure, without any discernible qualities, whether of taste or of odor. It is cold in summer and warm in winter. There is also rainwater, which is also excellent, though it quickly becomes putrid. Water from snow and ice is the coarsest in substance, the worst, and the coldest. Some water contains pharmacological potencies, such as salt water, which has the same power as salt; the water that comes from tar pits and whose temper and potency resembles that of tar; sulfurous water, which resembles sulfur in its potency; and the water that comes from alum mines, whose potency also resembles that of alum.

(20) Some foods dry the body—for example, dry bread, millet grain, and rice. Some foods moisten, such as succulent herbs and meats. Some foods warm, such as the foods taken with mustard and pepper; and some cool, such as fruit and the two kinds of cucumber.[48] Some drinks, such as cold water, cool and moisten. Some drinks warm and moisten, such as new wine. Some warm and dry, such as spiced honeyed wine;[49] and some cool and dry, such as drinks made from vinegar and water.

---

47. "Necessarily" is used in the sense of *The Small Art*, B23, K 1:367, where it refers to the six non-naturals—the six categories of things that affect health but that cannot be avoided; see pp. 18–19, n. 43, above. Two MSS put a section heading before this paragraph: "On the kinds of waters."

48. *Al-qithā*ʾ and *al-khiyār*, respectively the Arabic and Persian words for cucumber. While it is possible that only one Greek word is being translated, several kinds of cucumber were known and used medicinally from ancient times; see *Encyclopaedia Iranica*, s.v. "Cucumber."

49. *Khandīqūn*. The readers of this text were not sure what it meant. Ibn al-Tilmīdh glosses it as "A drink of honey with spices." Another MS has the gloss: "A plant put into a drink." It is mentioned in the epitome of *The Small Art* as a warming drink, where the MSS read it variously as *khadīqūn*, *fandīqūn*, *khandīqūn*, and *qandīṭūn*; see paragraph 70, p. 21, below. The justification for my translation is Ibn Sīnā, *Qānūn* (Bulaq ed.), 3:368, which gives two recipes in which various spices were mixed with honeyed wine.

(١٩) الماء يستعمل في بعض المواضع ضرورةً بمنزلة ما يستعمل في الشّرب[١١٣] وفي بعضها غير ضرورة بمنزلة ما يستعمل في الاستحمام[١١٤]، وأصناف الماء تختلف، وذلك أنّ منه أجاميًّ[١١٥] لا يجرى، وهو غليظ رديّ، ومنه ماء العيون، وهو جيّد وأفضله ما كان يخرج من عيون غائرة ويسخن ويبرّد سريعاً، وهو خفيف الوزن، وإذا نظرت إليه رأيته صافياً نقياً، وليس فيه شيء من الكيفيّات ظاهر لا في المذاق ولا في الرّائحة ويكون في الصّيف بارداً وفي الشّتاء حارّاً، ومنه ماء الأمطار، وهو أيضاً جيّد غير أنّ العفونة تسرع إليه، ومنه ماء الثّلج والجمد، وهو أغلظ جوهراً وأرداً وأشدّ برداً، ومنه ماء توجد فيه قوى دوائيّة بمنزلة ماء الملح[١١٦] الذي قوّته قوّة الملح[١١٧] والماء الذي يخرج من عيون القير[١١٨] الذي يشبه مزاجه وقوّته قوّة القير[١١٩]، والماء الكبريتيّ، وهو الذي يشبه الكبريت في قوّته، والماء الذي يخرج من معدن الشّب وقوّته أيضاً قوّة الشّب.

(٢٠) والأطعمة[١٢٠] منها ما يجفّف البدن بمنزلة الخبز اليابس والعدس الجاورس والأرز ومنها ما يرطّب بمنزلة البقول الرّطبة واللّحوم[١٢١]، ومنها ما يسخّن بمنزلة ما يتّخذ من الأطعمة بالخردل والفلفل، ومنها ما يبرّد بمنزلة الفاكهة والقثاء[١٢٢] والخيار، والأشربة منها ما يبرّد ويرطّب بمنزلة الماء البارد، ومنها ما يسخّن وترطّب بمنزلة الشّراب الحديث[١٢٣]، ومنها ما يسخّن ويجفّف بمنزلة الخنديقون[١٢٤]، ومنها ما يبرّد ويجفّف بمنزلة الأشربة التي يتّخذ بالخلّ والماء.

---

١١٣ DSY: الشّراب | ١١٤ S: في الحمام للاستحمام | ١١٥ ADMY: أجاميًّا | ١١٦ DY: الماء المالح؛ M^2: القار | ١١٧ M^2: القار | ١١٨ A^2DY: القار؛ F: + وهو | ١١٩ A^2DY: القار؛ S: القفر | ١٢٠ DFY: الأطعمة | ١٢١ DY: + الدّسمة | ١٢٢ F: والقيء | ١٢٣ S^h: لأنّ المائيّة [؟] فيه بعد موجود | ١٢٤ A^hM^h: حـ[اشية] له هو شراب عسل بأفاوية؛ S^h: نبات يطرح في الشّراب

(21) Sleep and wakefulness have various effects in the body, since sleep strengthens the natural faculty and weakens the psychic faculty. Wakefulness does the opposite, strengthening the psychic faculty and weakening the natural faculty. Sleep is followed by the retention of what is evacuated from the body, whereas wakefulness results in the evacuation of what had been retained in the body. The exact effect of sleep varies in accordance with what it encounters in the body. If it encounters matter in the body that has not been processed or food that has not been assimilated, it processes the matter, digests the food, and heats and moistens; whereas, if it encounters the body pure and empty, the heat is drawn to the innate moisture by which it subsists and destroys it, after which the coldness is left in the body. If sleep encounters the body when there is matter in it whose quantity does not overpower the faculty, it is also beneficial and strengthens the natural heat. If it encounters the body when the quantity of the matter in it overpowers the faculty, it quenches the natural heat. This occurs, for example, in the beginning of periodic fevers at the time when the physician will come in to the sick person and command him to stay awake.

(22) One knows the quality of the things by which health is acquired from the species of the disease that they are intended to treat, and one knows their quantity from the things that are called the "daughters of

(٢١) النّوم واليقظة يختلف فعلهما في البدن، وذلك أنّ النّوم يقوّي القوّة الطّبيعيّة ويرخي القوّة النّفسيّة واليقظة بخلاف ذلك تقوّي القوّة النّفسيّة وترخي[١٢٥] القوّة الطّبيعيّة[١٢٦]، والنّوم يتبعه احتباس ما يستفرغ[١٢٧] من البدن، واليقظة يتبعها استفراغ ما هو محتبس في البدن، وفعل النّوم خاصّةً يختلف بقدر ما يصادف في البدن، وذلك أنّه إن صادف في البدن مادّةً لم تنضج وغذاء لم يستمرئ، فأنضج المادّة، وهضم الغذاء وأسخن[١٢٨] و رطبُ، وإن[١٢٩] صادف البدن نقيّا حاويًا[١٣٠] فعطفت[١٣١] الحرارة على الرّطوبة الغريزيّة التي قوامها بها فأفنتها أعقب ذلك برودة[١٣٢] البدن، وإذا[١٣٣] صادف النّوم البدن أيضاً وليس فيه مادّة مقدارها قاهر للقوّة نفع وقوّى الحرارة الطّبيعيّة، فإن صادفه وفيه مادّة مقدارها قاهر للقوّة[١٣٤] طفأ الحرارة الطّبيعيّة بمنزلة ما يعرض ذلك في ابتداء نوائب الحميات النّائبة في الوقت الذي يتقدّم فيه إلى المريض وتأمره أن يكون يقظانًا[١٣٥].

(٢٢) كيفيّة الأشياء[١٣٦] التي تستفاد بها الصّحّة تعرف من نوع العلّة التي يقصد بها لمداواتها[١٣٧]، ومقدارها يعرف من الأشياء التي تقال لها بنات[١٣٨] الأركان،

---

١٢٥ DFY: تقوي. . . وترخي > ترخي | ١٢٦ D: + القوى النّفسانيّة؛ S: –الطّبيعيّة. . . الطّبيعيّة > النّفسانيّة؛ S^2: يقوي القوّة الطّبيعيّة . . . بخلاف ذلك يرخى القوّة الطّبيعيّة ويقوي القوّة النّفسانيّة؛ Y: القوّة الطّبيعيّة > الطّبيعيّة وتقوّي النّفسانيّة | ١٢٧ S^h: [إ]نّ قوّة الدّافعة تحتاج في [فعـ]ـلها إلى . . م بالاستعانة بالقوّة المحرّكة حركة الإراديّة، وهذه هي متعطّلة في وقت النّوم. | ١٢٨ DSY: أسخن | ١٢٩ ADMY: فإن | ١٣٠ DY: خاويًا؛ S: خاليًا | ١٣١ A: تعظّمت؛ F: انعطف؛ DM^2SY: فعطفت؛ M: تعطّفت | ١٣٢ ADMY: + في | ١٣٣ DSY: فإذا | ١٣٤ FS: –نفع. . . القوّة؛ F^1S^2: نفع وقوّى الحرارة الطّبيعيّة، وإن صادف [F^1: فإن صادفه] وفيه [F^1: + مادّة] مقدارها قاهر للقوّة [F^1: + صحّ] | ١٣٥ ADMY: يقظان | ١٣٦ S^h: الأدوية | ١٣٧ A: بها مداواتها [كذ]؛ M: المداواتها [كذ]؛ FS: بها لمداواتها | ١٣٨ A^2: بزاق [؟]

the elements,"[50] the evidences upon which the decision on treatment is based. These are, for example, the age of the patient, which can be either childhood, adolescence, middle age, or old age; the temperament of the patient; the present time of year; the weather that day; the place to which he resorts;[51] his customary habits; and the trade he engages in. There are nine kinds of temperament: four simple kinds, which are hot, cold, moist, and dry; four composite, which are hot and dry, hot and moist, cold and dry, and cold and moist; and one moderate. There are also various kinds of habits; for some men are in the habit of drinking wine, some are in the habit of drinking water, some are in the habit of eating once a day, some are in the habit of eating twice or three times a day, some are accustomed to having evacuation, and some are not. There are four times of the year: first, spring, which is moderate; second, summer, which is hot and dry; third, autumn, which is dry, moderately cool, but with uncertain weather; and fourth, winter, which is cold and wet.

[The differences between the Empiricists and the Rationalists][52]

(23) The Empiricists use the symptoms following from diseases to deduce what it is that they have done many times that has been beneficial. The Rationalists use them to deduce what they ought to do. Take, for example, swelling.[53] All swelling is generated from matter belonging to one of the humors that flows into one of the organs. There are four

50. K 1:70; H 5; G[a] 25. Galen mentions these as factors that the Rationalist physician must consider when determining the proper treatment for a particular patient; but his text here, at least, is not the source for the term "daughters of the elements," as a term for the humors.

51. One MS reads: "for the patient to live in."

52. Commentary on the fourth chapter, according to the modern editions, begins here.

53. K 1:70; H 5; G[a] 24–5: An elaboration of Galen's example, showing how the Rationalist's methods would cause him to diagnose and treat a hematoma differently than would the Empiricist. Galen gives a similar argument involving rabies; see K 1:73; H 7–8; G[a] 30–31.

وهي الشّواهد التي عليها مبنى الأمر بمنزلة سنّ المريض التي إمّا أن يكون سنّ الصّبيّ وإمّا سنّ الشّباب وإمّا سنّ الكهول وإمّا سنّ الشّيوخ، ومزاج المريض والوقت الحاضر من أوقات السّنة، وحال الهواء في ذلك اليوم[١٣٩] والبلد الذي يأويه[١٤٠] والعادة التي جرى عليها والصّناعة التي يعالجها، وأصناف المزاج[١٤١] تسعة منها أربعة بسيطة، وهي الحارّ والبارد والرّطب واليابس، ومنها أربعة مركّبة، وهي الحارّ اليابس والحارّ الرّطب والبارد اليابس والبارد الرّطب، ومنها واحد معتدل، والعادة أيضاً أصناف، وذلك أنّ من النّاس من عادته شرب الشّراب، ومنهم من عادته شرب الماء[١٤٢]، ومنهم من قد اعتاد أن يأكل مرّةً في اليوم[١٤٣]، ومنهم من قد اعتاد أن يأكل[١٤٤] مرّتين أو ثلث[١٤٥]، ومن النّاس من قد اعتاد الاستفراغ، ومنهم من لم يعتد ذلك، وأوقات السّنة أربعة، أحدها الرّبيع، وهو معتدل، والآخر الصّيف، وهو حارّ يابس، والثّالث الخريف، وهو يابس معتدل البرودة مضطرب الحال، والرّابع الشّتاء، وهو بارد رطب.

[الفرق بين أصحاب التّجربة وأصحاب القياس]

(٢٣) الأعراض التّابعة للأمراض يستدلّون[١٤٦] بها أصحاب التّجارب على ما قد فعلوه مراراً كثيرة فنفع، ويستدلّون[١٤٧] بها أصحاب القياس على الشّيء الذي ينبغي أن يفعل، مثال ذلك الورم، فإنّ كلّ ورم إنّما[١٤٨] يتولّد من مادّة تنصبّ إلى واحد من

---

١٣٩ ADMY: الوقت | ١٤٠ A^1M^2: + المريض؛ S^h: ليسكنه المريض | ١٤١ S^h: ذكر المزاج ها[هنا] لأجل ذكره له بقو[ى؟] مزاج المريض | ١٤٢ F: + البارد | ١٤٣ ADMY: في اليوم مرّةً | ١٤٤ AM: + في اليوم | ١٤٥ ADM: ثلثاً؛ S: + مرّات | ١٤٦ ADMY: ويستدلّ | ١٤٧ ADMY: ويستدلّ | ١٤٨ A^2M^2: فإنّما

humors: blood, which is hot and moist; phlegm, which is cold and moist; yellow bile, which is hot and dry; and black bile, which is cold and dry. Four kinds of swellings are generated from these four humors. The hematoma known as a phlegmona arises from blood. The soft swelling known as an edema arises from phlegm. There is no swelling arising from yellow bile, but the swelling known as erysipelas results from blood tainted with yellow bile. Herpes, not erysipelas, results from yellow bile alone. Finally, from black bile arises the firm swelling, known by its hardness, which is called scirrhus. Suppose that this swelling that we are using as an example is the one that comes from blood and is thus the one known as a phlegmona. When this swelling occurs, the symptoms that follow from it are the inflation of the swollen organ, redness, distention, painfulness, hardness, and resistance to touch.[54] When these symptoms appear, the Empiricist is reminded by them that he has treated swellings like this many times in early stages with garden nightshade and in later stages with chamomile, fenugreek, and linseed, and that these drugs were helpful and beneficial. The Rationalist will deduce from these symptoms that he needs to evacuate the matter that has accumulated in that organ and the excess[55] that has flowed into it and that the organ must be strengthened so that thereafter it will not allow them to flow into it. These two approaches differ from each other. That is because strengthening the organ so that it will not accept anything excessive can only be accomplished by things that restrain and by things that return the organ from excess in temperament to moderation. On the other

54. Reading *li-al-mass* (to touch) for *li-al-ḥiss* (to sensation).

55. A gloss of IT reads: "That is, excessive, meaning additional to what is sufficient."

الأعضاء من واحد من الأخلاط، والأخلاط أربعة، الدّم وهو حارّ رطب، والبلغم وهو بارد رطب، والمرّة الصّفراء وهي حارّة يابسة، والمرّة السّوداء وهي باردة يابسة، ويتولّد عن هذه الأربعة الأخلاط أربعة أجناس من الأورام، فيحدث عن الدّم الورم الدّمويّ الذي يقال له فلغمونيّ ويحدث عن البلغم الورم الرّخو الذي يقال له أوذيما، ويحدث عن المرّة الصّفراء لا، بل عن[١٤٩] الدّم الذي ضربت فيه الصّفراء الورم المعروف بالحمرة لأن الصّفراء وحدها إنّما تحدث عنها النّملة لا الحمرة، ويحدث عن المرّة السّوداء الورم الجاسيّ المعروف بالصّلابة، وهو الذي يقال له سقيروس، فأنزل[١٥٠] أنّ هذا الورم الذي تمثّلنا به هو ورم من دمٍ، وهو الذي يقال له فلغمونيّ، فإنّ هذا ورم إذا حدث تبعته[١٥١] هذه الأعراض، وهي انتفاخ العضو الوارم وحمرته وتمدّده ووجعه وصلابته ومدافعته للحسّ، وإذا ظهرت[١٥٢] هذه الأعراض تذكر بها صاحب التّجارب أنّه قد عالج مثل هذا الورم مراراً كثيرةً في مبدء أمره بعنب[١٥٣] الثّعلب، وفي آخره بالبابونج والحلبة وبزر الكتّان، فنفعه ذلك وأنجع[١٥٤] فيه، وأمّا صاحب القياس، فإنّه يستدلّ بها على أنّه يحتاج إلى استفراغ ما قد حصل في ذلك العضو من المادّة والفضل[١٥٥] التي أنصبّت إليه وتقوية العضو حتّى لا يقبل ما ينصبّ إليه منها بعد ذلك، والسّبيل في كلّ واحدة[١٥٦] من هاتين الخصلتين[١٥٧] غيره في الأخرى، وذلك أنّ تقوية العضو حتّى لا تقبل شيئاً من الفضل إنّما يكون بالأشياء القابضة، والأشياء التي تردّ العضو عن إفراط المزاج إلى اعتداله، وأمّا

١٤٩ D: لا بدّ؛ S²: النّملة وعن | ١٥٠ Ga؛ FM: فأقول | ١٥١ DFY: تبعته | ١٥٢ AM: + فيه | ١٥٣ AM: لعب | ١٥٤ A: المجع [؟]؛ A¹DM²SY: وأنجح | ١٥٥ AʰMʰ: حاشية له أي الفاضلة يعني الزايدة عن الكفاء [Aʰ: الكفاف]؛ F: مادّة الفضل؛ DSY: – والفضل | ١٥٦ DFY: واحد | ١٥٧ S: القضيّتين

hand, the evacuation of what flowed into it can be accomplished by two means. First, the matter may be blocked and sent back, which is done at first by astringents. Second, what has accumulated and is not going back and is not being restrained or expelled must be evacuated. In later stages that is accomplished by things that warm and relax, for the drugs that hinder and repel are those that are astringent and cooling, while the ones that evacuate are those that warm and relax.[56]

(24) There are three genera of faculties in the body.[57] First, there is the genus of the psychic faculty, which is in the brain. Inferences can be made about it on the basis of the health or weakness of the volitional actions. Second, there is the genus of the vital faculty, which is in the heart. Inferences can be made about it by means of the pulse. Third, there is the genus of the natural faculty. The source of this faculty is the liver. Inferences can be made about it on the basis of the urine and the feces, similar to the wash water of fresh meat just after slaughter.

(25) If the occurrence of the hematoma is from an antecedent cause, such as a blow or stroke, it must be treated at first with things that warm and relax it in order to evacuate the humor that is causing it. If it occurs due to a preceding cause that existed before—that is, from a plethora occurring in the body—it is first necessary to restrain and block the matter until it is evacuated from the body. After that, one returns to things that warm and relax, and the swelling is treated with them. The signs indicating plethora are that the patient is afflicted with sluggishness and torpor, so that he does not move and feels a heaviness in his entire body; he is flushed; his arteries inflate more than is natural; and

---

56. A gloss of IT reads: "Clear in the manuscript. That is, because some of the drugs that expel this excess from the organ do so by repelling; but if the excess has become such that it cannot be expelled, the expulsive drugs will not be used. Other drugs are evacuant—the ones that warm and relax—so, therefore, they are the ones that are used in the later stages, since, in expelling the excess, one does not wish to damage the essence." The last two words differ in the MSS, but the sense seems clear.

57. It is not clear why the faculties are discussed here.

استفراغ ما قد حصل ممّا انصبّ إليه، فيكون بأمرين، أحدهما قمع المادّة وردّها إلى خلف، وذلك يكون في مبدأ الأمر بالأشياء القابضة، والآخر استفراغ ما قد حصل وصار لا يرجع[١٥٨] ولا ينقمع ولا يندفع، وذلك يكون في آخر الأمر بالأشياء التي تسخن وترخى لأنّ الأدوية[١٥٩] منها قامعة[١٦٠] دافعة، وهي التي تقبّض وتبرّد، ومنها ما تستفرغ، وهي التي تسخن وترخى.

(٢٤) أجناس ما في البدن من القوى ثلثة، أحدها جنس القوّة النّفسانيّة، وهي التي في الدّماغ، ويستدلّ عليها بصحّة الأفعال الإراديّة وضعفها، والآخر جنس القوّة الحيوانيّة، وهي التي في القلب ويستدلّ عليها بالنّبض، والثّالث جنس القوّة الطّبيعيّة، ومبدأ هذه القوّة الكبد، ويستدلّ عليها بالبول والبراز الشّبيه بغسالة اللّحم[١٦١] الطّري القريب العهد بالذّبح ـه

(٢٥) إن كان حدوث الورم الدّمويّ من سبب بادٍ بمنزلة الضّربة والصّدمة، فينبغي أن يعالج في أوّل الأمر بالأشياء التي تسخن وترخى كيما يستفرغ الخلط الفاعل له، وإن كان حدوثه من سبب سابق متقدّم، أعني من امتلاء حاصل في البدن، فينبغي أوّلاً أن تقمع المادّة وتمنع حتّى إذا استفرغ البدن رجع إلى الأشياء التي تسخن[١٦٢] وترخي فعولج بها، وعلامات الدّالّة على[١٦٣] الامتلاء هي أن يكون الإنسان يعتريه كسل وفتور عن الحركة ويجد ثقلاً في جميع بدنه، ويصير لونه أحمر وينتفخ عروقه

١٥٨ A^1M^2S: صار ولا يرجع > وليس يرجع | ١٥٩ A^hM^h: حاشية له ظ في خ أي لأنّ الأدوية التي تخرج هذا الفضل عن العضو منها دافعة، فإذا صار الفضل بحيث لا يندفع فلن تستعمل الدّافعة، ومنها مستفرغة وهي التي تسخن وترخي، فهي إذن التي بحيث استعمالها بآخره إذا لم يطمع في اندفاع الفضل بردع المانع [M^h: تردّ الماهيّة] | ١٦٠ F: مانعة | ١٦١ M^hY^h: حـ[اشية] له أي ذلك دليل خاصّ بضعفها | ١٦٢ F: + يستفرغ | ١٦٣ ADMY: ـ الدّالّة على

his body is distended. Treatment varies in the quantity of that by which it is treated, as occurs when a greater or smaller quantity of blood is removed; in the means that are employed, as occurs when the thing is expelled several times or only once; or else in the entire genus altogether, as occurs when a constricting poultice is employed instead of evacuation.

(26) The things indicating evacuation[58] are plethora; the health of the faculty; adolescence; the season of spring; temperate air;[59] the habit of evacuation; and a trade whose practitioners need evacuation, such as those trades not involving physical labor.[60] The things indicating that evacuation ought not to be employed include weakness, whether of the psychic, vital, or natural faculties. In the case of the psychic faculty, this is whether it involves the faculties of sensation, motion, or governing and deliberation; and in the case of the natural faculty, whether it involves the attractive faculty, the retentive faculty, the transformative faculty, or the expulsive faculty. These factors [indicating against evacuation] also include age, specifically childhood or old age; the season of the year, specifically if it is summer or winter; the place, if it is extremely

58. K 1:70–71; H 5–6; G[a] 26–27: Galen is discussing the circumstances under which a patient with a plethora of blood should be treated by bloodletting.

59. One MS adds: "and place." A gloss of IT reads: "A manuscript reads 'having matter, temperate air, and place.'"

60. A gloss of IT reads: "There is something to consider. He ought to have mentioned 'in the place'; and, in fact, it is that way in Galen's original text. When he says, 'and temperate air,' that is not already implied when he says, 'and the season of spring'; for they are in the habit of saying, 'and the season of the year and the state of the air that day.' If that had been added, it ought to have said, 'and temperate air and place,' which is what is actually found in some manuscripts. When he later mentions the opposite case—the things that indicate that there ought not to be evacuation—he only mentions the place if it is extremely hot or cold. Therefore, the first 'and a temperate place' is correct in Galen's original text, for he said, 'and the place is temperate.'" The point is that while Galen mentions the season of spring as appropriate for bloodletting, he does not specifically mention the weather being mild; while, on the other hand, Galen specifically mentions that the place, as well as the season, should be temperate, which the epitomist omits to mention. The syntax of the gloss is murky.

أكثر ممّا كانت عليه بالطّبع وتمدّد جلده، المداواة تختلف إمّا في مقدار الشّيء الذي يداوى به بمنزلة ما يعرض ذلك إذا استفرغ من الدّم مقدار أكثر أو أقلّ، وإمّا في الوجه الذي يستعمل به ذلك الشّيء بمنزلة ما يعرض ذلك عند ما يستفرغ الشّيء مراراً كثيرة أو مرّةً[١٦٤]، وإمّا في الجنس كلّه جملةً بمنزلة ما يعرض ذلك إذا استعمل مكان الاستفراغ ضماد قابض[١٦٥].

(٢٦) الأشياء التي تدلّ على[١٦٦] الاستفراغ هي الامتلاء وصحّة القوّة وسنّ الشّباب ووقت الرّبيع واعتدال الهواء[١٦٧] وعادة الاستفراغ والصّناعة التي يحتاج المعالج لها إلى الاستفراغ بمنزلة الصّناعات التي لا تعب فيها[١٦٨]. وأمّا الأشياء التي تدلّ على أنّه لا ينبغي أن يستعمل الاستفراغ، فهي ضعف القوّة إن كانت نفسانيّةً[١٦٩] وإن كانت حيوانيّة وإن كانت طبيعيّة، ومن النّفسانيّة[١٧٠] أيضاً إن كانت قوّة الحسّ وإن كانت قوّة الحركة وإن كانت قوّة التّدبير والسّياسة ومن الطّبيعيّة إن كانت القوّة الجاذبة وإن كانت القوّة الماسكة وإن كانت القوّة المغيّرة وإن كانت القوّة الدّافعة، والسّنّ إذا كان صغيراً وشيخاً[١٧١]، والوقت الحاضر من أوقات السّنة إذا كان صيفاً

---

١٦٤ F²؛ DFY: – أو مرّة | ١٦٥ DSY: ضماداً قابضاً | ١٦٦ ADMY: + الحاجة إلى | ١٦٧ F: + والبلد؛ AʰMʰ: حاشية له خ له موادّ واعتدال الهواء واعتدال والبلد صح | ١٦٨ AʰMʰ: له النظر عمّا ذكر في البلد واجب، وكذلك هو في الفصّ، وليس قوله واعتدال الهواء بداخل في قوله ووقت الرّبيع كان عادتهم أن يقولوا وفصل السّنة وحال الهواء اليوميّ، فإن كان أزاد ذلك، فينبغي أن يقال، واعتدال الهواء والبلد، وقد وجد كذلك في بعض النسخ، ولمّا ذكر بعد مقابل ذلك، وهو الأشياء التي تدلّ على أنّه لا ينبغي أن يستفرغ لم يذكر الا البلد إذا كان شديد الحرّ أو البرد، فبحسب ذلك يكون الأوّل واعتدال البلد صحّ من الفصّ، فإنّه قال، وكان البلد معتدلاً | ١٦٩ AM: نفسيّة | ١٧٠ AM: النفسيّة | ١٧١ F: كان سنّ صبيّ صغير أو شيخ فانٍ؛ DY: أو شيخاً؛ S: كان صبيّ أو شيخ

cold, like the land of the Slavs,[61] or extremely hot, like the land of the Ethiopians; and lack of habituation to evacuation.

(27) The Empiricists and the Rationalists know precisely the same things—that is, sickness and the evidences[62] that determine how the treatment is to be performed—but the Empiricists know them by memory and observation and the Rationalists by deduction.[63] The Rationalists deduce from the state of affairs what they ought to do in that case. Thus, they deduce from each thing that is according to nature that it ought to be preserved and continued and from each thing that is outside nature that it ought to be rooted out and extirpated. There are three things that are unnatural:[64] disease, its cause, and the symptom correlated with it.

(28) Some causes, such as blows and bites, afflict the body from outside and are called antecedent causes. Some, such as plethora and putrefaction,[65] move in the body from inside and are called preceding causes. Some have other causes prior to themselves and are the causes most nearly associated with the occurrence of diseases; for example, the heating of the heart in a fever. These are called cohesive causes.[66]

---

61. Galen's original text—and presumably the Greek original of the epitome—mentions Scythians, the Slavs not yet having appeared in history, whereas the Arabic translation of *The Medical Sects* and the present epitome mention Slavs, the Scythians having disappeared by then.

62. One MS glosses this as "the daughters of the elements"; see pp. 22–23, paragraph 22, above.

63. K 1:72–74; H 7–9; G[a] 27–31. Galen demonstrates that the Rationalists differ from the Empiricists in seeking to know the cause of the disease, since it is sometimes necessary to know the cause in order to determine the correct treatment.

64. Or "outside of nature," which are to be distinguished from the natural things, such as air and exercise, and which were later known as the non-naturals.

65. A gloss of IT reads: "Here he gives plethora and putrefaction as two examples of preceding causes. Ḥunayn said in his *Questions* that putrefaction is the cohesive cause of septic fever. You should know that plethora and putrefaction can occur without heating the heart; and so as long as the putrefaction does not heat the heart, the putrefaction is also a prior cause. When it begins to heat the heart and brings about a fever, it is a cohesive cause. That is why he confines himself to mentioning it here, enumerating putrefaction as a preceding cause, even though he said, 'and putrefaction.' When he mentioned putrefaction, Ḥunayn said that putrefaction is one of the cohesive causes, giving as an example the putrefaction that causes fever. He did not say 'like putrefaction' without qualification." Regarding the passage from *Questions* mentioned in this gloss, see Ḥunayn ibn Isḥāq, *Masāʾil*, 261–63; Ḥunayn ibn Isḥāq, *Questions on Medicine for Scholars*, 84, which translates *sabab sābiq* as "antecedent cause" rather than "preceding cause."

66. On these three causes, see p. 2 above with n. 7.

أو شتاءً، والبلد إذا كان شديد البرد بمنزلة بلاد الصّقالبة، أو شديد الحرّ بمنزلة بلاد الحبشة، وقلّة الاعتياد للاستفراغ.

(٢٧) أصحاب التّجارب وأصحاب القياس يعرفون أشياء واحدة بأعيانها— أعني المرض والشّواهد[١٧٢] التي عليها مبنيّ الأمر في الأشياء التي يداوى بها إلاّ أنّ أصحاب التّجارب يعرفون ذلك بالحفظ والرّصد وأصحاب القياس بالاستدلال[١٧٣]، وأصحاب القياس يستدلّون من نفس الأمر على ما ينبغي أن يفعل فيه، فيستدلّون من كلّ شيء هو في الطّبع على أنّه ينبغي أن يحفظ ويستبقي ومن كلّ شيء هو خارج عن الطّبع على أنّه ينبغي أن يقلع[١٧٤] ويستأصل، والأشياء الخارجة عن الطّبيعة ثلثة، المرض وسببه والعرض اللاّزم له.

(٢٨) والأسباب منها ما يرد على البدن من خارج، ويقال لها أسباب بادية بمنزلة الضّربة والنّهشة، ومنها ما يتحرّك في البدن من داخل، ويقال لها أسباب سابقة بمنزلة الامتلاء والعفونة[١٧٥]، ومنها ما يتقدّمها أسباب أُخر، وتكون هي أقرب الأسباب إلى حدوث الأمراض، ويقال لها أسباب واصلة بمنزلة سخونة القلب في الحمّى.

١٧٢ S^h: بنات الإسطقسات | ١٧٣ F: –وأصحاب القياس بالاستدلال | ١٧٤ A^2M^2: يقطع | ١٧٥ $A^hM^hY^h$: حاشية له ضرب هاهنا الامتلاء والعفونة مثالين للأسباب السّابقة، وقد قال حنين في مسائله أنّ العفونة هي سبب واصل للحمّى العفينة، ويجب أن تعلم أنّه يمكن أن يؤخذ [A^h: يوجد] امتلاء وعفونة لم يبلغ أن يسخن القلب فما دامت [؟] لم تسخنه، فهي سبب متقادم أيضاً، وإذا [A^hY^h: فإذا] رقت [A^hY^h: وفت] باسخانه وإصدار [A^hY^h: إحداث] الحمّى ، فهي سبب واصل، وكذلك [Y^h: لذلك] اقتصر [A^hY^h: أقبض] هاهنا في عدّه العفونة سبباً سابقاً على أن قال والعفونة، وقال حنين عند ذكر العفونة [Y^h: فالعفونة] في الأسباب الواصلة مثل العفونة المحدثة للحمّى ولم يقل [Y^h: يقبل] مثل العفونة فقط.

(29) Venomous animals have different kinds of poison.[67] One kind causes excessive dryness, resulting eventually in convulsions—as is the case with rabies, which does most of its damage in the brain. Some poison chills excessively, so that the patient imagines that he has been struck with a freezing stone—as happens with a scorpion's venom, which does most of its damage to the heart. Some heats excessively, as is the case with snake and viper venom, which putrefies the organ and consumes it with its sharp burning.

(30) Some symptoms indicate the disease itself, some indicate the cause of the disease, and some indicate the location of the disease.[68] For example, the variation of the speed of the pulse during fever indicates the fever itself. The signs indicating plethora indicate the cause of the fever—for example, sluggishness, heaviness of the body, inflation of the blood vessels, and flushing. The symptoms occurring in the patient with pleurisy indicate the location of the disease. They are acute fever, coughing, shortness of breath, and prickly pain.

(31) A lesion may occur from an internal cause from within or from an observable cause manifest from without.[69] An internal cause from within is, for example, a pungent humor that consumes and burns or a plethora that distends and separates. An external cause originating from without is either some body that has a soul, which is a body that grows, or a body that does not have a soul, which is one that does not grow. A lesion that occurs from a body that does not have a soul is either from a body that stretches, such as a rope; from a body that cuts, such as a sword; from a body that burns, such as fire; from a body that bruises, such as a stone; or from a body that pierces, such as a spear. The lesion that results from a body that has a soul is, for example, the

67. K 1:74; H 7–8; G[a] 29–31: Galen mentions the bite of a rabid dog or poisonous snake as an example of when a Rationalist would need to know the antecedent cause—that is, the external cause of the disease—since the symptom itself would not be sufficient to indicate the proper course of treatment. See also pp. 41–42, paragraph 46, below.

68. K 1:72–73; H 7; G[a] 27–28: Galen refers vaguely to the use of symptoms to indicate appropriate treatment or the cause of a disease. This paragraph gives a more precise account of the use of symptoms in diagnosis.

69. K 1:73–74, 88–90; H 8, 18–19; G[a] 30–31, 62–63: *Qarḥa* (lesion) can refer to a wound, ulcer, sore, or lesion—the common meaning being an injury or disease of some sort visible on the exterior of the body. "Lesion" seems the most neutral translation, so long as it is remembered that it does not refer to internal abnormalities.

(٢٩) سموم الحيوان ذوات السّمّ تختلف، فمنها ما يجفّف تجفيفاً مفرطاً حتّى أنّه يحدث تشنّجاً بمنزلة سمّ الكلب الكَلِب الذي أكثر مضرّته للدّماغ[١٧٦]، ومنها ما يبرّد تبريداً مفرطاً حتّى يظنّ المريض أنّه يرمى بحجارة[١٧٧] البرد بمنزلة سمّ العقرب الذي أكثر مضرّته القلب[١٧٨]، ومنها ما يسخن إسخاناً مفرطاً بمنزلة سمّ الثّعبان، وسمّ الأفاعي الذي يعفن العضو ويأكله ويحرقه بحدّته.

(٣٠) ومن[١٧٩] الأعراض أشياء تدلّ على نفس المرض، ومنها أشياء تدلّ على سبب المرض، ومنها أشياء تدلّ على موضع المرض، مثال ذلك أنّ اختلاف النّبض في وقت[١٨٠] الحمّى في السّرعة يدلّ على نفس الحمّى، والعلامات الدّالة على الامتلاء تدلّ على سبب الحمّى بمنزلة الكسل عن الحركة وثقل البدن وانتفاخ العروق وحمرة اللّون، والأعراض الحادثة بصاحب ذات الجنب تدلّ على موضع العلّة والمرض، وهي الحمّى الحادّة والسّعال وضيق النّفَس والوجع النّاخس.

(٣١) حدوث القرحة يكون إمّا من سبب باطن من داخل وإمّا من سبب ظاهر من خارج، والسّبب الباطن من داخل بمنزلة خلطٍ حادٍّ يأكل ويحرق أو امتلاء يمدِّد ويفرّق، وأمّا السّبب الظّاهر من خارج، فإمّا أن يكون جسماً من الأجسام ذوات النّفوس، وهي الأجسام النّامية، وإمّا من جسم[١٨١] لا نفس له، أي غير نامٍ، والقرحة الحادثة عن جسم لا نفس له يكون إمّا من جسم يمدّ بمنزلة الحبل، وإمّا من جسم يقطع بمنزلة السّيف، وإمّا من جسم تحرق بمنزلة النّار، و إمّا من جسم يرضّ بمنزلة الحجر، وإمّا من جسم يثقب بمنزلة السّهم، فأمّا القرحة التي تحدث عن جسم ذي نفس، فبمنزلة

---

١٧٦ ADM: بالدّماغ؛ Y: في الدّماغ | ١٧٧ ADMY: + من | ١٧٨ ADY: بالقلب؛ M: في القلب | ١٧٩ ADSY: من | ١٨٠ ADMY: – وقت | ١٨١ AM: + ما

lesion resulting from the bite of an animal. Animals that bite either have venom or do not have venom. The lesion resulting from the bite of an animal not having venom is always similar to the lesion resulting from a body not having a soul, there being no distinction or difference between them. On the other hand, the bite of a venomous animal is inevitably followed by injurious symptoms whose like does not occur from bodies not having souls. These injurious symptoms may occur only in the later stages, so that, in the beginning, there is no difference from the lesion produced by a body without a soul; or they may occur immediately. The lesion in which the injurious symptoms occur in the later stages is like the bite of a rabid dog, for in the first days this lesion is like other lesions, but in the later stages injurious and deadly symptoms appear, such as hydrophobia and convulsions. In the case of some of the lesions in which the injurious symptoms occur immediately, the resulting symptoms may be only in the lesion itself—for example, lesions accompanied by putrefaction, the destruction of some organ, or blackness appearing in the lesion. In other cases, the symptoms resulting from the lesion affect the entire body—for example, lesions that cause convulsions because the venom damages the brain, fainting resulting from damage to the heart, or loss of color and jaundice resulting from damage to the liver.

(32) Those who are bitten by venomous animals are treated externally by pungent, hot, absorbent drugs placed upon the wound to draw out the venom. They are treated internally by drugs that dry and absorb the venom, such as theriac[70] and its equivalents.

70. A master antidote containing specific antidotes to a large number of poisons.

القرحة الحادثة عن نهشة حيوان[١٨٢]، والحيوان الذي ينهش لا يخلوا من أن يكون إمّا حيوان[١٨٣] له سمّ وأمّا حيوان[١٨٤] لا سمّ له، والقرحة الحادثة عن نهشة حيوان لا سمّ له لا يزال دائمًا شبيهة بالقرحة الحادثة عن جسم لا نفس له لا خلاف بينهما ولا فرق، فأمّا القرحة التي تحدث عن نهشة حيوان ذي سمٍّ، فلا بدّ من أن يتبعها لا محالة أعراض رديّة لا يكون مثلها في القرحة الحادثة عن الأجسام التي لا نفس لها، وهذه الأعراض الرّديّة إمّا أن تتبع القرحة في آخر الأمر حتّى تكون في مبدأ أمرها لا فرق بينها وبين القرحة الحادثة عن جسم لا نفس له، وإمّا أن تتبعها في أوّل الأمر، والقرحة التي يتبعها الأعراض الرّديّة في آخر الأمر هي مثل القرحة الحادثة عن نهشة الكلب الكَلِب، فإنّ هذه القرحة تكون في الأيّام الأوّل شبيهة بسائر القروح، ثمّ إنّها في آخر الأمر تحدث أعراضًا رديّةً مهلكةً بمنزلة التّفزّع من الماء والتّشنّج، وأمّا القرحة التي يتبعها الأعراض الرّديّة في أوّل الأمر، فمنها ما يكون الأعراض التّابعة له في القرحة وحدها بمنزلة القرحة التي تكون معها عفونة[١٨٥] أو عطب عضوًا[١٨٦] من الأعضاء أو من[١٨٧] سواد يظهر فيه، ومنها ما يكون الأعراض التّابعة له تعمّ البدن كلّه بمنزلة القروح التي تحدث عنها التّشنّج عندما يضرّ السّمّ بالدّماغ أو الغشي عندما يضرّ بالقلب أو إحالة اللّون واليرقان عندما يضرّ بالكبد ــهـ[١٨٨]

(٣٢) الذين تنهشهم الحيوانات ذوات السّموم يداوون من خارج بأدوية حادّة حارّة جاذبة[١٨٩] توضع على القرحة كيما تجذب السّمّ، ومن داخل بأدوية تجفّف وتنشف السّمّ بمنزلة التّرياق وما أشبهه ــهـ[١٩٠]

---

١٨٢ AM: الحيوان | ١٨٣ ADS: حيوانًا | ١٨٤ ASY: حيوانًا | ١٨٥ S: + وتشنّج | ١٨٦ ADY: عضو | ١٨٧ ADMY: - من | ١٨٨ AS: - ــهـ | ١٨٩ F: بأدوية حارّة مادّته؛ F[1]: جاذبة؛ S: - حارّة جاذبة؛ DY: - جاذبة | ١٩٠ AS: - ــهـ

Commentary on chapter 4 of Galen's book *On the Medical Sects*[71]

[The Rationalists' criticism of the Empiricists]

(33) The Rationalists condemn the Empiricists on three grounds. First, some—for example, Asclepiades—say that experience has no stability.[72] He said that since bodies are constantly changing, never remaining in a single state, it is impossible to remember what was beneficial in a great many cases.[73] It is not fair to criticize them on this ground, since even though bodies are constantly changing, their change is not so great that a drug will not be helpful twice; for physicians seek sensible change, not the change as it exists in the nature [of the body].[74] Second, they say that experience is insufficient for what is needed, as Erasistratus has said. He was convinced that simple, noncomposite things could be found on the basis of experience by which simple, noncomposite diseases could be treated; but he denied the existence of treatments using composite drugs for composite disease based on experience. This, too, is not a fair criticism of them. Just as experience can certainly discover treatments for simple, noncomposite diseases, likewise it is possible for it to find and discover treatments for composite diseases. The reason is that diseases can be simple and noncomposite—for example, throbbing,[75] tertian fever, and phlegmatic fever. Others are composite—for example, the hematoma that throbs

71. Corresponding to the first half of chapter 5 of the modern edition; K 1:75–76; H 9; G[a] 32–35. The epitome follows Galen's text closely but fills in the arguments.

72. Galen gives this as the first objection of the Rationalists to the Empiricists. Galen disapprovingly quotes at length arguments of Asclepiades against the possibility of coherent medical experience, especially in *On Medical Experience*; see Galen, *Three Treatises*, passim. See appendix 1, s.v. "Asclepiades."

73. This misses the point that Galen attributes to Asclepiades here, which is not that there is too much experience to remember, but, rather, that things do not recur in exactly the same way. The fault is probably in the Arabic translation, since both John of Alexandria (*Commentaria*, 4va4–10) and Agnellus (*Lectures* 36r) are clear about the fluidity of experience.

74. A gloss reads: "That is, the nature of man, which is manifest to sense."

75. The scribes seem to have read the uncommon word *ḍarabān* (throbbing) as the dual *ḍarbān* and inflected it accordingly.

شرح الباب الرّابع
من كتاب جالينوس في فرق الطّبّ[١٩١]

[احتجاج أصحاب القياس على أصحاب التّجربة]

(٣٣) أصحاب القياس يطعنون على أصحاب التّجربة من ثلثة وجوهٍ، أحدها أنّهم قالوا إنّ التّجربة لا ثبات لها بمنزلة أسقليباذس[١٩٢]، فإنّ هذا قال إنّه لما كانت الأبدان دائمة التّغيّر لا تقف على حال واحدة بتّةً صار حفظ ما قد نفع مراراً كثيرةً ممّا لا يمكن ولم ينصف هذا في طعنه على القوم، وذلك أنّ الأبدان وإن كانت دائمة التّغيّر فليس[١٩٣] يبلغ من تغيّرها أن يكون الدّواء لا ينفع مرّتين لأنّ الأطبّاء إنّما يطلبون التّغيّر المحسوس[١٩٤]، لا التّغيّر الموجود في الطّبع[١٩٥]. والوجه الثّاني أنّهم قالوا إنّ التّجربة ليس تكمل[١٩٦] لما يحتاج إليه بمنزلة إراسسطراطس، فإنّ هذا يقرّ بانّ العلل البسيطة المفردة توجد مداواتها بأشياء بسيطة مفردة[١٩٧] بطريق التّجارب، فأمّا وجود مداواة علل مركّبة بأدوية[١٩٨] مركّبة بطريق التّجارب، فذلك عنده ممّا ينكره، وهذا أيضاً لم ينصف في الطّعن عليهم، وذلك أنّه كما أنّ التّجربة تستخرج وتجد مداواة الأمراض البسيطة المفردة، كذلك قد يجوز أن تجد وتستخرج مداواة الأمراض المركّبة، لأنّ الأمراض منها بسيطة مفردة بمنزلة الضّربين[١٩٩] وحمّى الغبّ، وحمّى البلغم، ومنها مركّبة بمنزلة الورم الدّمويّ الذي يضرب فيه الحمرة والحمّى المركّبة من حمّى الغبّ ومن حمّى البلغم،

---

١٩١ $F^{h}S$؛ ADMY: – من كتاب جالينوس في فرق الطّبّ | ١٩٢ M: أسقليبيذس | ١٩٣ F: التّحلّل والتّغيّر | ١٩٤ F: المحبوس | ١٩٥ S^{h}: أي طبع الإنسان ويظهر حسّا | ١٩٦ AM: حـ[اشية] له أي ليس تلق | ١٩٧ F: توجد. . . مفردة > قد يجوز أن يداوا؛ DSY: – مفردة | ١٩٨ F: وجود. . . أدوية > أن يكون أمراضاً توجد مداواتها بأشياء | ١٩٩ S^{2}؛ $F^{2}S$: النقرس

due to erysipelas, the fever compounded from tertian fever and phlegmatic fever, and the fever known as hectic when it is accompanied by a putrid fever. The same is true of the things by which diseases are treated. Some are simple and noncomposite, such as purslane and garden nightshade; and others are composite, such as collyria and electuaries. Third, they say that experience has no technical method. In this respect, Athenaeus said that it is not in accordance with the technical method by which the practitioners arrive at the principles of their art.[76] This criticism of them is justified, since anything that has no reason is not part of an art, as Plato said.[77]

76. As a Pneumatist, Athenaeus held a variant of the humoral theory and, in issues of epistemology, would have sided with the Rationalists. Galen does not mention Athenaeus here, nor does he give his view of this criticism. See appendix 1, s.v "Athenaeus."

77. Perhaps this refers to the passage on mimetic arts at the end of the *Sophist*. A gloss in one MS reads: "Any man who has not comprehended the cause of his craft is not a craftsman."

والحمّى المعروفة بالدِّقّ إذا كانت معها حمّى من عفونة، وكذلك الأشياء التي يداوى بها الأمراض، منها أشياء بسيطة مفردة بمنزلة البقلة الحمقاء وعنب الثّعلب، ومنها أشياء مركّبة بمنزلة الأكحال والمعجونات، والوجه الثّالث أنّهم قالوا إنّ التّجربة ليس لها مذهب صناعيّ بمنزلة ما قال أثيناوس، فإنّ هذا قال إنّها غير لازمة للطّريق الصّناعي الذي به يصلون[٢٠٠] أصحاب الصّناعات إلى أحكام صناعاتهم[٢٠١]، وقد أنصف هذا في طعنه عليهم، وذلك أنّ كلّ أمر لا قياس معه، فهو غير صناعيّ[٢٠٢] كما قال أفلاطن[٢٠٣] هـ

٢٠٠ AM: يصل | ٢٠١ DFY: صناعتهم | ٢٠٢ F^2: كلّ أمر لا يؤتى سبب ما هو صانعه لا يستحقّ منع أن أسمّيه صانعًا | ٢٠٣ AD: فلاطن؛ F: أفلاطون

Commentary on chapter 5[78] of Galen's book *On the Medical Sects*

[The Empiricists' criticism of the Rationalists[79]]

(34) The Empiricists condemn the Rationalists on three grounds. First, some of them say that inference can only affirm something insofar as it is better known, clearer, and more convincing; but as for its being able to discover the truth itself and the entity existing in nature—no. Second, one group of them says that, even if inference could be used to discover an existent entity in nature—which they deny is possible—it is nevertheless useless. Third, yet another group of them says that, even if something useful can be discovered by inference, it is not something that is really necessary but only something extra, since what can be discovered by experience is sufficient for what is needed.

(35) The Rationalists particularly seek to know three things that the Empiricists are not interested in.[80] First, there is the nature of the body—by which I mean here every nature within the scope of this investigation. Second, there are the causes of diseases—that is, the preceding causes and the cohesive causes, since the antecedent causes might equally well be investigated and considered by the Empiricists. Third, there are the powers of the things by which health can be acquired, for the Empiricists do not investigate the primary effects of drugs, nor do they study the power by which each drug has the effect that it has.

78. K 1:76–79; H 9–12; G[a] 36–43; corresponding to the second half of chapter 5 of modern Greek editions. A gloss in two MSS notes that "the chapter heading is omitted in many manuscripts," as is the case with **F**, though it is added by a collator. Some MSS omit: "Of Galen's Book on the Sects of Medicine." This section of Galen's text deals mainly with the dispute between the two groups about the legitimacy of the Rationalists' use of anatomy, deduction, and logic.

79. K 1:76–77; H 10; G[a] 37–39, of which this paragraph is a close paraphrase.

80. K 1:76; H 10; G[a] 37.

شرح الباب الخامس
من كتاب جالينوس في فرق الطّبّ[٢٠٤]

[احتجاج أصحاب التّجربة على أصحاب القياس]

(٣٤) وأصحاب التّجارب يطعنون على أصحاب القياس من ثلثة وجوه، أحدها أنّ بعضهم قال إنّ القياس إنّما يوجب الشّيء من طريق ما هو أولى وأشبه وأقنع، فأمّا أن يكون يقدر على استخراج نفس الحقّ، والأمر الموجود في الطّبع، فلا[٢٠٥]، والآخر أنّ قوماً منهم قالوا إنّه وإن كان القياس يمكن أن يستخرج به ما أنكروه[٢٠٦] أولئك من الأمر الموجود في الطّبع، فإنّه ليس ينتفع بذلك. والثّالث أنّ قوماً آخر[٢٠٧] منهم قالوا إنّه وإن كان ما يستخرج بالقياس ممّا ينتفع به، فليس ممّا لا بدّ منه ضرورةً لكن هو شيء فضل إذ كان ما يستخرج بالتّجارب يفي بما يحتاج إليه.

(٣٥) أصحاب القياس خاصّةً يطلبون معرفة ثلثة أشياء لا يطلبونها[٢٠٨] أصحاب التّجارب، أحدها طبيعة البدن، أعني بقولي ها هنا طبيعة جميع باب النّظر في الطّبائع، والثّاني أسباب الأمراض، أعني الأسباب السّابقة والأسباب الواصلة لأنّ الأسباب البادئة قد ينظر فيها ويطلبها أصحاب التّجارب، والثّالث قوى الأشياء[٢٠٩] التي تستفاد بها الصّحّة، وذلك لأنّ أصحاب التّجارب لا ينظرون في فعل الأدوية الذي هو فعل أوّل ولا يطلبونه ولا يبحثون عن القوّة التي بها يفعل كلّ واحد من الأدوية ما يفعل.

---

٢٠٤ ADMY: – شرح . . . الطّبّ؛ A^hM^h: رؤوس الأبواب ملغاة [؟] في كثير من النّسخ: | ٢٠٥ S: + ويستدلّون على ذلك باختلاف القياسيّين في إدراك الحقّ في كلّ شيء | ٢٠٦ ADMY: أنكره | ٢٠٧ ADMY: آخرين | ٢٠٨ ADMY: يطلبها | ٢٠٩ F^2S: الأدوية

(36) The sect of Rationalists has three tools that they particularly use and that are not employed by the sect of the Empiricists: anatomy, deduction from the thing itself to what ought to be done with it, and the science of logic.[81] There are two approaches to anatomy, for there is what occurs by chance, such as the opportunities that occur in war, and there is what occurs by art, using either a living animal or a dead animal. By means of anatomy learned from a living animal, it is possible to know the actions and uses of the organs, while by means of the anatomy learned from a dead animal, it is possible to know the substance of each organ specific to it; its structure, size, number, and orientation; and its commonalities with other organs. The Empiricists condemn the study of anatomy on two grounds: first, they claim that it cannot be used to discover that which is actually needed; and, second, if it does discover something, it is not something that really must be known in the art of medicine.

(37) Something that is not manifest may be not manifest by nature, as is the case with everything not accessible to the senses but known by the mind.[82] The method by which it is known and the inference that leads to it are called analogism, which is inference to the hidden by means of the manifest.[83] On the other hand, what is not manifest may not be hidden by nature nor to the art, but, rather, be hidden at some particular time. This is the case, for example, with something that is by nature sensible but that is concealed because it is too far away, is very small in quantity,[84] or is being sought with a sense foreign to it. The

81. K 1:77; H 10; G[a] 38–40, ἀνατομή, ἔνδειξις, and διαλεκτικὴ θεορία. The discussion of the methods of anatomical research is not based on Galen's text here, but these methods are discussed in detail in Galen's *On Anatomical Procedures* 1.2, 2.1–2. Human dissection and vivesection were rarely, if ever, practiced in Roman or medieval times.

82. K 1:77; H 11–12; G[a] 40–42. Galen's discussion concerns whether a special kind of scientific deduction is needed in medicine or whether only the informal inductive inference used in everyday life is necessary.

83. Ἀναλογισμός. Similar definitions for this word and for ἐπιλογισμός (epilogism) are found in Galen, *On Hippocrates' Prognostics* I, in K 18B:26. In *An Outline of Empiricism*, Galen remarks that the Empiricists "call their own form of reasoning 'epilogism,' and the form of reasoning characteristic of the Rationalists 'analogism,' since they do not care to agree even in their terminology. In the same way, they also call their most concise accounts not 'definitions' but 'descriptions.'" The term "epilogism" apparently arose in Hellenistic philosophical debates about inductive inference, particularly in ordinary life; see Michael Frede, in Galen, *Three Treatises*, xxiii, 62–63.

84. Some MSS have the gloss: "or because there is an obstruction in front of it."

(٣٦) ولفرقة أصحاب القياس ثلث آلات يستعملونها خاصّةً ولا تستعملها فرقة أصحاب التّجارب، وهي التّشريح والاستدلال من نفس الشّيء على ما ينبغي أن يفعل به وعلم المنطق، والتّشريح يكون على ضربين، وذلك أنّ منه ما يقع بالاتّفاق بمنزلة ما يعرض من ذلك في الحرب، ومنه ما يكون من فعل الصّناعة إمّا في حيوان حيّ وإمّا في حيوان ميّت، والذي يكون في حيوان حيّ يعرف به أفعال الأعضاء ومنافعها، والذي يكون في حيوان ميّت يعرف به جوهر كلّ واحد من الأعضاء المخصوص به وخلقته ومقداره وعدده ووضعه ومشاركته لما يشاركه، وأصحاب التّجارب يطعنون على التّشريح من وجهين، أحدهما أنّهم يزعمون أنّه ليس يستخرج به ما يحتاج إليه، والثّاني أنّه وإن استخرج به شيء فليس هو ممّا لا بدّ[٢١٠] منه ضرورةً في الصّناعة.

(٣٧) الشّيء الذي ليس بظاهر إمّا أن يكون في طبعه غير ظاهر بمنزلة كلّ شيء لا يقع عليه الحسّ، وإنّما يعرف بالعقل، والباب الذي به يعرف هذا والقياس[٢١١] الذي يدلّ عليه يقال له أنالوجسموس[٢١٢]، وهو القياس على الخفيّ بالظّاهر، وإمّا أن يكون ليس بخفي في الطّبع ولا في الصّناعة، لكنّه ممّا يخفي وقت من الأوقات بمنزلة كلّ شيء هو في طبعه محسوس إلاّ أنّه لبعد شقّته[٢١٣] أو لصغر مقداره أو لأنّه يطلب بحاسّة غريبة منه[٢١٤] قد

٢١٠ M: + له | ٢١١ AM: القياس | ٢١٢ F: أفالوجسموس | ٢١٣ DSY: مسافته

method by which such a thing is deduced is called epilogism, which is inference to the manifest by the manifest. The Rationalists use inference from the manifest to the hidden, while the Empiricists use inference from the manifest to the manifest, claiming that this is beneficial in refuting those who condemn what can be seen by sense, revealing what is hidden and concealed, and showing the truth about those who seek to accuse the Empiricists of errors and involve them in fallacies. The hidden thing may be hidden from us by nature, as, for example, the substance of God—may He be blessed and exalted—and the substance of the mind, soul, and nature. Things of that sort can be known only by inference from the manifest to the hidden. The hidden thing may be hidden from sense, for something can escape the senses due to four causes: being too far away, like the ship that is in the middle of the sea and is thus hidden from someone on shore; small size, such as the dust that floats in the air and is invisible to us unless it enters a sunbeam from a skylight or window; its belonging to the genus of another sense, such as sounds, which are not apparent to taste; or its being concealed and veiled, such as a stone in the depths of the sea, which is concealed by the seawater. The Empiricists dislike inference from the manifest to the hidden because irresolvable disagreement can occur. They praise inference from the manifest to the manifest on the grounds that it is a method that tends not to result in disagreement; but if disagreement does occur, it is easy to resolve. They claim that disagreement is something that indicates that the thing concerning which there is disagreement has not been comprehended,[85] nor has its reality been grasped. It is the failure to comprehend the thing and reach its essence that is the cause of the disagreement.

85. K 1:78–79; H 11–12; G[a] 42: Κατάληψις (comprehension) is a technical term of the Empiricists, defined by Galen as "true and certain knowledge," and is used for primary cognition in Stoic epistemology.

صار خفيّا، والباب الذي يدلّ على هذا يقال له أفيلوجسموس[٢١٥]، وهو القياس بالظّاهر على الظّاهر. فأمّا أصحاب[٢١٦] القياس يستعملون القياس بالظّاهر على الخفي، وأمّا أصحاب التّجارب فيستعملون القياس من الظّاهر على الظّاهر، ويزعمون أنّ هذا نافع في الرّدّ على من يطعن على ما يرى حسّاً وفي كشف ما قد توارى وغاب وفي كشف أمر القوم الذين يلتمسون أن يغالطوا أصحاب التّجارب ويخدعوهم[٢١٧] بالأغاليط[٢١٨]. الشّيء الخفيّ إمّا أن يكون[٢١٩] في طبعه[٢٢٠] خفيّا بمنزلة جوهر[٢٢١] الله تبارك وتعالى، وجوهر العقل والنّفس والطّبيعة، وما كان كذلك فإنّما يعرف بالقياس من الظّاهر على الخفيّ، وإمّا أن يكون إنّما هو خفيّ عند الحسّ، وهذا يفوت الحسّ لواحد من أربعة أسباب، إمّا لبعد الشّقّة[٢٢٢] بمنزلة السّفينة التي تكون في لجّة البحر[٢٢٣]، فتخفى على من في شاطئ البحر، وإمّا لصغر مقداره بمنزلة الهباء الذي يطير[٢٢٤] في الهواء، فإنّ هذا متى لم يدخل شعاع الشّمس من كوّة أو من روزنة لم يتبيّن لنا، وإمّا لأنّه من جنس حاسّة أخرى بمنزلة الصّوت الذي لا يتبيّن للمذاق، وإمّا لأنّ شيئاً يغطيه ويستره بمنزلة حجر في قعر البحر يغطيه ماء البحر. أصحاب التّجارب يكرهون القياس من الظّاهر على الخفيّ لأنّه أمر يقع فيه اختلاف ولا يقع عليه الحكم، ويحمدون القياس من الظّاهر على الظّاهر من طريق أنّه أمر لا يقع فيه اختلاف، وإن وقع كان الحكم فيه سهلاً[٢٢٥]، والاختلاف زعموا أنّه أمر يدلّ على أنّ الشّيء الذي فيه الاختلاف[٢٢٦] لم يدرك ولم يوقف على حقيقته والقصور على[٢٢٧] إدراك الشّيء وبلوغ حقيقته، هو سبب الاختلاف.

---

٢١٤ $A^2F^2M^2S$: + أولأنّ دونه حائلاً؛ S: – منه | ٢١٥ AM: + أيضاً؛ F: أنالوجسموس؛ S: أفيلوجسيموس | ٢١٦ A^hFM^h: وأصحاب | ٢١٧ A^1؛ ADY: ويخدعونهم | ٢١٨ F^2؛ F: بالأغلوطات؛ DY: بالأغلاط | ٢١٩ AM: + عندنا | ٢٢٠ A^1M^1؛ AM: بالطّبع؛ S: في نفسه | ٢٢١ M: ذات | ٢٢٢ S: الطّرف؛ DY: المسافة | ٢٢٣ S: اليمّ | ٢٢٤ S: الهباء الطّائر | ٢٢٥ F: سهل؛ F^h: والاختلاف إذكان في خفيّ كان الحكم فيه غير سهل والاختلاف صح | ٢٢٦ ADMY: اختلاف | ٢٢٧ ADMY: عن

Commentary on chapter 6 of Galen's book *On the Medical Sects*[86]

[The opinions of the Methodists]

(38) The followers of the third sect, who are the Methodists, excuse themselves from investigating causes, habits, ages, times of year, temperaments, countries, faculties, and organs of the body. When they turn to diseases, they also excuse themselves from investigating diseases in their specificity as individuals, since they are infinite. Rather, they try to discover the general communities of diseases, since these are easier to find and less trouble for the mind to acquire. They assume that there are three communities of diseases: first, flux, which is flowing; second, costiveness, which is adhesion; and, third, a compound of the two. One acquires health through therapy by regimen, surgery, or the use of drugs. Some adherents of the third sect think that these communities are communities common to all disease, including those treated by regimen, those treated by surgery, and those treated by drugs. Others among them think that the communities existent in diseases treated by regimen are flux, costiveness, and their combination, whereas they think that the diseases treated by surgery are other communities.[87] This is to say that the thing that is treated can be foreign and harmful

86. K 1:79–81; H 12–13; G[a] 44–47; corresponding to the first half of chapter 6 of the modern Greek edition. Galen here gives a summary of the views of the Methodists, which the epitome follows closely and elaborates.

87. A "community" (κοινότης) is the pattern of perceptible signs and symptoms by which the Methodist physician recognizes the inward state of costiveness or flux; see p. xlvii above. Galen does not mention the third subsect of the Methodists. Nutton (*Ancient Medicine*, 192) mentions the communities treated by surgery and says that the later Methodists devised ever-finer distinctions among communities.

شرح الباب السّادس
من كتاب جالينوس في فرق الطّبّ[٢٢٨]

(٣٨) وأمّا[٢٢٩] أهل الفرقة الثّالثة، وهم أصحاب الحِيَل، فإنّهم يستعفون من النّظر في الأسباب والعادات والأسنان وأوقات السّنة والأمزاج[٢٣٠] والبلدان والقوى وأعضاء البدن، وإذا صار وا[٢٣١] إلى الأمراض استعفوا أيضاً من النّظر في الخاصّيّة[٢٣٢] الأفراد منها لأنّها ممّا لا نهاية له، ويقصدون لجمل[٢٣٣] الأمراض العامّيّة من قِبَل أنّها أسهل وجوداً وأهون تحصيلاً في العقل، و وضعوا أنّ جمل[٢٣٤] الأمراض العامّيّة ثلث، إحدىهنّ[٢٣٥] الانبعاث، أي الاسترسال، والأخرى الاحتقان، أي الاستمساك، والثّالثة التّركيب منهما. اجتلاب الصّحّة، وهو[٢٣٦] المداواة تكون إمّا بالتّدبير وإمّا بعلاج اليد وإمّا باستعمال الأدوية، وبعض أهل الفرقة الثّالثة يجعلون هذه الجمل جملاً تعمّ جميع الأمراض ما كان منها يداوى بالتّدبير، وما كان منها[٢٣٧] يداوي بعلاج اليد، وما كان يداوى بالأدوية، وبعضهم جعل الجمل الموجودة في الأمراض التي تداوى بالتّدبير الانبعاث والاحتقان، والتّركيب[٢٣٨] منهما، وجعل الأمراض التي تداوى بعلاج اليد جملاً أُخَر، وهي أن يكون الشّيء الذي يعالج إمّا

٢٢٨ F^{h}S؛ ADMY: – من كتاب جالينوس في فرق الطّبّ | ٢٢٩ DFY: فأمّا | ٢٣٠ F: والمزاج؛ S: والأمزجة | ٢٣١ AM: صار | ٢٣٢ M^2: الخاصّة؛ SY: خاصّيّة | ٢٣٣ A: لحمل؛ DSY: بجمل | ٢٣٤ MS: حملى | ٢٣٥ F: أحدهنّ؛ DY: أحدها | ٢٣٦ DFY: وهي | ٢٣٧ ADMY: – منها؛ | ٢٣٨ F: الانبعاث. . . وللتّركيب > الاسترسال والتّركيب

by nature, such as the stones that form in the bladder, or foreign and harmful in its place, such as intestinal rupture, which is the hernia resulting from a rupture. They are all in agreement that the diseases treated by drugs are to be treated in accordance with the communities that we have mentioned.

(39) When the followers of this sect say "costive disease," they mean the blockage of the excesses that are continually being produced—for example, retention of the urine, constipation of the bowels, and retention of perspiration. When they say "fluent disease," they mean an excessive evacuation of these excesses—for example, diarrhea, urine so excessive that the patient cannot retain it, and excessive sweat. When they refer to "a disease compounded of the two diseases," they mean a disease that includes both conditions—for example, the eye when it is both swollen and running with tears. They try to treat costive diseases by inducing relaxation and flow—for example, the treatment of a swollen knee with a poultice compounded from fenugreek, sweet melilot, linseed, barley meal, and chamomile. They treat fluent diseases by retention and constriction—for example, treating diarrhea with quince. In the case of composite diseases, they pay more attention to the more serious and acute than they do to the other part. For example, in the case of a swollen eye, if the tearing is more serious than the swelling, they treat it with salves that restrain and block the matter; but if the swelling is more serious, they use resolvent salves.

غريباً منكرًا في طبعه بمنزلة الحصاة التي تتولّد في المثانة وإمّا غريباً منكرًا في موضعه بمنزلة فتق[٢٣٩] الأمعاء، وهو القيلة الحادثة عن الفتق، وأمّا[٢٤٠] الأمراض التي تداوى بالأدوية، فإنّهم متّفقون مشتركون[٢٤١] على جميع ما يداوى به من الجمل[٢٤٢] التي ذكرناها.

(٣٩) وأهل هذه الفرقة يعنون بقولهم علّة احتقانيّة[٢٤٣] احتباس الفضول التي لم تزل تجري بمنزلة أسر[٢٤٤] البول[٢٤٥] وحصر البطن واحتباس العرق، ويعنون بقولهم علّة انبعاثيّة[٢٤٦] الإفراط في استفراغ هذه الفضول بمنزلة الخلفة وكثرة البول الذي لا يقدر صاحبه على حبسه وكثرة العرق، ويعنون بقولهم علّة مركّبة من العلّتينِ العلّة الجامعة للأمرينِ بمنزلة العين إذا كانت وارمة كثيرة الدّموع معاً، ويقصدون لمداواة العلل الاحتقانيّة بالأرخاء والتّسليس بمنزلة ما تداوي الرّكبة الوارمة بالضّماد المؤلّف من الحلبة[٢٤٧] وإكليل الملك وبزر الكتّان ودقيق الشّعير والبابونج، ولمداواة العلل الانبعاثيّة بالإمساك والسّدّ بمنزلة ما يداوي الخلفة بالسّفرجل، فأمّا العلل المركّبة، فإنّهم يقصدون فيها إلى الأهمّ والأشدّ أكثر منهم إلى غير ذلك بمنزلة ما تداوى العين الوارمة إن كانت دموعها أكثر من ورمها بالشّيافات التي[٢٤٨] تقمع[٢٤٩] وتردع المادّة، وإن كان ورمها أكثر فبالشّيافات المحلّلة ـــه

---

٢٣٩ $A^{2}FM^{2}$؛ AS: فرو؛ DY: قرو | ٢٤٠ S: – وأمّا؛ DY: فأمّا | ٢٤١ AM: مشتركون متّفقون؛ F: متّفقين مشتركين | ٢٤٢ F^{2}: العلل | ٢٤٣ $A^{2}DM^{2}SY$: استمساكيّة؛ S^{2}: اعتقاليّة | ٢٤٤ DFY: حصر | ٢٤٥ F: + واستمساكه | ٢٤٦ $A^{2}DM^{2}SY$: استرساليّة؛ S^{2}: انبعاثيّة؛ F: استرساليّة وانبعاثيّة | ٢٤٧ F: الحلية | ٢٤٨ F: بالشّياف الذي | ٢٤٩ ADMY: + وتدفع

Commentary on chapter 7[88] of Galen's book *On the Medical Sects*[89]

[The differences among the sects]

(40) Each of the three sects seeks something in particular. The sect of Empiricists has two things that they seek to acquire: first, to remember what they have apprehended by experience; and, second, to follow what is manifest to sensation. The sect of Rationalists has two things that they particularly seek to acquire: first, to deduce from the thing itself what ought to be done in that case; and, second, to know that which is not manifest to sense. The sect of Methodists has two things that they seek to acquire: first, to deduce from the thing itself what ought to be done in that case; and, second, to follow what is manifest to sense. From these four things—memory, deduction, what is manifest to sense, and what is not manifest to sense—there are six combinations, two of which cannot occur, one of which is not held by any of the sects, and three that are held by these three sects. They are in this pattern:

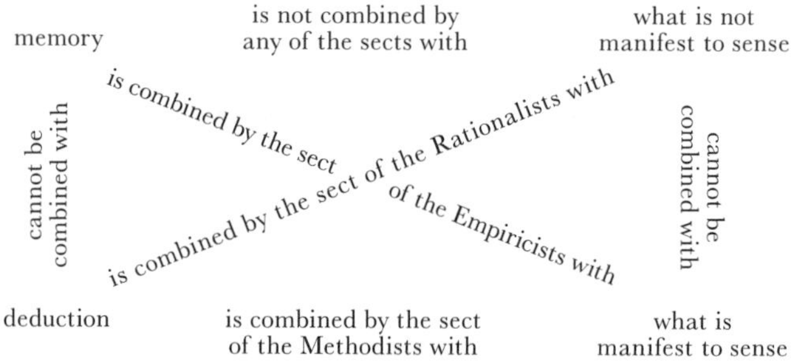

88. K 1:81–83, H13–15; G^a 48–55; corresponding to the second part of chapter 6 of the modern Greek edition: Galen explains how the Methodists distinguish their school from those of the Rationalists and Empiricists. The epitome gives a good summary of Galen's discussion. Galen's chapter ends with his caustic comments on the famous statement attributed to the Methodists that, while Hippocrates had said that life was short and art long, medicine can actually be learned in six months.

89. Omitted in most MSS.

شرح الباب السّابع
من كتاب جالينوس في فرق الطّبّ[٢٥٠]

[الفرق بين الفرق الثّلاث]

(٤٠) كلّ واحد[٢٥١] من الثّلث الفرق يقصد لشيء، ففرقة أصحاب التّجارب تقصد لأمرَيْن تطلبهما[٢٥٢]، أحدهما حفظ ما قد استدرك بالتّجارب والآخر اتّباع ما هو ظاهر للحسّ، وفرقة أصحاب القياس تقصد لأمرَيْن تطلبهما خاصّةً، أحدهما الاستدلال من نفس الشّيء على ما ينبغي أن يفعل في أمره، والآخر معرفة[٢٥٣] ما لا يظهر للحسّ، وفرقة أصحاب الحيل تقصد لأمرَيْن تطلبهما أحدهما الاستدلال من نفس الشّيء على ما ينبغي أن تفعل في أمره، والآخر[٢٥٤] اتّباع ما هو ظاهر للحسّ، فهذه الأربعة الأشياء، أعني الحفظ والاستدلال، والأمر الظّاهر للحسّ، والأمر الذي لا يظهر للحسّ يكون منها ستّة تراكيب، اثنان منها لا يثبتانِ و واحد ليس ينتحله أحد من أهل الفرق وثلثة تنتحلها أصحاب هذه الفِرَق الثّلثة[٢٥٥] على هذا المثال ـــه

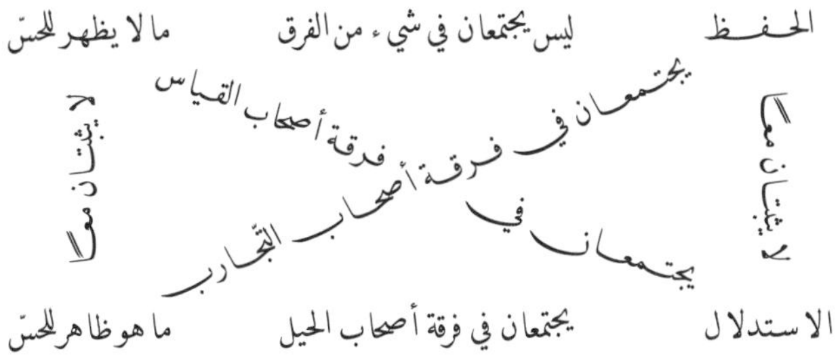

٢٥٠ ADMY: – من كتاب جالينوس في فرق الطّبّ | ٢٥١ AS: واحدة | ٢٥٢ F: مطلبا نهما؛ S: – تطلبهما | ٢٥٣ F: – والآخر معرفة؛ DF²Y: ومعرفة | ٢٥٤ DFY: – والآخر | ٢٥٥ ADMY: – الثّلثة

(41) The adherents of the third sect disagree about its definition, for one of them says that it is the knowledge of the communities that are manifest to sense and that relate to the end of medicine; another says that it is the knowledge of manifest communities according with the end of medicine; another group says that it is the knowledge of the manifest communities according with and following from the end of medicine; and Thessalus says that it is the knowledge of the communities proximate to the end of medicine that are absolutely necessary to it.[90]

(42) The Empiricists and the Methodists disapprove of investigating entities that are not manifest to sensation.[91] However, the Empiricists avoid doing so because they consider them to be unknowable, whereas the Methodists consider them to be unprofitable. The Methodists disagree with the Empiricists about memory and about why they feel no need for that which is not manifest to sense, and even disapprove of investigating it. They disagree with the Rationalists about sensible entities. The Methodists disagree with both in another way, for they dispense with entities that they claim are profitless. These are the causes, ages, seasons, places, and affected parts.[92] Despite this, however, the Empiricists and the Rationalists agree about practice, for their treatment uses the very same drugs.[93] They disagree, however, about how to discover the knowledge of the things they use in their treatments, for the Empiricists want to discover those things that they discover by means of observation and memory, while the Rationalists seek to use deduction. On the other hand, the Methodists and the Rationalists do not agree in their practice,[94] for the Methodists investigate neither the causes nor the daughters of the elements, nor the means by which one discovers the knowledge of the drugs used in treatment, since they disregard anything that is not manifest to sense.

---

90. K 1:81; H 13–14; G[a] 48–49: This paragraph is a near translation of Galen's original, although he also stresses that the Methodists define the method in terms of the communities.

91. K 1:82; H 14; G[a] 49–51.

92. The Methodists believed that it was only necessary to know whether a disease was a state of costiveness or flux, so there was no need to know its causes. Galen does not specifically mention either causes or affected parts.

93. K 1:79; H 12; G[a] 43: Galen remarks that the differences between Rationalists and Empiricists are mainly theoretical, since their treatments are usually the same.

94. Some MSS read "agree in neither practice nor doctrine."

(٤١) قد اختلف أهل الفرقة الثّالثة في حدّها، فقال بعضهم إنّها هي معرفة الجمل الظّاهرة للحسّ التّابعة لتمام[٢٥٦] الطّبّ، وقال بعضهم إنّها معرفة الجمل الظّاهرة الموافقة لتمام الطّبّ، وقال قوم آخرون إنّها معرفة الجمل الظّاهرة الموافقة التّابعة لتمام الطّبّ، وقال ثاسلس إنّها معرفة الجمل القريبة من تمام الطّبّ التي لا بدّ منها فيه ضرورةً.

(٤٢) أصحاب التّجارب وأصحاب الحيل يكرهون النّظر في الأمر الذي لا يظهر للحسّ ويتجنّبونه[٢٥٧] إلاّ أنّ أصحاب التّجارب يفعلون ذلك على أنّه أمر مجهول، وأصحاب الحيل على أنّه أمر لا ينفع. أصحاب الحيل مخالفون لأصحاب التّجارب في الحفظ وفي جهة الاستعفاء ممّا لا يظهر للحسّ والكراهة له ولأصحاب القياس في الأمر الظّاهر للحسّ ولهما جميعاً من جهة أخرى، وذاك[٢٥٨] أنّهم يحذفون أشياءً يزعمون أنّه لا ينتفع بها، وهي الأسباب والأسنان وأوقات السّنة والبلدان والأعضاء[٢٥٩] الآلمة، ومع هذا أيضاً، فإنّ أصحاب التّجارب وأصحاب القياس متّفقون[٢٦٠] في العمل، وذاك[٢٦١] أنّهم يداوون بأشياء واحدة بأعيانها، وهم مختلفون في الوجه الذي به يستخرج معرفة الأشياء التي بها يداوون من طريق[٢٦٢] أصحاب التّجارب يريدون أن يستخرجوا ما يستخرجونه من ذلك بالرّصد والحفظ، وأصحاب القياس بالاستدلال، فأمّا أصحاب الحيل وأصحاب القياس، فغير متّفقين في العمل[٢٦٣] لأنهم لا ينظرون في الأسباب ولا في بنات الأركان ولا في الوجه الذي تستخرج به معرفة الأشياء التي بها تكون المداواة لأنّهم يهربون من الأمر الذي لا يظهر للحسّ.

---

٢٥٦ A^2M^2: لغاية | ٢٥٧ F: يجتنبونه؛ S: ويجتلبون [؟] | ٢٥٨ A^2M^2: وهي؛ DFY: – وذاك؛ S: فهو | ٢٥٩ A^hM^h: حاشية له مثل ما يستدلّ على مرض الكبد بألم الجانب الأيمن | ٢٦٠ F: متّفقين؛ DY: يتّفقون | ٢٦١ DFY: وذلك | ٢٦٢ ADSY: + أنّ | ٢٦٣ AM: + ولا في القول

(43) The sect of Methodists cannot avoid a dilemma, for either they are correct in their claim and so do great good for the art of medicine when they omit what is extraneous and unprofitable, or else they are wrong in their claim and do great harm to the art of medicine by omitting what is truly necessary to the art.[95]

95. K 1:83–84; H 15–16; G^a 53.

(٤٣) ليس تخلوا فرقة أصحاب الحيل من أحد أمريْنِ، إمّا أن تكون صادقة في دعواها، فتكون صناعة الطّبّ تنتفع بها منفعة عظيمة، إذ كانت تحذف منها ما هو فضل لا ينتفع به، وإمّا أن تكون كاذبةً في الدّعوى، فتكون مضرّتها لصناعة الطّبّ مضرّة عظيمة إذ كانت تحذف الأشياء التي لا بدّ منها في الصّناعة ضرورةً.

Commentary on chapter 8[96] of Galen's book *On the Medical Sects*[97]

[Galen's criticism of the Methodists]

(44) There are two ways to investigate entities and make judgments about them. That is because entities can be discovered and their truth known from their falsehood either by inference, if they are intelligible entities, or by sense, if they are sensible entities. In his controversy here with the adherents of this third sect, Galen uses the second of the two methods of investigation. That is because it is easier for beginners in the science of medicine and because the adherents of this sect also use it and prefer it.

(45) The adherents of the third sect fail to consider the matter of ages; fail to consider causes, whether antecedent causes, preceding causes, or cohesive causes; and fail to consider the seasons of the year, countries, or organs of the body. These last are diverse because some of them have to do with nerves, such as the stomach and womb; some of them are venous, such as the tongue; and some are arterial, such as the lungs and spleen. Thus, let us first consider the investigation of the causes. There are three causes: antecedent, preceding, and cohesive. The Rationalists seek to know all three, and so they investigate them. The Empiricists seek to know the antecedent causes insofar as they are manifest to sense. For this reason, Galen thought that he ought to frame the debate between the Empiricist and the Methodist as being first about antecedent causes, since neither of them investigates or seeks to know the preceding or cohesive causes—the Empiricists because they are not manifest to sense, and the Methodists because they are useless.

96. K 1:83–87, H 15–18; G[a] 53–60; corresponding to chapter 7 of the modern Greek edition and the remainder of chapters 7 and 8 of the Arabic translation. Galen has the Methodist argue that consideration of antecedent and cohesive causes is unnecessary. The epitome is a summary of Galen's text with some clarification of his examples.

97. Omitted in most MSS.

شرح الباب الثّامن
من كتاب جالينوس في فرق الطّبّ[٢٦٤]

[احتجاج جالينوس على أصحاب الحيل]

(٤٤) النّظر في الأمور والحكم عليها تكون على ضربين، وذلك أنّ الأمور تختبر[٢٦٥] وتعرف حقّها من باطلها، إمّا بالقياس إن كانت من الأمور المعقولة، وإمّا بالحسّ إن كانت من الأمور المحسوسة، وجالينوس يستعمل في مناظرة أصحاب هذه الفرقة الثّالثة هاهنا أوّلاً هذا الباب الثّاني من بابي[٢٦٦] النّظر، وذلك أنّه[٢٦٧] أسهل على الدّاخلين في علم الطّبّ، ولأنّ أهل هذه الفرقة أيضاً يقدّمونه ويؤثّرونه.

(٤٥) وأهل هذه الفرقة الثّالثة يحذفون النّظر في أمر الأسنان والنّظر في أمر الأسباب، ماكان منها من الأسباب البادية وماكان منها من الأسباب السّابقة وماكان منها من الأسباب الواصلة، والنّظر في أوقات السّنة وفي البلدان وفي أعضاء البدن التي هي مختلفة لأنّ بعضها عصبيّة بمنزلة المعدّة والرّحم، وبعضها عروقيّة بمنزلة اللّسان، وبعضها شريانيّة بمنزلة الرّية والطّحال. فليكن النّظر أوّلاً في أمر الأسباب، والأسباب ثلثة، البادية والسّابقة والواصلة، وأصحاب القياس يطلبون معرفة هذه الثّلثة كلّها، وينظرون فيها، وأصحاب التّجارب يطلبون معرفة الأسباب البادية[٢٦٨] من طريق أنّها ظاهرة للحسّ، ولذلك رأى أن يجعل المناظرة بين صاحب التّجربة وصاحب الحيل أوّلاً في هذه الأسباب البادية[٢٦٩]، فأمّا الأسباب السّابقة والأسباب الواصلة، فليس ينظرون[٢٧٠] فيها ولا يطلبون[٢٧١] معرفتها أصحاب التّجارب من طريق أنّها لا تظهر للحسّ ولا أصحاب الحيل من طريق أنّها لا تنفع.

---

٢٦٤ ADMY: – من. . . الطّبّ | ٢٦٥ F^2؛ F: تعتبر | ٢٦٦ M: تأتي | ٢٦٧ FS: لأنه | ٢٦٨ S^h: صاحبه وملكه هو العبد الفقير جار الله الإبرهاتيّ [؟] | ٢٦٩ DF^2Y: + من طريق أنّها ظاهرة | ٢٧٠ AM: ينظر | ٢٧١ AM: تطلب

Commentary on chapter 9[98] of Galen's book *On the Medical Sects*[99]

[The Empiricists' criticism of the Methodists]

(46) The antecedent cause may be a cause of one of the systemic diseases, as when hot winds are a cause of fever; it may be a cause of one of the organic diseases, in the way that a blow can be the cause of swelling; or it may be a cause of dieresis,[100] in the way that a sword, spear, or animal bite may be the cause of a lesion. Because the case of dieresis is the clearest, and because it is obvious that it occurs from the antecedent cause, we will explain that, so long as this cause is not known, it is not possible to treat the dieresis occurring from it with a treatment that will lead to a cure. Thus, suppose that a man has received a bite from a rabid dog[101] that wounds him somewhere on his body, and he recognizes that the dog was rabid from the signs indicating rabies. These are that the dog's eyes protrude,[102] his tongue hangs from his mouth, his tail is limp, and he is thirsty but does not drink water. Those who are bitten by rabid dogs are treated in the same way as those bitten by other poisonous animals: externally by hot, pungent drugs placed upon the wound to expand and open the head of the wound and draw up and expel the poison, and internally by things that destroy and dry the poison, such as theriac

98. K 1:87–92; H 18–22; G[a] 61–70; corresponding to chapter 8 of the modern Greek text and chapter 9 of the Arabic translation. The Empiricist defends the use of antecedent causes by pointing out that the bite of a rabid dog needs to be treated differently from other small puncture wounds and that factors such as age, season, and the organ affected need to be taken into account in treatment. The epitome follows the text closely, systematizing Galen's polemical exposition.

99. Omitted in most MSS.

100. Literally, "separation of connection."

101. K 1.88–89; H 18–19; G[a] 62–64: Galen presents this example as a parable of two men bitten by rabid dogs. The foolish man goes to a Methodist physician, who treats him for a minor puncture wound and thus kills him by his negligence. The wise patient goes to an Empiricist physician, who determines the antecedent cause—a rabid dog—treats the man for rabies, and thus saves his life. The point is that the treatments are opposite, and the physician needs to know the antecedent cause to determine which is appropriate to a particular patient.

102. Two MSS contain the gloss: "In the Sheikh's commentary on Galen's original text, it says 'and his eyes are sunken, red, and disturbed.' The text in

شرح الباب التّاسع
من كتاب جالينوس في الطّبّ[٢٧٢]

[احتجاج أصحاب التّجربة على أصحاب الحيل]

(٤٦) السّبب البادي إمّا أن يكون سبباً لمرض من الأمراض المتشابهة الأجزاء بمنزلة ما تكون السّمائم سبباً للحمّى، وإمّا أن يكون سبباً لمرض من الأمراض الآليّة بمنزلة ما تكون الضّربة[٢٧٣] سبباً للورم، وإمّا أن يكون سبباً لتفرّق الاتّصال بمنزلة ما يكون السّيف أو السّهم أو نهشة الحيوان سبباً للقرحة، ولأنّ الأمر في تفرّق الاتّصال أبين وحدوثه[٢٧٤] عن السّبب البادي ظاهر[٢٧٥]، فنحن[٢٧٦] نبيّن أنّه متّى لم يعرف هذا السّبب[٢٧٧] ما هو لم يمكن أن يداوى تفرّق الاتّصال الحادث عنه مداواة يبرّأ بها، فأنزل[٢٧٨] أنّ إنساناً نهشة كلب كَلِب نهشةً خرق بها موضعاً من بدنه وعرف أنّه كلب كلب من هذه العلامات الدّالّة عليه، وهي أن عينيهِ تكونانِ ناتئتيْنِ[٢٧٩] ولسانه يكون خارجاً عن فيه وذنبه يكون مسترخياً ويعطش ولا يشرب الماء، والذين ينهشهم كلب كلب يداوون كما يداوى سائر من نهشة حيوان من الحيوانات الأُخرذوات السّموم إمّا من خارج، فبالأدوية الحارّة الحادّة[٢٨٠] التي توضع على القرحة ممّا توسع وتفتح رأس القرحة وتجتذب[٢٨١] السّمّ وتخرجه، وأمّا من داخل، فبالأشياء التي

---

٢٧٢ ADMY: – من. . . الطّبّ | ٢٧٣ A^hM^h: أي الصّدمة أو الصّكّة | ٢٧٤ DF^1Y: وأنّ حدوثه | ٢٧٥ DF^1Y: + للحسّ | ٢٧٦ F: نحن؛ DY: – فنحن | ٢٧٧ F: + البادي | ٢٧٨ A^2M^2: فإن رأينا | ٢٧٩ A^hM^h: قال في تفسير الشّيخ لفص الكتاب: وعيناه غائرتان حمراوانِ مضطربتان للصّورة التي في متن هذه النّسخة، وهي نائتينتين هي بمعنى مضطربتين فتكون حينئذٍ الغلط في إعجام البلد بتاء صحّ المعنى | ٢٨٠ A^2FM^2S: الجاذبة | ٢٨١ ADMY: وتجذب

and the like, since, if the dieresis occurs from the bite of a poisonous animal like a rabid dog or viper, it needs to be widened and opened, but if it occurs from a sword, fire, or spear that is not poisoned, it needs only to be closed and scarred over. The situation with dieresis being as we have described it, we know that it is useful to consider causes, for the treatment can change or vary in accordance with the cause. Treatment can change or vary in its quantity, its quality, or its entire genus, as we have described here, for we can treat one and the same disease differently if the efficient cause is different.

(47) It can also be shown that consideration of the patient's age is beneficial, as I will explain.[103] If we wish to draw blood from a vein, we will not do so from a small child, because the body of a small child is subject to rapid dissolution by reason of its heat and moisture, because most of his blood is used in growth and the increase of his body, and because he has little strength. We also do not draw blood from an old man, for he has less blood than before, his body has become cold and dry, and his strength has weakened. We draw blood only from someone in the prime of youth, since there are none of the contraindicants found in children and the old. He does not resemble the child in the rapid dissolution of the body, lack of strength, or the amount of blood used up in the growth of the body; nor is he like the old man with his lack of blood, predominance of coldness and dryness, and lack of strength.

(48) It can also be known in the following way that consideration of countries is beneficial. I say that some places are moderate in climate; in those that are, we draw blood when necessary. Some are northerly and very cold, such as the lands of the Slavs; in such places, we do not

---

this manuscript is 'staggering,' in the sense of disturbed. Thus, the error is [the placement of] the *tāʾ* in the dialect of the city, but the meaning is correct." The author of the gloss is explaining why the epitomist has written that the dog's eyes are *nātiʾatayn* (bulging), whereas a commentator has written *ghāʾiratān* (sunken). The gloss explains that the word is a dialectical misspelling of *nāʾitatayn* (staggering), which can be taken to mean disturbed. "The Sheikh" is presumably Abū al-Faraj ibn al-Ṭayyib (d. 435/1053), who wrote the only attested Arabic commentary on *The Medical Sects*. The passage referred to is found in MS Manisa 1772 f. 30a, which does indeed read "sunken," though not "red and disturbed."

103. K 1:89–92; H 19–22; G[a] 64–70: The epitome elaborates on Galen's brief discussion of the relevance of age and country in determining whether to let blood. The original text also cites Hippocrates on the relevance of the time of year in prescribing drugs. The epitome modifies Galen's treatments somewhat, probably to bring them into line with Islamicate medical practices.

تفني[٢٨٢] السمّ وتجفّفه بمنزلة التّرياق وما أشبهه، لأنّ تفرّق الاتّصال إن كان حدوثه عن نهشة حيوان ذي سمّ بمنزلة كلب كلب أو أفعى، فهو يحتاج إلى أن يوسع ويفتح، وإن كان حدوثه عن سيف أو نار أو سهم غير مسموم، فهو يحتاج إلى أن يلحم ويدمل فقط، وإذا[٢٨٣] كان الأمر في تفرّق الاتّصال على ما وصَفتُ، فقد علم منه أنّ النّظر في أمر الأسباب[٢٨٤] ممّا ينتفع به إذ[٢٨٥] كانت المداواة تختلف وتتغيّر بحسب السّببّ، واختلاف المداواة وتغيّرها تكون إمّا في مقدارها وإمّا في كيفيّتها وإمّا في جملة جنسها على ما وصفنا هاهنا، فإنّا نداوي العلّة الواحدة بعينها إذا اختلف السّبب الفاعل لها مداواة مختلفة.

(٤٧) وقد يعلم أيضاً أنّ النّظر في أمر الأسنان ممّا ينتفع به ممّا أصف. أقول إنّا إذا أردنا أن نفصد عُرقًا لم نفصده لصبيّ صغير لأنّ الصّبيّ الصّغير بدنه يتحلّل سريعًا بسبب حرارته و رطوبته ولأنّ أكثر الدّم يتصرّف في نمائه و زيادة بدنه، ولأنّ قوّته ضعيفة، ولا نفصد العرق أيضاً لشيخ، فإن لان دمه قد قلّ وبدنه قد برد وجفّ وقوّته قد ضعفت، وإنّما نفصد العرق لمن كان في عنفوان سنّ الشّباب فقط، لأن[٢٨٦] ليس فيه من الموانع التي في الصّبيان و في الشّيوخ شيء إذ كان ليس هو مثل الصّبيّ في سرعة تحلّل البدن وضعف القوّة وكثرة ما ينصرف من الدّم في نماء بدنه، ولا مثل الشّيخ في قلّة الدّم وغلبة البرد[٢٨٧] واليبس وضعف القوّة،

(٤٨) ويعلم أيضاً أنّ النّظر في أمر البلدان ممّا ينتفع به من هذا الوجه. أقول إنّ البلدان منها معتدل المزاج، وما كان منها كذلك، فنحن نفصد فيه العرق في وقت الحاجة، وبعضها شماليّ كثير البرد بمنزلة بلاد الصّقالبة، وما كان كذلك فلسنا نفصد

---

٢٨٢ F^2؛ F: نفس | ٢٨٣ AM: وإذ | ٢٨٤ F: + البادية | ٢٨٥ DFY: إذا | ٢٨٦ ADMY: لأنّه | ٢٨٧ DFY: وغلبة البرد > والبرد

draw blood. Some are southerly and burn like fire; in such places, we also do not draw blood due to the strong tendency of the air's heat to cause the dissolution of the body, as happens in the country of Abyssinia.[104]

(49) In the following way, it can also be known that consideration of the organs of the body is necessary and beneficial.[105] I say that a hematoma is treated by things that differ in accordance with the organ that is swollen. If the swelling is in the eye, it is treated with collyria.[106] If it is in the ear, it is treated with wine vinegar and rose oil. If it is in the uvula, it is treated with mulberry pulp.[107] If it is in the leg, it is treated with poultices.[108] If it is in the stomach, it is treated with douches of oil and warm water. From the examples I have given of organic diseases, it is also clear that the investigation of causes is both necessary and useful. That is because a swelling may be of itself—that is, without an antecedent cause—or from an antecedent cause. If it is of itself, then it is from a preceding cause and is thus from a plethora in the body. A swelling that is of this sort does not need relaxing substances before one evacuates the body entirely. If it previously needed things to restrain and block it, or if it was from an antecedent cause such as what results from a blow or shock, it initially needs things to relax and dissolve it.

(50) What we have described can be investigated in two ways. First, what can be discovered from sensation suits the Empiricists, being in accord with their position. Thus, the words of the Empiricists can be used to criticize the Methodists.[109] Second, what results from inference suits the Rationalists since it conforms with their position; so the Rationalist debates the Methodist here.

---

104. Galen's examples are the Arctic ("under the Bears") and Egypt.

105. K 1:90–92; H 21–22; G[a] 67–70: The epitome continues to give systematic expositions of the examples that Galen gives in his polemic.

106. Galen says "astringents," στούφοντες. G[a] reads *al-akḥāl al-qābiḍah* (astringent collyria).

107. Galen's remedy is the fruit of the Egyptian thorn, ἀκάνθης Αἰγυπτίας ὁ κάρπος. G[a] reads *al-shawkah al-Miṣrīyah*, which the editor identifies as the sant tree, *acacia nilotica*. G[a] also suggests fissile alum—*al-shabb al-yamānī* (Yemeni alum).

108. Galen's remedy is lancing with a knife and soaking with olive oil.

109. Galen cast the latter half of *The Medical Sects* as a debate between the Methodists and their opponents. This paragraph summarizes the positions expounded by the Empiricists and Rationalists in chapters 9 and 10 of the epitome, corresponding to chapters 8 and 9 of the modern Greek editions.

فيه العرق، ومنها جنوبيّ يلتهب ناراً، وما كان كذلك أيضاً فليس نقصد فيه العرق لكثرة ما يحلّله الهواء من البدن بحرارته بمنزلة ما يعرض ذلك في بلاد الحبشة.

(٤٩) ويعلم أيضاً أنّ النّظر في أمر أعضاء البدن ممّا يحتاج إليه وينتفع به من هذا الوجه، أقول أنّ الورم الدّمويّ يداوى بأشياء تختلف بحسب العضو الوارم، فإن كان ذلك الورم في العين فدواؤه الأكحال، وإن كان في الأُذن فدواؤه خلّ الخمر ودهن الورد، وإن كان في اللّهاة، فدواؤه رُبّ التوت، وإن كان في السّاق فدواؤه الضّماد، وإن كان في البطن فدواؤه[٢٨٨] النّطول[٢٨٩] بالدّهن والماء[٢٩٠] الحارّ، وقد تبيّن[٢٩١] أيضاً أنّ النّظر في أمر الأسباب ممّا يحتاج إليه وينتفع به من هذا المثال الذي أمثله لك من الأمراض الآليّة، وذلك أنّ الورم يكون إمّا من قِبَل نفسه، أعني من غير سبب بادٍ، وإمّا من سبب بادٍ، وإذا كان من قِبَل نفسه، فهو من سبب سابق، وهو من امتلاء يكون في البدن، وما كان من الورم كذلك، فليس يحتاج إلى الأشياء المرخية[٢٩٢] دون أن يستفرغ البدن[٢٩٣] كلّه، فأمّا قبل ذلك فيحتاج[٢٩٤] إلى أشياء تقمع وتمنع، وأمّا إذا كان من سبب بادٍ بمنزلة ما يعرض من ضربة أو من صدمة، فهو يحتاج منذ أوّل أمره[٢٩٥] إلى أشياء ترخي وتحلّل.

(٥٠) والنّظر على ما وصفنا بابانِ، أحدهما من الحسّ، وهو باب يشاكل ويوافق أصحاب التّجارب، وقد نوظر[٢٩٦] به أصحاب الحيل عن[٢٩٧] لسان أصحاب التّجارب، والآخر من القياس، وهو باب يوافق ويشاكل[٢٩٨] أصحاب القياس، فصاحب القياس يناظر صاحب الحيل بهذه المناظرة ها هنا.

---

٢٨٨ F: – وإن. . . فدواؤه؛ M: فداؤه؛ DY: – فدواؤه | ٢٨٩ F: والنّطول؛ DY: فالنّطول | ٢٩٠ ADSY: بالماء | ٢٩١ DSY: يتبيّن | ٢٩٢ A: + ظنّ؛ D: المرضيّة | ٢٩٣ AM: – البدن | ٢٩٤ AM: يحتاج | ٢٩٥ AM: مرّة؛ DY: الأمر | ٢٩٦ F^2؛ F: ينظر | ٢٩٧ F^2؛ AFM: على | ٢٩٨ ADMY: يشاكل و يوافق

Commentary on chapter 10[110] of Galen's book *On the Medical Sects*[111]

[The Rationalists' criticism of the Methodists]

(51) One group of the Methodists says that the communities are in the bodily organs themselves, and another group of them says that they are in the humors contained by the organs of the body.[112] The view of those who hold that the communities are in the organs themselves can be refuted by pointing out that, while they say that the communities are things that are manifest to sense, they also say that they are in the bodies of the organs themselves. Now, the organs are not all manifest to sense—only those that are near[113] the exterior of the body. Those that are inside the body are concealed and so are not manifest to sense. If the organ is not manifest to sense, such of these communities as occur in it are not manifest to sense. As for those who say that the communities occur in the humors contained in those organs and who claim that diseases occur only when these humors are subject to excessive retention or evacuation, their theory can be refuted by pointing out that the quantity of evacuation is often immoderate[114] and that this can be beneficial, not harmful—for example, in a crisis.

110. K 1:93–105, H 22–32; G^a 71–90; corresponding to chapter 9 of the modern Greek edition and chapter 10 of the Arabic text: Galen's text contains the Rationalists' (and Galen's) refutation of the Methodists. His main arguments are that the Methodists themselves disagree about the nature of communities, which is their fundamental theoretical concept, and that their concepts of flux and costiveness are unclear, as shown by their treatment of swelling or inflammation.

111. Omitted in most MSS.

112. K 1:93–96; H 22–25; G^a 72–76. Galen describes these two views slightly differently: According to the Methodists, the communities are "the dispositions of the body themselves" (ἐν αὐταῖς τῶν σωμάτων ταῖς διαθέσεςι, *ḥālāt al-abdān*), which, according to others, are "the natural secretions" (ταῖς κατὰ φύσιν ἐκκρίσσεσι, *al-istifrāghāt al-ṭabīʿiyah*).

113. Some MSS read "on" for "near."

114. Gloss of IT: "There is a point to investigate here, for going beyond moderation—that is, the amount that ought to evacuated—is harmful. An opponent might object that so long as it is beneficial, it is not immoderate—that is, beyond

شرح الباب العاشر[٢٩٩]
من كتاب جالينوس في فرق الطّبّ[٣٠٠]

[احتجاج أصحاب القياس على أصحاب الحيل]

(٥١) إنّ أصحاب الحيل منهم قوم يقولون إنّ هذه الجمل[٣٠١] تكون في نفس أعضاء البدن، ومنهم قوم يقولون إنها تكون في الأخلاط التي يحتوي عليها أعضاء البدن، فمن قال منهم إنّ هذه الجمل تكون في الأعضاء أنفسها، فقوله ينتقض من أنّه بعد ما قال إنّ الجمل أشياء تظهر للحسّ قال إنّها موجودة في أجسام الأعضاء أنفسها، وليس الأعضاء كلّها ظاهرة للحسّ، بل إنّما يظهر منها للحسّ ما كان يلي[٣٠٢] في ظاهر البدن، وأمّا ما[٣٠٣] كان في باطن البدن مستوراً، فليس هو بظاهرٍ للحسّ، وإذا لم يكن العضو ظاهراً للحسّ، فما يحدث فيه أيضاً من هذه الجمل ليس بظاهر للحسّ، فأمّا من قال منهم بأنّ الجمل تحدث[٣٠٤] في الأخلاط التي تحتوي عليها الأعضاء، وكان يزعم أنّ الأمراض إنّما هي إفراط هذه الأخلاط في الاحتباس أو في الاستفراغ، فقوله ينتقص من أنّه قد يعرض مراراً كثيرة من الاستفراغ مقدار يجاوز[٣٠٥] الاعتدال، فلا يضرّ بل ينفع بمنزلة ما يكون ذلك في البحران.

---

٢٩٩ F^1؛ F: – شرح الباب العاشر | ٣٠٠ ADMY: – من. . . الطّبّ | ٣٠١ S: + الثّلث؛ S^h: الاستفراغ الاحـ. . . والمركّب منهما | ٣٠٢ ADMY: – يلي | ٣٠٣ ADMY: وأمّا ما > وما | ٣٠٤ A^2M^2S: توجد | ٣٠٥ A^hM^h: حـ[اشية] له، ينظر فيه فإنّ المجاوز للاعتدال، أي لما قيس حقّه أن يستفرغ يضرّ، والخصم يقول ما دام نافعاً، فليس بمجاوز للاعتدال، أي لما قيس العدل أن يستفرغ ولما من حقّه أن يستفرغ إلاّ أن يقاس ذلك بالمعتدل في وقت الصّحّة ويلتزمه الخصم، فيصحّ الحجّة عليه.

(52) Some of the things that are evacuated from the body are unnatural in their quantity, such as feces, urine, sweat, and vomit[115] when an excessive quantity of any one of these is expelled.[116] Some evacuations are unnatural in their quality, such as feces that are excessively moist and burning hot, urine that is red or black, and sweat that is cold. Some of them are unnatural in every respect, such as hemorrhage, for blood is not naturally evacuated from the body. However, we do find quite often that when it is evacuated, it is a cause of health, not of disease.

(53) The small intestines have three parts.[117] The first is called the pylorus; and the second, the jejunum. After that comes the rest of the ileum, which is collectively called the small intestine. These are the bowels. Then there is the large intestine, which also has three parts: first, the caecum; second, the colon; and, third, the rectum.[118]

(54) If a patient has diarrhea, the Methodists will say that it is one of the flowing diseases—which is to say, a flux. One might then ask them: "Since it is possible for this fluent disease to occur in each of these intestines, not one of which is manifest to sense, how do you know which one this disease has occurred in?" That is because, if someone

---

what rightly ought to be evacuated and what is the correct amount to evacuate. However, that is in comparison with what is evacuated during a time of health. This follows from the opponent's position, so the argument against him is valid."

115. A gloss of IT: "Perhaps vomit ought not to be included here because it is not natural to man. However, one might mention the vomit of nursing babies, so perhaps that could be given as an example here, since it is prevalent in them."

116. K 1:94–95, H 23–24; G[a] 73–74: Galen, arguing against those Methodists who hold that unnatural evacuation is a fluent disease, points out that unnatural evacuation is often helpful. This paragraph clarifies what unnatural evacuation is, distinguishing between unnatural quantity and unnatural quality, which Galen does not do.

117. K 1:95–99; H 24–26; G[a] 74–78. This is Galen's response to the Methodists who held that the communities are states in the organs. Galen sets up a dilemma: either they cannot know the evacuation that takes place within the body, especially in the various parts of the intestinal tract, or else they need to know the medical theory that would tell them the states of the inner organs.

118. Galen mentions the parts of the intestines in passing, but does not list them. In Arabic, names of the three parts of the small intestine are, literally, "the gatekeeper," "the one that fasts," and "circles" or "coils." The parts of the large intestine are the "one-eyed," "colon," and "straight intestine." Except for "colon," which is a loanword from Greek, all the terms are calques of the Greek terms and are thus identical in root meaning with the modern scientific terms. A gloss by IT says of the ileum: "These coils are called fascial [pertaining to wrappings] and are the third part of the small intestine."

(٥٢) الأشياء التي تستفرغ من البدن منها ما هو خارج عن الطّبيعة[٣٠٦] في مقداره بمنزلة الثّفل والبول والعرق والقيء[٣٠٧] إذا خرج من كلّ واحد منها مقدار مفرط، ومنها ما هو خارج عن الطّبيعة[٣٠٨] في كيفيّته بمنزلة الثّفل الذي يكون كثير الرّطوبة ومتشيّطاً[٣٠٩] محترقاً[٣١٠] والبول إذا كان أحمر أو أسود والعرق إذا كان بارداً[٣١١]، ومنها ما استفراغه خارج عن الطّبيعة في جميع جهاته بمنزلة انفجار الدّم لأنّ الدّم ليس له في الطّبع حدّ يستفرغ به من البدن، وقد نجده مراراً كثيرة يستفرغ، فلا يكون ذلك سبباً للمرض بل سبباً للصّحّة.

(٥٣) الأمعاء منها دقاق، وهي ثلثة، أحدها يقال له البوّاب، والآخر يقال له الصّائم، ومن بعد ذلك سائر الاستدارات[٣١٢] التي يقال لها[٣١٣] جملةً[٣١٤] أمعاء دقاق، وهي المصارين[٣١٥]، ومنها غلاظ، وهي أيضاً ثلثة، أحدها الأعور، والآخر القولن والثّالث المعاء المستقيم.

(٥٤) فإذا حدث بإنسان خلفة، ثمّ قال أصحاب الحيل إنّ ذلك علّة من العلل الاسترساليّة، أي الانبعاث، قيل لهم إذ[٣١٦] كان[٣١٧] قد[٣١٨] يمكن هذه العلّة الاسترساليّة أن تحدث في كلّ واحد من هذه الأمعاء التي ليس منها ولا واحد يظهر للحسّ من أين تعلمون في أي هذه حدثت تلك العلّة، وذلك أنّ من أراد أن

٣٠٦ A^2؛ AM: الطّبع | ٣٠٧ A^hM^h: حاشية له. عسى القيء لا يصلح للأنسان [A: للمنال] هاهنا أنّه [A: لأنّه] ليس في الطّبع إلاّ أن يحتيج بقيء الأطفال المرضعين، فعسى ذلك ممّا يجوز التّمثيل به هاهنا لأنّه أكثريّ فيهم | ٣٠٨ A^2؛ AM: الطّبع | ٣٠٩ ADY: أو متشيّطاً؛ F: أو منبسط | ٣١٠ F: محترق؛ S: محرقاً | ٣١١ F: أردلة رائحة | ٣١٢ A^hM^h: حـ[اشية] له هذه الاستدارات يسمّى اللفائفي، وهو الثّالث من الدّقّاق | ٣١٣ Y: له | ٣١٤ DY: – جملة | ٣١٥ AM: المصران؛ M^h: المصارين المصران هو الجيد لأنّه جمع مصير | ٣١٦ DFY: إذا | ٣١٧ F^1: كذلك؛ S: لأن | ٣١٨ ADMY: – قد

wants to know the diseases of the inner organs, he must be knowledgeable about the natures of the organs, the meaning of signs, the theory of the elements according to the school of the physicists, anatomy, and the science of logic. No one who knows logic will fall victim to fallacies about names and attributes in the way that the Methodists do; but, rather, he will oppose distended to soft, tender to hard, rarefied to dense, and evacuation to the retention of what is to be evacuated.

(55) Someone who knows well the things that we have mentioned will know this: if something that had been retained in some organ prior to its evacuation is then evacuated, it will have been evacuated by one of six causes.[119] It might be because it became finer and subtler, as is the case with the bleeding that happens to women. It might be because it became larger in quantity, as happens with vomit in those who are drunk. It might be because the vessels expanded, as happens in sexual intercourse. It might be because something attracts it from inside or from outside—from inside, as when some organ becomes warm; or from outside, as is the case with hot air or hot medicine. It might be because the repulsive power has strengthened, as is the case of a patient who drinks cold water at the height of his fever, it then being quenched from that hour. It might be because the retentive power weakens, as is the case with the patient who loses consciousness when he passes feces. We also know that the retention of what ought to be evacuated is due to

119. K 1:99–101; H 27–29; G[a] 82–84: Galen contrasts the various causes of evacuation and retention recognized by the ancient Hippocratics with the simplistic Methodist theory of larger and smaller pores. The epitomist converts Galen's allusions to various causes of evacuations and retention into six clearly defined categories and then adds examples.

يتعرّف علل الأعضاء الباطنة، فإنّه يحتاج أن يكون عالماً بطبائع الأعضاء وبقوّة العلامات وبالنّظر في أمر الإسطقسات على مذهب[٣١٩] علم الطّبائع وبالتّشريح[٣٢٠] وبعلم المنطق، فإن من هو عالم بالمنطق لا يغلط كغلط أصحاب الحيل في الأسماء والصّفات لكنّه يجعل بأزاء الرخو المتمدّد وبإزاء الصّلب اللّيّن وبإزاء المتخلخل الكثيف وبإزاء الاستفراغ امتناع ما يستفرغ.

(٥٥) ومن كان عالماً بالأشياء التي ذكرناها قبل علماً جيّداً، فهو يعلم أنّ الشّيء الذي قد كان قبل استفراغه محتبساً في عضو من الأعضاء ثمّ استفرغ، فإنّما استفرغ بواحد من ستّة أسباب، إمّا لأنّه قد رقّ ولطف بمنزلة ما يعرض للنّساء من النّزف، وإمّا لأنّه قد كثر بمنزلة ما يعرض القيء[٣٢١] للسكران[٣٢٢]، وإمّا لأنّ المجاري قد اتّسعت بمنزلة ما يعرض في الجماع، وإمّا لأنّ شيئاً[٣٢٣] يجذبه من داخل أو من خارج، أمّا من داخل فبمنزلة واحد من الأعضاء إذا كان قد سخن، وأمّا من خارج فبمنزلة الهواء الحارّ والدّواء الحارّ، وإمّا لأنّ القوّة الدّافعة قد قوّيت بمنزلة ما يعرض لمن يشرب ماءً بارداً في منتهى حمّاه، فيغرق من ساعته، وإمّا لأنّ القوّة الماسكة تضعف بمنزلة ما يعرض لمن يغشى عليه عندما[٣٢٤] يخرج منه الغائط. ونعلم أيضاً أنّ احتباس

---

٣١٩ ADMY: + أصحاب | ٣٢٠ DSY: والتّشريح | ٣٢١ F: – للنّساء. . . القيّ؛ S: – القيّ | ٣٢٢ F: + أنّ القيّ؛ DY: للسّكران القيّ | ٣٢٣ S: الأشياء | ٣٢٤ AM: إن

six causes opposite to these: because it becomes either more or less gross, or because the vessels are compressed, or because there is nothing external or internal attracting it, or because the retentive power has weakened, or because the cohesive power has strengthened.

(56) Some of the organs are loose and spongy, so if some particular matter flows to one of them, the thinner matter will ooze from it, while the coarser matter will be retained.[120] The Methodists think that this is a compound disease, as occurs in the eye, the nose, and the mouth. Other organs are dense and have no pores. If matter flows to them, they do not ooze, and nothing flows from them, as happens with the thigh, the arm, and the calf. The Methodists are mistaken when they say that a swelling with nothing oozing from it is a simple, noncomposite disease, while the swelling with something oozing from it is compound. They do not know that, when the swelling also occurs in a spongy organ like the eye, if its matter is fine, it oozes and some of the matter flows from the organ; but if the matter of the swelling is coarse, the organ does not ooze and nothing flows from it.

120. K 1:102–5; H 29–31; G[a] 86–90: Galen is criticizing the Methodist account of "mixed communities"—diseases where there is both swelling, indicating costiveness; and discharge, indicating flux. One MS glosses *mutakhalkhal* ("spongy" or "rarefied") as "in Syriac, *mukhalkhal*." This is the Syriac *mḥalḥal* (hollow or perforated). The note is probably because the word usually means "rarefied" in philosophical Arabic. Galen mentions wool and sponges as examples of materials that retain fluids despite having large pores. This undermines the Methodist use of hypothetical larger and smaller pores to explain evacuation and retention.

ما قد كان يستفرغ إنّما يكون لستّة أسباب مخالفة لتلك إمّا لأنه قد غلظ وإمّا لأنه قد قلّ وإمّا لأنّ المجاري قد تكاثفت[٣٢٥] وإمّا لأنّه[٣٢٦] ليس شيء[٣٢٧] يجذبه من داخلٍ ولا من خارجٍ وإمّا لأنّ القوّة الدّافعة قد ضعفت وإمّا لأنّ القوّة الماسكة قد قويت.

(٥٦) الأعضاء منها ما هو سلس متخلخل[٣٢٨]، وما كان كذلك فهو إذا انصبّ إليه شيء من الموادّ شيء[٣٢٩] رشح[٣٣٠] منه الرّقيق من تلك المادّة واحتبس الغليظ. وأصحاب الحيل يظنّون أنّ هذا هو علّة مركّبة بمنزلة ما يعرض ذلك في العين وفي الأنف وفي الفم، ومنها ما هو كثيف لا مسامّ له، وما كان كذلك فهو إذا انصبّت إليه مادّة لم يرشح ولم يجر منها شيء بمنزلة ما يعرض ذلك في الفخذ والعضد والسّاق. وأصحاب الحيل يغلطون ويقولون إنّ الورم الذي لا يرشح منه شيء هو مرض بسيط مفردٌ، والورم الذي يرشح منه شيء هو مركّب، ولا يعلمون أنّ الورم إذا حدث في عضو متخلخل أيضاً[٣٣١] بمنزلة العين إن كانت مادّته رقيقة رشح وجرى منه بعضها، وإن كانت غليظة لم يرشح ولم يجر منه شيء منها.

---

٣٢٥ DY: تكاثف؛ F: ضاقت | ٣٢٦ DSY: لأنّ؛ F: إلا | ٣٢٧ S: – شيء؛ DY: شيئاً | ٣٢٨ Y: فتخلخل؛ Mʰ: في السّرياني مخلخل | ٣٢٩ ADMY: – شيء | ٣٣٠ F: وشح؛ F²: شيء رشح | ٣٣١ AM: – أيضاً

This ends the Alexandrians' epitomes of Galen's book

On the Medical Sects

using the method of commentary and abridgment,

translated by Ḥunayn ibn Isḥāq[121]

---

121. The texts of the colophon vary somewhat among the manuscripts. An early MS adds the blessing: "Praise be to God, the Lord of the worlds, and may He be glorified forever and ever. May His mercy be upon us." A late MS adds an Islamic blessing: "May God have mercy upon him [Ḥunayn], for he is worthy of it. Blessing be upon our Prophet Muḥammad and all his family. God is sufficient for us and a generous guardian."

An earlier MS has the colophon: "Written by Sallām ibn Ṣāliḥ, the teacher in Shafarᶜām, on Thursday, the 26th of April 6748 of Adam [1240 CE]." Shafarᶜām is a village near Acre, in present-day Israel. The date is according to the Byzantine era of the world, so this scribe was presumably a Christian.

An early MS contains a gloss with two tables, formatted slightly differently, explaining the composition of the simple flavors. I have translated the first here and noted differences in brackets: "The qualities of simple flavors are compounded from subtle, dense, and intermediate and in their temperament from hot, cold, and intermediate. This is a diagram of their composition:

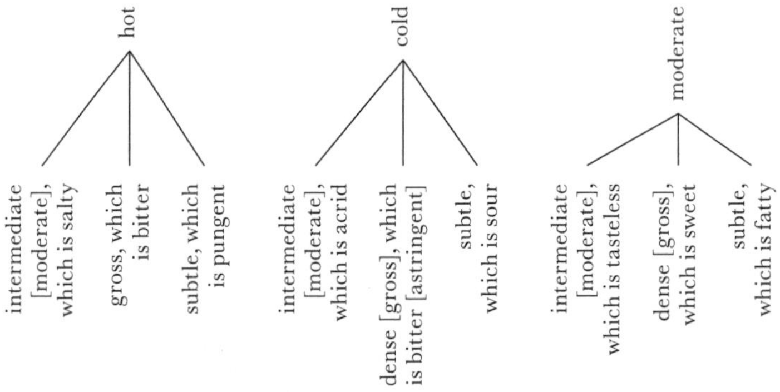

The caption of the second form of the table of simple flavors reads: "The table explaining the temperaments of the foods compounded from the hot, cold, and moderate mixture, and from subtle, gross, and moderate substance. They are the simple foods, which are of nine species."

An early MS has the gloss: "Whenever we add to impure bodies, we increase them in power [?], as Hippocrates said." The same MS has a note in an official Ottoman hand written by one Aḥmad Shaykhzāda, the trustee of the endowments of Mecca and Medina, recording its donation by Sultan Maḥmūd I (r. 1730–59) as *waqf*, presumably when he established the Aya Sofya library.

An early MS adds: "It is followed by *The Small Art*."

تمّت[٣٣٢] جوامع الإسكندرانيّين[٣٣٣] لكتاب[٣٣٤] جالينوس في فرق الطّبّ على الشّرح والتّلخيص[٣٣٥] ترجمة حنين بن إسحق[٣٣٦]

---

٣٣٢ ADMY: تمّ؛ Y: + كتاب | ٣٣٣ M: الإسكندارانيّين؛ F: - الإسكندرانيّين | ٣٣٤ F: كتاب | ٣٣٥ F: في . . . التّلخيص > وجمل معانيه مشروحة ملخّصة ولله الحمد | ٣٣٦ DF: - ترجمة . . . إسحق؛ DY: + رحمه الله والحمد لله حقّ حمده، كما هو أهله ومستحقّه، وصلواته على نبيّه محمّد وآله آجمعين، وحسبنا الله ونعم الوكيل. || M: +

وكتب سلاّم بن صالح المعلّم بشفرعام
في يوم الخميس سادس عشرين نيسان
سنة ست ألف سبع مائة وثمان وأربعين لأدم.

S: على الشّرح . . . إسحق > ويتلوه كتاب الصّناعة الصّغيرة؛ S^h: لمّا كانت الأجسام الغير نقيّة كلّما عدّدناها زدنا شرًّا كما قال إبقراط. || A: + والحمد لله ربّ العالمين والتّسبيح لله دائماً أبداً وعليها رحمته أمين. || A: + مفردات الطّعوم تتركّب في كيفيّاتها من لطيف وكثيف ومتوسّط وفي قوامها من حارّ وبارد ومعتدل، وهذه صورة تركيبها:

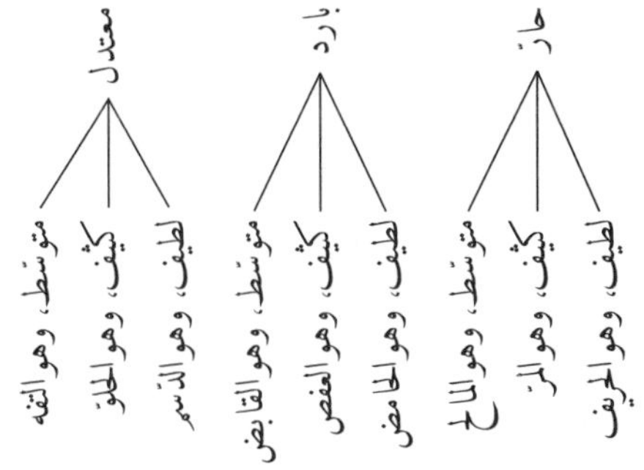

هذا الجدول في معرفة أمزاج الطّعوم المركّبة من المزاج الحارّ والبارد والمعتدل، ومن الجوهر اللّطيف والغليظ والمعتدل، وهي مفردات الطّعوم التّسعة الأنواع:

	حار	بارد	معتدل
غليظ	المرّ	العفص	الحلوّ
لطيف	الحريف	الحامض	الدّسم
معتدل	المالح	القابض	المسيح

THE ALEXANDRIAN EPITOMES OF GALEN
VOLUME 1

The Alexandrian Epitome of Galen's Book *The Small Art of Medicine*

In the Name of God, the Merciful, the Compassionate[1]

The Alexandrian Epitome of Galen's Book Known as

The Small Art [of Medicine]

Using the method of commentary and abridgment

Translated by [Abū Zayd] Ḥunayn ibn Isḥāq [al-ᶜIbādī][2]

[Introduction]

[Methods of instruction][3]

(1) Some people think that there are five kinds of method in instruction:[4] first, the method that proceeds by analysis and conversion; second, the method that proceeds by synthesis; third, the method that proceeds by dialysis of the definition; fourth, the method that proceeds by

1. Some MSS add: "God is the master of success" or "My trust is in God on high alone."

2. Some MSS omit the author's name, and only one gives it in full.

3. B i.1–5; K 1:305–7; G[a] 3–7: I have added the subheadings, which, with a few exceptions, are not part of the text and have no MS authority.

4. An early MS reads: "kinds of instruction and methods therein." Galen begins the *The Small Art of Medicine* with an account of the three methods of instruction (διδασκαλία) and the methods he uses in his works—in this case, the dialysis of the definition of medicine. The account given here is much more elaborate and includes alternative accounts and many examples.

بسم الله الرحمن الرحيم[١]

جوامع الإسكندرانيّن لكتاب[٢] جالينوس المعروف بـ

الصّناعة الصّغيرة الطّبيّة[٣]

على الشّرح والتّلخيص[٤]

ترجمة[٥] حنين ابن إسحق[٦]

[مقدمة]

[ذكر المسالك في التّعليم]

(١) أنحاء المسالك في التّعاليم[٧] بحسب رأي بعض النّاس خمسة. أحدها المسلك الذي يجري الأمر فيه على طريق التّحليل والعكس، والثّاني المسلك الذي يكون على طريق التّركيب، والثّالث المسلك الذي يكون على طريق تحليل الحدّ، والرّابع المسلك

١ AM: + الله وليّ التّوفيق؛ Y: + ثقتي بالله تعالى وحده | ٢ AM: جوامع كتاب؛ S: جوامع كتاب لكتاب | ٣ المعروف . . . الصّغيرة > AM: في الصّناعة الطّبيّة؛ A²DM²Y: في الصّناعة الطّبيّة الصّغيرة؛ S: المعروفة بالصّناعة الصّغيرة | ٤ F: – على . . . والتّلخيص | ٥ F: نقل أبي زيد | ٦ F: + العبّاديّ؛ DSY: – ترجمة . . . إسحق | ٧ F: أنحاء التّعاليم والمسالك فيها

division; and, fifth, the method that proceeds by attributes and descriptions. Three of these five methods proceed in order: the analysis that is called conversion, synthesis, and dialysis of the definition.[5] Two of them proceed without order: division and description.[6] Division tends to proceed without order because the thing being divided can be subject to division in various ways: division of the genus into the species, as when animal is divided into man, horse, and ox; division of the species into individuals, as when man is divided into Plato, Asclepiades, and Socrates; division of the whole into parts, as when the body is divided into head, hands, and feet; division of the substance into accidents, as when man is divided into white, red, and black; division of accidents into substances, as when white is divided into snow, milk, and swan; or the division of the ambiguous word into the various meanings that it shares, as when the name "dog" is applied to the dogfish, the domestic dog, the dog star, and the philosopher of the school known as Cynics.[7] The whole can be divided into parts in two ways: either into homoeomerous parts,[8] as when nerve is divided into nerves or when vein is divided into veins; or into parts that are not homoeomerous, as when the body as a whole is divided into such parts as the head, the hands, the feet, and the chest. Description and attribution tend to proceed without order just because the one who assigns attributes to the thing and describes it may do so with respect to its quantity, its quality, or its relation to something else.

---

5. These are Galen's three methods: ἀνάλυσις, σύνθεσις, and ὅρου διάλυσις. The terms are rendered identically in the Arabic translation of *The Small Art*. Galen's explanation is clearer: "There are three types of definition, each with its place in the order. First is that which derives from the notion of an end, by analysis. Second is that from the putting together of the analysis. Third is that from the decomposition of the definition." (Trans. Singer with slight modifications.) Galen explains that in other works he had used the method of analysis of the goals of medicine. Other authors have attempted to proceed by synthesizing the results of medicine. This work will use the method of dialysis of the definition as being the easiest method for acquiring an overview of medicine, though it is scientifically inferior to analysis.

6. These are the unsystematic equivalents of analysis and dialysis of the definition. Division differs from analysis in that the distinctions made need not be essential. Description is definition that uses accidental rather than essential properties. The classic example is the definition of man as a featherless biped, which correctly identifies man but does not reveal his essence as a rational animal. Galen mentions but dismisses the fourth method.

7. Whose name means "of the dog."

8. That is, parts of the same nature as what is being divided, such as the parts of a quantity of water, which are also water.

الذي يكون على طريق القسمة، والخامس المسلك الذي يكون على طريق الصّفات والرّسوم، وهذه الخمسة المسالك منها ثلثة تجري على ترتيب، وهي التّحليل الذي يقال له العكس والتّركيب وتحليل الحدّ، ومنها اثنانِ يجريانِ على غير تربيب، وهما القسمة والرّسم[٨]، أمّا القسمة فصارت تجري على غير ترتيب لأنّ الشّيء الذي يقسم تقع عليه القسمة على وجوه شتّى، إمّا على جهة قسمة الجنس إلى الأنواع بمنزلة الحيوان إلى الإنسان والفرس والثّور وإمّا على جهة قسمة النّوع إلى الأشخاص بمنزلة الإنسان إلى فلاطون وأسقليبياذس وسقراطيس، وإمّا على جهة قسمة الكلّ إلى الأجزاء بمنزلة[٩] البدن إلى الرّأس واليديْنِ والرّجليْنِ، وإمّا على جهة قسمة الجوهر إلى الأعراض بمنزلة الإنسان إلى الأبيض والأحمر والأسود، وإمّا على جهة قسمة الأعراض إلى الجواهر[١٠] بمنزلة الأبيض إلى الثّلج واللّبن والحيوان المسمّى قُقْنُس، وإمّا على جهة قسمة اللّفظ المشترك إلى المعاني المختلفة المشتركة[١١] بمنزلة اسم الكلب الواقع على كلب البحر وكلب البرّ والكوكب المسمّى كلباً والفيلسوف الذي من حزب الكلبيّين، وقسمة الكلّ إلى الأجزاء تكون على ضربيْنِ، إمّا إلى أجزاء متشابهة بمنزلة العصبة إذا قسمت أعصاباً والعرق إذا قسم عروقاً، وإمّا إلى أجزاء غير متشابهة بمنزلة جملة البدن إذا قسم أجزاء منها الرّأس ومنها اليدانِ ومنها الرّجلانِ ومنها الصّدر، فأمّا الرّسم والوصف، فإنّما صارا على غير ترتيب لأنّ الواصف والرّاسم إنّما يصف الشّيء ويرسمه إمّا من كميّته وإمّا من كيفيّته وإمّا من إضافته ونسبته إلى غيره[١٢].

---

٨ AM: والرّسوم | ٩ AM: + قسمة | ١٠ AD: الجوهر | ١١ AM: - المشتركة [D: + فيه] | ١٢ F: + وذلك أجمع غير محصور ولا مرتّب

(2) According to other people there are three kinds of instruction, for they say that all instruction must be either by way of conversion and analysis, by way of synthesis, or by way of dialysis of the definition.[9] Each of these may be conducted either in order or without order. Analysis and conversion are ordered whenever they omit nothing in the middle, since it is a return from the last to the first. This is also the case in division, for division is included in analysis. In analysis, for example, we might say that the body is analyzed into the compound organs, the compound organs into the tissues, the tissues into the humors, the humors into the nutriments, and the nutriments into the elements, which are fire, air, water, and earth. The corresponding example of division is to say that the body is divided into the head, the arms, the trunk, and the legs. Each of these is divided into the parts from which it is compounded, as the head is divided into bone, brain, and sense organs.[10] Then these, in turn, are divided into those things that they are compounded from until the division ends in the tissues, where it stops. Unordered analysis and division are when one omits something in the middle and[11] alters their order and arrangement. In analysis, for example, you might say that the body is analyzed into the tissues, but you would have overlooked the compound organs. You might also say that the body is analyzed into the organs, the organs into the tissues, the tissues into the nutriments, and the nutriments into the humors, thus altering their order and arrangement. The corresponding

9. These are the three methods mentioned by Galen. They correspond to the three ordered methods mentioned in the previous paragraph.

10. Literally, "the senses." Two late MSS read "spinal cord."

11. Some MSS read: "or."

(٢) وأنحاء التّعاليم بحسب رأي قوم آخرين ثلثة، وذلك أنّهم قالوا إنّ كلّ تعليم لا يخلوا من أن يكون إمّا على جهة العكس والتّحليل وإمّا على جهة التّركيب وإمّا على جهة تحليل الحدّ، وكلّ واحد من هذه تجري إمّا على ترتيب وإمّا على غير ترتيب، والتّحليل والعكس يكونانِ[١٣] على ترتيب[١٤] عندما لا يدع شيئًا ممّا في[١٥] الوسط إذا هو رجع من الأخير إلى الأوّل، وكذلك الأمر في القسمة لأنّ القسمة داخلة في التّحليل، مثال ذلك في التّحليل أن تقول إنّ البدن يحلّ إلى الأعضاء المركّبة والأعضاء المركّبة إلى الأعضاء المتشابهة[١٦] الأجزاء والأعضاء المتشابهة الأجزاء إلى الأخلاط، والأخلاط إلى الأغذية، والأغذية إلى الإسطقسات، وهي النّار والهواء والماء والأرض، ونظير ذلك في القسمة أن تقول إنّ البدن ينقسم إلى الرّأس واليدينِ والصّدر[١٧] والرّجلينِ، وكلّ واحد من هذه ينقسم إلى الأجزاء التي هو منها مركّب بمنزلة الرّأس إلى العظم والدّماغ والحواسّ، ثمّ إنّ هذه أيضاً تنقسم إلى الأجزاء التي هي مركّبة منها إلى أن تنتهي القسمة إلى الأعضاء المتشابهة الأجزاء، فتقف عندها، وأمّا التّحليل والقسمة على غير ترتيب فيكونانِ عندما يدع الإنسان بعضما في الوسط وتغيّر[١٨] ترتيب الأشياء ونظامها[١٩]، مثال ذلك في التّحليل أن تقول إنّ البدن يحلّ إلى الأعضاء المتشابهة الأجزاء، فقد تركنا[٢٠] الأعضاء المركّبة، أو تقول إنّ البدن يحلّ إلى الأعضاء المركّبة والمركّبة إلى المتشابهة الأجزاء والمتشابهة الأجزاء إلى الأغذية والأغذية إلى الأخلاط، فتكون قد غيّرت التّرتيب والنّظام، ونظير ذلك

---

١٣ A^2DM^2Y: يكون؛ F: في العكس يكون | ١٤ A^2M^2: رتبته | ١٥ F^h: هوصح | ١٦ F: المشابهة F consistently uses this spelling, so it is not noted hereafter in the variants | ١٧ DFY: –والصّدر | ١٨ ADM: أوتغيّر | ١٩ Y^1: +الأشياء ونظامها | ٢٠ ADY: تركت؛ M: تركّبت

example of division is to say that substance is divided into body and what is not body, and body[12] into rational and irrational;[13] but you would have omitted breathing and living in the middle. You might also say that substance is divided into animate and inanimate, and animal into body and non-body, in which case you would have altered the order and arrangement.

(3) Synthesis is also sometimes conducted with order and arrangement. That occurs when you proceed from the first to the last without omitting anything or altering the arrangement. For example, you might say that plants come to be when the elements are combined and that humors come to be from plants,[14] tissues from humors, organs from tissues, and the whole body from the organs. Synthesis is sometimes conducted without order, as when you omit or alter some things. For example, you might say that when the elements are combined, the humors arise from them, but you would have omitted the nutriments. You might also say that when the elements are combined, plants come to be from them and that humors come to be from plants, organs from the humors, and tissues from the organs, in which case you would have altered the order and arrangement.

(4) Dialysis of the definition may be conducted with arrangement and order. That is when you omit nothing and alter nothing.[15] For example, you might say that medicine is the knowledge of healthy and diseased states and of states that are neither healthy nor diseased. Each of these three states can apply to three things: first, the body; second,

12. Two late MSS insert: "is divided into growing and not growing, and growing is divided into living and dead, and living is divided . . . "

13. One of the two late MSS adds: "and the rational is divided into the mortal and the immortal, so if you say that substance is divided into body and non-body, and body is divided into growing and not growing . . . "

14. One early MS reads: "nutriments result from plants, humors from nutriments."

15. This is the method supposedly used in *The Small Art* and thus in this epitome.

في القسمة أن تقول إنّ الجوهر ينقسم إلى جسم ولا[٢١] جسم والجسم[٢٢] إلى النّاطق وغير النّاطق[٢٣]، فتكون قد تركت في الوسط المتنفّس والحيّ، أو تقول إنّ الجوهر ينقسم إلى الحيوان وغير الحيوان، والحيوان منه جسم ومنه[٢٤] لا جسم، فتكون قد غيّرت التّرتيب والنّظام.

(٣) وأمّا التّركيب فمنه أيضاً ما يجري على التّرتيب والنّظام، وذلك عند ما تسلك[٢٥] من الأوّل إلى الأخير من غير أن تدع شيئاً أو تغيّر نظاماً[٢٦]، مثال ذلك أن تقول إنّ الإسطقسات إذا تركّبت حدث عنها النّبات، ويحدث عن النّبات[٢٧] الأخلاط وعن الأخلاط الأعضاء المتشابهة الأجزاء وعن المتشابهة الأجزاء الأعضاء المركّبة وعن المركّبة جملة البدن، ومنه ما يجري على غير ترتيب، وذلك عند ما تدع بعض الأشياء أو تغيّر بعضها، مثال ذلك أن تقول إنّ الإسطقسات إذا تركّبت حدث عنها الأخلاط، فتكون قد تركت الأغذية، أو تقول إنّ الأسطقسات إذا تركّبت حدث عنها النّبات وعن النّبات الأخلاط وعن الأخلاط الأعضاء المركّبة وعن[٢٨] المركّبة المتشابهة الأجزاء، فتكون قد غيّرت التّرتيب والنّظام.

(٤) وأمّا تحليل الحدّ، فمنه ما يجري على نظام وترتيب، وذلك عند ما لا تدع شيئاً ولا تغيّر شيئاً، مثال ذلك أن تقول إنّ الطّبّ هو معرفة الأمور الصّحيّة والمرضيّة، والتي ليست بصحيّة ولا مرضيّة، وإنّ كلّ واحد من هذه الأمور الثّلثة يقع على ثلثة

---

٢١ AM: وغير | ٢٢ DY: + ينقسم إلى نامي وغير نامي والنّامي ينقسم إلى حيّ وموات والحيّ ينقسم | ٢٣ Y: ناطق والنّاطق ينقسم إلى مائت وغير مائت فإن قلت إنّ الجوهر ينقسم إلى جسم ولا جسم والجسم ينقسم إلى نامي وغير نامي | ٢٤ F: – ومنه؛ S: فيه | ٢٥ A^2DM^2Y: يمضى؛ F: يمضي يسلك | ٢٦ M: نظام؛ DY: بغير نظام | ٢٧ F: + الأغذية وعن الأغذية | ٢٨ AM: + الأعضاء

the prophylactic and efficient cause; third, the sign that either indicates the present condition, informs about the past condition so that it can be recollected, or precedes a condition so as to give information about what will be. Each of these three signs either does so in the present time or does so constantly or does so usually. Dialysis of the definition also sometimes proceeds without order or arrangement, which is when you omit or alter some things. This occurs when it is not conducted according to the order we have mentioned and there is some alteration or omission in the middle.

(5) Analysis and division have in common the property that they both begin from one and end in many, but they differ in that analysis and conversion take as their starting point something that is actual and is sensibly one and end in many things that are potential and intelligible. For example, beginning from body, one ends up with the elements, whose existence in the body is intelligible and potential, not actual and sensible. Division, on the other hand, takes one thing that is one potentially and intelligibly and divides it into many things that are actual, such as the division of the genus into species and the species into individuals. It can also take something that is actually one and go from that to what is actually many, such as the division of the whole into parts, the substance into accidents, or the accident into substances.

(6) These are the three methods of instruction when studying:[16] analysis, synthesis, and dialysis of the definition. Definitions, descriptions, and demonstrations[17] are not methods of instruction but, rather,

16. That is, the three listed by Galen.

17. They are, respectively, true definitions yielding knowledge of the essence ("man is a rational animal"), definitions distinguishing the thing from other things by means of nonessential attributes ("man is a featherless biped"), and scientific proof yielding certainty, as in geometry. All are standard logical terms.

أشياء، أحدها البدن والآخر السّبب الحافظ والفاعل والثّالث العلاّمة التي تدلّ على[٢٩] الحاضر والتي تنبئ عن الماضي، وتذكر[٣٠] به، والتي تتقدّم فتنبئ عمّا سيكون، وكلّ واحدة[٣١] من هذه العلاّمات الثّلاث إمّا أن تفعل ذلك في الوقت الحاضر وإمّا أن تفعله دائماً وإمّا على الأكثر. ومنه ما يجري على غير ترتيب ولا نظام، وذلك عندما تدع بعض الأشياء أو تغيّر بعضها متى لم يجر الأمر في التّرتيب على ما ذكرنا وكان فيه تغيّراً[٣٢] أو نقصاناً[٣٣] في الوسط.

(٥) التّحليل والقسمة يشتركانِ في أنّهما جميعاً يبتدئان من واحد وينتهيانِ إلى كثير ويختلفانِ في أنّ التّحليل والعكس إنّما يأخذ شيئاً هو بالفعل والحسّ واحد، وينتهي[٣٤] به إلى أشياء كثيرة، هي بالقوّة والمعقول، مثال ذلك أن ينتهي بالبدن إلى الإسطقسات التي إنّما وجودها في البدن بالعقل والقوّة لا بالفعل والحسّ. فأمّا القسمة فتأخذ واحداً[٣٥] هو بالقوّة والمعقول[٣٦] واحد فتقسمه إلى أشياء هي[٣٧] كثيرة بالفعل بمنزلة قسمة الجنس إلى الأنواع والنّوع إلى الأشخاص، أو تأخذ شيئاً هو بالفعل واحد، فتصير[٣٨] به إلى أشياء هي بالفعل كثيرة بمنزلة قسمة الكلّ إلى الأجزاء أو قسمة الجوهر إلى الأعراض أو العرض[٣٩] إلى الجواهر.

(٦) وعند التّحصيل يقال إنّ أنحاء التّعاليم هي هذه الثّلثة، التّحليل[٤٠] والتّركيب وتحليل الحدّ. فأمّا الحدود والرّسوم والبراهين فليست من أنحاء التّعاليم، لكنّها علوم

---

٢٩ F: + الوقت | ٣٠ AM: فتذكر | ٣١ DFY: واحد | ٣٢ DS: تغيراً | ٣٣ YA^2M^2: ونقصان؛ A^hM^h حـ[اشية] له، أي وجد فيه تغير أو نقصان وهي كان التامة [؟] | ٣٤ AM: فينتهي | ٣٥ DFM: واحد | ٣٦ F: - والمعقول | ٣٧ DSY: - هي | ٣٨ DSY: ويضرّ ٣٩ DMS: والعرض؛ Y: الأعراض | ٤٠ S: + والعكس

are logical sciences. The instruction that proceeds by means of dialysis of the definition[18] is called by various names: dialysis of the definition, conversion of the definition, opening up of the definition, division of the definition, expansion of the definition, explanation of the definition, and summary of the definition.

(7) Three groups of physicians have sought to employ these three methods of instruction. Dialysis of the definition was employed by the followers of Herophilus and Heraclides the Erythrean. Synthesis was also employed by the followers of Herophilus and by Athenaeus, who was from Attaleia of Antioch.[19] Analysis and conversion were used by Galen alone among the physicians.[20] Both analysis and conversion and dialysis of the definition give their practitioner an advantage in some respect. Dialysis of the definition is superior to conversion in that it deals concisely with everything in the lesson and makes it easy for the students to remember what has been taught.[21] Conversion is superior to dialysis of the definition in that it is of higher rank and in the rigor of its technical method.[22]

(8) Some definitions are essential.[23] These are the ones that are taken from the substance of the thing defined, as when man is defined as a mortal rational animal capable of reason and culture. Others are descriptive.[24] These are the definitions taken from the accidents attaching to the things, as when man is defined as an animal with broad fingernails and erect posture whose body is covered with bare skin.

---

18. The method used by Galen in *The Small Art.*

19. Following Galen's Greek text, since the names are mangled in the Arabic. On these three physicians, see appendix 1.

20. Galen boasts that he was the first to use this method to expound medicine from its first principles; see B i.2; K 1:306; G[a] 4.

21. B i.4; K 1:306; G[a] 5–6.

22. That is, analysis, which deduces the art of medicine from first principles.

23. Οὐσιώδης: That is, essential, revealing the essence of what is defined.

24. Ἐννοηματικοῖς: That is, descriptive definitions as mentioned on p. 54, n. 17, above.

منطقيّة، والتّعليم الذي يجري على طريق تحليل الحدّ يسمّى بأسماء كثيرة، وهي تحليل الحدّ وعكس الحدّ وتفتيح[٤١] الحدّ وقسمة الحدّ وبسط الحدّ وتفسير الحدّ وتلخيص الحدّ.

(٧) وقد التمس تعاطي[٤٢] هذه التّعاليم الثّلثة قوم من الأطبّاء. تعاطى[٤٣] تحليل الحدّ أصحاب إيروفليس وإيراقليدس الأرثاريّ[٤٤]، وتعاطى التّركيب أصحاب إيروفليس[٤٥] أيضاً وأثيناوس المعروف بأطالوس الأنطاكيّ[٤٦]. فأمّا التّحليل والعكس فجالينوس[٤٧] دون سائر الأطبّاء[٤٨] استعمله، وكلّ واحد من التّحليل والعكس وتحليل الحدّ يفوق صاحبه في شيء، أمّا تحليل الحدّ فيفوق العكس في أنّه يأتي بكلّ شيء من التّعليم باختصار وإيجاز وفي أنّه يسهل حفظ ما يُأتى به على المتعلّمين، وأمّا العكس فإنّه يفوق في[٤٩] تحليل الحدّ في جلالة القدر وفي لزوم الطّريق الصّناعيّ.

(٨) والحدود منها جوهريّة وهي التي تؤخذ من جوهر الشّيء المحدود بمنزلة حدّ الإنسان أنّه حيوان ناطق مائت قابل للعقل[٥٠] والأدب، ومنها رسوميّة وهي الحدود التي تؤخذ من الأعراض التّابعة[٥١] للأشياء بمنزلة ما يحدّ الإنسان بأنّه حيوان عريض الأظفار منتصب القامة[٥٢] يعلوا بدنه جلد[٥٣].

٤١ F^{2}: + أيضاً، location uncertain؛ DY: وقسمة | ٤٢ DFY: استعمال بعض | ٤٣ ADMY: فتعاطى | ٤٤ $A^{2}M^{2}$: ايراقليدس والارومري؛ D: إبيروقليس وايرقليدس الأويريّ؛ F: إيراقليدس الأيري؛ S: إيروفيلس وإبرافليدس الأرثاريّ؛ Y: وإيرقليدس الأويري | ٤٥ D: إبيروفيلس؛ S: إيروفيلس | ٤٦ $A^{h}M^{h}$: في الفصّ المعروف بارثراس واثيناوس المعروف باطالوس؛ F: أثيناوس الأرظاليّ؛ S: وآيثانوس الأنطاليّ؛ DY: أثيناوس الأنطاكيّ | ٤٧ F: + وحده | ٤٨ F: + والنّاس | ٤٩ ADMY: فإنّه يفوق في > فيفوق | ٥٠ F: للعلم؛ DY: العقل | ٥١ AM: اللّاحقة؛ $A^{2}M^{2}$: التّابعة اللّاحقة | ٥٢ F: + مُعشّي | ٥٣ F: جلده

(9) Galen employs each of these methods of instruction in various places in his books. He uses the method that proceeds by means of analysis and conversion in his books *On Diseases and Symptoms*, *On the Method of Healing*, and *On Affected Parts*. He uses synthesis in the books *On the Natural Faculties*, *On the Elements [According to Hippocrates]*, and *On the Temperament*. He uses dialysis of the definition in the present book and in his book *On the Affirmation of Medicine*,[25] since his goal in both books is to be brief and concise.

---

25. Probably *On the Constitution of the Art of Medicine*, K 1:224–304, of which no Arabic translation is known but which does follow the same method as *The Small Art*.

(٩) وقد استعمل جالينوس كلّ واحد من هذه[٥٤] الثّلثة التّعاليم في مواضع من كتبه، فاستعمل[٥٥] التّعليم الذي يجري على طريق التّحليل والعكس في «كتاب العلل والأعراض» وفي «كتاب المواضع الآلمة» و«كتاب حيلة البرء[٥٦]»، واستعمل التّركيب في «كتاب القوى الطّبيعيّة» وفي «كتاب الإسطقسات» وفي «كتاب المزاج[٥٧]»، واستعمل تحليل الحدّ في هذا الكتاب الحاضر وفي كتابه «في إثبات الطّبّ» من طريق أنّه قصد فيهما إلى الاختصار والإيجاز.

٥٤ DFY: – هذه | ٥٥ F: استعمل؛ DY: واستعمل | ٥٦ AM: حيلة البرء وكتاب المواضع الألمة | ٥٧ ADMY: الإسطقسات وفي كتاب المزاج [DY: الأمزاج] وفي كتاب القوى الطّبيعيّة

[Chapter 1]

[The definition of medicine]

(10) Medicine is defined as being the knowledge of entities relating to health, entities relating to disease, and entities that are neither healthy nor diseased.[26] In this definition, knowledge is mentioned where the genus of the thing would be mentioned; and where the differentia would be mentioned, entities relating to health, to disease, and to neither health nor disease are mentioned.[27]

(11) Three things relate to health:[28] first, the body receptive to health;[29] second, the sign indicating health; and third, the cause effecting and preserving health. There are three things relating to disease: first, the body receptive to disease, which is the body called "diseased"; second, the sign indicating disease; and third, the cause that brings the disease into being or preserves it. There are three things relating neither to health nor to disease: first, the body that is receptive to a state

---

26. B 1.1; K 1:307; G^a 7: Galen gives exactly this definition, though Ḥunayn renders it quite differently in G^a. It originated with Herophilus; cf. *Epitome of the Medical Sects*, p. 54, paragraph 5 and p. 55, n. 23, above; see B 396–98 and von Staden, *Herophilus*, 89–114, for detailed discussions of the history of this definition. Von Staden believes that the third category of neutrals has a connection with Stoic thought.

27. B 1.2–3; K 1:307; G^a 8–9: Galen notes that "healthy" and "diseased" are used somewhat ambiguously as applying to the body, the cause, and the sign. Hereafter, I will omit the awkward "entities"—which are not in the Greek anyway—and render the terms as seems natural in English in each context.

One MS contains the gloss: "Ḥunayn said, 'He means by what he says here that knowledge [or science] is the genus of the different kinds of knowledge. That is because it includes the sciences of medicine, astronomy, logic, and geometry. Knowledge includes these sciences and so, in this respect, is their genus. Mention of the healthy states, the diseased states, and the states that are neither healthy nor diseased takes the place of reference to the differentia. That is because health is part of the science of medicine, but knowledge of disease is one of the sciences of medicine. Thus, they are specifically mentioned among the sciences. This takes the place of reference to the differentia." On the other hand, Galen says (B 1b.1; K 1:307; G^a 8), "The term 'knowledge' is to be understood in its common, not its technical, sense." His point is probably that medicine is not a science in the sense of a formal demonstrative system like geometry. The point made by Galen and Ḥunayn is that this is not a proper, essential definition.

28. B 1.1–5; K 1:307–8; G^a 8–10.

29. A gloss in one MS reads: "That is, the body that is called healthy."

[الفصل الأوّل]

[ذكر حدّ الطّبّ]

(١٠) الطّبّ يحدّ بأنّه العلم بالأمور الصّحيّة والمرضيّة والتي ليست بصحيّة ولا مرضيّة[٥٨]، وذِكر العلم في هذا الحدّ[٥٩] يقوم مقام ذكر جنس الشّيء وذكر الأمور الصّحيّة والمرضيّة والتي ليست بصحيّة ولا مرضيّة تقوم مقام ذكر الفصول[٦٠].

(١١) والأمور الصّحيّة ثلثة، أحدها البدن القابل للصّحّة[٦١]، والثّاني العلامة الدّالّة على الصّحّة، والثّالث السّبب الفاعل والحافظ للصّحّة، والأمور المرضيّة ثلثة، أحدها البدن القابل للمرض، وهو البدن الذي يسمّى مريضاً[٦٢]، والثّاني العلامة الدّالّة على المرض، والثّالث السّبب الذي يحدث المرض أو يحفظه[٦٣]، والأمور التي ليست بصحيّة ولا مرضيّة ثلثة، أحدها البدن القابل للحال[٦٤] التي ليست بصحّة[٦٥] ولا مرض[٦٦]،

---

٥٨ ^{a}G: إنّ الطّبّ هو معرفة الأشياء المنسوبة المتّصلة بالصّحّة والمرض وبالحال التي لم يخلص للإنسان فيها صحّة ولا مرض | ٥٩ DSY: – الحدّ | ٦٠ ^{h}A^{h}M: قال حنين: أراد بقوله هاهنا أنّ العلم جنس العلوم، وذلك أنّ منه علم الطّبّ وعلم النّجوم وعلم المنطق وعلم الهندسة، والعلم يحوي هذه العلوم، فصار جنسها [^{h}A: جنساً] من قِبَل هذه، وذكر الأمور الصّحيّة والمرضيّة والتي ليست بصحيّة ولا مرضيّة يقوم مقام ذكر الفصول، وذلك لأنّ الصّحّة هو من علم الطّبّ والمرض هو علم من علوم الطّبّ، فقد انفصلت منه من العلوم هذا يقوم مقام ذكر الفصول. | ٦١ ^{h}S: وهو البدن الذي يدعى صحيحاً | ٦٢ DY: بمريض؛ F: مرض | ٦٣ F: يبقيه | ٦٤ DY: للحالتيْن | ٦٥ DSY: صحيّة | ٦٦ DSY: مرضيّة؛ ^{h}S: وهو البدن الذي يدعى ليس بصحيح ولا مريض

relating neither to health nor to disease;[30] second, the sign that indicates this state; and, third, the cause that effects this state or that preserves it.[31] If these are combined, nine combinations result, as follows:[32]

healthy body	diseased body	body neither healthy nor diseased
sign of health	sign of disease	sign of neither health nor disease
cause of health	cause of disease	cause of neither health nor disease

(12) Each of these—body, sign, and cause, whether healthy, diseased, or neither healthy nor diseased—is said to be so in two senses: first, being so at the present moment and, second, being so absolutely—that is, not[33] at the present moment.[34] If each of these three <that is, body, cause, and sign[35]> is combined with each of these three states <that is, health, disease, and neither health nor disease[36]> and with these two times, eighteen combinations result in the following manner:

<1> a healthy body at the present moment	<2> a sign of health at the present moment	<3> a cause of health at the present moment
<4> a healthy body absolutely	<5> a sign of health absolutely	<6> a cause of health absolutely

30. A gloss in one MS reads: "That is, the body that is called neither healthy nor diseased."

31. B 1b.5–6; K 1:307–8; G[a] 10: Galen mentions these nine categories, but his point is different; medicine, he explains, is primarily about causes—in the first instance, causes of health, and then causes of diseases—and only secondarily about bodies. In the practice of medicine, however, diagnosis of the body comes first, which is done by signs; and only then can the cause be determined.

32. Here and in the other tables in this text, the MSS differ in format, with some presenting the items as simple lists, some as tables, and some as numbered lists. Two of the earliest MSS consistently use the table format, and only the youngest MSS consistently number the items.

33. An old MS adds: "only."

34. B 1.7; K 1:308–9; G[a] 9–10.

35. A gloss in one MS.

36. A gloss in one MS.

والثّاني العلامة التي تدلّ على هذه الحال، والثّالث السّبب الذي يفعل هذه الحال أو يحفظها، وإذا ألّفت هذه تولّد منها تسعة تراكيب على هذه الصّفة[٦٧]:

بدن صحيح	بدن مريض	بدن لا صحيح ولا مريض
علامة صحيّة	علامة مرضيّة	علامة لا صحيّة ولا مرضيّة
سبب صحيّ	سبب مرضيّ	سبب لا صحيّ ولا مرضيّ

(١٢) وكلّ[٦٨] واحد من البدن والعلامة والسّبب الصّحيّ منها والمرضيّ والذي ليس بصحيّ ولا مرضيّ يقال على ضربيْن، أحدهما أن يكون كذلك في الوقت الحاضر، والآخر أن يكون كذلك مطلقاً، أي ليس في الوقت الحاضر[٦٩]، وإذا ألّفت كلّ واحدة[٧٠] من هذه الثّلثة[٧١] مع كلّ واحد من هذه الثّلث[٧٢] الحالات مع كلّ واحد من هذيْنِ الوقتيْنِ تولّد من ذلك ثمانية عشر تركيباً[٧٣] على هذه الصّفة[٧٤]:

‹آ› بدن صحيح في الوقت الحاضر	‹ب› علامة صحيّة في الوقت الحاضر	‹ج› سبب صحيّ في الوقت الحاضر
‹د› بدن صحيح مطلق	‹ه› علامة صحيّة مطلقة	‹و› سبب صحيّ مطلق

٦٧ AM: arranged as plain text; DY: arranged as an abjad-numbered list | ٦٨ AM: فكلّ | ٦٩ S: + فقط | ٧٠ DY: واحد | ٧١ M^h: [حاشية] له له، أي البدن والسّبب والعلامة | ٧٢ M^h: [حاشية] له، أعني الصّحّة والمرض والحال التي ليست بصحّة ولا مرض. | ٧٣ M: كيباً [كذ] | ٧٤ S: المثال؛ ADMY: Arranged as an abjad-numbered list

<7> a diseased body at the present moment	<8> a sign of disease at the present moment	<9> a cause of disease at the present moment
<10> a diseased body absolutely	<11> a sign of disease absolutely	<12> a cause of disease absolutely
<13> a body that is neither healthy nor diseased at the present moment	<14> a sign of neither health nor disease at the present moment	<15> a cause of neither health nor disease at the present moment
<16> a body that is neither healthy nor diseased absolutely	<17> a sign of neither health nor disease absolutely	<18> a cause of neither health nor disease absolutely

(13) The absolute—that which is not in the present moment—is divided into two classes: first, the chronic[37] and, second, the usual.[38] If these are combined, twenty-seven combinations result, as follows:

<1> a healthy body at the present time	<2> a sign of health at the present time	<3> a cause of health at the present time
<4> a chronically healthy body	<5> a chronic sign of health	<6> a chronic cause of health
<7> a usually healthy body	<8> a usual sign of health	<9> a usual cause of health
<10> a diseased body at the present time	<11> a sign of disease at the present time	<12> a cause of disease at the present time
<13> a chronically diseased body	<14> a chronic sign of disease	<15> a chronic cause of disease
<16> a usually diseased body	<17> a usual sign of disease	<18> a usual cause of disease

37. Or "continual."
38. B 1.7; K 1.308; G^{a} 11.

‹ز› بدن مريض في الوقت الحاضر	‹ح› علامة مرضيّة في الوقت الحاضر	‹ط› سبب مرضيّ في الوقت الحاضر
‹ي› بدن مريض مطلق	‹يا› علامة مرضيّة مطلقة[٧٥]	‹يب› سبب مرضيّ مطلق
‹يج› بدن لا صحيح ولا مريض في الوقت الحاضر	‹يد› علامة لا صحيّة ولا مرضيّة في الوقت الحاضر	‹يه› سبب لا صحيّ ولا مرضيّ في الوقت الحاضر
‹يو› بدن لا صحيح ولا مريض مطلق	‹يز› علامة لا صحيّة ولا مرضيّة مطلقة	‹يح› سبب لا صحيّ ولا مرضيّ مطلق

(١٣) والمطلق، وهو ما ليس في الوقت الحاضر، ينقسم قسمين، أحدهما الدّائم والآخر الذي على الأكثر، وإذا ألّفت ذلك تولّد منه سبعة وعشرين[٧٦] تركيباً على هذه الصّفة[٧٧]:

‹آ› بدن صحيح في الوقت الحاضر	‹ب› علامة صحيّة في الوقت الحاضر	‹ج› سبب صحيّ في الوقت الحاضر
‹د› بدن صحيح دائم الصّحّة	‹ه› علامة صحيّة دائماً	‹و› سبب صحيّ دائماً
‹ز› بدن صحيح في أكثر الأمر[٧٨]	‹ح› علامة صحيّة في أكثر الأمر[٧٩]	‹ط› سبب صحيّ في أكثر الأمر[٨٠]
‹ي› بدن مريض في الوقت الحاضر	‹يا› علامة مرضيّة في الوقت الحاضر	‹يب› سبب مرضيّ في الوقت الحاضر
‹يج› بدن مريض[٨١] دائم المرض	‹يد› علامة مرضيّة دائماً	‹يه› سبب مرضيّ دائماً
‹يو› بدن مريض في أكثر الأمر[٨٢]	‹يز› علامة مرضيّة في أكثر الأمر[٨٣]	‹يح› سبب مرضيّ في أكثر الأمر[٨٤]

٧٥ MS: مريضة مطلقة | ٧٦ AS: وعشرون؛ D: وعشر | ٧٧ ADMY: Arranged as an abjad-numbered list | ٧٨ ADMY: على الأكثر | ٧٩ ADMY: على الأكثر | ٨٠ ADMY: على الأكثر | ٨١ F: مرضيّ | ٨٢ ADMY: على الأكثر | ٨٣ ADMY: على الأكثر | ٨٤ ADMY: على الأكثر

<19> a body neither healthy nor diseased at the present time	<20> a sign of neither health nor disease at the present time	<21> a cause of neither health nor disease at the present time
<22> a body chronically neither healthy nor diseased	<23> a chronic sign of neither health nor disease	<24> a chronic cause of neither health nor disease
<25> a body usually neither healthy nor diseased	<26> a usual sign of neither health nor disease	<27> a usual cause of neither health nor disease

(14) That which is neither healthy nor diseased—whether it is a body, a sign, or a cause—is said to be so in three respects.[39] First, it may have neither health in the extreme nor disease in the extreme; but, rather, it may be in between, as is the case with the diseases affecting the bodies of the old and convalescents. Second, they may both apply but in different organs, as is the case with someone whose foot or hand is chronically ill but the rest of whose body is healthy.[40] Third, they may both apply but at different times, as is the case with someone who is healthy in winter but diseased in summer. Thus, if these are combined with the first pattern, twenty-seven combinations result, as follows:[41]

39. B 1.8; K 1:308; G[a] 11–12. The explanation in the epitome is clearer and narrower in scope that Galen's.

40. This differs from Galen's definition of the second class: "participating in both."

41. A gloss in two MSS reads: "That is, these nine combined with 'at the present time,' 'chronically,' and 'usually' result in twenty-seven combinations." The examples given in angle brackets in some of the following entries are not in most MSS—a fact noted in a gloss to the older MS that contains them.

‹يط› بدن لا صحيح ولا مريض في الوقت الحاضر	‹ك› علامة لا صحيّة ولا مرضيّة في الوقت الحاضر	‹كا› سبب لا صحيّ ولا مرضيّ في الوقت الحاضر
‹كب› بدن لا صحيح ولا مريض دائمًا	‹كج› علامة لا صحيحة[٨٥] ولا مرضيّة دائمًا	‹كد› سبب لا صحيّ ولا مرضيّ دائمًا
‹كه› بدن لا صحيح ولا مريض في أكثر الأمر[٨٦]	‹كو› علامة لا صحيّة ولا مرضيّة في أكثر الأمر[٨٧]	‹كز› سبب لا صحيّ ولا مرضيّ في أكثر الأمر[٨٨]

(١٤) والذي ليس بصحيح ولا مريض بدنًا[٨٩] كان أو علامةً أو سببًا[٩٠] يقال على ثلثة أنحاء، أحدها أن يكون ليس له صحّة في الغاية ولا مرض في الغاية، لكنّه في الوسط بمنزلة أبدان الشّيوخ وأبدان النّاقهين من العلل[٩١]، والثّاني أن يكون جامعًا للأمرين في أعضاء مختلفة بمنزلة من تكون رجله أو يده زمنة، ويكون سائر بدنه صحيحًا، والثّالث أن يكون جامعًا للأمرين في أوقات مختلفة بمنزلة إنسان يكون في الشّتاء صحيحًا و في الصّيف مريضًا، فإذا[٩٢] ألّفت[٩٣] هذه تولّد منها في التّأليف[٩٤] الأوّل[٩٥] سبعة وعشرين[٩٦] تركيبًا على هذه الصّفة[٩٧]:

٨٥ ADSY: صحيّة | ٨٦ ADMY: على الأكثر | ٨٧ ADMY: على الأكثر | ٨٨ ADMY: على الأكثر | ٨٩ DFY: بدن؛ $A^hM^hY^h$: حاشية له: أي والذي ليس بصحيّ ولا مرضيّ بدنًا كان | ٩٠ M^1؛ M: – سببًا؛ DY: سبب | ٩١ F: – من العلل؛ DY: الأمراض | ٩٢ AM: وإذا | ٩٣ A^2: اتّفقت؛ M^2: تألّفت | ٩٤ M^2؛ DY: التّأليفات | ٩٥ A^1: تسعة؛ D: – الأول؛ S: – التّأليف الأوّل | ٩٦ A^1SY: عشرون؛ A^hM^h: حاشية له: أي هذه التّسعة [Y^h: سبعة] مع الوقت الحاضر ومع الدّائم ومع الأكثر يولّد منها سبعة وعشرون [A^h: وعشرين] تركيبًا. | ٩٧ M^h: الأمثلة ليس في أكثر النّسخ؛ ADMY: Arranged as an abjad-numbered list

<1> A body that, at the present moment, is neither healthy nor diseased to either extreme, <for example, the convalescent, since the convalescent is neither healthy in the extreme nor diseased in the extreme>	<2> A body that is chronically neither healthy nor diseased to either extreme, <such as the blind man, who is neither healthy in the extreme nor diseased in the extreme>	<3> A body that is usually neither healthy nor diseased to either extreme, <such as someone who is diseased for six months of the year and healthy the rest of it>
<4> A body that, at the present moment, is neither healthy nor diseased, since it is in both conditions but in different organs	<5> A body that is chronically neither healthy nor diseased, since it is in both conditions but in different organs <such as the blind man, since the eyes of the blind man are chronically diseased but his body is healthy>	<6> A body that is neither healthy nor diseased, since it is usually in both conditions but in different organs
<7> A body that, at the present moment, is neither healthy nor diseased, since it is in both conditions but at different times	<8> A body that is chronically neither healthy nor diseased, since it is in both conditions but at different times	<9> A body that is usually neither healthy nor diseased, since it is in both conditions but at different times
<10> A sign of neither health nor disease to either extreme at the present moment	<11> A sign always indicating neither health nor disease to either extreme	<12> A sign usually indicating neither health nor disease to either extreme
<13> A sign of neither health nor disease, since it indicates both conditions at the present moment but in different organs	<14> A sign of neither health nor disease, since it always indicates both conditions but in different organs	<15> A sign of neither health nor disease, since it usually indicates both conditions but in different organs

‹آ› بدن لا صحيح ولا مريض ليس فيه ولا[٩٨] في واحد[٩٩] من الغايتيْنِ في الوقت الحاضر ‹مثل النّاقه لأنّ النّاقه ليس هو صحيح في الغاية ولا سقيم في الغاية[١٠٠]›	‹ب› بدن لا صحيح ولا مريض ليس فيه ولا في واحد في واحد[١٠١] من الغايتيْنِ دائماً ‹مثل الأعمى لأنّ الأعمى ليس هو صحيح في الغاية ولا سقيم في الغاية[١٠٢]›	‹ج› بدن لا صحيح ولا مريض ليس فيه ولا[١٠٣] واحد من الغايتيْنِ على الأكثر[١٠٤] ‹مثل أن يكون في ستّة أشهر مريضاً والباقي صحيح[١٠٥]›
‹د› بدن لا صحيح ولا مريض جامع للأمريْنِ في أعضاء مختلفة في الوقت الحاضر	‹ه› بدن لا صحيح ولا مريض جامع للأمريْنِ[١٠٦] في أعضاء مختلفة دائماً ‹مثل الأعمى لأنّ الأعمى هو أبداً مريض من عينيْه صحيح في بدنه[١٠٧]›	‹و› بدن لا صحيح ولا مريض جامع للأمريْنِ في أعضاء مختلفة في أكثر الأمر[١٠٨]
‹ز› بدن لا صحيح ولا مريض جامع للأمريْنِ في أوقات مختلفة في الوقت الحاضر	‹ح› بدن لا صحيح ولا مريض جامع للأمريْنِ في أوقات مختلفة دائماً	‹ط› بدن لا صحيح ولا مريض جامع للأمريْنِ في أوقات مختلفة على الأكثر[١٠٩]
‹ي› علامة لا صحيّة ولا مرضيّة ليست ولا في واحدة[١١٠] من الغايتيْنِ في الوقت الحاضر	‹يا› علامة لا صحيّة ولا مرضيّة ليست ولا في واحدة من الغايتيْنِ دائماً	‹يب› علامة لا صحيّة ولا مرضيّة ليست ولا في واحدة من الغايتيْنِ على الأكثر[١١١]
‹يج› علامة لا صحيّة ولا مرضيّة جامعة للأمريْنِ في أعضاء مختلفة في الوقت الحاضر	‹يد› علامة لا صحيّة ولا مرضيّة جامعة للأمريْنِ في أعضاء مختلفة دائماً	‹يه› علامة لا صحيّة ولا مرضيّة جامعة للأمريْنِ في أعضاء مختلفة في أكثر الأمر[١١٢]

---

٩٨ S: فيه ولا > هو ولا | ٩٩ DMS: واحدة | ١٠٠ AM | ١٠١ ADMY: واحدة | ١٠٢ AM | ١٠٣ F: – ولا؛ S: فيه ولا > هو ولا | ١٠٤ F: في أكثر الأمر | ١٠٥ AM | ١٠٦ S: + جميعاً | ١٠٧ AM | ١٠٨ ADMY: على الأكثر | ١٠٩ S: في أكثر الأمر | ١١٠ FY: واحد | ١١١ F: في أكثر الأمر | ١١٢ ADMY: على الأكثر

<16> A sign of neither health nor disease at the present moment, since it indicates both conditions but at different times	<17> A sign of neither health nor disease, since it always indicates both conditions but at different times	<18> A sign of neither health nor disease, since it usually indicates both conditions but at different times
<19> A cause, at the present moment, of neither health nor disease to either extreme	<20> A chronic cause of neither health nor disease to either extreme	<21> Usually a cause of neither health nor disease to either extreme
<22> A cause of neither health nor disease, since, at the present moment, it causes both but in different organs	<23> A chronic cause of neither health nor disease, since it always causes both but in different organs	<24> A cause of neither health nor disease, since it usually causes both but in different organs
<25> A cause of neither health nor disease at the present moment, since it causes both but at different times	<26> A chronic cause of neither health nor disease, since it causes both but at different times	<27> A usual cause of neither health nor disease, since it causes both but at different times

(15) The body, sign, or cause that is neither healthy nor diseased and that is in various organs is said to be so in two ways: first, equally and, second, unequally.[42] The following eighteen combinations are generated from this second of the aspects implied by what is neither healthy nor diseased.[43]

42. B 1.8; K 1:308–9; G^{a} 11–12: A gloss in an old MS reads: "To have equal disease in the body is when half the body is diseased and half is healthy."

43. The second sense is when some organs are healthy and others diseased. A gloss in several MSS reads: "That is, if it is taken at the present time, chronically, and usually."

‹يو› علامة لا صحيّة ولا مرضيّة جامعة للأمرَين في أوقات مختلفة في الحاضر	‹يز› علامة لا صحيّة ولا مرضيّة جامعة للأمرَين في أوقات مختلفة دائمًا	‹يح› علامة لا صحيّة ولا مرضيّة جامعة للأمرَين في أوقات مختلفة في أكثر الأمر[١١٣]
‹يط› سبب لا صحيّ ولا مرضيّ ليس هو ولا في واحدة[١١٤] من الغايتَين في الوقت الحاضر	‹ك› سبب لا صحيّ ولا مرضيّ ليس هو ولا في واحدة من الغايتَين دائمًا	‹كا› سبب لا صحيّ ولا مرضيّ ليس هو ولا في واحدة من الغايتَين في أكثر الأمر[١١٥]
‹كب› سبب لا صحيّ ولا مرضيّ جامع للأمرَين في أعضاء مختلفة في الوقت الحاضر	‹كج› سبب لا صحيّ ولا مرضيّ جامع للأمرَين في أعضاء مختلفة دائمًا	‹كد› سبب لا صحيّ ولا مرضيّ جامع للأمرَين في أعضاء مختلفة في أكثر الأمر[١١٦]
‹كه› سبب لا صحيّ ولا مرضيّ جامع للأمرَين في أوقات مختلفة في الوقت الحاضر	‹كو› سبب لا صحيّ ولا مرضيّ جامع للأمرَين في أوقات مختلفة دائمًا	‹كز› سبب لا صحيّ ولا مرضيّ جامع للأمرَين في أوقات مختلفة في أكثر الأمر[١١٧]

(١٥) وما ليس بصحيّ ولا مرضيّ ممّا يكون كذلك في أعضاء مختلفة بدن[١١٨] كان أو[١١٩] علامة أو[١٢٠] سبب[١٢١]، فهو يقال على ضربَين، أحدهما على المساواة[١٢٢] والآخر على غير مساواة، فيتولّد من ذلك الوجه الثاني من الوجوه التي يدلّ عليها ما ليس بصحيح ولا مريض[١٢٣] تأليف[١٢٤] فيه ثمانية عشر تركيباً على هذه الصّفة[١٢٥]:

١١٣ ADMY: على الأكثر | ١١٤ F: واحد | ١١٥ ADM: على الأكثر | ١١٦ ADMY: على الأكثر | ١١٧ ADMY: على الأكثر | ١١٨ AM: بدنًا | ١١٩ DSY: أم | ١٢٠ DSY: أم | ١٢١ ADMY: سببًا | ١٢٢ S^h: على مساواة المرض في البدن إن يكون نصف البدن مريضاً ونصفه صحيحًا | ١٢٣ $A^hD^hM^h$: حاشية له: أي إذا أخذ في الوقت الحاضر ودائمًا وعلى [D^h: على] الأكثر. | ١٢٤ SY: يتألّف
١٢٥ ADMY: Arranged as an eighteen-item abjad-numbered list

<1> A body that, at the present moment, is neither healthy nor diseased, since it is in both conditions but in different organs equally	<2> A body that is chronically neither healthy nor diseased, since it is in both conditions but in different organs equally	<3> A body that is usually neither healthy nor diseased, since it is in both conditions but in different organs equally
<4> A body that, at the present moment, is neither healthy nor diseased, since it is in both conditions but in different organs unequally	<5> A body that is chronically neither healthy nor diseased, since it is in both conditions but in different organs unequally	<6> A body that is usually neither healthy nor diseased, since it is in both conditions but in different organs unequally
<7> A sign of neither health nor disease, since it indicates both conditions at the present moment but in different organs equally	<8> A sign of neither health nor disease, since it indicates both conditions chronically but in different organs equally	<9> A sign of neither health nor disease, since it usually indicates both conditions but in different organs equally
<10> A sign of neither health nor disease, since it indicates both conditions at the present moment in different organs unequally	<11> A sign of neither health nor disease, since it indicates both conditions chronically but in different organs unequally	<12> A sign of neither health nor disease, since it usually indicates both conditions but in different organs unequally

‹آ› بدن لا صحيح ولا مريض جامع للأمرَين في أعضاء مختلفة على التّساوي[١٢٦] في الوقت الحاضر	‹ب› بدن لا صحيح ولا مريض جامع للأمرَين في أعضاء مختلفة على التساوي[١٢٧] دائمًا	‹ج› بدن لا صحيح ولا مريض جامع للأمرَين في أعضاء مختلفة على التّساوي[١٢٨] في أكثر الأمر[١٢٩]
‹د› بدن لا صحيح ولا مريض جامع للأمرَين في أعضاء مختلفة على غير مساواة[١٣٠] في الوقت الحاضر	‹ه› بدن لا صحيح ولا مريض جامع للأمرَين في أعضاء مختلفة على غير التّساوي[١٣١] دائمًا	‹و› بدن لا صحيح ولا مريض جامع للأمرَين في أعضاء مختلفة على غير مساواة[١٣٢] في أكثر الأمر[١٣٣]
‹ز› علامة لا صحيّة ولا مرضيّة جامعة للأمرَين في أعضاء مختلفة على التّساوي[١٣٤] في الوقت الحاضر	‹ح› علامة لا صحيّة ولا مرضيّة جامعة للأمرَين في أعضاء مختلفة على التّساوي[١٣٥] دائمًا	‹ط› علامة لا صحيّة ولا مرضيّة جامعة للأمرَين في أعضاء مختلفة على التّساوي[١٣٦] في أكثر الأمر[١٣٧]
‹ي› علامة لا صحيّة ولا مرضيّة جامعة للأمرَين في أعضاء مختلفة على غير التّساوي[١٣٨] في الوقت الحاضر	‹يا› علامة لا صحيّة ولا مرضيّة جامعة للأمرَين في أعضاء مختلفة على غير التّساوي[١٣٩] دائمًا	‹يب› علامة لا صحيّة ولا مرضيّة جامعة للأمرَين في أعضاء مختلفة على غير التّساوي[١٤٠] في أكثر الأمر[١٤١]

١٢٦ AM: المساواة | ١٢٧ AM: المساواة | ١٢٨ AM: المساواة | ١٢٩ ADMY: على الأكثر | ١٣٠ D: التّساوي؛ FY: على التّساوي؛ S: المتساوي | ١٣١ AM: المساواة؛ F: التّسا [كذا] | ١٣٢ D: على غير التّساوي؛ FSY: على التّساوي | ١٣٣ ADMY: على الأكثر | ١٣٤ D: على غير التّساوي؛ AM: على المساواة | ١٣٥ AM: المساواة | ١٣٦ AM: المساواة | ١٣٧ ADMY: على الأكثر | ١٣٨ AM: المساواة | ١٣٩ AM: المساواة | ١٤٠ AM: مساواة؛ Y: + على غير التّساوي | ١٤١ ADMY: على الأكثر

<13> A cause neither healthful nor causing disease, since, at the present moment, it causes both but in different organs equally	<14> A cause neither healthful nor causing disease, since it chronically causes both but in different organs equally	<15> A cause neither healthful nor causing disease, since it usually causes both but in different organs equally
<16> A cause neither healthful nor causing disease, since, at the present moment, it causes both but in different organs unequally	<17> A cause neither healthful nor causing disease, since it chronically causes both but in different organs unequally	<18> A cause neither healthful nor causing disease, since it usually causes both but in different organs unequally

(16) Medicine is the knowledge of matters relating to health, to disease, and to neither health nor disease.[44] This statement of ours does not imply knowledge of all cases, since that is impossible and cannot be completely attained; nor does it imply knowing some of them, since that does not resemble the path of the art. Rather, it means which [of the three categories] each is from. People have differed about how this point is to be explained. Some say that the meaning of "whichever of them each is from" is the knowledge [of the varying conditions] of the majority of people; but this is absurd because it is impossible for someone to know all the people in a single city, let alone know a majority of those in the entire world. Moreover, this statement also implies that medicine is the individual knowledge of these matters [case by case], and Galen rejected and refuted this view.[45] Some people say that, by

44. B 1.9; K 1:309; G^a 12–13, directly quoting the definition given by Galen at B 1.1; K 1:307; G^a 7 and p. 57, paragraph 10, above: Galen points out the ambiguity of the definition: It can refer to all things relating to health, disease, and the neutral; to some things; or to the kind of things (τὸ ὁποίων, a variant of the usual word for quality; G^a *ay shayʾin iltamisat maʿrifatufu minhā*). Galen rejects the first as impossible and the second as deficient and unscientific (οὐ τεχνικόν; literally, "not of the art"). The third, though, is both scientific and sufficient for all parts of the art of medicine. "The knowledge of what kind of things fall into each category is both scientific and sufficient for all the individual parts of the art" (trans. Singer). The point is that the art of medicine does not consist in knowing all particular cases, nor in knowing only some particulars, but, rather, in knowing how to categorize the various matters relating to health and disease. The last point seems to have been obscure both to the epitomist and the translator, since it is explained at length in the text yet is still vague in the Arabic.

45. In his attacks on the Empiricists; see pp. 33–34, paragraph 33, above.

‹يج› سبب لا صحيّ ولا مرضيّ جامع للأمريْنِ في أعضاء مختلفة على التّساوي[١٤٢] في الوقت الحاضر	‹يد› سبب لا صحيّ ولا مرضيّ جامع للأمريْنِ في أعضاء مختلفة على التّساوي[١٤٣] دائمًا	‹يه› سبب لا صحيّ ولا مرضيّ جامع للأمريْنِ في أعضاء مختلفة على التّساوي[١٤٤] في أكثر الأمر[١٤٥]
‹يو› سبب لا صحيّ ولا مرضيّ جامع للأمريْنِ في أعضاء مختلفة على غير التّساوي[١٤٦] في الوقت الحاضر	‹يز› سبب لا صحيّ ولا مرضيّ جامع للأمريْنِ في أعضاء مختلفة على غير[١٤٧] التّساوي[١٤٨] دائمًا	‹يح› سبب لا صحيّ ولا مرضيّ جامع للأمريْنِ في أعضاء مختلفة على غير التّساوي[١٤٩] في أكثر الأمر[١٥٠]

(١٦) الطّبّ هو معرفة الأمور الصّحيّة والمرضيّة والتي ليست بصحيّة ولا مرضيّة، وقولنا هذا ليس يدلّ على الجميع لأنّ ذلك ممّا لا يمكن ولا يوقف على منتهاه ولا يدلّ أيضاً على البعض لأنّ ذلك لا يشبه الطّريق الصّناعيّ، لكنّه يدلّ على أيّ شيء كان منها. وقد اختلف النّاس في تفسير هذا المعنى، فقال بعضهم إنّ معنى قولنا[١٥١] أي شيء كان منها إنّما هو معرفة جلّ النّاس، وهذا شنع[١٥٢] لأنّ الإنسان لا يقدر أن يعرف جلّ من في مدينة واحدة فضلاً[١٥٣] أن يعرف جلّ من في العالم كلّه، على أنّه يجب بحسب هذا القول أيضاً إن يكون الطّبّ إنّما هو معرفة بالبعض من هذه الأمور، وجالينوس قد هرب[١٥٤] من هذا ودفعه، وقال بعض النّاس[١٥٥] إنّ قوله أي شيء كان

١٤٢ AM: المساواة | ١٤٣ AM: المساواة | ١٤٤ AM: المساواة | ١٤٥ ADMY: على الأكثر | ١٤٦ AM: المساواة | ١٤٧ S: – غير | ١٤٨ AM: المساواة | ١٤٩ ADM: المساواة | ١٥٠ ADMY: على الأكثر | ١٥١ DFY: قوله | ١٥٢ M: أشنع؛ S: شنيع | ١٥٣ M: فضل؛ DY: فدع | ١٥٤ F¹؛ F: يجبّ | ١٥٥ F¹؛ F: – النّاس: DY: وبعض قال

"whichever of them each is from," he means whoever comes to the physician; but this statement also implies knowledge of the individual. Another group says that, when he said "whichever of them each is from," he means the [patient's] impaired functions; but this is clearly false, since it cannot ever be said that the functions of healthy people are impaired. The most accurate thing that can be said about the meaning of his statement "whichever of them each is from" is this: that this statement of his implies that the physician possesses knowledge of the methods and means universally and generically, and that, by means of these universal generic methods, he grasps individual particular matters and thereby knows them.

منها إنّما أراد به من يأتي الطّبيب، وهذا القول أيضاً يوجب معرفة البعض، وقال قوم آخرون[١٥٦] إنّ معناه في قوله أي شيء كان منها إنّما[١٥٧] هو من أفعاله مضرورة وهذا[١٥٨] كذب صراح لأنّ الأصحّاء بأسرهم ليس يجوز أن يكون أفعالهم مضرورة، وأصدق الأقاويل في معنى قوله أي شيء كان منها هو أنّ قوله هذا يدلّ على أنّه يكون عند الطّبيب[١٥٩] علم بالطّرق والمذاهب الكلّيّة[١٦٠] الجنسيّة، فيقف بهذه الطّرق الكلّيّة الجنسيّة[١٦١] على الأشياء المفردة الجزئيّة، فيعرفها بها[١٦٢].

١٥٦ F: آخر؛ DY: وقوم آخر قالوا | ١٥٧ F: عني ضرر الأفعال | ١٥٨ S: + أيضاً هو؛ DY: + أيضاً | ١٥٩ D^1S^1؛ DSY: الإنسان | ١٦٠ S: – الكلّيّة؛ S^1: الكلّيّة والعامّيّة | ١٦١ D: – فيقف. . . الجنسيّة؛ F: – الجنسيّة | ١٦٢ F^1: + إمّا أن يكون كذلك دائماً وإمّا أن يكون كذلك في الوقت الحاضر وإمّا على الأمر الأكثر والبدن الصّحيح صحّ الوقت

[Chapter 2][46]

[Bodies]

(17) The body that is healthy at the present time is the one that, at the present time, has a moderate temperament and symmetrical structure. "Moderate" and "symmetrical" are used in two ways:[47] first, when there are equal parts of the things by which the temperament is moderate and of the things by whose symmetry the symmetrical parts are symmetrical; and, second, when the things in it are of unequal quantities but the quantities correspond to what is needed <by the body.>[48] The body that is chronically healthy is the one that has a moderate temperament and symmetrical structure in all years. The body that is usually healthy is the one that falls short of the optimal states of health,[49] but only falls short by a small quantity.[50] Likewise, the diseased body is diseased either at the present time, or chronically, or for the most part. The bad temperament may be in all the tissues, as happens in the case of fever; in one, as happens in gout; or in the best and highest ranking of them, as happens in melancholia.[51] In addition, the asymmetry and lack of moderation in the composite organs may also

46. B 2.1; K 1:309–10; G^a 15. One early MS begins this paragraph with "The healthy body is either always so, or is so at the present time, or is so for the most part. The body . . . "

47. B 4.3; K 1:314–15; G^a 25–26.

48. Added in one early MS.

49. One MS adds the gloss: "Which are chronic."

50. B 2.1–7; K 1:310–13; G^a 15–21.

51. One MS has the gloss: "Both examples—gout and melancholia—involve a single organ."

[الفصل الثّاني]

[ذكر الأبدان]

(١٧) البدن الصّحيح في الوقت الحاضر هو الذي يكون في ذلك الوقت الحاضر معتدل المزاج مستوي التّركيب، والمعتدل والمستوي تقال على ضربيْنِ، أحدهما أن يكون فيه من الأشياء التي بامتزاجها اعتدل[١٦٣] والأشياء التي باستوائها استوى[١٦٤] أجزاء متساوية، والآخر أن يكون فيه من تلك الأشياء مقادير غير متساوية إلاّ أنّ تلك المقادير موافقة لما يحتاج إليه[١٦٥]. وأمّا البدن الصّحيح دائماً فهو الذي يكون معتدل المزاج مستوي التّركيب في جميع الأسنان، وأمّا البدن الصّحيح على الأكثر[١٦٦] فهو الذي يكون ناقصاً عن أفضل حالات الصّحّة إلاّ أنّ مقدار نقصانه مقدار[١٦٧] يسير، وكذلك البدن المريض لا يخلوا من أن يكون إمّا في الوقت الحاضر وإمّا دائماً وإمّا على الأكثر، وسوء المزاج إمّا أن يكون في جميع الأعضاء المتشابهة الأجزاء بمنزلة ما يعرض ذلك في الحمّى، وإمّا أن يكون في بعضها بمنزلة ما يعرض في النّقرس، وإمّا أن يكون في أشرفها وأجلّها قدراً بمنزلة ما يعرض في الوسواس السّوداويّ[١٦٨]، وخروج الأعضاء المركّبة عن الاستواء والاعتدال في التّركيب

---

١٦٣ AY: اعتدال | ١٦٤ ADFY: استواء | ١٦٥ F: + الأبدان الصّحيحة | ١٦٦ F: أكثر الأمر | ١٦٧ AM: – مقدار | ١٦٨ M^h: المثالانِ هما في بعض، أعني النّقرس والوسواس.

exist either in all the organs, as was the case with Thersites;[52] in one of them, as is the case with a long, narrow head; or in the best and highest ranking of them. The best and highest ranking, in this sense, may be either with respect to what is needed for the continuance of life, as is the case with one who has constricted blood vessels in the liver; or with respect to the good quality of life, as with someone whose fingers are joined together. The body that is neither healthy nor sick in that it combines both—that is, extreme health and extreme sickness—either exhibits both but in different organs, as is the case with the head and feet, or exhibits both in a single organ. If it exhibits both in a single organ, then either the structure of that organ is symmetrical but its temperament is not moderate, as when the head is well formed, but it is hotter or colder than it ought to be; or it is the opposite of that, so that its temperament is moderate but its structure is asymmetrical, as is the case when the head has a moderate temperament but is extremely protuberant to the front or back; or its structure is symmetrical, but its temperament is simultaneously moderate in some qualities and immoderate in others, as is the case in what is moderate in heat and coldness but immoderate in moisture and dryness or vice versa; or else the

---

52. This is the Thersites of the *Iliad*, trans. Fagels, B:216–19:

> Here was the ugliest man who ever came to Troy.
> Bandy-legged he was, with one foot clubbed,
> both shoulders humped together, curving over
> his caved-in chest, and bobbing above them
> his skull warped to a point,
> sprouting clumps of scraggly, woolly hair.

The MSS have something like “Eustes.” A gloss in several MSS reads, with minor variations: “Eustes [Thersites] was a man mentioned by the poet Homer. He was in the army and was hunched in his breast and back and had a long head that resembled a boat. His hair was thin, and his legs were crooked and lame.” This reference is evidence that the epitomes were composed in Greek. Thersites is not mentioned in the corresponding text of *The Small Art*, though he does appear in three other works of Galen (K 3.469, 5.15, 18a.253, 289). The gloss probably goes back to a gloss in the Greek text, since it is closely based on Homer, who lists the same four deformities. The corruption of the name must also be very early in the Arabic textual tradition, but it is easily explained in Arabic—a *thāʾ-rāʾ* read as *yāʾ-wāw*. A recent article identifies his condition as cleidocranial dysplasia, a genetic bone deformity; Simms, “Missing Bones, 33–40. *BNP* 14 col. 556. See appendix 1, s.v. “Thersites.”

إمّا أن يكون أيضاً في جميع الأعضاء بمنزلة ما كان ذلك في ثرسطس[١٦٩]، وإمّا في بعضها بمنزلة ما يكون في المسفّط الرّأس، وإمّا في أشرفها وأجلّها قدراً، وشرف هذه وجلالة قدرها إمّا أن يكون فيما يحتاج إليه لقوام[١٧٠] الحياة بمنزلة من يكون عروق كبده ضيّقة، وإمّا فيما يحتاج إليه من[١٧١] جودة[١٧٢] الحياة بمنزلة من يلتحم أصابعه بعض[١٧٣] ببعض، والبدن الذي ليس بصحيح ولا سقيم[١٧٤] الجامع للأمرين، أعني الصّحّة[١٧٥] في الغاية والسّقم[١٧٦] في الغاية إمّا أن يكون جامعاً للأمرين في أعضاء مختلفة بمنزلة الرّأس والرّجلين، وإمّا في عضو واحد، والجامع لهما في عضو واحد إمّا أن يكون تركيب ذلك العضو منه مستوياً، ومزاجه غير معتدل بمنزلة ما يكون الرّأس حسن الشّكل إلاّ أنّه أحرّ[١٧٧] وأبرد ممّا ينتغي، وإمّا أن يكون على خلاف ذلك، فيكون مزاجه معتدلاً وتركيبه غير مستوٍ بمنزلة ما يكون الرّأس معتدل المزاج إلاّ أنّه أشدّ نتوًا إمّا إلى قدّام وإمّا إلى خلف وإمّا أن يكون تركيبه مستوٍ[١٧٨] ومزاجه معتدل[١٧٩] وغير معتدل معاً في كيفيّات مختلفة بمنزلة ما يكون معتدل[١٨٠] في الحرارة والبرودة غير معتدل في الرّطوبة واليبس[١٨١] أو على خلاف ذلك، وإمّا أن يكون بخلاف هذا[١٨٢]

---

١٦٩ AM: يوسطيس؛ FS: يوسطس؛ DY: بوسيطس؛ $A^{h}D^{h}F^{h}M^{h}S^{h}Y^{h}$: حاشية له [$D^{h}$: –]: يوسطيس [$Y^{h}$: – حاشية له يوسطيس] هو رجل ذكر أوميروس [A^{h}: أرميرس؛ S^{1}: أميروس؛ Y^{h}: آوميرس] الشّاعر أنّه كان في العسكر وكان أحدب في صدره وظهره [F^{h}: قال أمروس الشّاعر إنّ يوسطس كان رجلاً أحدب في ظهره وصدره] وكان رأسه [A^{h}: ورأسه] مستطيلاً يشبه [F^{h}: بسببها] الزّورق أزعر الشّعر سوقه [$A^{h}S^{h}$:منشوفه؛ D^{h}: مشرقة؛ Y^{h}: مشرفه] أزور أعرج [$D^{h}Y^{h}$: + بهنّ شبيه]. | ١٧٠ ADMY: في قوام | ١٧١ ADMY: في | ١٧٢ F: + صلاح؛ S: إصلاح | ١٧٣ ADMY: بعضها | ١٧٤ S: ينضم؛ DY: مريض | ١٧٥ DFY: صحّة | ١٧٦ F: وسقم؛ DY: سقماً | ١٧٧ S^{1}؛ S: –أحرّ | ١٧٨ ADMY: مستوياً | ١٧٩ ADMY: معتدلاً | ١٨٠ AY: معتدلاً | ١٨١ ADMY: واليبوسة | ١٨٢ F: ذلك

opposite of the previous state, so that it is moderate with respect to its temperament and both symmetrical and asymmetrical in structure in different respects, as is the case with that which is symmetrical in form but asymmetrical in position, size, or number, or the opposite.

(18) The active primary qualities are those that have an effect, and the passive primary qualities are those in which the effect occurs.[53] These are four in all. Two of them are active, which are those in which activity is greater; these are heat and coldness. Two are passive, which are those in which it is more usual to receive an effect; these are dryness and moisture. There are two kinds of opposition among primary qualities. The first is called the opposition of the predominantly active qualities, which are heat and coldness. The second is called the opposition of the predominantly passive qualities, those that are more likely to have effects occur in them. These are dryness and moisture.

(19) The term "now"—that is, the present moment—is used in two senses:[54] first, the durationless point of time; and second, a duration of time, as when we say, "It is summer now."

53. B 2.5; K 1:312–13; G[a] 19–20: Galen mentions the active and passive qualities but does not identify them.

54. B 2.6; K 1:313; G[a] 21: Galen mentions without explanation that "now" has two meanings. The Arabic translation of *The Small Art* adds in some MSS: "The first having no parts and the second having parts"; cf. B, p. 281, n. 1, which cites this passage of the epitome and the commentary of ʿAlī ibn Riḍwān.

فيكون في جملة مزاجه معتدل[١٨٣] ويكون في تركيبه مستوي[١٨٤] وغير مستوٍ معاً في أنحاء مختلفة بمنزلة ما يكون مستوي[١٨٥] في الخلقة غير مستوٍ في الوضع أو في المقدار أو في العدد أو على خلاف ذلك ـــه

(١٨) الكيفيّات الأوَل الفاعلة التي تفعل والمنفعلة التي يقع بها الفعل هي أربع، اثنتانِ[١٨٦] منها فاعلتانِ يكون بهما الفعل أكثر[١٨٧]، وهما الحرارة والبرودة، واثنتانِ منفعلتانِ يقع بهما الفعل أكثر[١٨٨]، وهما اليبس[١٨٩] والرّطوبة[١٩٠]. أصناف التّضادّ في الكيفيّات الأوَل صنفانِ، أحدهما يقال له تضادّ الكيفيتَيْن اللتيْن[١٩١] هما[١٩٢] أكثر فعلاً، وهما الحرارة والبرودة، والآخر يقال له تضادّ الكيفيّات اللتيْن هما أكثر انفعالاً[١٩٣]، ووقوع الفعل بهما أكثر، وهما اليبس[١٩٤] والرّطوبة.

(١٩) معنى الآن وهو الوقت الحاضر يقع على أمريْن، أحدهما النّقطة التي لا عرض لها من الزّمان، والآخر الوقت الذي له عرض بمنزلة ما نقول إنّ الآن صيف.

---

١٨٣ ADMY: معتدلاً | ١٨٤ ADMY: مستوياً | ١٨٥ ADMY: مستوياً | ١٨٦ F: اثنيْنِ؛ M: اثنانِ | ١٨٧ DY: فاعلتانِ. . . أكثر > الفعل لهما أكثر أي أكثر فعلاً | ١٨٨ F: واثنيْنِ منهما أكثر انفعالاً؛ S: –أكثر؛ DY: يكون. . . أكثر > يقع بهما الفعل أكثر أي أكثر انفعالاً | ١٨٩ ADMY: اليبوسة | ١٩٠ A^2M^2: الكيفيّات الأوَل الفاعلة والمنفعلة هي أربع، اثنتانِ منها أكثر فعلاً، وهما الحرارة والبرودة، واثنتانِ أكثر انفعالاً، وهما اليبوسة [A^2: اليبس] والرّطوبة. | ١٩١ DY: + الفعل؛ S^2: الكيفيّتيْنِ الفاعلتيْنِ | ١٩٢ F: اللتيْنِ هما > الفاعلتيْنِ وهما؛ S: وهما؛ D: بما | ١٩٣ F: نفعاً؛ M: فعلاً؛ F^2: انفعالاً | ١٩٤ ADMY: اليبوسة

[Chapter 3]

[Signs]

(20) Some signs are of health and some of disease.[55] In each class, some are diagnostic of the present state, some are prognostic of the future, and some are mnemonic of what is past. Some of the signs that are of neither health nor disease are diagnostic of a body that combines extremes of health and disease; some are diagnostic of a body that does not have extremes of either health or disease; and some are diagnostic of a body that is healthy at one moment and diseased at another. Each of these three kinds of sign is either diagnostic of the present, prognostic of the future, or mnemonic of what is past. If these signs are combined, the following fifteen combinations result.[56]

<1> A sign diagnostic of health	<2> A sign prognostic of health	<3> A sign mnemonic of health
<4> A sign diagnostic of disease	<5> A sign prognostic of disease	<6> A sign mnemonic of disease
<7> A sign diagnostic of a body that is neither healthy nor diseased in the extreme	<8> A sign prognostic of a state that is neither healthy nor diseased in the extreme	<9> A sign mnemonic of a state that is neither healthy nor diseased in the extreme
<10> A sign diagnostic of a body combining both conditions	<11> A sign prognostic of a state combining both conditions	<12> A sign mnemonic of a state combining both conditions
<13> A sign diagnostic of a body that is healthy at one moment and diseased at another	<14> A sign prognostic of a state that is healthy at one moment and diseased at another	<15> A sign mnemonic of a state that is healthy at one moment and diseased at another

55. B 3.1–3; K 1:313; G^a 22–24: This passage follows Galen's text closely.
56. Other MSS give a numbered list rather than a table.

[الفصل الثّالث]

[ذكر العلامات]

(٢٠) العلامات منها صحيّة ومنها مرضيّة، وكلّ واحد من هذينِ الصّنفينِ منه ما هو دالّ على الأمر الحاضر، ومنه ما هو منذر بما يستأنف، ومنه مذكّر بما قد مضى، وأمّا العلامات التي ليست بصحيّة ولا مرضيّة، فبعضها يدلّ على البدن الجامع للغايتيْنِ من الصّحّة والمرض[١٩٥]، وبعضها يدلّ على البدن الذي لا صحّة له في الغاية ولا مرض في الغاية[١٩٦]، وبعضها يدلّ على البدن الذي يكون في وقت من الأوقات صحيحًا وفي وقت آخر مريضاً، وكلّ[١٩٧] واحد من هذه الثّلثة الأصناف من العلامات إمّا أن يدلّ على الحاضر وإمّا أن ينذر بالمستأنف وإمّا أن يذكّر بالسّلف[١٩٨]، وإذا[١٩٩] ألّفت هذه العلامات صار منها خمسة عشر تركيباً على هذه الصّفة[٢٠٠]:

‹آ› علامة تدلّ على الصّحّة	‹ب› علامة تنذر بالصّحّة	‹ج› علامة تذكّر بالصّحّة
‹د› علامة تدلّ على المرض	‹هـ› علامة تنذر بالمرض	‹و› علامة تذكّر بالمرض
‹ز› علامة تدلّ على بدن لا صحيح ولا مريض في الغاية	‹ح› علامة تنذر بحال لا صحيّة ولا مرضيّة في الغاية	‹ط› علامة تذكّر بحال لا صحيّة ولا مرضيّة في الغاية
‹ي› علامة تدلّ على بدن جامع للأمريْنِ	‹يا› علامة تنذر بحال جامعة للأمريْنِ	‹يب› علامة تذكّر بحال جامعة
‹يج› علامة تدلّ على بدن صحيح في وقت مريض في وقت آخر	‹يد› علامة تنذر بحال صحيّة في وقت مرضيّة[٢٠١] في آخر	‹يه› علامة تذكّر بحال صحيّة في وقت مرضيّة في وقت[٢٠٢] آخر ـــهـ

١٩٥ F^h: + في أعضاء مختلفة | ١٩٦ ADMY: – في الغاية | ١٩٧ DF: فإذاً كلّ | ١٩٨ AM: بما قد مضى | ١٩٩ S: فإذا | ٢٠٠ ADMY: Presented as a fifteen-item abjad-numbered list |
٢٠١ DSY: ومرضيّة | ٢٠٢ ADMY: – وقت

(21) Some signs—those diagnostic of the present state of affairs—are things that are beneficial only to the patient. Other signs—those mnemonic of what has already happened—are beneficial only to the physician, since these may be beneficial in garnering praise for the physician and manifesting his skill and acuity to men.[57] Still others—those prognostic of what will be—are of use to both patient and physician. Signs may be analyzed in another way in that the benefit of those of them that are diagnostic of the present redounds, in the first place, to the patient in the form of the treatment he needs but accrues accidentally to the physician in that his treatment is successful. The benefit of those that are mnemonic of what has already happened redounds primarily to the physician in that he needs to acquire praise and reputation but also redounds accidentally to the patient in that, if he has confidence in the skill and acuity of the physician, he will follow his instruction and trust what he says, thus making it more likely that the treatment will have a good outcome. The benefit of the signs prognostic of the future redounds to both of them, both primarily and accidentally.

57. B 3.4; K 1:314; G[a] 24: Galen's text lacks this intrusion of marketing considerations and simply observes that mnemonic signs are less useful than the diagnostic and prognostic.

(٢١) العلامات منها أشياء ينتفع بها المريض فقط، وهي العلامات الدّالّة على ما هو حاضر، ومنها أشياء ينتفع بها الطّبيب فقط، وهي العلامات المذكّرة بما قد سلف لأنّ هذه إنّما ينتفع بها في أن يمدح الطّبيب ويظهر للنّاس[٢٠٣] حذقه وفراهته[٢٠٤]، ومنها ما[٢٠٥] ينتفع بها المريض والطّبيب معاً، وهي العلامات التي تنذر بما سيكون. وللعلامات تصريف آخر، وهو أنّ ما هو منها دالّ على[٢٠٦] الحاضر، فنفعه[٢٠٧] أوّلاً يعود[٢٠٨] على المريض فيما يحتاج إليه من المداواة ومن طريق العرض يعود نفعها على الطّبيب في أن ينجح عمله، وما هو منها مذكّر بما قد سلف، فَنَفْعه يعود أوّلاً على الطّبيب فيما يحتاج إليه من اكتساب الحمد والمدح ويعود نفعه من طريق العرض على المريض أيضاً في أنّه إذا وثق بحذق الطّبيب وجودة نظره[٢٠٩] استسلم[٢١٠] إليه[٢١١] وركن إلى قوله، وذلك ممّا يتبعه حسن العاقبة في المداواة، وما هو منها منذر بما يستأنف، فنفعه يعود عليهما جميعاً أوّلاً وبالعرض.

٢٠٣ S: النّاس؛ DY: + أنّه | ٢٠٤ ADMY: – وفراهته | ٢٠٥ S: وممّا | ٢٠٦ S: + الوقت | ٢٠٧ F: ونفعها | ٢٠٨ AM: يعود أوّلاً | ٢٠٩ A^2M^2؛ AM: بصره | ٢١٠ D: أسلم؛ D^1Y: استنام | ٢١١ ADMY: له

[Chapter 4]

[The best states of health]

(22) The body that is in the most excellent state of health combines moderation in the temperament of the tissues, symmetry in the structure of the functional organs, and a laudable state of continuity in the entire body.[58] Some of the signs indicating moderation of temperament in the tissues are essential, such as moderation of the primary qualities of heat, coldness, dryness, and moisture. Others are accidental. Some such accidental signs are palpable, such as moderate hardness and moderate softness. Some are visible, such as moderate pallor and moderate ruddiness. Some are both palpable and visible, such as the hair and its varieties—shaggy, thin, curly, lank, and so forth. The excellent body is moderate in all these respects. Another sign is completeness, as in unimpaired functions. Likewise, some of the signs indicating ideal states in the structure of functional organs are essential; these are construction, dimensions, number, and position. The five aspects of structure are apertures, concavity, shape, roughness, and smoothness. Position includes two things: erect position and contiguity. If the structure of the functional organs is in these ideal states, then the body is moderate with respect to

58. B 4.3–5; K 1:314–15; G[a] 25–26: Continuity refers to the absence of physical damage and lesions in the tissues and organs; see paragraphs 23, 61, 81, pp. 72–73, 108, 125, below.

[الفصل الرّابع]

[ذكر أفضل الحالات في الصّحّة]

(٢٢) ما كان من الأبدان على أفضل الحالات في الصّحّة، فهو جامع للاعتدال في مزاج الأعضاء المتشابهة الأجزاء واستوائها[٢١٢] في تركيب الأعضاء الآليّة، والحال المحمودة في اتّصال[٢١٣] جملة البدن[٢١٤]. العلامات الدّالّة على اعتدال مزاج الأعضاء المتشابهة الأجزاء منها ما هو جوهريّ بمنزلة الاعتدال في الكيفيّات الأوَل، وهي الحرارة والبرودة والرّطوبة[٢١٥] واليبوسة، ومنها ما هو عرضيّ، وهذه العرضيّة منها ملموسة وهي الصّلابة المعتدلة واللّين المعتدل، ومنها مبصورة[٢١٦]، وهي البياض المعتدل والحمرة المعتدلة، ومنها ملموسة ومبصورة معاً بمنزلة الشّعر وتوابعه، وهي الأزبّ والأزعر والجعد والسّبط[٢١٧] وما أشبه ذلك، فإنّ البدن الفاضل يكون معتدلاً في هذه الأحوال كلّها، ومنها تماميّة بمنزلة تمام الأفعال، وكذلك أيضاً العلامات الدّالّة على أفضل الحالات في تركيب الأعضاء الآليّة، منها ما هي جوهريّة، وهي الخلقة والمقادير والعدد والوضع، والخلقة تجمع خمسة أشياء، وهي التقعير[٢١٨] والتّجويف والشّكل والخشونة والملاسة، والوضع يجمع شيئيْن، وهما الوضع[٢١٩] المستقيم[٢٢٠] والمشاركة، والبدن الذي حاله أفضل الحالات في تركيب

---

٢١٢ ADMY: والاستواء | ٢١٣ F: إيصال | ٢١٤ F: + في الاستدلال على اعتدال مزاج الأعضاء | ٢١٥ AMY¹: واليبوسة والرّطوبة؛ Y: – والرّطوبة | ٢١٦ A²M²: مبصرة؛ S: المبصرة | ٢١٧ A: البسيط؛ DY: الزّيب والزّعارة والجعودة والسّبوطة | ٢١٨ ADMY: الثّقب | ٢١٩ ADM: الموضع | ٢٢٠ F: + والمائل

each of them. Some states are accidental, such as comeliness and beauty; and some have to do with completeness, such as unimpaired functions. The defect in a usually healthy body by reason of which it is not always healthy is either in the tissues or in the functional organs. That which is in the tissues is either in one quality or in two. That which is in the functional organs is either in one species or in two.

(23) Of the things by which the tissues and organs acquire excellence in their forms, two are, as it were, the genera.[59] One of them is the temperament and the other the size. Others are like the species. There are also two of these: first, moderation in temperament and,

59. B 4.3–6; K 1:315–16; G[a] 25–27.

الأعضاء الآليّة هو معتدل في هذه كلّها، ومنها عرضيّة بمنزلة الحسن والجمال، ومنها تماميّة بمنزلة تمام الأفعال، والآفة في الأبدان الصّحيحة على أكثر الأمر التي بسببها صارت ليست بدائمة الصّحّة إمّا أن تكون في الأعضاء المتشابهة الأجزاء وإمّا أن تكون في الأعضاء الآليّة، والذي[٢٢١] في الأعضاء المتشابهة الأجزاء[٢٢٢] إمّا في كيفيّة واحدة وإمّا في اثنتين[٢٢٣]، والتي[٢٢٤] في الأعضاء الآليّة إمّا في نوع واحد وإمّا في نوعين.

(٢٣) الأشياء[٢٢٥] التي بها تتمّ[٢٢٦] الفضيلة في هيئة الأعضاء المتشابهة الأجزاء وفي هيئة الأعضاء الآليّة، بعضها يجري مجرى الجنس[٢٢٧]، وهما شيئان، أحدهما

---

٢٢١ F^{1}: الذي في > وفي؛ S^{1}: التي | ٢٢٢ $F^{1}S^{1}$؛ FSY: –وإمّا. . . الأجزاء | ٢٢٣ DFY: كيفيّتين | ٢٢٤ ADMY: والذي | ٢٢٥ $A^{h}M^{h}$: + الله وليّ التّوفيق] 'هذا الموضع الذي أثبت على الحاشية من نسخة آخرى من الجوامع، وهو الذي أوّله الأشياء التي بها يتمّ الفضيلة في هيئة الأعضاء المتشابهة الأجزاء، وفي هيئة الأعضاء الآليّة هو في الفصّ هكذا، وأجناس الآفات هي أجناس الأشياء التي تتمّ بها فضيلتها، وهي في الأعضاء المتشابهة الأجزاء المزاج، وفي الأعضاء الآليّة العدد والمقادير والخلق والوضع والاتّصال مشترك بينهما، وفي هذه الأجناس بأعيانهما [M: بأعيانهما] تكون آفة الأبدان السّقيمة على أي المعنيين اللذين ينتظمهما هذا الاسم [Ga: + فقد فهمت أمرها [Ga: أمرهما] .

حاشية له: هذا الكلام في الفصّ موجز واضح، وهو شاهد النّسخة [A: للنّسخة] التي أثبت الفضل منها على الحاشية لأنّ [A: لأنّه] في تلك النّسخة أبدل لفظة المقدار بلفظة الهيئة في الكلام بأسره، والهيئة يمكن بوجه مّا أن تدلّ بها جملة على أحوال الأعضاء الآليّة بأسرها التي ذكرت في الفصّ مفصّلة ، وبدل [A: فقيل] العدد والمقادير [A: للعدد والمقدار] والخلق والوضع، فأمّا ما في متن، فنسختي التي كتبت على حاشيتها وفي عدّة نسخ كان في جميعها مكان الهيئة المقادير فقط، فإنّ لفظة المقادير لا تدلّ دلالة الهيئة التي قد يكفي في إجمال ما فصّله [A: + الفصّ] بقوله العدد والمقادير والخلق والوضع لأنّ لفظة المقادير المستعملة على الجميع هي أحد تفاصيل ذلك الجميع، فليعتمد على ما في الحاشية، أعني الذي أبدل فيه لفظة المقادير بلفظة الهيئة لصحّة دلالته وشهادة الفصّ له.

٢٢٦ F: يتمّ بها؛ D: تتمّ | ٢٢٧ AM: بعضها يجري مجرى الجنس > منها ما هي الأجناس

second, moderation in quantity.[60] There are also two genera of the things by which defects occur in the tissues and organs: first, the temperament and, second, the size.[61] There are also two things that are species: first, deviation from moderation in the temperament and, second, deviation from symmetry in the structure. Moreover, the organs need to admit of continuity in order to have an excellent form. By this I mean both the tissues, such as nerves, arteries, and veins; and the compound organs, such as the head, arm, and leg.

(24) The difference between the sickly and the diseased person is that the functions of the sickly person are not perceptibly impaired, whereas the impairment of the functions of the diseased person is perceptible.[62] The difference between the sickly person and the person who is usually healthy is that the sickly person is predisposed to fall victim to the causes of disease and his functions are usually impaired, whereas

---

60. Some MSS read "form." A gloss appears in two MSS: "This passage, which has been inscribed in the margin from another manuscript of the epitomes, is that which begins 'Of the things by which the tissues and organs acquire excellence in their forms' and is as follows in [Galen's] original text [B 4.6–7; K 1:315–16; G[a] 27]: 'The genera of defects are the same as the genera of the things by which [the body] acquires excellence. They are the temperament in the tissues and the number, dimensions, structure, and position. Continuity is common to both [tissues and organs]. The defects of diseased bodies fall into these same genera in whichever of the two meanings of the word you understand.'

"A gloss to this: This explanation in the original text is concise and clear and corroborates the manuscript whose reading is recorded in the margin because, in that manuscript, the word 'dimension' is replaced with the word 'form' throughout the discussion. 'Form' can, in a certain sense, indicate in general all the states of the functional organs that are mentioned in detail in the original text. It replaces 'number,' 'dimensions,' 'structure,' and 'position.' As for what is in this text—in my manuscript upon whose margin I am writing and in a number of other manuscripts—there is always 'dimensions' in place of 'form,' for the word 'dimensions' does mean something like 'form' in such a way that it could sum up what is specified in detail [in Galen's text] by the terms 'number,' 'quantities,' 'form,' and 'position.' That is because the word 'quantities,' when used for all of them, is also one of the items within that whole. Thus, reliance ought to be placed on what is in the gloss—that is, the one in which the word 'quantities' is replaced with the word 'form'—due to its correct denotation and the corroboration of the underlying text."

The problem is textual. While the majority of the MSS read *miqdār* ("size" or "dimension"), some record *hayʾah* ("form") as a variant, which the writer of the gloss—probably Ibn al-Tilmīdh—prefers since it can include the four categories of number, dimensions, structure, and position listed in Galen's text.

61. Some MSS read "form."

62. B 4.7–12; K 1:316–18; G[a] 27–31: Galen is trying to establish criteria to distinguish degrees of health and disease while avoiding conflating fragile health

المزاج والآخر المقدار[٢٢٨]، وبعضها يجري مجرى النّوع، وهما شيئانِ[٢٢٩]، أحدهما الاعتدال في المزاج والآخر الاعتدال في المقدار[٢٣٠]، والأشياء أيضاً التي من قِبَلها تدخل الآفة على[٢٣١] الأعضاء المتشابهة الأجزاء، والآليّة[٢٣٢] منها ما هي أجناس وهما شيآنِ أحدهما المزاج، والآخر المقدار[٢٣٣]، ومنها ما هي أنواع، وهما أيضاً شيئانِ أحدهما الخروج عن اعتدال المزاج والآخر الخروج عن استواء التّركيب[٢٣٤]، والأعضاء أيضاً تحتاج إلى أن يسلم لها اتّصالها[٢٣٥] لتكون هيئتها الهيئة الفاضلة، أعني المتشابهة الأجزاء منها، وهي العصب والعروق الضّوارب وغير الضّوارب وما أشبه ذلك، والمركّبة، وهي الرّأس واليد والرّجل وما أشبه ذلك.

(٢٤) الفرق بين المسقام والمريض أنّ المسقام ليس ما بأفعاله من المضارّ محسوس[٢٣٦] والمريض مضارّ أفعاله محسوسة، والفرق بين المسقام والذي هو[٢٣٧] صحيح على الأكثر أنّ المسقام[٢٣٨] مستعدّ ليقهر[٢٣٩] من[٢٤٠] الأسباب المرضيّة[٢٤١] وأفعاله[٢٤٢] مضرورة أكثر، والذي هو صحيح على الأكثر ليس هو بمستعدّ لينقهر[٢٤٣] الأسباب المرضيّة[٢٤٤] وأفعاله[٢٤٥]

---

٢٢٨ D^1؛ A^2DM^2Y: الهيئة | ٢٢٩ F: شيئَيْنِ | ٢٣٠ D^1؛ A^2DY: الهيئة | ٢٣١ DFY: أيضاً. . . على > التي يتمّ بها آفة | ٢٣٢ F: والأعضاء الآليّة | ٢٣٣ $A^2D^2M^2Y^2$: الهيئة | ٢٣٤ M^2: الهيئة؛ DY: + الهيئة | ٢٣٥ F: – اتّصالها؛ S: اتّصاله | ٢٣٦ AM: محسوساً | ٢٣٧ F^1: المسقام والذي هو > الممراض والمصحّ وهو الذي قال فيه إنّه | ٢٣٨ F^1: الممراض | ٢٣٩ ADMY: ينقهر | ٢٤٠ F^1: للانقهار عن | ٢٤١ ADF^1M^1Y: الممرضة | ٢٤٢ F^1: وتضرّ أفعاله | ٢٤٣ ADY: + من؛ F^1: للانقهار عن | ٢٤٤ ADY: الممرضة؛ M^1: الممرضة، خ يعني البادية | ٢٤٥ F^1: وتضرّ أفعاله

the one who is usually healthy is not predisposed to fall victim to the causes of disease and his functions are less impaired. The degrees between the most excellent degree of health and the extreme[63] degree of disease are the following:

<First>	<Second>	<Third>	<Fourth>	<Fifth>
One of the extremes, health that is in the extreme of excellence	After that, the lower levels of health	Then intermediate health, being the state of the one who is neither healthy nor diseased	Then the degrees of sickliness, which are the states of the healthy when he is in health but sickly, which is bad	The other extreme, established disease, which is the worst of all

(25) The bodies in whose functions some defect occurs can have that happen to them either by the entire loss of the function—as when organs lose sensation and motion—or by a deficiency in the function. If the deficiency is small in quantity, it is very difficult to diagnose; but if it is considerable, then it is easy to diagnose. The function may also be improper, in which case the diagnosis is easy, as with organs in which pain or convulsions occur. Impairments occurring in the body are of

with actual disease. In particular, he wishes to avoid the doctrine of ἀειπάθεια, the theory that we are constantly diseased to some greater or lesser degree. While this may be true in some sense, it is of no practical value to the practicing physician; cf. *On the Temperament* 3, K 1:676–77. Either the epitomist or the translator has introduced a terminological distinction that does not exist in the corresponding passage of Galen's text between "diseased" (νοσώδης, *marīḍ*) and "sickly" (*misqām*). In place of "sickly," Galen has "neither" (οὐδέτερος, *laysa . . . wa-lā*), which is not a very satisfactory term to describe bodies prone to disease. Galen does, however, mention that "diseased" (νοσώδης) has two senses (B 4.7; K1:316; G[a] 27).

63. Two MSS have the gloss: "Another gloss by him. If what is in the text is corrected, then its interpretation is that the most excellent class of health is part of the excellence, and the most excellent class of sickness is from the excess—that is, the superfluity." The problem is that some MSS read, literally: "the most excellent level of disease." The author of the gloss points out that the root *F-Ḍ-L* means both "excellence" and "excess." The table reflects Galen's categories at B 4.10–12; K 1:317–18; G[a] 30–31.

مضرورة أقلّ، وفيما بين أفضل[٢٤٦] طبقات الصّحّة وأقصى[٢٤٧] طبقات المرض مراتب على هذا المثال:[٢٤٨]

‹الأوّل[٢٤٩]›	‹الثّاني[٢٥٠]›	‹الثّالث[٢٥١]›	‹الرّابع[٢٥٢]›	‹الخامس[٢٥٣]›
أحد الطرفين الصّحّة التي هي[٢٥٤] في غاية[٢٥٥] الفضيلة	وبعد ذلك مراتب الصّحّة الدّون	ثمّ الصّحّة الوسطى، وهي حال من ليس بصحيح ولا مريض[٢٥٦]	ثمّ مراتب السّقم وهي حالات الأصحّاء، صحّة مسقامة وذلك رديء	والطّرف الآخر المرض المستحكم، وهو أردئ الجميع

(٢٥) الأبدان التي يحدث بها في أفعالها[٢٥٧] آفة لا تخلوا من أن يكون ذلك يعرض لها إمّا عند ذهاب الفعل جملة[٢٥٨] بمنزلة الأعضاء التي يبطل حسّها وحركتها[٢٥٩] وإمّا عند ما ينقص الفعل، ونقصانه إن كان يسير المقدار فتعرّفه يكون عسراً شاقّاً[٢٦٠]، وإن كان كثيراً فتعرّفه يكون سهل، وإمّا عند ما يجري الفعل مجرى رديّاً منكراً[٢٦١] وتعرّف ذلك يكون سهلاً[٢٦٢] بمنزلة الأعضاء التي يحدث بها وجع أو تشنّج. المضارّ الحادثة

---

٢٤٦ M^{1}: في يعني من الفضيلة | ٢٤٧ $DF^{1}SY$: أفضل؛ D^{1}: أقصى؛ $A^{h}M^{h}$: يعني من الزّيادة، أي الفضل؛ $A^{h}M^{h}$: حاشية له، إن يصحّ [A: صحّ] ما في المتن، فتأويله أنّ أفضل طبقات الصّحّة هو من الفضيلة وأفضل طبقات المرض هو من الفضل، أي الزّيادة. | ٢٤٨ DY: Table arranged as continuous text. $A^{h}M^{h}$: حاشية له: الجيّد أن يكتب في الثّالث ثمّ الوسطى، وهي حال من ليس بصحيح ولا مريض إلاّ أنّ ذكره في الرّابع أنّه حال الأصحّاء صحّته [A^{h}: صحّة] مسقامة يجوز أن يكون كلّ ما ذكر غير الخامس داخلة في صحّةٍ مّا. | ٢٤٩ DFSY: -الأوّل | ٢٥٠ DFSY: -الثّاني | ٢٥١ DFSY: -االثّالث | ٢٥٢ DFSY: -الرّابع | ٢٥٣ DFSY: -الخامس | ٢٥٤ AMY: -هي | ٢٥٥ F: الغاية من | ٢٥٦ DY: + وذلك وسط | ٢٥٧ F: بها في أفعالها > بأفعالها | ٢٥٨ F^{1}؛ S: + وهذا ظاهر | ٢٥٩ AM: أو حركتها | ٢٦٠ F: شاقّ؛ S: -شاقّاً | ٢٦١ S: رديّ؛ DY: رديّ منكراً | ٢٦٢ F: سهل

two kinds. First, the impairment may be concealed from the senses and exist by nature. In that case, the body is not said to be diseased. Second, the impairment may be perceptible. In this case, the body is said to be of intermediate health if the impairment is slight—that is, the body is neither healthy nor diseased. If it is greater than that, the body is said to be in a sickly state of health. If the impairment is even greater than that, the body is said to be diseased.

(26) Some bodies that fall short of the most excellent state do so only slightly, others more so, others yet more, and still others by a great deal. Likewise, some signs also fall slightly short of indicating the optimal form, others more so, others yet more, and still others fall short by a great deal.[64] The combination of the levels of these states results in the following.

First, the state of healthy bodies in the most excellent health
Second, the state of healthy bodies whose health is less than excellent
Third, the state of bodies that are neither healthy nor diseased
Fourth, the state of the sickly body
Fifth, the state of the mildly diseased body
Sixth, the state of the body in the extreme of disease

64. Some MSS include the gloss: "A gloss of his: The addition of the digression about signs is not acceptable—not because it is not true, but, rather, because the discussion here is confined to bodies." But the author of the gloss is wrong, since Galen discusses both subjects; see B 4.7–5.1; K 1:316–18; G[a] 27–32.

في البدن ضربانِ، أحدهما أن تكون المضرّة تخفى عن الحسّ وتوجد في الطّبع، وإذا كان ذلك لم يقل إنّ البدن مريض، والآخر أن تكون المضرّة موجودة حسّاً، وإذا كانت كذلك فإنّها إن كانت مضرّة يسيرة قيل إنّ ذلك البدن صحيح صحّة وسطاً، أي ليس هو بصحيح ولا مريض إن كانت أكثر من ذلك[٢٦٣]، قيل إنّ البدن صحيح صحّة مسقامة، وإن كانت أكثر من ذلك أيضاً قيل إنّه مريض.

(٢٦) والأبدان النّاقصة عن أفضل الهيآت، بعضها ينقص نقصاناً يسيراً، وبعضها نقصانه[٢٦٤] أكثر، وبعضها أكثر من ذلك أيضاً، وبعضها نقصانه[٢٦٥] أكثر بكثير، وكذلك[٢٦٦] أيضاً العلامات بعضها تنقص عن الدّلالة على الهيأة الفاضلة مقداراً يسيراً، وبعضها نقصانه عن ذلك أكثر، وبعضها أكثر من ذلك، وبعضها كثير النّقصان جدّاً، وتصنيف مراتب هذه الحالات تجري على هذا مثال[٢٦٧]:

‹آ›	‹ب›	‹ج›	‹د›	‹ه›	‹و›
أوّلها حال الأبدان الصّحيحة[٢٦٨] صحّة[٢٦٩] في غاية الفضيلة	والثّاني حال الأبدان التي صحّتها[٢٧٠] دون الفضيلة[٢٧١]	والثالث حال الأبدان التي ليست بصحيحة ولا مريضة[٢٧٢]	والرّابع حال البدن الذي هو[٢٧٣] المسقام	والخامس حال البدن المريض مرض[٢٧٤] دون	والسّادس حال من هو في غاية المرض

٢٦٣ DFY: + كثيراً | ٢٦٤ AM: نقصاناً | ٢٦٥ AM: نقصاناً | ٢٦٦ $A^{h}M^{h}$: حاشية له: لا استصوب إضافة هذه الزّيادة في العلامات لا لأنّها ليست بحقّ، ولكن لأنّ الكلام هاهنا مقصور على الأبدان. | ٢٦٧ S^{2}؛ S: – مقداراً. . .مثال؛ F: + وهي ستّة ADMY: Table of six rows with abjad numerals | ٢٦٨ DFY: – الأبدان الصّحيحة | ٢٦٩ DY: الصّحّة حال صحّة؛ F: الصّحّة | ٢٧٠ ADMY: التي صحّتها > الصّحيحة صحّة | ٢٧١ F: – الفضيلة | ٢٧٢ DFY: بصحيّة ولا مرضيّة | ٢٧٣ ADMY: – الذي هو | ٢٧٤ AS: مرضاً

We have already described in a general way how to diagnose sickly bodies, but now we will describe in detail how to diagnose each of the organs individually. The sickly body may be sickly absolutely, which is what we mention here; or it may be sickly at the present moment, which we will mention later.[65]

65. Perhaps p. 118, paragraph 73, below.

وجملة الأمر في تعرّف الأبدان المسقامة[٢٧٥] قد وصفناها فيما تقدّم، فأمّا على التّفصيل، أعني[٢٧٦] تعرّف كلّ واحد من الأعضاء على الانفراد، فنحن نصفه ها هنا، فالبدن المسقام لا يخلوا من أن يكون مسقاماً مطلقاً[٢٧٧]، وهو هذا[٢٧٨] الذي نذكره في هذا الموضع أو مسقاماً[٢٧٩] في الوقت الحاضر، وهو الذي نذكره في آخر الأمر.

---

٢٧٥ M^1 : + أمّا الجنسيّ منها؛ S: + أمّا الجنسيّ | ٢٧٦ AM: بمعنى | ٢٧٧ AM: مطلق | ٢٧٨ DSY: – هذا | ٢٧٩ DFY: مسقام

[Chapter 5][66]

[The genera of the organs]

(27) There are four genera of organs in the body. First, there are the principal organs, which are the brain, liver, heart, and testicles. Then there are the organs subordinate to the principal organs. These are the nerves, which are subordinate to[67] the brain; the arteries, which are subordinate to the heart; the veins, which are subordinate to the liver; and the sperm ducts, which are subordinate to the testicles. There are also the organs that have, of themselves, innate faculties. These are the bones, cartilage, ligaments, soft flesh, and so forth. Finally, there are organs that have both innate faculties and faculties reaching them from other organs, such as the arms, legs, trunk, and the like.

---

66. B 5.2–3; K 1:319–20; G[a] 32–35: This chapter is an introduction to the main part of the book, which deals with the diagnosis and treatment of the diseases of the major organs.

67. Literally, "dependents and servants of."

[الفصل الخامس]

[ذكر أجناس الأعضاء]

(٢٧) أجناس ما في البدن من الأعضاء أربعة، وذلك أنّ منها رئيسة[٢٨٠]، وهي الدّماغ والقلب والكبد[٢٨١] والأنثيان[٢٨٢]، ومنها خول وخدم للرّؤساء، وهي العصب، وهي[٢٨٣] خول وخدم للدّماغ، والعروق الضّوارب، وهي خول وخدم للقلب، والعروق الغير الضّوارب[٢٨٤]، وهي خول وخدم للكبد، وأوعية المني، وهي خول وخدم للأنثيين، ومنها ما لها في أنفسها قوى غريزيّة، وهي العظام والغضاريف[٢٨٥] والرّباطات واللّحم الرّخو وما أشبه ذلك، ومنها ما لها قوى غريزيّة وقوى تجري إليها من غيرها، وهي اليدانِ والرّجلانِ والصّدر وما أشبه ذلك.

---

٢٨٠ A: ما هي رئيسيّة؛ DY: رؤساء؛ D¹M: ما هي رئيسة؛ S: أعضاء رئيسة | ٢٨١ AM: والكبد والقلب | ٢٨٢ FS: والأنثيين | ٢٨٣ S: أنّ الأعصاب هي | ٢٨٤ S¹؛ ADMY: التي ليست بضوارب؛ S: الضّوارب؛ S¹: غير الضّوارب | ٢٨٥ DY: + والأغشية

[Chapter 6][68]

[The diagnosis of the brain]

(28) There are five genera of signs indicating the states of the brain essentially. There are also accidental signs, such as the speed—or lack thereof—with which injury caused by external causes occurs. The first of the five essential genera is the condition of the head; the second is the condition of the sensory functions; the third is the condition of the motor functions; the fourth is the state of the deliberative—that is, the governing—functions; and the fifth is the state of the natural functions. The condition of the head embraces three things: first, its size; second, its shape; and, third, its hair. In size, the head may be either large or small. Its shape may be either well formed or badly formed. The hair has three aspects: first, its size; second, its shape; and, third, its color. The size of hair may be either coarse or fine in texture. Its shape may be curly or straight. Its color may be fire-colored, red, blond, white, or black. There are five sensory functions: vision, hearing, smell, taste, and touch. The motor functions are those brought to completion by the muscles.

68. B 6.1; K 1:319–20; G[a] 35–36: A close summary of Galen's text, except that Galen mentions "change coming about as a result of external influences," instead of accidental signs. The epitomist's reference to accidental signs probably comes from B 6.11; K 1:323; G[a] 41, where Galen mentions that "if the brain is well balanced in respect to the four qualities, it . . . will be very little harmed by any external influence—things which heat, cool, dry, or moisten."

[الفصل السّادس]

‹الاستدلال على الدّماغ[٢٨٦]›

(٢٨) العلامات الدّالّة على حالات الدّماغ بعضها جوهريّة، وهي خمسة أجناس، وبعضها عرضيّة بمنزلة ما يتّفق له أن يكون تسرع إليه المضرّة من الأسباب التي تحدث من خارج أو لا تسرع إليه[٢٨٧]، فأمّا الخمسة الجوهريّة[٢٨٨]، فأحدها حال الرّأس والثّاني حال الأفعال الحسّاسة والثّالث حال الأفعال المحرّكة والرّابع حال الأفعال السّياسيّة،[٢٨٩] المدبّرة[٢٩٠]، والخامس حال الأفعال الطّبيعيّة، وحال الرّأس تجمع ثلثة أشياء، أحدها مقداره والثّاني شكله والثّالث شعره، أمّا مقدار الرّأس فإنّه إمّا أن يكون كبيراً وإمّا أن يكون[٢٩١] صغيراً، وأمّا شكله فإنّه إمّا أن يكون حسن الشّكل وإمّا أن يكون رديّ الشّكل، وأمّا شعره فيلزمه[٢٩٢] ثلثة أشياء، أحدها مقداره والثّاني شكله والثّالث لونه، أمّا مقدار الشّعر فإنّه لا يخلوا من أن يكون إمّا غليظ الطّاقة وإمّا دقيق الطّاقة، وأمّا شكله فإنّه لا يخلوا من أن يكون إمّا جعداً وإمّا سبطاً، وأمّا لونه فإنّه يكون إمّا بلون النّار وإمّا أحمر وإمّا أشقر وإمّا أبيض وإمّا أسود. وأمّا[٢٩٣] الأفعال الحسّاسة، فهي خمسة، البصر والسّمع والشّمّ والمذاق[٢٩٤] واللّمس، وأمّا الأفعال المحرّكة[٢٩٥]، فهي الأفعال التي تتمّ بالعضل[٢٩٦]، وأمّا الأفعال السّياسيّة،

---

٢٨٦ ADMSY: –الاستدلال على الدّماغ | ٢٨٧ F: إليها؛ S: + كالحرارة والبرودة | ٢٨٨ AM: الجوهريّة الخمسة | ٢٨٩ ADMY:+أي | ٢٩٠ F: –المدبّرة | ٢٩١ S: –أن يكون؛ DY: وإمّا أن يكون> أو | ٢٩٢ ADMY: ففيه | ٢٩٣ DFY: فأمّا | ٢٩٤ DSY: والذّوق | ٢٩٥ DY: الإراديّة؛ D[1]Y[1]: المحرّكة | ٢٩٦ F: + والحركة الإراديّة؛ DY: + المحرّكة

The deliberative—that is, the managing—functions are imagination, thought, and memory. The natural functions are the absorption of nutriment, its retention, its coction, and the expulsion of what remains of it.

(29) The head may be either large or small.[69] Under all conditions, a small head indicates that the structure and form of the brain are poor. A large head caused by abundant matter and a weak faculty indicates that the structure and form of the brain are poor. When its size is caused by the amount of matter and a healthy faculty, then a large head indicates that the structure and form of the brain are excellent. A large head has signs and indications by which its condition can be diagnosed. Did it occur due to the quantity of matter combined with weakness of the faculty, or was it because of the quantity of matter combined with a healthy faculty? These signs and indications are in the things that grow from it and from its shape. In the case of things growing from it, if the nerves, spinal cord, and neck are extremely thick, then the cause of its size is that there is a great deal of matter and a strong faculty. If the parts we mentioned are thin and weak, the cause of that is the quantity of matter combined with a weakness of the faculty.

(30) As for the indication given by the shape of the head,[70] if the head is well formed, its large size is caused by the large quantity of matter and the strength of the faculty. If it is badly shaped, the cause is the quantity of matter combined with a weakness in the faculty. The head whose large size indicates that the structure and form of the brain are excellent can be identified by the things that originate from it and from its shape—from the things that originate from it if the neck is strong,

69. B 6.2–3; K 1:320; G[a] 36–37.
70. B 6.3–6; K 1:320–21; G[a] 37–38.

أي[٢٩٧] المدبّرة، فهي التّخيّل والفكر[٢٩٨] والذكر، وأمّا الأفعال الطّبيعيّة، فهي جذب الغذاء وإمساكه وإنضاجه ودفع ما يبقى منه.

(٢٩) والرّأس يكون إمّا كبيراً و إمّا صغيراً[٢٩٩]، والصّغير يدلّ على كلّ حال أنّ بنية الدّماغ[٣٠٠] وهيئته[٣٠١] رديّة، وأمّا الكبير فإنّه إن كان السّبب في كبره كثرة المادّة وضعف القوّة، فهو يدلّ على أنّ الدّماغ رديّ البنية والهيئة، فإن[٣٠٢] كان السّبب في ذلك كثرة المادّة وصحّة القوّة، فهو يدلّ على أنّ بنية الدّماغ وهيئته جيّدة فاضلة، وللرّأس الكبير علامات ودلائل يتعرّف بها أمره، هل عرض له ذلك بسبب كثرة المادّة مع ضعف من[٣٠٣] القوّة أو[٣٠٤] بسبب كثرة المادّة مع صحّة[٣٠٥] من القوّة، وهذه العلامات والدّلائل تكون في[٣٠٦] الأشياء التي تنبت منه ومن شكله، أمّا من الأشياء التي تنبت منه فإنّه إن[٣٠٧] كان العصب والنّخاع والعنق غليظاً قويّاً[٣٠٨]، فالسّبب في كبره أنّ المادّة كانت كثيرة والقوّة قويّة، وإن كان ما سمّيناه[٣٠٩] دقيقاً ضعيفاً، فالسّبب في ذلك كثرة المادّة مع ضعف من[٣١٠] القوّة.

(٣٠) وأمّا الدّلالة من شكل الرّأس فإنّه إن كان الرّأس حسن الشّكل، فالسّبب في كبره أنّ المادّة كانت كثيرة والقوّة قويّة، وإن كان رديّ الشّكل، فالسّبب في ذلك كثرة المادّة مع ضعف من القوّة، والرّأس الكبير الذي يدلّ على أنّ الدّماغ فاضل البنية والهيئة تعرف من الأشياء التي منشأها منه ومن شكله، أمّا من الأشياء

---

٢٩٧ DFY: – أي | ٢٩٨ F: في الفكر؛ DY: والتّفكّر | ٢٩٩ ADMY: صغيراً وإمّا كبيراً | ٣٠٠ FY[1]: الرّأس | ٣٠١ S: بنية؛ Y: وهيئة | ٣٠٢ DFY: وإن | ٣٠٣ S: ضعف من < صحّة | ٣٠٤ AM: أم | ٣٠٥ S: ضعف | ٣٠٦ ADMY: من | ٣٠٧ F: إذا | ٣٠٨ F: غليظ قويّ؛ F[2]: غليظانِ قويّانِ | ٣٠٩ DFY: سمّينا؛ F[1]: ينبت | ٣١٠ S: – من

the arms and legs perfectly shaped, and all of the genera of nerves thick and strong; and from its shape if it is well shaped. The head has two protrusions. The one in the front is because the sensory nerve grows from the front of the brain. The other, which is in the back, results from the motor nerve and spinal cord growing from the back of the brain. Each of these two protrusions can be deficient or excessive. If it is deficient and small, that is due either to a deficiency in the matter, in which case the deficiency is less serious than otherwise, or it is due to a weakness of the faculty, which is more harmful. They are both diagnosed from things whose origin is in the brain. The excess is when the head is narrow, which is either because of an abundance of matter combined with a weakness of the faculty, which is harmful, or because of an abundance of matter combined with a healthy faculty, which is excellent.

(31) There are five seams, or sutures, in the head.[71] Three of them are true sutures, and two of them are not. One of the three true sutures is the coronal suture, which runs from the front of the brain to where it reaches the crown of the head.[72] The second is the skewerlike suture, which is the one that bisects the head vertically in a straight line from front to back.[73] The third [the lambdoid suture] resembles the Greek letter lambda and is the suture that is in back, in this shape: Λ. The two that are not true sutures are on the sides of the head and are called the

71. B 6.6–7; K 1: 321–22; G[a] 38–40; cf. pseudo-Galen, *Introduction to Medicine*, K 14:720: Galen alludes to the lambdoid suture while discussing the relation of the brain and the spinal cord.

72. It runs from the temples, slanting slightly backwards over the top of the skull, and defines the frontal bone.

73. Several MSS have the gloss: "A gloss of his: They give the name 'skewerlike' to the combination of the sagittal [arrowlike] and lambdoid sutures, but the one that divides the head vertically into two halves is the sagittal suture." Our name "sagittal" comes from the Latin word for arrow.

التي منشأها منه[٣١١]، فإن[٣١٢] يكون العنق قويًّا، واليدانِ والرّجلانِ بأفضل هيئة، وجميع أجناس العصب غليظة قويّة، وأمّا من شكله، فبأن يكون حسن الشّكل، وللرّأس نتوآنِ، أحدهما من قدّام، وذلك لأنّ العصب الحسّيّ ينبت من مقدّم الدّماغ، والآخر من خلف، وذلك لأنّ العصب المحرّك والنّخاع ينبت من مؤخّر الدّماغ، وكلّ واحد من هذينِ النّتوينِ ينقص ويزيد[٣١٣] إلاّ أنّ نقصانه وصغره تكون إمّا من قِبَل نقصان في المادّة، وهذا النّقصان أقلّ رداءةً من غيره، وإمّا من قِبَل ضعف من القوّة، وهذا أعظم شرًّا، وجميعاً يتعرّفانِ من الأشياء التي منشأها من الدّماغ، وأمّا[٣١٤] الزّيادة، فهي[٣١٥] أن يصير[٣١٦] الرّأس مسفّطاً، فيكون إمّا بسبب كثرة المادّة مع ضعف من القوّة، وذلك رديّ، وإمّا بسبب كثرة المادّة مع صحّة من القوّة، وذلك أفضل.

(٣١) شؤون الرّأس، وهي دروزه، خمسة، ثلثة منها هي دروز بالحقيقة، واثنانِ ليسا دروزه[٣١٧] بالحقيقة، أمّا الثّلثة الحقيقيّة[٣١٨]، فواحد منها الدّرز[٣١٩] الإكليليّ[٣٢٠]، وهو من مقدّم الدّماغ حيث يوضع الإكليل من الرّأس، والآخر السّفوديّ، وهو الذي يقطع الرّأس في طوله بنصفينِ على الاستقامة من قدّام إلى خلف، والثّالث الشّبيه باللاّم في حروف اليونانيّين، وهذا[٣٢١] الدّرز الذي من خلف على هذا المثال ٨[٣٢٢]، وأمّا[٣٢٣] الاثنانِ اللذانِ ليسا بحقيقيّينِ، فهما على[٣٢٤] جنبتيَ الرّأس، ويقال لهما

٣١١ F: + وأمّا من شكله فبأن يكون حسن الشكل؛ S: – منه | ٣١٢ ADSY: فبأن | ٣١٣ F: + وعند نقصانه يكون صغيراً وعند زيادته يكون عظيماً | ٣١٤ DY: + من | ٣١٥ A: وهو؛ A^2: فهو؛ DMY: وهي | ٣١٦ A^2F^2S: يكون | ٣١٧ A^2M^2: ظ دروزاً؛ DY: دروز | ٣١٨ A^1؛ AM: بالحقيقة | ٣١٩ DF: الدّروز | ٣٢٠ S: الأعلى | ٣٢١ ADMY: وهو | ٣٢٢ F: – ٨؛ DSY: < | ٣٢٣ AM: فأمّا | ٣٢٤ AM: عن

two squamous sutures because they connect bone to bone, one on the right side and the other on the left.[74] This is a diagram of the five sutures. The brain is divided into two parts in accordance with the suture resembling a lambda. The front part of it is called the anterior of the brain, and the back part is called the posterior of the brain.

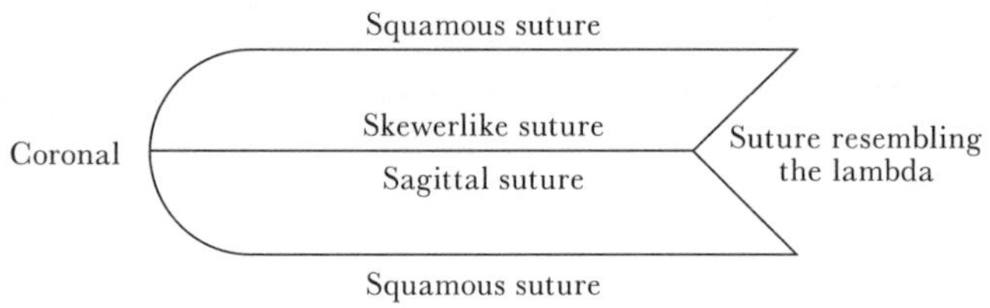

(32) Nerves are either motor or sensory.[75] The motor nerves must originate in the posterior of the brain, for many motor nerves grow from there, but few sensory nerves. The sensory nerves must originate in the anterior of the brain, since many sensory nerves grow from there, but few motor nerves.

(33) There are three deliberative—that is, managing—functions: imagination, thought, and memory.[76] Each of these three has a desirable excellence and an undesirable defect. Thus, imagination's excellence is accepting knowledge quickly. This is one of the indications that the anterior part of the brain is receptive and quick to be imprinted,

74. They separate the temporal bones around the ear and get their name from the fact that they resemble scales.

75. B 6.7; K 1:321–22; G^a 39–40.

76. B 6.9–10; K 1:322; G^a 40–41: Galen describes intellectual virtues and vices, but less systematically and without specific references to imagination and thought.

الدّرزيْنِ القشريّيْن لأنهما التزاق[٣٢٥] عظم بعظم، واحد من الجانب الأيمن، والآخر من الجانب الأيسر، وهذا مثال الخمسة الدّروز[٣٢٦]:

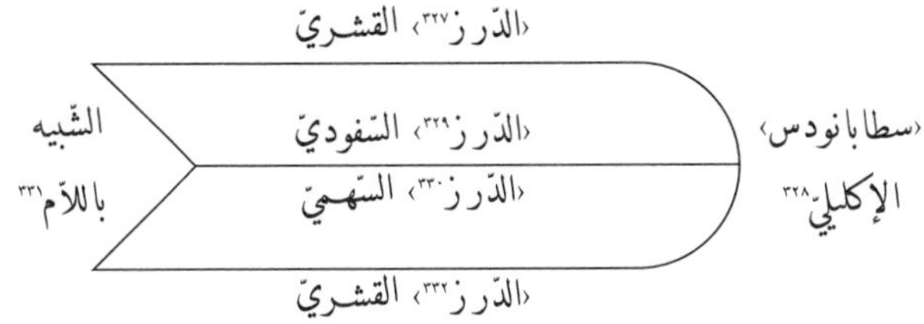

الدّماغ[٣٣٣] مقسوم بجزئيْنِ[٣٣٤] حيث الدّرز الشّبيه باللاّم والجزء المقدّم منه يقال له مقدّم الدّماغ، والجزء الخلف منه[٣٣٥] يقال له مؤخّر الدّماغ.

(٣٢) العصب[٣٣٦] منه محرّك ومنشأه من مؤخّر الدّماغ لأنّ هذا الجزء ينبت منه عصب محرّك كثير وعصب حسّاس يسير، ومنه حسّاس ومنشأه من مقدّم الدّماغ لأنّ هذا الجزء ينبت منه عصب حسّاس كثير وعصب محرّك يسير.

(٣٣) الأفعال السّياسيّة، أي المدبّرة، ثلثة، وهي التّخيّل والفكر والذّكر، ولكلّ واحدة من هذه الثّلثة فضيلة تحمد وآفة تذمّ. ففضيلة التّخيّل سرعة قبول العلم[٣٣٧]، وهذا ممّا يدلّ على أنّ الجزء المقدّم من أجزاء الدّماغ سريع الانطباع والقبول لأنّ

---

٣٢٥ S: التصاق | ٣٢٦ M^h: + خ ط ب؛ $A^hD^hM^hY^h$: حـ[اشية] له، إنّما يسمّى السّفّوديّ مجموع السّهميّ واللاّميّ، فأمّا القاسم للرّأس طولا بنصفيْنِ [D^hY^h: نصفيْنِ طولاً] فقط، فهو السّهميّ. | ٣٢٧ DFSY: – الدّرز | ٣٢٨ AM: سطابانودس الإكليليّ؛ F: – الإكليليّ | ٣٢٩ DFSY: – الدّرز | ٣٣٠ DFSY: – الدّرز | ٣٣١ S^h: في كتابة اليونانيّين | ٣٣٢ DFSY: – الدّرز | ٣٣٣ S: والدّماغ | ٣٣٤ F: + أحدهما مقدّمه والآخر مؤخّره تقسامه | ٣٣٥ ADMY: – منه | ٣٣٦ F: أفعال العصب هي الحسّ والحركة والعصب | ٣٣٧ F^2: التعلّم

since imagination is in this particular part of the brain. The most suitable temperament for that is the moderately moist temperament, since it is impossible for something that is dry and hard to be imprinted quickly and easily receive the form, nor is it possible for this to happen with something that is extremely moist. What is moderate and between these two extremes is best. The defect of imagination—its bad state—is difficulty in receiving instruction. This is one of the things that indicates that the anterior part of the brain is hard and is imprinted and receives the form only with difficulty. Hardness is a consequence of one of two things, either dryness or coldness. It is most likely, in this context, to be associated with dryness. The excellence and perfection of thought is subtlety of mind and quick-wittedness. These are among the things that indicate the subtlety of the psychic spirit. The subtlety of the psychic spirit depends upon attaining complete coction. The attainment of complete coction is a result of moderate heat and limited moisture. The defect of thought—its bad state—is the backwardness and slowness of understanding. These indicate coarseness of the psychic spirit. Coarseness of the psychic spirit results from one of two things: either excessive moisture or coldness. The excellence of memory is ease of recollection, which indicates that the posterior part of the brain is dry. That is where memory is, and memory requires something that has stability and permanence, attributes of the dry thing. The defect of memory—its bad state—is forgetfulness, which indicates that the posterior portion of the brain is moister than it ought to be. Something that is moist runs and flows, having neither rest nor permanence.

التّخيّل إنّما يكون بهذا الجزء من أجزاء الدّماغ، وأوفق الأمزاج لذلك المزاج المعتدل الرّطوبة إذ كان ليس يمكن سرعة الانطباع وسهولة قبول الصّورة في الشّيء اليابس الصّلب، ولا في الشّيء الكثير الرّطوبة، بل في الشّيء المعتدل فيما بينهما، وآفة التّخيّل وسوء حاله عسر القبول للتّعليم[٣٣٨]، وذلك ممّا يدلّ على أنّ الجزء المقدّم[٣٣٩] من أجزاء الدّماغ عسر الانطباع والقبول للصّورة لأنّه صلب، والصّلابة[٣٤٠] تابعة لأحدى[٣٤١] أمرَينِ[٣٤٢] إمّا لليبس[٣٤٣] وإمّا[٣٤٤] للبرد[٣٤٥]، والأولى[٣٤٦] بها في هذا الموضع أن تكون تابعة لليبس، وأمّا الفكر ففضيلته وكماله لطافة الذّهن وسرعة الفهم، وذلك ممّا يدلّ على لطافة الرّوح النّفسانيّ[٣٤٧] ولطافة الرّوح النّفسانيّ[٣٤٨] تابعة لبلوغ الغاية في النّضج، وبلوغ الغاية في النّضج تابعة[٣٤٩] لاعتدال الحرارة وقلّة الرّطوبة، وآفة الفكر وسوء حاله التّخلّف والإبطاء في الفهم، وذلك ممّا يدلّ على غلظ الرّوح النّفسانيّ[٣٥٠]، وغلظ الرّوح النّفسانيّ[٣٥١] تابع لأحد أمرَينِ إمّا لكثرة الرّطوبة وإمّا للبرودة. وأمّا الذكر ففضيلته[٣٥٢] جودة الحفظ، وذلك يدلّ على يبس[٣٥٣] الجزء المؤخّر من أجزاء الدّماغ لأنّ الذكر إنّما يكون هناك، والحفظ يحتاج إلى شيء له ثبات وبقاء، واليابس هو على هذه الصّفة، وآفة[٣٥٤] الذّكر وسوء حاله النّسيان، وذلك[٣٥٥] يدلّ على انّ الجزء المؤخّر من أجزاء الدّماغ أرطب ممّا ينبغي، وإذا كان الشّيء رطباً فهو سيّال جارٍ لا لبث له ولا بقاء.

---

٣٣٨ F: التّعليم؛ S: للتّعلّم | ٣٣٩ S: المؤخّر | ٣٤٠ M: وللصّلابة؛ F: الصّلابة | ٣٤١ DFY: لأحد | ٣٤٢ M: الأمرَين؛ S: لأمرَينِ | ٣٤٣ AD^{1}MY: لليبوسة؛ F: اليبس | ٣٤٤ D^1؛ DY: أمّا | ٣٤٥ ADMY: للبرودة؛ | ٣٤٦ D: والأوّل؛ F: + والأولى | ٣٤٧ AM: النفسيّ | ٣٤٨ DSY: – النّفسانيّ | ٣٤٩ ADMY: تابع | ٣٥٠ AM: النفسي | ٣٥١ F^1؛ F: وغلظ. . . النّفسانيّ؛ AM: النّفسيّ؛ | ٣٥٢ DF: ففضيلة | ٣٥٣ DFY: يبوسة؛ DY: + في | ٣٥٤ F: آفة | ٣٥٥ F: + ممّا

[Chapter 7][77]

[The moderate temperament of the brain]

(34) Rapid changes of opinions and lack of stability in resolution are among the things that indicate that the temperament of the brain is hot. The proof of this is the condition of children. Stability of resolution indicates that the temperament of the brain is cold, as is proven by the condition of the old. If the temperament of the brain is moderate, then its states are moderate, whether in the sensory functions of vision, hearing, smell, taste, and touch; in the voluntary functions, which are all the motions; in the deliberative or governing functions, which are imagination, thought, and memory; or in the natural functions, which are known by the superfluities evacuated through the nose, the two apertures in the palate, and the ears. It is not quickly harmed by external things that it encounters, such as the things that heat, cool, moisten, or dry. Moreover, if a person's brain is moderate in temperament, his hair will be inclined to be the color of fire when he is a baby, light red when he is a child, and when he is an adult, reddish-blond[78] midway between straight and curly—providing that he lives in a country with a moderate climate. Hair is black and curly due to either the heat of the brain, the heat of the country, or the heat of the humors and their dominance by bile, the result of the liver's heat.

---

77. B 6.11–7.6; K 1:323–26; G[a] 41–52.

78. B 6.12; K 1:323; G[a] 42: A gloss in several MSS reads, "[Galen's] original text says, 'When they reach maturity, their hair is blond inclining to red.' Many MSS of the epitomes read 'reddish-blond'"—which all of ours do. Galen's Greek is ὑπόξανθος—"*yellowish* or *lightish brown*," according to Liddell-Scott-Jones.

[الفصل السّابع]

[ذكر مزاج الدّماغ المعتدل]

(٣٤) وممّا يستدلّ به على أنّ مزاج الدّماغ حارّ سرعة الاستحالة إلى الآراء وقلّة الثّبات على العزيمة، والدّليل على ذلك حال الصّبيان، وممّا يستدلّ به على أنّ مزاج الدّماغ بارد الثّبات على العزيمة، والدّليل على ذلك حال الشّيوخ. إذا كان مزاج الدّماغ معتدلاً كانت أحواله معتدلة في الأفعال الحسّاسة[٣٥٦]، وهي البصر والسّمع والشّمّ والمذاق واللّمس ، وفي الأفعال الإراديّة، وهي الحركات أجمع[٣٥٧] وفي الأفعال السّياسيّة[٣٥٨] المدبّرة، وهي التّخيّل والفكر والذّكر، وفي الأفعال الطّبيعيّة، وهي التي تعرف بالفضول التي تستفرغ من الأنف ومن ثقبَي الحنك ومن الأذنيْن، ويكون لا يسرع إليه المضارّ من الأشياء التي تلقاه من خارج بمنزلة الأشياء التي تسخن أو تبرّد أو ترطّب أو تيبّس[٣٥٩]، وإذا كان الدّماغ أيضاً[٣٦٠] معتدل المزاج كان شعره في وقت ما يكون الإنسان طفلاً يضرب إلى لون النّار وفي وقت ما يكون صبيّاً يضرب إلى الحمرة النّاصعة، وفي وقت المنتهى والكمال يكون شعره أصهب أشقر[٣٦١] وسط[٣٦٢] فيما بين السّبط والجعد، وذلك في البلد[٣٦٣] المعتدل. الشّعر يكون جعداً أسود إمّا بسبب حرارة الدّماغ وإمّا بسبب حرارة البلد وإمّا بسبب حرارة الأخلاط وغلبة المرار عليها، وذلك تابع لحرارة الكبد.

---

٣٥٦ AM: الحسّيّة؛ S: الحاسّة | ٣٥٧ DSY: –أجمع | ٣٥٨ M: السّياسة؛ AM: +أي | ٣٥٩ M: يبس؛ $A^2F^2M^2S$: تجفّف؛ | ٣٦٠ DSY: أيضاً الدّماغ | ٣٦١ $A^hM^hY^h$: حاشية له: في الفصّ: وإذا صار إلى حال التّمام، فإنّ شعره يصير أشقر إلى الحمرة وفي كثير من نسخ الجوامع أصهب أشقر. | ٣٦٢ DSY: وسطاً | ٣٦٣ S: الوقت؛ S^2: البلد؛ DY: البدن

[Chapter 8][79]

[Immoderate temperaments of the brain]

(35) If the brain ceases to have a moderate temperament, its temperament will either diverge slightly from the natural temperament or diverge greatly from it. The sign of a slight divergence is weak and hidden; but if the divergence is large, its sign is strong and obvious. The brain can have eight kinds of bad temperaments, just as is the case for the kinds of bad temperament in the other organs. There are four singular, simple kinds and four compound kinds. The singular, simple kinds are the hot, the cold, the moist, and the dry. The compound kinds are the hot and dry, the hot and moist, the cold and dry, and the cold and moist. The divergence of each of these kinds from the natural state may be slight, in which case their signs are not obvious; or their divergence may be great, in which case their signs are obvious.

79. B 6.13–14; K 1:323–24; G^a 42–44.

[الفصل الثّامن]

[ذكر أمزجة الدّماغ الغير المعتدلة]

(٣٥) وإذا كان الدّماغ زائلاً عن المزاج المعتدل، فليس يخلوا مزاجه الزّائل من أن يكون إنّما زال عن المزاج الطّبيعيّ زوالاً يسيراً أو يكون قد زال عنه زوالاً كبيراً، فإن كان زواله يسيراً، كانت علامته ضعيفة خفيّة، وإن كان زواله كبيراً كانت علامته قويّة بيّنة. وأصناف المزاج الرّديّ من[٣٦٤] مزاج[٣٦٥] الدّماغ ثمانية كمثل ما عليه أصناف المزاج الرّديّ في سائر الأعضاء، منها أربعة أصناف مفردة بسيطة، ومنها أربعة مركّبة، أمّا البسيطة المفردة[٣٦٦]، فالحارّ والبارد والرّطب واليابس، وأمّا المركّبة فالحارّ اليابس والحارّ الرّطب والبارد اليابس والبارد الرّطب، وجميع هذه الأصناف إمّا أن يكون زوالها عن الحال الطّبيعيّة زوالاً يسيراً وتكون علاماتها غير بيّنة، وإمّا أن يكون زوالها كثيراً وتكون علاماتها بيّنة.

---

٣٦٤ F^2S: الغالب على | ٣٦٥ F: أمزاج | ٣٦٦ AM: المفردة البسيطة

[Chapter 9][80]

[The temperament of the eye]

(36) It is possible to tell whether the temperament of the eye is hot or cold by its motions, the condition of the blood vessels in it, and the way that the eye feels to the touch. It is possible to tell whether the temperament of the eye is moist or dry by the fact that, if the temperament is moist, the eye will be soft to the touch and will be filled with moisture. If its temperament is dry, the eye will be hard and will be dry and dull. There is a general sign for the eye by which it is possible to tell whether some quality is in excess in it, just as it is possible to tell this about any other organ. This is that things similar to the quality, if in excess, harm the eye; and the things contrary to the quality help it. The condition of the size of the eye can be known from its structure and its functioning. A large eye with good structure indicates that the matter from which it is composed is abundant and of moderate temperament. If it is large but its structure is not good, that indicates that the matter is abundant but that it is not moderate. As for its functioning, if it functions perfectly well, that indicates that the matter from which it is composed is excellent. A deficiency in its function indicates that there is something wrong with its temperament. The condition of a small eye can also be discovered from its structure and its functioning. As for its structure, if it is small and its shape is good, that is an indication that the matter from which it is composed is slight in quantity but that it is excellent and of moderate temperament. If it is small and its shape is not good, that indicates that the matter is slight in quantity and its

80. B 9.1–3; K 1:329–30; G[a] 52–53: The epitome skips over several pages in which Galen discusses how to diagnose the eight deficient temperaments of the brain, but the epitomist's account of the eye is much more elaborate than Galen's; see appendix 2, pp. 203–8, below.

[الفصل التّاسع]

ذكر مزاج العين

(٣٦) الذي يستدلّ به على مزاج العين أحارّ هو أم بارد حال العين في حركاتها[٣٦٧] وحال العروق التي فيها وحال ما يتبيّن للمس منها، والذي يستدلّ به على مزاجها أرطب هو أم يابس[٣٦٨]، أنّ المزاج الرّطب تكون معه العين[٣٦٩] ليّنة الملمس وتكون مملوءة رطوبة، واليابس تكون معه العين صلبة وتكون يابسة جافّة، وللعين علامة عامّيّة يستدلّ بها على[٣٧٠] كيفيّة تفرط عليها كمثل ما يستدلّ به على كلّ عضو آخر، أي عضو[٣٧١] كان، وهي أنّ الأشياء المشبهة[٣٧٢] للكيفيّة المفرطة عليها تضرّها، والأشياء المخالفة لها تنفعها. الحال في كبر العين تعرف من خلقتها ومن فعلها، أمّا من خلقتها فإنّها إن كانت كبيرة وكانت خلقتها حسنة دلّ ذلك على أنّ المادّة التي منها خلقت كانت كثيرة معتدلة المزاج، وإن كانت كبيرة ولم تكن خلقتها حسنة دلّ ذلك على أنّ المادّة كانت كثيرة إلّا أنّها لم تكن متعدلة، وأمّا من فعلها فإنّها إن كانت تفعل فعلها على التّمام حسناً دلّ ذلك على أنّ المادّة التي منها خلقت كانت جيّدة، وإن كان في فعلها تقصير دلّ ذلك على سوء مزاجها، وكذلك الحال أيضاً في صغر العين تعرف من خلقتها ومن فعلها، أمّا من خلقتها فإنّها إن كانت صغيرة وكان شكلها حسناً كان ذلك دليلاً[٣٧٣] على أنّ المادّة التي منها خلقت كانت يسيرة إلّا أنّها كانت[٣٧٤] معتدلة المزاج جيّدة[٣٧٥]، فأمّا إن كانت صغيرة ولم يكن شكلها حسناً، فذلك منها يدلّ على أنّ المادّة

٣٦٧ A¹M¹: ذكائها؛ F: حركته | ٣٦٨ AM: أيبس | ٣٦٩ DSY: العين معه | ٣٧٠ DSY: + كلّ | ٣٧١ ADMY: الأعضاء | ٣٧٢ DM²Y: المتشابهة | ٣٧٣ DFY: دليل | ٣٧٤ AM: + جيّدة | ٣٧٥ AM: - جيّدة

temperament bad. As for its functioning, if it functions well, that is an indication that the matter of which it is composed is excellent. If it does not function well, that indicates that its temperament is bad.

(37) Blue eyes are due to a lack of the aqueous humor found in the eyes, this humor's clarity and purity, the luminosity or large size of the lens, or the eye being protuberant.[81] Blackness in the eye is due to the quantity of the aqueous humor or its coarseness and turbidity, or to the lens not being luminous, its being too small, or its being sunken.

(38) This topic can be divided in another way, as follows: The eye is blue or black by reason of either the aqueous humor or the lens. If it is the aqueous humor, it is because of either its quantity or its quality. If its quantity is small, the eye becomes blue; and if it is large, the eye becomes black. If it is due to its quality, then if it is clear, the eye becomes blue; and if it is coarse and turbid, the eye becomes black. When the cause is the lens, that is due to either its quantity, its quality, or its position. If the cause is its quality, then if it is luminous, the eye will for that reason be blue; but if not, the eye will for that reason be black. If the cause is its quantity, the eye will become blue if it is abundant; but if it is little, the eye will be something other than blue, <which is to say, black.>[82] If the cause is its position, then, if the eye is protuberant, it will be blue; but if

81. B 9.4–5; K 1:330–31; G^{a} 53–55: Two MSS have the gloss: "A gloss of his: One ought to know that the primary cause for blueness and blackness is the color of the uvea [the iris and ciliary body] and that the rest of what he mentions are factors increasing the blueness or blackness."

82. Inserted in some MSS.

كانت يسيرة ومزاجها كان رديًّا، وأمّا من فعلها فإنّها إن كانت تفعل فعلها حسنًا دلّ ذلك على أنّ المادّة التي منها خلقت كانت جيّدة، وإن كانت لا تفعل فعلها حسنًا دلّ ذلك على أنّ مزاجها[٣٧٦] رديًّا[٣٧٧].

(٣٧) الزّرقة[٣٧٨] تكون في العين إمّا بسبب نقصان الرّطوبة الشّبيهة ببياض البيض التي في العين وإمّا بسبب صفائها ونقائها، وإمّا بسبب ضياء[٣٧٩] الرّطوبة الشّبيهة بالجليد، وإمّا بسبب كثرتها[٣٨٠]، وإمّا بسبب أنّها موضوعة ممّا يلي خارج. الكحولة[٣٨١] تكون في العين إمّا بسبب كثرة الرّطوبة الشّبيهة ببياض البيض، وإمّا بسبب غلظها وكدورتها، وإمّا بسبب أنّ الرّطوبة الجليديّة ليست بمضيئة، وإمّا بسبب أنّها صغيرة، وإمّا بسبب أنّها موضوعة[٣٨٢] ممّا يلي داخل.

(٣٨) وقد يقسم هذا المعنى بقسمة[٣٨٣] أخرى على هذه الحكاية. العين تكون زرقاء أو كحلاء[٣٨٤] إمّا بسبب الرّطوبة الشّبيهة ببياض البيض، وإمّا بسبب الرطوبة الشّبيهة بالجليد، ومن أجل[٣٨٥] الرّطوبة الشّبيهة ببياض البيض إمّا لكميّتها وإمّا لكيفيّتها، أمّا بسبب كميّتها فإنّها إن كانت يسيرة صارت العين بها زرقاء، وإن كانت كثيرة صارت العين بها كحلاء، وأمّا بسبب كيفيّتها فإنّها إن كانت صافية صارت العين بها زرقاء، وإن كانت غليظة كدرة صارت العين بها كحلاء، وأمّا بسبب الرّطوبة الجليديّة فيكون ذلك إمّا لكمّيّتها[٣٨٦] وإمّا لكيفيّتها[٣٨٧] وإمّا لوضعها، أمّا بسبب كيفيّتها فإنّها إن كانت

٣٧٦ ADMY: + كان | ٣٧٧ ADMY: رديّ | ٣٧٨ A^hM^h: حاشية [له]: ينبغي أن تعلم أنّ أولى الأسباب بالزّرقة والكهولة هو لون العنبيّة وسائر ما ذكره ها هنا هو من مؤكّدان الزّرقة والكهولة. | ٣٧٩ FY: صفاء | ٣٨٠ A^2DFM^2Y: كبرها | ٣٨١ F: والحكومة؛ DF^1Y: والكحولة | ٣٨٢ S^2: أنّ وضعها غايره ؟ | ٣٨٣ S: – بسبب أنّها صغيرة. . . بقسمة؛ مضاف في S^2؛ S^2: على قسمة | ٣٨٤ F: زرقها أو كحكتها؛ DY: وكحلاء | ٣٨٥ AM: ومن أجل> وسبب؛ DY: فسبب | ٣٨٦ ADMY: لكيفيّتها | ٣٨٧ $ADMY^1$: لكمّيّتها؛ Y: – وإمّا لكميّتها

it is sunken, it will be black. The finer the aqueous humor and the greater its quantity, the more moist the eye is.[83] The coarser it is and the less its quantity, the drier the eye will be. This is also the case with the lens, which can differ and change either with respect to temperament or with respect to texture. If the difference and change so alters the temperament as to make it dry, the eye will become dry. If they make the temperament moister, then the eye will become moist. If the difference and change so alters the texture that it becomes finer than it ought to be, the eye will then become moist. If they alter the texture so that it becomes coarser than it ought to be, the eye will become dry.

(39) This same topic can be divided in yet another way, as follows: The eye is moist or dry by reason of the aqueous humor or by reason of the lens. This can be caused by the quantity of the aqueous humor or by its quality—that is, its texture. If the quantity is large, the eye will thereby become moist; but if it is small in quantity, the eye will thereby become dry. If the quality and texture are fine, the eye will thereby become moist;

83. B 9.6; K 1:331; G[a] 55.

مضيئة تصير[٣٨٨] العين بها زرقاء، وإن لم تكن مضيئة تصير[٣٨٩] العين بها كحلاء، وأمّا بسبب كمّيّتها فإنّها[٣٩٠] إن كانت كبيرة تصير[٣٩١] العين[٣٩٢] زرقاء، وإن كانت صغيرة تصير[٣٩٣] العين[٣٩٤] غير زرقاء[٣٩٥]، وأمّا بسبب وضعها، فإنّها إن كانت موضوعة ممّا يلي خارج صارت العين زرقاء، وإن كانت موضوعة ممّا يلي داخل صارت العين كحلاء. كلّما كانت الرّطوبة الشّبيهة ببياض البيض أرقّ وأكثر ممّا ينبغي كانت العين أرطب، وكلّما كانت أغلظ وأقلّ كانت العين أيبس، وكذلك الرّطوبة الشّبيهة بالجليد، وقد[٣٩٦] تختلف وتتغيّر[٣٩٧]، إمّا من طريق مزاجها وإمّا من طريق قوامها، واختلافها وتغيّرها من طريق المزاج إن كان ميلها إلى اليبس تصير العين[٣٩٨] يابسة، أو ميلها إلى الرّطوبة تصير العين[٣٩٩] رطبة[٤٠٠]، وأمّا اختلافها وتغيّرها من طريق القوام، فإنّه إن كان ميلها إلى الرّقّة بأكثر[٤٠١] ممّا ينبغي تصير[٤٠٢] العين بها[٤٠٣] رطبة، وميلها إلى الغلظ بأكثر[٤٠٤] ممّا ينبغي تصير[٤٠٥] العين[٤٠٦] يابسة.

(٣٩) وقد يقسم هذا المعنى بعينه بقسمة أخرى على هذه[٤٠٧] الحكاية. العين تكون رطبة أو يابسة، إمّا بسبب الرّطوبة الشّبيهة ببياض البيض وإمّا بسبب الرّطوبة الجليديّة. أمّا بسبب الرّطوبة الشّبيهة ببياض البيض، إمّا لكمّيّتها وإمّا لكيفيّتها، أعني

---

٣٨٨ ADMY: صارت؛ F: + كانت | ٣٨٩ F^{1}S؛ ADMY: صارت | ٣٩٠ F^{1}: فإن؛ S: - فإنّها | ٣٩١ F^{1}S؛ ADFMY: صارت | ٣٩٢ ADM: + بها ؛ F: - مضيئة . . . زرقاء | ٣٩٣ ADMY: صارت | ٣٩٤ ADMY: + بها | ٣٩٥ AM: + أي كحلاء؛ DF^{1}Y: كحلاء؛ DY: - غير كحلاء | ٣٩٦ ADY: قد | ٣٩٧ ADM: + وتغيّرها؛ Y: وتتغيّر | ٣٩٨ ADMY: ميلها. . . العين> يخرج بها إلى أن يكون أيبس ممّا ينبغي صارت العين بها | ٣٩٩ ADMY: أو. . . العين> وإن كان يخرج بها إلى الرّطوبة بصير العين بها | ٤٠٠ DY: + وإن كان يخرج بها إلى أن يكون أيبس صارت العين بها يابسة | ٤٠١ ADMY: خرج بها إلى أن تكون أرقّ | ٤٠٢ ADMY: صارت | ٤٠٣ F: مضيئة؛ SY: - بها | ٤٠٤ ADMY: وإن خرج بها [DY: + اكثر ممّا ينبغي] إلى أن يكون أغلظ | ٤٠٥ ADMY: صارت | ٤٠٦ ADMY: + بها | ٤٠٧ ADMY: على هذه> بهذه

but if they are coarse, the eye will thereby become dry. If it is caused by the lens, then it is due to either its temperament or its texture. If the temperament is dry, the eye will thereby become dry; and if it is moist, the eye will become moist. If its texture is coarse, that is the cause of the eye becoming dry; but if the texture is fine, the eye thereby becomes moist.

The structure of the eye[84]

(40) The eye is composed of three fluids and two layers. The first of the three fluids resembles liquid glass [the vitreous humor], the second resembles ice [the crystalline humor], and the third resembles egg white [the aqueous humor]. The first of the two layers originates from the solid covering [dura mater] of the brain, and the second from the delicate covering [pia mater]. The portion of the layer that originates

84. Galen's text does not have an account of the anatomy of the eye. In the previous paragraphs, I have rendered these terms using standard anatomical terms rather than the descriptive terms that I have translated literally here. "Fluids" are *ruṭūbāt*, meaning "moistures" or "humors," and are not to be confused with *akhlāṭ*, indicating the four humors. "Layers" are *ṭabaqāt*, meaning "cloaks" in older terminology. See appendix 2, pp. 203–8, for a detailed account of the confusing terminology used here for the anatomy of the eye.

قوامها. أمّا[٤٠٨] من طريق كمّيتها، فإنّها إن كانت كبيرة المقدار صارت العين بها رطبة، وإن كانت يسيرة المقدار صارت العين بها[٤٠٩] يابسة. وأمّا من طريق كيفيّتها، وقوامها[٤١٠]، فإنّها إن كانت رقيقة صارت العين بها رطبة، وإن كانت غليظة صارت العين بها يابسة. وبسبب الرّطوبة الجليديّة، إمّا لمزاجها وإمّا لقوامها، أمّا من طريق مزاجها، فإنّها إن كانت يابسة صارت العين بها يابسة، وإن كانت رطبة صارت العين بها رطبة، وأمّا من طريق قوامها فإنّها إن كانت غليظة صارت العين[٤١١] يابسة، وإن كانت[٤١٢] رقيقة صارت العين بسببها[٤١٣] رطبة.

ذكر تركيب العين

(٤٠) العين مركّبة من ثلث رطوبات وطبقتين، إمّا الثّلث الرّطوبات[٤١٤]، فالواحدة[٤١٥] منهنّ شبيهة[٤١٦] بالزّجاج الذّائب[٤١٧]، والأخرى شبيهة[٤١٨] بالجليد، والثّالثة شبيهة[٤١٩] ببياض البيض[٤٢٠]، وأمّا[٤٢١] الطّبقتان، فأحدهما[٤٢٢] منشأها من الغشاء الصّلب من غشاء الدّماغ، والأخرى[٤٢٣] من الغشاء الرّقيق[٤٢٤]، والطّبقة[٤٢٥] التي منشأها من الغشاء

٤٠٨ F: فأمّا؛ S: وأمّا | ٤٠٩ DSY: –بها | ٤١٠ AM: أعني قوامها | ٤١١ ADMY: + بسببها | ٤١٢ AM: + ليّنة | ٤١٣ S: –بسببها؛ DY: –يابسة. . . بسببها | ٤١٤ DFY: رطوبات | ٤١٥ F: فواحدة | ٤١٦ ADMY: الشّبيهة | ٤١٧ S^{l}: إلى الدّماغ يميل | ٤١٨ ADMY: الشّبيهة | ٤١٩ ADMY: الشّبيهة | ٤٢٠ S^{h}: إلى القدام يميل | ٤٢١ DFY: فأمّا | ٤٢٢ AM: فواحدة منهنّ؛ S: فإحداهما؛ DY: فالواحدة منهما | ٤٢٣ F: والثّانية | ٤٢٤ M: الدّقيق | ٤٢٥ $A^{h}M^{h}Y^{h}$: حاشية له: ذكر الطّبقات [A^{h}: + فيه] هاهنا ليس في الفصّ، وذلك عساه [Y^{l}: لأنّه] آخذ من شرح الفصّ، والمسمّى [$A^{h}Y^{h}$: + فيه] الشّبكة [$A^{h}Y^{h}$: الشّبكيّة هاهنا هو [Y^{h}: هي] المسمّى في كتاب حنين الطّبقة الصّلبة، ولفظ الشّبكة في كتاب حنين [M^{h}: –الطّبقة. . . حنين] اسم لنفس العصبة [A^{h}: العضلة] الجوفاء الباصرة إذا اتّسعت [Y^{h}: انبعثت] في حجاج العين واحتوت على نصف الجليديّة من خلف احتواء الشّبكة [Y^{h}: الشّبكيّة] على الصّيد.

from the solid membrane that is behind the lens is called the layer resembling a net [retina, but actually the sclera] because that part of it contains all that is in the eye, like a net.[85] That which is in front of the lens is called the layer resembling horn [cornea] because of the fineness and clarity of this part. The part of the layer growing from the fine membrane that is behind the lens is called the layer resembling the placenta [choroids] due to the large number of arteries and veins in this part. The portion of it that is in front of the lens is called the layer resembling a grape [uvea, which is the combination of the iris and the ciliary body] because this part of it is similar to a single grape. The covering connected to the layer resembling horn, which is the one that originates from the covering above the skull, is a sort of protection and mantle for these things that are behind it. In Greek, it is called ἐπιπεφυκώς [the conjunctiva].

85. Several MSS have the gloss: "A gloss of his: The layers here are not in [Galen's] underlying text, but [the epitomist] might have taken them from commentaries on the text. What is here called the retina is called 'the hard layer' [sclera] in Ḥunayn's book. The word 'retina' in Ḥunayn's book is used for the hollow optic nerve, since it spreads out around the ball of the eye and contains half of the crystalline moisture from behind, embracing it like a net on its prey." See appendix 3, pp. 209–11.

الصّلب يسمّى[٤٢٦] ما هو منها من وراء الرّطوبة الجليديّة الطّبقة الشّبيهة بالشّبكة لأنّ ذلك الجزء منها يحتوي على جميع ما في العين بمنزلة الشّبكة، وأمّا ما هو منها من قدّام[٤٢٧] الجليديّة، فيسمّى الطّبقة الشّبيهة بالقرن[٤٢٨]، وذلك لرقّة هذا الجزء وصفائه. فأمّا الطّبقة التي[٤٢٩] من الغشاء الرّقيق، فيسمّى ما هو منها من وراء الرّطوبة الجليديّة الطّبقة الشّبيهة بالمشيمة[٤٣٠]، وذلك لكثرة ما في هذا الجزء من العروق الضّوارب وغير الضّوارب، وأمّا ما هو منها من قدام الجليديّة، فيسمّى الطّبقة الشّبيهة بالعنبيّة[٤٣١] لأنّ هذا الجزء منها شبيه بحبّة العنب، وأمّا[٤٣٢] الغشاء المتّصل بالطّبقة الشّبيهة بالقرن، وهو الذي منشأه من الغشاء الذي فوق القحف، فإنّما هو بمنزلة الوقاية[٤٣٣] واللّباس لهذه الأشياء التي وراءه، ويقال له باليونانيّة افيفافوقوس[٤٣٤].

٤٢٦ F: ليس | ٤٢٧ M: قوام | ٤٢٨ AM: بالفرق؛ F: بالقرنيّ | ٤٢٩ AM: + تنبت | ٤٣٠ M^{1}؛ A: المشيمية؛ M: – بالمشيمة | ٤٣١ AM: بالعنبة؛ $F^{1}Y$: العنبيّة؛ S: الطّبقة. . . بالعنبة > العنبية | ٤٣٢ A: فأمّا؛ DY: أمّا | ٤٣٣ $A^{2}M^{2}S$: الواقية | ٤٣٤ A^{1}: افيفاقوس؛ D: افيفافيوس؛ F: افيفافوفوس؛ M^{1}: ظ أفينافوس؛ S: أفنفافوس؛ Y: افيغافيوس

[Chapter 10][86]

<The temperament of the heart>

(41) The heart has eight species of bad temperament, just like the species of bad temperament in the rest of the organs. Four of these eight are simple and singular, and four compound. The simple species are heat, coldness, dryness, and moisture. The compound species are heat and dryness, heat and moisture, coldness and dryness, and coldness and moisture.

(42) There are three species of signs indicating that the bad temperament of the heart is hot: first, the species of signs peculiar to this temperament and inseparable from it; second, the species of signs not peculiar to this temperament and not inseparable from it; and, third, the species of the signs intermediate between these two.[87] The signs peculiar to and inseparable from the hot temperament are a large volume of breath and a rapid and continuous pulse. The signs that are not peculiar to this temperament and are separable from it are irascibility and a broad chest; for the philosophical dispositions can alter

86. B 10.1–11.7; K 1:331–37; G^a 56–64: Galen begins by reminding the reader that, when we speak of a greater heat, coldness, dryness, or moisture of a part, these terms are relative to that part, not to some other object. Thus, if a heart is comparatively cold by nature, its mixture will still be much hotter than that of the hottest brain. Galen then discusses these eight temperaments in turn. The epitome follows him closely. The chapter title is found in only one of the old MSS.

87. B 10.1–6; K 1:332–33; G^a 56–59: Galen mentions only the first of these three categories, contrasting it to signs indicating heat in the body as a whole. The following gloss found in two MSS explains the intermediate category not mentioned as such by Galen, though he does mention the examples: "A gloss by him: He thinks it best to place intermediate signs between the signs that are peculiar to this temperament and inseparable from it and those that are not peculiar to it but are still inseparable from it. These are the association of the entire body with the heart and the quantity of hair on the chest. These two signs are correlates of their cause—which is the temperament of the heart—in the same way that the first class of signs are, except that they are not continuously apparent, since it is possible that the temperament of the liver interferes, counteracting the temperament of the heart with respect to these two signs, thereby concealing what they indicate in a way similar to the signs that are not peculiar to this temperament and which may be separated from it. Due to the existence

[الفصل العاشر]

‹ذكر مزاج القلب[٤٣٥]›

(٤١) أنواع سوء مزاج القلب ثمانية كمثل ما يكون عليه أنواع سوء مزاج كلّ واحد من سائر الأعضاء، ومن هذه الثّمانية أربعة مفردة بسيطة وأربعة مركّبة، أمّا البسيطة المفردة فالحرارة والبرودة واليبوسة[٤٣٦] والرّطوبة[٤٣٧]، وأمّا المركّبة فالحرارة واليبوسة والحرارة والرّطوبة والبرودة واليبوسة والبرودة والرّطوبة[٤٣٨].

(٤٢) والعلامات الدّالّة على سوء مزاج القلب الحارّ ثلثة أنواع، أحدها نوع العلامات الخاصّيّة بهذا المزاج الحارّ التي لا يزايله، والثّاني[٤٣٩] نوع العلامات التي ليست بخاصّيّة لهذا[٤٤٠] المزاج ولا غير مفارقة له، والثّالث نوع العلامات المتوسّطة فيما بين هذينِ. أمّا العلامات[٤٤١] الخاصّيّة التي لا يزايل المزاج الحارّ، فعظم التّنفّس وسرعة النّبض وتواتره، وأمّا العلامات التي ليست بخاصّيّة لهذا المزاج ولا غير مفارقة له، فالغضب وسعة الصّدر، أمّا الغضب فإنّه ينتقل ويتغيّر[٤٤٢] عمّا يوجبه هذا المزاج

---

٤٣٥ DF¹Y؛ AFMS: – ذكر مزاج القلب | ٤٣٦ DFY: – واليبوسة | ٤٣٧ F: + اليبس؛ DY: + واليبوسة | ٤٣٨ S: – وأمّا. . . الرّطوبة؛ S²: وأربعة مركّبة وهو الحرارة مع الرّطوبة والحرارة مع اليبس والبرودة مع الرّطوبة والبرودة مع اليبس | ٤٣٩ ADMY: والآخر | ٤٤٠ AM: بهذا | ٤٤١ AʰMʰ: حاشية له: إنّما استخار أن يجعل بين العلامات الخاصّيّة التي لا تزايل وبين العلامات التي ليست بخاصّيّة ولا غير مفارقة علامات متوسّطة بينهما، وهي مشاركة البدن كلّه للقلب وكثرة الشّعر في الصّدر لأنّ هاتينِ العلامتينِ يلزمانِ سببهما، أعني مزاج القلب لزوم القسم الأوّل من العلامات إلاّ أنّ ظهورهما لا يدوم بسبب إمكان معارضة مزاج الكبد المخالف لمزاج القلب لهما فليخفى ما يدلاّنِ عليه تشبهانِ العلامات غير الخاصّيّة التي قد يفارق ولوجود ملزومهما، وهو المزاج القلبيّ وتباعهما له تشبهانِ العلامات الخاصّيّة التي لا تزايل خلا أنّ هذه المتوسّطة وإن لم تزايل فقد يزايل ظهورها بسبب اعتراض خلاف المزاج من المبدأ الآخر، أعني الكبد. | ٤٤٢ ADMY: يتغيّر وينتقل

irascibility and change what this temperament would normally necessitate, and the size of the brain can affect the breadth of the chest and change it from what this temperament would ordinarily necessitate.[88] The intermediate signs that are between these two species of signs are the association in this temperament of the entire body with the heart and a hairy chest. That is because these two signs are changed by coldness and moisture in the liver. When we say that the breadth of the chest is altered by the size of the brain, we mean that, if the brain is large in size, there is the correlate effect that the spinal cord is also large, since it originates in the brain. A large spinal cord requires the vertebrae that contain it to be large; and if the vertebrae are large, the ribs linked with them are necessarily large. If the ribs are large, the chest that is composed from them must be large and broad. A broad chest is thus a consequence of one of three things: either the heat of the heart, the large size of the brain, or both together. The signs indicating that the temperament of the heart is cold are a weak pulse, a narrow chest, extreme timidity, a hairless chest, and coldness in the entire body.[89] The signs indicating that the temperament of the heart is dry are a hard pulse, a brutish character,[90] and dryness of the entire body.[91] The signs indicating that the temperament of the heart is moist are a soft pulse, anger that is easily roused and quickly placated, and moisture in the entire body, provided that it is not counteracted by the liver.[92]

---

of what is correlate to them—which is the temperament of the heart and the fact that they follow it—they resemble the signs that are peculiar to that temperament and are inseparable from it. However, in the case of these intermediate signs, even though they are inseparable, their appearance may be separable due to the interference of the contrary temperament from another source—by which I mean the liver." See p. 94, paragraph 45, below. The point is that both heat in the heart and heat in the liver cause overall heat in the body and hair on the chest, so that heat or cold in one can imitate or mask heat or cold in the other.

88. B 10.7; K1:336–37; G[a] 64. On "philosophical dispositions" see p. 92, n. 98, below.

89. B 10.7; K 1:333–34; G[a] 59–60.

90. Three MSS contain the gloss: "Another MS reads: 'of a bestial temperament.' Understand that this means without hasty anger but if roused, slow to be placated." Cf. B 11.1; K 1:335; G[a] 61: "The spirit not readily roused, but fierce and implacable."

91. B 10.8; K 1:334; G[a] 60.

92. B 10.9; K 1:334; G[a] 61.

بالأخلاق الفلسفيّة، وأمّا سعة الصّدر فإنّها تتغيّر[٤٤٣] وتخالف ما يوجبه هذا المزاج بمقدار الدّماغ، وأمّا العلامات المتوسّطة فيما بين ذينك النّوعين، فبمشاركة[٤٤٤] البدن كلّه للقلب في هذا المزاج وكثرة الشّعر في[٤٤٥] الصّدر، وذلك أنّ[٤٤٦] هاتين العلامتين تتغيّران ببرودة[٤٤٧] الكبد ورطوبتها، وقولنا إنّ سعة الصّدر تتغيّر بسبب[٤٤٨] مقدار الدّماغ إنّما نريد[٤٤٩] أنّ الدّماغ إذا كان عظيم[٤٥٠] المقدار[٤٥١] لزم من ذلك ووجب عنه أن يكون النّخاع أيضاً عظيم المقدار إذ كان منشأه منه، وإذا كان النّخاع عظيماً وجب أن يكون الفقارات[٤٥٢] المحتوية عليه[٤٥٣] كباراً، وإذا كانت الفقار كباراً وجب أن يكون الأضلاع النّاشئة منها[٤٥٤] كباراً، وإذا كانت الأضلاع كباراً وجب أن يكون الصّدر المؤلّف منها كبيراً واسعاً، وسعة الصّدر تابعة لأحد ثلثة أشياء، إمّا لحرارة القلب وإمّا لعظم مقدار الدّماغ وإمّا لهما جميعاً، والعلامات الدّالّة على مزاج القلب البارد صغر النّبض وضيق الصّدر وإفراط[٤٥٥] الجبن وقلّة الشّعر في[٤٥٦] الصّدر وبرد[٤٥٧] جميع البدن[٤٥٨]. والعلامات الدّالّة على مزاج القلب اليابس صلابة النّبض وسبعيّة[٤٥٩] الخلق ويبس جميع البدن[٤٦٠]، والعلامات الدّالّة على مزاج القلب الرّطب لين النّبض وسرعة الغضب وسهولة سكونه ورطوبة البدن إن لم يقاومه[٤٦١] الكبد.

---

٤٤٣ S: + وينتقل | ٤٤٤ ADMY: فمشاركة | ٤٤٥ AM: على | ٤٤٦ ADMY: لأنّ | ٤٤٧ AM: بمقدار برودة | ٤٤٨ A^2M^2: بحسب | ٤٤٩ ADMY: + به؛ DY: أريد | ٤٥٠ F: أعظم؛ S: عظيماً | ٤٥١ F: مقداراً؛ S: – المقدار | ٤٥٢ AM: الفقار | ٤٥٣ ADMY: الحاوية له | ٤٥٤ ADM: المؤلّفة معها؛ Y: المؤلّفة منها | ٤٥٥ A: وكثرة؛ M: وكثر | ٤٥٦ AM: على | ٤٥٧ AM: وبرودة | ٤٥٨ S: + وصغر النّفس | ٤٥٩ M^1، FM: وسعة؛ $A^hD^hM^hY^h$: وجد في نسخة على سبعيّة الخلق، افهم [وجد. . . فمنهم > D^h: إنّهم؛ Y^h: افهم] قلّة المسارعة إلى الغضب، وأمّا إذا غضبت فيبطء [Y^h؟: فبطي] ما يزول حاشية له. | ٤٦٠ F: + والعلامات الدّالّة على مزاج القلب البارد صغر النّبض وضيق الصّدر وإفراط الجبن وقلّة الشّعر في الصّدر وبرد جميع البدن | ٤٦١ DY: يقاوم

[Chapter 11][93]

[Compound temperaments of the heart]

(43) The signs indicating that the temperament of the heart is hot and dry are a large and hard pulse, a large volume of breath, a large chest, and a daring and violent character,[94] quick anger,[95] and implacability. If they have this character, they are insolent. They also have a large quantity of coarse hair on their chests and are hot and dry in their entire bodies. The signs indicating that the temperament of the heart is hot and moist[96] are a large and soft pulse, irascibility that is quickly placated, a large chest, a large volume of breath, frequent putrid diseases if the moisture is excessive, and a strong heat in the entire body. If this temperament is not counteracted by the liver, the body is moist as well as hot. The signs indicating that the temperament of the heart is cold and moist[97] are a soft, small pulse, excessive timidity, sluggishness, placidity, lack of hair on the chest, and coldness in the entire body. If this temperament is not counteracted by the liver, the body is moist as well as cold. The signs indicating that the temperament of the heart is cold and dry are a small and hard pulse, a small and narrow chest with little hair, shallow respiration, and coldness and dryness in the entire body.[98]

---

93. B 11.1–2; K 1:334–35; G^a 61–62.

94. There are various readings of these words. Two MSS contain the gloss: "He means that dryness is associated with difficulty in being roused to anger but is also associated with difficulty in subduing the anger once it is roused, which is brutishness. This is not false." Galen's own text reads: "Quick to action, spirited, and speedy; fierce, unkind, reckless, shameless; tyrannical in character; bad-tempered and implacable."

95. One MS has the gloss: "Because if heat is dominant, motion is easy."

96. B 11.3–4; K 1:335–36; G^a 62–63.

97. B 11.5; K 1:336; G^a 63.

98. B 11.6–7; K 1:336; G^a 63–64. Galen ends this section by remarking that innate natural characteristics caused by the temperaments are not to be confused with good and bad character traits brought on by philosophical attitudes.

[الصل الحادي عشر]

[ذكر الأمزجة المركّبة للقلب]

(٤٣) والعلامات الدّالّة على مزاج القلب الحارّ اليابس عظم النّبض وصلابته وعظم التّنفّس وسعة الصّدر والجرأة وحدّة[٤٦٢] الأخلاق وسرعة الحركة[٤٦٣] إلى الغضب وإبطاء[٤٦٤] سكونه[٤٦٥]، وإذا كان الخلق كذلك، فهو من أخلاق العتاة وكثرة الشّعر وتكاثفه في الصّدر[٤٦٦] وحرارة جميع البدن ويبسه. والعلامات الدّالّة على أنّ مزاج القلب حارّ رطب عظم النّبض[٤٦٧] ولينه[٤٦٨] وسهولة انبعاث[٤٦٩] الغضب جدّاً وسرعة سكونه[٤٧٠] وكبر الصّدر وعظم التّنفّس وكثرة الأمراض العفونيّة[٤٧١] إن أفرطت الرّطوبة وقوّة[٤٧٢] حرارة جميع البدن، وإن لم يخالف[٤٧٣] الكبد كان البدن مع حرارته رطباً أيضاً[٤٧٤]. والعلامات الدّالّة على أنّ مزاج القلب بارد رطب لين النّبض وضغره وإفراط الجبن والكسل وقلّة الغضب وقلّة الشّعر في الصّدر وبرد[٤٧٥] جميع البدن، وإن[٤٧٦] لم يخالفه الكبد كان البدن مع برده رطباً. والعلامات الدّالّة على أنّ مزاج القلب بارد يابس صغر النّبض وصلابته وصغر الصّدر وضيقه وقلّة الشّعر في الصّدر وصغر التّنفّس[٤٧٧] وبرد جميع البدن ويبسه.

---

٤٦٢ F:والجرأة وحدّة > والحرارة والنّجدة في؛ S: > ونجدة؛ DY: > والجراة والنّجدة في؛ $A^{1}M^{1}$: والحدة والحرارة؛ $A^{h}D^{h}M^{h}Y^{h}$: يريد أن اليبس يتبعه عسر الافتعال [D^{1}: الانفعال] عن الغضب ويتبعه أيضاً إذا انفعل بعد العسر عن الغضب عسر زوال الغضب [$D^{1}Y^{1}$: – الغضب؛ D^{1}: + فإن سمّي عسر زوال الغضب] سبعيّة فليس بكذب | ٤٦٣ S^{h}: لأنّ الحرارة إذا غلبت سهلت الحركة | ٤٦٤ ADMY: وعسر؛ M: سرعة عسر | ٤٦٥ S^{h}: لأجل اليبوسة | ٤٦٦ AM: + وحدّة الأخلاق | ٤٦٧ S^{h}: بحرارته | ٤٦٨ S^{h}: للرّطوبة | ٤٦٩ ADMY: الخروج إلى | ٤٧٠ F: جموده | ٤٧١ F: العفنة | ٤٧٢ F: الرّطوبة وقوّة > وقوّيت الرّطوبة؛ S: وقويت | ٤٧٣ AM: يخالفه | ٤٧٤ AM: أيضاً رطباً | ٤٧٥ AM: وبرودة | ٤٧٦ DY: فإن؛ F: إن | ٤٧٧ DY^{1}: وضعف النّفس؛ D^{1}: وصغر النّفس؛ S^{h}: لعلّة الطّبع[؟] ولصغر الصّدر

[Chapter 12][99]

<The temperament of the liver>

(44) The signs indicating that the temperament of the liver is hot are thick blood vessels, abundant yellow bile and, at the end of youth, also abundant black bile,[100] hairiness on the soft parts of the belly, and heat throughout the body if it is not counteracted by the heart. The signs indicating that the temperament of the liver is cold are narrow blood vessels, excess phlegm in the blood, lack of hair on the soft parts of the belly, and coldness throughout the body if it is not counteracted by the heart. The signs indicating that the temperament of the liver is moist are softness of the blood vessels, moisture of the blood, and moisture throughout the body. The signs indicating that the temperament of the liver is dry are hardness of the blood vessels, coarseness of the blood, and dryness throughout the body.[101] The signs indicating that the temperament of the liver is hot and dry are thick, coarse hair on the soft parts of the belly,[102] broad and hard blood vessels, scanty and coarse blood, and dryness and heat throughout the body if it is not counteracted by the heart.

---

99. B 12.1–5; K 1:337–38; G^a 64–66: A slightly abridged translation of Galen's text. One old MS contains the gloss: "You ought to know that the pains of the liver can be used to indicate whether the humors are in excess or deficiency. That is because the liver is the source where the humors are generated."

100. One old MS contains the gloss: "Because the more heat corrupts, the more it inflames the humors."

101. One MS adds the gloss: "Because dryness is such as to destroy, transform, and coarsen moisture."

102. One MS adds the gloss: "Because of the vapor generated."

[الفصل الثّاني عشر]

‹ذكر مزاج الكبد[٤٧٨]›

(٤٤) والعلامات الدّالّة على أنّ مزاج الكبد حارّة[٤٧٩] سعة العروق وكثرة المرّة الصّفراء، وفي منتهى الشّباب[٤٨٠] الزّيادة في[٤٨١] السّوداء وكثرة الشّعر في مراقّ البطن وحرارة جميع البدن إن لم يخالفها القلب. والعلامات الدّالّة على أنّ مزاج الكبد بارد ضيق العروق وإفراط البلغم في الدّم وقلّة الشّعر في مراقّ البطن وبرد[٤٨٢] البدن إن لم يخالفها القلب. والعلامات الدّالّة على أنّ مزاج الكبد رطب لين العروق و رطوبة الدّم و رطوبة جميع البدن[٤٨٣]. والعلامات الدّالّة على أنّ مزاج الكبد يابس[٤٨٤] صلابة العروق وغلظ الدّم ويبس جميع البدن[٤٨٥]. والعلامات الدّالّة على أنّ مزاج الكبد حارّ يابس كثرة الشّعر[٤٨٦] وكثافته في مراقّ البطن وسعة العروق وصلابتها وقلّة الدّم وغلظه ويبس جميع البدن وحرارته إن لم يخالفها القلب.

---

٤٧٨ AMS: – ذكر مزاج الكبد؛ D: ذكر الكبد | ٤٧٩ ADMY: الحارّ؛ Sʰ: ينبغي أن تعلم أنّ أوجاع الكبد تستدلّ عليها من الأخلاط في زيادتها ونقصانها، وذاك أنّ الكبد هو معدن تولّد الأخلاط. | ٤٨٠ A: الشّبابيّ؛ Sʰ: لأنّ الحرارة كلّما أفسدت أحرقت الأخلاط | ٤٨١ F¹S؛ ADMY: الزّيادة في> كثرة | ٤٨٢ AM: وبرودة؛ DY: + جميع؛ Sʰ: لضعفهم لقلّة الحرارة | ٤٨٣ ADMY: والعلامات. . . اليابسة. . . والعلامات. . . الرّطبة. . . رطوبة جميع البدن | ٤٨٤ F: اليابسة؛ M: اليابس | ٤٨٥ Sʰ: لأنّ اليبوسة شأنها أن تفني الرّطوبة وتقلبها وتغلظها | ٤٨٦ Sʰ: لأجل البخار المتولّد

(45) The way in which the heart and the liver counteract and oppose each other's temperaments is this:[103] The heat of the heart strongly overcomes the coldness of the liver, and the coldness of the heart overcomes the heat of the liver more weakly. The moisture of the heart does not overcome the dryness of the liver at all, and the heart's dryness overcomes the moisture of the liver only weakly. The heat of the liver overcomes the coldness of the heart weakly, and the moisture of the liver strongly overcomes the dryness of the heart; but the liver's coldness is less powerful in overcoming the heat of the heart. The liver's dryness always overcomes the moisture of the heart.

(46) The signs indicating that the temperament of the liver is hot and moist[104] are the large size of the blood vessels,[105] abundant hair in the soft parts of the belly (though less than is the case with the hot and dry temperament), frequent putrid diseases, and moisture and heat throughout the body. The signs indicating that the temperament of the liver is cold and dry are narrow blood vessels, paucity of blood, scanty hair on the soft parts of the belly, and dryness throughout the body. The signs indicating that the temperament of the liver is cold and moist are narrow blood vessels, extremely scanty hair on the soft parts of the belly, an excess of phlegm in the blood, and moisture throughout the body.

103. B 12.6–7; K 1:338; G^a 66–67: Following Galen's text closely. One old MS has the following table and gloss:

	heat counteracts the coldness of			heat weakly counteracts the coldness of	
	coldness slightly counteracts the heat of			coldness counteracts the heat of	
The heart's	moisture does not counteract the dryness of	the liver.	The liver's	moisture counteracts to an intermediate degree the dryness of	the heart.
	dryness slightly counteracts the moisture of			dryness always counteracts the moisture of	

104. B 12.8–11; K 1:338–39; G^a 68–69.

105. One MS adds the gloss: "Due to the heat, since the heat that occurs with moisture has less effect."

(٤٥) والحال في مخالفة كلّ واحد من القلب والكبد مزاج الآخر ومقاومته إيّاه أنّ حرارة القلب تغلب[٤٨٧] برودة الكبد غلبةً قويّةً[٤٨٨] وبرودته أضعف[٤٨٩] قوّة في غلبة[٤٩٠] حرارتها، ورطوبته لا تغلب[٤٩١] يبسها أصلاً، ويبسه يغلب[٤٩٢] رطوبتها غلبةً[٤٩٣] ضعيفةً[٤٩٤]، وحرارة الكبد تغلب[٤٩٥] برودة القلب غلبةً ضعيفةً[٤٩٦]، ورطوبتها تغلب[٤٩٧] يبسه غلبةً قويّةً[٤٩٨]، وبرودتها أضعف[٤٩٩] قوّة في الغلبة[٥٠٠] لحرارته، ويبسها لا يزال دائماً غالباً[٥٠١] لرطوبته[٥٠٢].

(٤٦) والعلامات الدّالّة على أنّ[٥٠٣] مزاج الكبد حارّ رطب عظم مقادير العروق وكثرة الشّعر في مراقّ البطن إلاّ أنّه على حال أقلّ منه في المزاج الحارّ[٥٠٤] اليابس وكثرة الأمراض العفونيّة ورطوبة جميع البدن وحرارته. والعلامات الدّالّة على أنّ مزاج الكبد بارد يابس ضيق العروق وقلّة الدّم وقلّة الشّعر في مراقّ البطن ويبس جميع البدن، والعلامات الدّالّة على أنّ مزاج الكبد بارد رطب ضيق العروق وغاية قلّة الشّعر في المراقّ وإفراط البلغم في الدّم ورطوبة جميع البدن.

٤٨٧ ADMY: تقهر؛ S: يقلب | ٤٨٨ ADMY: غلبةً قويةً > قهرًا قويًا | ٤٨٩ ADMY: أقلّ | ٤٩٠ ADMY: قهر | ٤٩١ ADMY: تقهر | ٤٩٢ ADMY : تقهر | ٤٩٣ ADMY : قهرًا | ٤٩٤ ADMY : ضعيفًا | ٤٩٥ ADMY: تقهر؛ F: + تقهر | ٤٩٦ ADMY : قهرًا ضعيفًا | ٤٩٧ ADMY : تقهر | ٤٩٨ ADMY: قهرًا قويًا | ٤٩٩ ADMY: أقلّ | ٥٠٠ ADMY: القهر | ٥٠١ AM: قاهرًا؛ DY: قاهر؛ S: غالب | ٥٠٢ S^h:

القلب	الحرارة تقاوم البوردة	الكبد	الكبد	الحرارة تقاوم البرودة ضعيفة	القلب
	البرودة تقاوم الحرارة أقلّ			البرودة تقاوم الحرارة	
	الرّطوبة لا تقاوم اليبوسة			الرّطوبة تقاوم اليبوسة مقاومة متوسّطة	
	اليبوسة تقاوم الرّطوبة يسير			اليبوسة تقاوم الرّطوبة دائمًا	

٥٠٣ FM: – أنّ | ٥٠٤ S^h: لأجل الحرارة لأنّ الحرارة التي تكون مع رطوبة يكون فعلها أقلّ

[Chapter 13][106]

<The temperament of the testicles>

(47) The signs indicating that the temperament of the testicles is hot are frequent sexual intercourse, the procreation of males, the emission of fertile sperm,[107] and abundance of pubic hair. The signs indicating that the temperament of the testicles is cold are the opposite of those things—that is, slowness to engage in sexual intercourse, scanty emissions,[108] procreating females, emitting infertile sperm, and a paucity of pubic hair.[109] The signs indicating that the temperament of the testicles is moist are the abundance and moisture of semen. The signs indicating that the temperament of the testicles is dry are a small quantity of thick semen. The signs indicating that the temperament of the testicles is hot and dry are the occurrence of a desire for sexual intercourse before a suitable age,[110] the thickness and paucity of semen, an abundance of pubic hair, a rapid climax and emission during sexual intercourse, and frequent procreation.[111] The signs indicating that the temperament of the testicles is hot and moist are the occurrence of desire for sexual intercourse before the arrival of the appointed time, the emission of sperm that seldom results in procreation, the abundance and moisture of semen, and harm caused by abstention from

106. B 12.1–8; K 1:339–41; G[a] 70–73.

107. Two MSS include the gloss: "That is, when he has sexual intercourse, the matter is suitable for procreation."

108. A gloss in one MS adds: "Due to a small quantity of semen."

109. One MS has the gloss: "Due to a lack of vapor."

110. A gloss in three MSS reads: "That is, before the proper time for puberty in the moderate temperament."

111. A MS has the gloss: "Due to the coarseness and ripeness of the semen and its lack of dampness."

[الفصل الثّالث عشر]

‹ذكر مزاج الأنثين[٥٠٥]›

(٤٧) والعلامات الدّالّة على أنّ مزاج الأنثيَيْنِ حارّ كثرة الجماع[٥٠٦] وتوليد الذّكورة وإنزال النّطفة المولّدة وكثرة الشّعر في الأعضاء التي حول[٥٠٧] العانة. والعلامات الدّالّة على أنّ مزاج الأنثيَيْنِ بارد هي من[٥٠٨] الأشياء المخالفة لهذه، أعني الإبطاء في الحركة إلى الجماع[٥٠٩] وقلّة الانتشار وتوليد الأناث وإنزال النّطفة التي لا يكون منها ولد وقلّة الشّعر في العانة[٥١٠]. والعلامات الدّالّة على أنّ مزاج الأنثيَيْنِ رطب كثرة المنى ورطوبته. والعلامات الدّالّة على أنّ مزاج الأنثيَيْنِ يابس قلّة المنى وغلظه، والعلامات الدّالّة على أنّ مزاج الأنثيَيْنِ حارّ يابس مسابقة الشّهوة[٥١١] للجماع[٥١٢] قبل[٥١٣] الوقت الموجب[٥١٤] وغلظ المنى وقلّته وكثرة الشّعر في العانة وسرعة الفراغ والإنزال عند الجماع وكثرة توليد الأولاد[٥١٥]. والعلامات الدّالّة على أنّ مزاج الأنثيين حارّ رطب مسابقة الشّهوة للجماع قبل وجوب[٥١٦] الوقت[٥١٧] وإنزال النّطفة التي قليل[٥١٨] ما يكون منها ولد[٥١٩] وكثرة المنى ورطوبته وأن يكون ذلك الإنسان إذا لم يباضع[٥٢٠]

---

٥٠٥ AMS: – ذكر مزاج الأنثيين | ٥٠٦ A^hM^h: أي متى جامع [A^h: جامعت] صلحت مادّته للتّوليد؛ S^h: لكثرة المنى | ٥٠٧ AM: – الأعضاء التي حول | ٥٠٨ ADMY: – من | ٥٠٩ F: للجماع؛ S^h: لقلّة المنى | ٥١٠ S^h: لقلّة البخار | ٥١١ $A^hD^hM^hY^h$: حاشية له: أي قبل الوقت المحدود للمراهقة في معتدل [A^h: والمعتدل؛ M: في المعتدل] المزاج | ٥١٢ M: الجماع؛ DY: إلى الجماع | ٥١٣ ADMY: + بلوغ | ٥١٤ ADMY: المحدود | ٥١٥ S^h: لغلظ المنى ونضجه وقلّة نداوته | ٥١٦ ADMY: حلول | ٥١٧ ADMY: + المحدود | ٥١٨ DFSY: قليلاً | ٥١٩ A: + وهو؛ M: ولداً | ٥٢٠ A^2M^2S؛ ADMY: يقرب الجماع؛ A^2M^2: يقدر على؛ F: يجامع

sexual intercourse. The signs indicating that the temperament of the testicles is cold and moist are that the person is slow to be roused in his desire for sexual intercourse,[112] that his pubic hair is scanty, and that his sperm is runny, like water. The signs indicating that the temperament of the testicles is cold and dry are that the person has sexual intercourse early, that his pubic hair is scanty, and that his sperm is thick and earthy.

112. Three MSS contain the gloss: "It is as though he means the opposite of the signs indicating the hot and moist temperament, which he says are a desire for sexual intercourse before its appointed time. Galen's text [B 13.7–8; K 1:340–41; G[a] 72–73] says in connection with the signs of the cold and moist temperament, 'Its possessor is slow to begin first engaging in sexual intercourse.' Then he says about the possessor of the cold and dry temperament that 'his other states are like those of the one before, except that his semen is coarser and smaller in quantity.'" These are signs mentioned by Galen but, for some reason, omitted by the epitomist.

ضرّه[٥٢١]. والعلامات الدّالّة على أنّ مزاج الأنثيينِ بارد رطب أن يكون الإنسان بطيًا[٥٢٢] ما ينهض شهوته للجماع[٥٢٣]، وإن يكون الشّعر التي[٥٢٤] في عانته قليلًا، وأن تكون نطفته رقيقة شبيهة بالماء. والعلامات الدّالّة على أنّ مزاج الأنثيين بارد يابس أن لا يسرع الإنسان[٥٢٥] في الجماع ويكون[٥٢٦] الشّعر في عانته قليلًا، ونطفته[٥٢٧] غليظةً أرضيّة.

٥٢١ A: طرّه؛ ADMY: + ذلك؛ A^2M^2: يضّرّه؛ F: يضرّه؛ S: – ضرّه | ٥٢٢ $A^hM^hY^h$: كأنّه يريد خلاف العلامات الدّالّة على المزاج الحارّ الرّطب، وهي قوله مسابقة الشّهوة للجماع قبل حلول الوقت المحدود في الفصّ في [A^h: – في] علامات مزاج الانثيين البارد الرّطب قال يبطئ صاحبهما في أوّل استعمال الجماع ثمّ قال في صاحب مزاجهما [Y^h: مزاجهم] البارد اليابس إنّ أمر صاحبهما في سائر أحواله كحال الذي [M^h: التي] قبله إلاّ أنّ منيه يكون [Y^h: – يكون] أغلظ ويكون قليلًا وتحت؛ DY: بطي | ٥٢٣ F: أن يكون. . . للجماع > قلّة شهوة الإنسان للجماع وإبطاء نهضته لها؛ S: بطئًا. . . للجماع > بطؤ شهوة الإنسان الجماع | ٥٢٤ ADSY: – التي | ٥٢٥ ADMY: أن يكون الإنسان لا يسرع | ٥٢٦ ADMY: وأن يكون | ٥٢٧ ADMY: ونطفته > وإن تكون نطفته

[Chapter 14][113]

<The temperament of the entire body—that is, the flesh>

(48) The signs indicating that the temperament of the entire body—that is, the temperament of the flesh in the entire body—is moderate are that the color of the body is mixed, being composed of red and white,[114] that the hair is intermediate between being curly and lank and blond,[115] that the body is intermediate to the touch in the various palpable qualities of heat and coldness, fatness and leanness, and softness and hardness.

---

113. B 14.1–16.8; K 1:341–46; G[a] 73–80: Galen's discussion is less systematic but more nuanced.

114. One MS adds the gloss: "Since neither of the two qualities of heat and coldness is dominant."

115. One MS adds the gloss: "Due to its moderation."

[الفصل الرّابع عشر]

‹ذكر مزاج البدن، يعني اللّحم[٥٢٨]›

(٤٨) والعلامات الدّالّة على أنّ مزاج جميع البدن معتدل، أعني مزاج اللّحم في[٥٢٩] جميع البدن، أن يكون لون البدن مختلطاً[٥٣٠] مركّبًا من حمرة وبياض[٥٣١]، وأن يكون شعره[٥٣٢] متوسّطاً فيما بين الجعد والسّبط أشقر[٥٣٣]، وأن يكون ملمسه معتدلاً[٥٣٤] فيما بين الكيفيّات الملموسة، وهي الحرارة والبرودة والسّمن والقضافة واللّين والصّلابة.

٥٢٨ AFMS: – ذكر. . . اللّحم | ٥٢٩ DFY: من | ٥٣٠ D: مختلطة؛ F: مخلطاً | ٥٣١ S^h: لعدم غلبة أحد الكيفيّتيْنِ الحرارة والبرودة | ٥٣٢ AM: الشّعر | ٥٣٣ S^h: <ا>لاعتداله | ٥٣٤ DM: معتدل

[Chapter 15]

[Its simple temperaments]

(49) The signs indicating that the temperament of the flesh is hot are that the body is hot to the touch and hairy, the complexion is ruddy,[116] and there is little fat. The signs indicating that the temperament of the flesh is cold are that the body is cold to the touch, the hair is scanty, the complexion is very pale, and there is abundant fat. The signs indicating that the temperament of the flesh is dry are that the body is lean and the skin is hard.[117] The signs indicating that the temperament of the flesh is moist are that the body is thick, yet is nonetheless soft.

116. One MS adds the gloss: "Because the heat causes the blood to rise to the surface."

117. One MS adds the gloss: "Due to the predominance of dryness."

[الفصل الخامس عشر]

[ذكر أمزجته البسيطة]

(٤٩) والعلامات الدّالّة على أنّ مزاج اللّحم حارّ أن يكون البدن حارّ الملمس كثير الشّعر الغالب على لونه الحمرة[٥٣٥] قليل الشّحم. والعلامات الدّالّة على أنّ مزاج اللّحم بارد أن يكون الغالب على البدن في ملمسه البرد[٥٣٦]، ويكون شعره قليلاً وبياضه[٥٣٧] وشحمه كثيرًا. والعلامات الدّالّة على أنّ مزاج اللّحم يابس أن يكون البدن قضيفًا والجلد منه صلبًا[٥٣٨]. والعلامات الدّالّة على أنّ مزاج اللّحم رطب أن يكون البدن غليظًا ويكون مع غلظه ليّنًا[٥٣٩].

---

٥٣٥ S^h: لبروزه الدّم إلى خارج لأجل الحرارة | ٥٣٦ AM: البرودة؛ DY: بارد | ٥٣٧ AM: + كثيرًا | ٥٣٨ S^h: لغلبة اليبوسة | ٥٣٩ DFY: ليّن؛ S: – والعلامات. . . ليّنًا

[Chapter 16]

[Its compound temperaments]

(50) The signs indicating that the temperament of the flesh is hot and dry are that the body is hot to the touch, the skin is hard, the hair is thick and curly, and the body is lean. The signs indicating that the temperament of the flesh is hot and moist are that the flesh is abundant, the body is soft and hot to the touch, and the hair is moderate. The signs indicating that the temperament of the flesh is cold and dry are hard skin, scanty hair, a pale complexion, and fat dispersed in the flesh. The signs indicating that the temperament of the flesh is cold and moist are ample flesh and fat,[118] a pale complexion, and scanty hair.

(51) Coldness reaches an organ quickly either because there is a coldness in the organ peculiar to it or because the organ is porous.[119] The organ receives coldness with difficulty either because it is hot or because it is dense. The organ appears dense either because of the quantity of the substance of flesh that is there[120] or because of the thickness of the bones that are under the flesh. The organ appears thin either because there is little muscle in it or because of the thinness of its bones.

---

118. Several MSS include the gloss: "By 'ample flesh and fat,' he would seem to mean the plumpness that is a sign of moisture. Coldness is associated with fat. [Galen's] text says [B 16.6; K 1:345; G[a] 79], 'In the case of the cold and moist temperament, if these two qualities do not diverge too greatly from moderation, then the body will have scanty hair and a pale complexion and be plump and fat. If there is a great divergence of these two qualities from moderation, then the other signs will be stronger in proportion to the increase of these qualities.'"

119. B 16.9; K 1:346; G[a] 80: A MS has the gloss: "So that the cold necessarily reaches it."

120. B 16.10–11; K 1: 346–47; G[a] 80–81: Galen is making a slightly different point than the epitome, that the apparent size of a part of the body may not be the result merely of the amount of flesh but also of the size of the bones beneath. Several MSS contain the gloss: "In two places—that is, in the flesh of muscle, not in pure flesh—that is, soft flesh."

[الفصل السّادس عشر]

[ذكر أمزجته المركّبة]

(٥٠) والعلامات الدّالّة على أنّ مزاج اللّحم حارّ يابس أن يكون البدن حارّ الملمس صلب الجلد ويكون شعره كثيرًا جعدًا[٥٤٠] ويكون بدنه قضيفًا. والعلامات الدّالّة على أنّ مزاج اللّحم حارّ رطب كثرة اللّحم ولين البدن وحرارة الملمس واعتدال الشّعر. والعلامات الدّالّة على أنّ مزاج اللّحم بارد يابس صلابة الجلد وقلّة الشّعر وبياض اللّون وتبدّد[٥٤١] الشّحم في اللّحم. والعلامات الدّالّة على أنّ مزاج اللّحم بارد رطب كثرة اللّحم وكثرة الشّحم[٥٤٢] وبياض اللّون وقلّة الشّعر.

(٥١) إسراع البرودة إلى العضو يكون إمّا لأنّ في العضو برودة خاصّةً له، وإمّا لأنّه متخلخل[٥٤٣]، وعسر قبول العضو[٥٤٤] للبرودة، إمّا لأنّه حارّ وإمّا لأنّه كثيف، ويكون العضو في منظره غليظًا، إمّا لكثرة ما فيه من جوهر اللّحم[٥٤٥]، وإمّا لغلظ العظام التي تحت اللّحم، ويكون العضو في منظره دقيقًا، إمّا لأنّ ما فيه من العضل[٥٤٦] قليل، وإمّا لدقّة ما فيه من العظام.

---

٥٤٠ S: جعدًا كثيرًا؛ DY: – جعدًا | ٥٤١ F: وتفرّق | ٥٤٢ $A^hM^hY^h$: حاشية له: كأنّه يريد بقوله كثرة [A^h: كثر] اللّحم وكثرة الشّحم كثرة العبول التي هي علامة الرّطوبة، وأمّا البرد فعاقد الشّحم، وفي الفصّ: فأمّا المزاج البارد الرّطب [Y^h: – البارد الرّطب] إذا كان فضله [A^h: فعله] في هاتيْنِ الكيفيّتيْنِ على المعتدل فضلًا يسيرًا، فإنّه يكون أزعر أبيض ليّنًا عبلا [Y^h: عبًا؟] سمينًا، فإن [M^h: وإن] كان فضل هاتيْنِ الكيفيّتيْنِ على المعتدل [M^h: لاعتدل] فضلًا كثيرًا، فإنّ سائر العلامات يزداد بحسب تزيّد [A^h: تزيلًا؟] بالكيفيّتيْنِ [Y^h: تزيّدًا لكيفيّته]. | ٥٤٣ S^h: فتصل إليه البرودة ضرورةً | ٥٤٤ AM: ويكون العضو عسر القبول | ٥٤٥ $A^hM^hY^h$: حاشية له: في الموضعيْنِ، يعني باللّحم العضل [M^h: العضد] لا اللّحم الخالص، أعني الرّخو. | ٥٤٦ AM: اللّحم؛ A^2: العضل؛ M^2: + الفضل

(52) There are four natural moistures in the organs of the body:[121] first, the moisture in the veins and arteries, which is blood; second, the moisture that is sprinkled through the organs like dew; third, the moisture that is in the moist organs recently coagulated and solidified, such as fat and flesh; and, fourth, the moisture that entered the organs at the time of the deposit of the sperm.

121. B 16.12–13; K 1:347; G^a 81–82: The epitome systematizes a passage in which Galen discusses the clinical aspects of the different ways that moisture can be present in the organs.

(٥٢) الرّطوبات الطّبيعيّة[٥٤٧] في أعضاء البدن أربع، إحديهنّ الرّطوبة التي في العروق، وهي الدّم، والثّانية الرّطوبة المبثوثة في الأعضاء بمنزلة الرّذاذ[٥٤٨]، والثّالثة الرّطوبة التي في الأعضاء الرّطبة[٥٤٩] القريبة الانعقاد[٥٥٠] وبالجمود وهي[٥٥١] الشّحم واللّحم[٥٥٢][٥٥٣]، والرّابعة الرّطوبة المداخلة[٥٥٤] للأعضاء[٥٥٥] منذ أوّل وقوع النّطفة.

---

٥٤٧ DFY: + التي | ٥٤٨ ADMY: الطّلّ؛ F: + والصّلّ | ٥٤٩ DFY: - الرّطبة | ٥٥٠ ADMY: العهد بالانعقاد؛ S: للانعقاد | ٥٥١ AM: بالجمود بمنزلة؛ DY: بالجمود وبمنزلة؛ S: والجمود وهي | ٥٥٢ AM: اللّحم والشّحم | ٥٥٣ AM: القريبة. . . واللّحم > الرّطبة [DY: - التي. . . الرّطبة] القريبة العهد بالانعقاد وبالجمود بمنزلة اللّحم والشّحم > ؛ DY: وبمنزلة الشّحم واللّحم | ٥٥٤ ADMY: التي هي مداخلة | ٥٥٥ S: + الأصليّة

[Chapter 17][122]

[The temperament of the stomach]

(53) The signs indicating that the stomach is dry are a strong thirst sated by a small quantity of drink, the occurrence of surfeit from a larger quantity, and an appetite for food in which dryness is predominant. The signs indicating that the stomach is moist are a small thirst, the ability to drink a large quantity without surfeit, and an appetite for foods in which moisture is predominant. The signs indicating that the stomach is hot are that it digests strong foods but destroys foods that are quickly transformed and that it has an appetite for foods that are extremely hot. The signs indicating that the stomach is cold are that the appetite is good but that it digests only easily digestible foods, digesting even these only with difficulty. The belch is sour, due to its coldness,[123] and the phlegm flows down to it from the head, provided that the brain is cold. This can be known by signs.

122. B 17.1–4; K 1:348–49; G[a] 83–85: Several MSS contain the gloss: "A gloss of his in a manuscript: So long as their functions operate, all organs are preserved by things similar to them, even if they have diverged a little. Likewise, as long as the stomach is dry and its functions operate, it is preserved by what is similar to it." Galen adds that, when the stomach is diseased, it desires opposites, not foods similar to its temperament, as is the case in times of health.

123. Two MSS have the marginal correction: "It is not due to its coldness."

[الفصل السّابع عشر]

[ذكر مزاج المعدة]

(٥٣) والعلامات الدّالّة على أنّ المعدة يابسة[٥٥٦] كثرة العطش والاكتفاء بالمقدار اليسير من الشّراب، وحدوث الكظّة[٥٥٧] من المقدار الكثير وحسن القبول للأطعمة التي اليبس عليها غالب. والعلامات الدّالّة على أنّ المعدة رطبة قلّة العطش والاحتمال للشّراب الكثير من غير كظّة وحسن القبول للأطعمة التي الرّطوبة عليها أغلب. والعلامات الدّالّة على أنّ المعدة حارّة أن تكون المعدة تهضم الأطعمة القويّة وتفسد فيها الأطعمة السّريعة التّغيّر، وتكون حسنة القبول لما كان من الأطعمة أشدّ حرارةً. والعلامات الدّالّة على أنّ المعدة باردة أن تكون الشّهوة جيّدة[٥٥٨]، وتكون المعدة تهضم الأشياء[٥٥٩] السّريعة الانهضام فقط، ولا تهضم هذه الأشياء[٥٦٠] إلاّ بكدٍّ[٥٦١] ويكون الجشاء حامض[٥٦٢]، إمّا[٥٦٣] لبردها، والبلغم ينصبّ إليها من الرّأس، وذلك إذا كان الدّماغ بارداً، وعرف بعلامات[٥٦٤].

---

٥٥٦ $A^hD^hM^hY^h$: حاشية له في نسخة: جميع الأعضاء ما دامت أفعالها جارية إنّما تحفّظها بالأشياء المتشابهة [D^h: المشابهة]، وإن كانت قد خرجت يسيراً، وكذلك المعدة ما دامت يابسة وفعلها يجري فحفظها [D^hY^h: وتحفّظها] بالشّبيه. | ٥٥٧ D^hY^h: الكظّة التّخمة | ٥٥٨ F: ناهضة؛ DY: جيّد؛ Y^2: تامّة | ٥٥٩ A^2M^2: الأطعمة | ٥٦٠ A: + أيضاً؛ DSY: أيضاً | ٥٦١ F: بكر؛ F^1: بكدّ؛ Y: – بكدّ | ٥٦٢ ADM: حامضاً | ٥٦٣ A^1M^1: ليس | ٥٦٤ DSY: – وإمّا لبردها. . . بعلامات

[Chapter 18][124]

[The temperament of the lungs]

(54) The signs that the lungs are hot are that the patient can quench thirst by inhaling cold air without drinking anything, that he feels a burning in his chest, that he breathes out strongly and breathes in a great deal of air, and that his voice is also loud. The signs that the lungs are cold are that they are quickly harmed by cold air, that they have a great deal of excess phlegm,[125] and that the voice is soft. The signs that the lungs are dry are that they do not generate many excretions and that the voice is like that of cranes. The signs that the lungs are

124. B 18.1–6; K 1:350–51; G[a] 87–88: Galen mostly discusses the causes of the types of voices. Two MSS include the gloss: "A gloss of his: That is, it is not by drink only, like thirst of the stomach. That is because cold drink is partly drunk toward the lung, which is thereby necessarily cooled, thus somewhat quenching the thirst. If his thirst is not quenched by air alone, it will continue. It says in Galen's text [B 17.7–10; K 1:349–50; G[a] 86–87], 'Someone whose thirst in these organs—that is, the heart and lung—is due to heat will be more likely to have his thirst quenched by air through expelling air by long and deep breathing. He feels a strong heat in his chest—not in his upper abdomen, as is perceived in the one who is thirsty due to heat. Moreover, if he drinks, his thirst is not quenched immediately. Drinking cold water will quench his thirst more, whereas a very hot drink will not quench it. Cold air will also quench the thirst of such a one if he takes deep breaths, but this will not quench the thirst of someone whose thirst is due to heat in his stomach.'"

125. A MS has the gloss: "Due to the coldness that traps the excretions in the windpipe and prevents air going out into it."

[الفصل الثّامن عشر]

[ذكر مزاج الرّئة]

(٥٤) والعلامات الدّالّة على أنّ[٥٦٥] الرّئة حارّة[٥٦٦] أن يكون العطش يسكن باستنشاق الهواء البارد لا بالشّراب[٥٦٧]، ويكون ذلك الأنسان يحسّ في صدره بالتهابٍ ويكون ما تخرجه النّخة عظيماً[٥٦٨] ويستنشق هواء كثير، ويكون صوته أيضاً عظيماً. والعلامات الدّالّة على أنّ[٥٦٩] الرّئة باردة[٥٧٠] أن يكون يسرع إليها الضّرر من الهواء البارد، ويكثر فيها الفضول البلغميّة[٥٧١]، ويكون الصّوت صغيراً. والعلامات الدّالّة على أنّ[٥٧٢] الرّئة يابسة[٥٧٣] أن يكون ما يتولّد فيها من الفضول يسيراً[٥٧٤]، ويكون الصّوت شبيهاً بصوت الكراكيّ. والعلامات الدّالّة على أنّ[٥٧٥] الرّئة رطبة[٥٧٦] كثرة

---

٥٦٥ AMY: + مزاج | ٥٦٦ AM: حارّ؛ F: أنّ. . . حارّة > الرّئة الحارّة؛ S: > أنّ الرّئة حارّة | ٥٦٧ ADMY: بالشرب؛ A^hM^h: حا[شية] له: أي لا بالشّرب [A^h: بالشّراب] فقط كعطش المعدة، وذلك لأنّ الشّراب البارد يشرب إلى الرّئة منه شيء، فيجب أن يبرد فيسكن العطش تسكيناً مّا، وإن لم يملأها ملأ الهواء البارد، ويدوم عليها دوامه، وفي الفصّ: ومن كان عطشه بسبب حرارة في هذه الآلات—يعني القلب والرّئة—فإنّه يستنشق من الهواء أكثر لإخراجه الهواء بالتّنفّس مدّة وطول ويحسّ في صدره بالالتهاب [A^h: بالتهاب] لا فيما دون الشّراسيف كما يحسّ من كان سبب عطشه حرارة في المعدة، وإذا شرب أيضاً ليس يسكن عطشه على المكان وشرب البارد يسكن عطشه أكثر ممّا يسكنه شرب الكثير الحارّ، وقد يسكن عطش من كانت هذه حاله الهواء البارد أيضاً إذا استنشقه، وليس يسكن عطش من كان سبب عطشه حرارة في المعدة. | ٥٦٨ AM: يخرج نفخة عظيمة | ٥٦٩ AM: + مزاج | ٥٧٠ A: بارد؛ M: بارداً | ٥٧١ S^h: لأجل البرودة تخلل الفضلات في قصبة الرّئة وخروج الهواء فيها | ٥٧٢ AM: + مزاج | ٥٧٣ AM: يابس

٥٧٤ D: يسير؛ F: يسيرة | ٥٧٥ AM: + مزاج | ٥٧٦ AM: رطب

moist are that a great many excretions are generated and the voice is hoarse. The causes of the voice [quality] can be supposed to differ in accordance with the differences in the voice. A loud voice is due to heat, a soft voice to coldness, a voice like a crane's to dryness, a hoarse voice to moisture, and a smooth voice to a moderate temperament. Likewise, a rough voice is due to dryness, a high-pitched voice to coldness, and a deep voice to heat.

(55) The temperament of each of the organs may be deduced from what each acquires from the things that it receives from outside the body and from its natural functions.[126] With respect to what it receives from outside, if the organ is warmed quickly, then heat is dominant in it. If it is cooled quickly, then coldness is dominant in it. With respect to the organ's natural functions,[127] if many excretions are generated in it, then it is colder, while if only a few excretions are generated in it, it is hotter.

126. B 18.8; K 1:351–52; G^{a} 89: A clarification by examples of Galen's comment that there are only faint external indications of the temperaments of the other internal organs, but that physicians should "attempt their diagnosis by observation both of the influences that benefit or damage them and of the actions of the natural faculties."

127. One MS contains the gloss: "These natural functions are taken in three respects; they are either nullified or reduced, or they operate in an undesirable way."

ما يتولّد فيها من الفضول[٥٧٧] وبحّة[٥٧٨] الصّوت[٥٧٩]. أسباب الصّوت تختلف بحسب اختلافه، فالصّوت العظيم يكون من قِبَل الحرارة والصّغير من قبل البرودة والشّبيه بصوت الكراكيّ من قبل اليبس[٥٨٠]، والأبحّ من قبل الرّطوبة، والأملس من قبل اعتدال المزاج، والخشن[٥٨١] من قبل اليبس[٥٨٢]، وكذلك أيضاً[٥٨٣] الصّوت الحادّ يكون من قبل البرودة، والثّقيل[٥٨٤] من قبل الحرارة.

(٥٥) وقد يستدلّ على مزاج كلّ واحد من الأعضاء بما يناله من الأشياء التي تلقاه من خارج، وبما تكون من أفعال[٥٨٥] الطّبيعة[٥٨٦]، أمّا من الأشياء التي تلقاه من خارج[٥٨٧]، فإنّه إن كان يسخن سريعاً فالغالب عليه الحرارة، وإن كان يبرد سريعاً، فالغالب عليه البرودة، وأمّا من أفعاله الطّبيعيّة[٥٨٨]، فإنّه إن كان يتولّد فيه فضل كثير، فهو أبرد وإن كان يتولّد فيه فضل يسير نضيج[٥٨٩]، فهو أسخن.

---

٥٧٧ S: الفضل؛ F: + فيها؛ | ٥٧٨ A^1؛ A: + يسيراً ويكون؛ A^1: بحة؛ A^2M^2: وبحوجه | ٥٧٩ A: + شبيهاً بصوت الكراكي والعلامة الدّالّة | ٥٨٠ ADMY: اليبوسة | ٥٨١ F: والحسّ | ٥٨٢ AM: اليبوسة | ٥٨٣ AM: – أيضاً | ٥٨٤ D: والثّقيل؛ F: والتّقبّل؛ Y: والثّقل | ٥٨٥ ADMY: أفعاله | ٥٨٦ AM: الطّبيعيّة | ٥٨٧ F: + بما يكون من أفعال الطّبيعة أمّا من الأشياء التي تلقاه من خارج | ٥٨٨ D: الطّبيعة؛ F: في أفعال الطّبيعة؛ S^h: هذه الأفعال الطّبيعيّة على ثلثة وجوه إمّا أن تبطل أو تنقص أو تجري مجرى منكر | ٥٨٩ AM: – نضيج؛ D: ينضج

[Chapter 19][128]

[Disorders]

(56) Some of the disorders occurring in the compound organs are perceptible to the senses and some are not. Examples of those perceptible to the senses are a long and narrow, fractured [?], large, small, or moderate head; a large, small, or moderate chest; or legs that are straight or bowed outward or inward.[129] Some of those not perceptible to the senses are diagnosed quickly, such as the disorders occurring in the stomach and bladder, while others are difficult to diagnose, such as the disorders occurring in the liver and bile ducts; and some cannot be diagnosed at all, such as the disorders occurring in the intestines and urinary ducts. Diseases occur in all of the organs that have been mentioned.

128. B 19.1–2; K 1:352–53; G[a] 90–91: At this point, Galen and the epitome move from discussing temperaments to discussing disease, beginning with a discussion of anatomical defects.

129. Two MSS include the gloss: "A gloss of his: It was not good to mention the moderate-sized head and trunk, since the section began by referring to disorders. It is not written in some of the manuscripts. When he says 'the legs being straight,' this is not equivalent to saying 'moderate,' because straightness in the legs—that is, in the two bones—is a defect; and being straight in shape, as Galen and others say, is when the upper part is like an animal's and the lower part inclines to the human. Therefore, it is permissible to mention straightness along with being bowed inward or outward among the examples of disorders; but it is not permissible to mention moderation in the sizes of the head and trunk among the examples of their defects."

[الفصل التّاسع عشر]

[ذكر الآفات]

(٥٦) الآفات الحادثة في الأعضاء المركّبة، بعضها يدرك حسّاً وبعضها لا يدرك حسّاً، أمّا الذي يدرك منها حسّاً، فبمنزلة الرّأس المسفّط[٥٩٠]، واللأطي[٥٩١] والكبير[٥٩٢] أو الصّغير أو المعتدل[٥٩٣]، وبمنزلة الصّدر الكبير أو الصّغير أو المعتدل[٥٩٤]، وبمنزلة السّاقيْنِ إذا كانتا[٥٩٥] على الاستقامة[٥٩٦] أو كانتا[٥٩٧] مقوستيْنِ إلى خارج أو إلى داخل، وأمّا الذي لا يدرك منها حسّاً، فبعضه يتعرّف[٥٩٨] سريعاً بمنزلة الآفات الحادثة في المعدة، وفي المثانة، وبعضه تعسرتعرّف مثله[٥٩٩] الآفات الحادثة في الكبد[٦٠٠] وفي مجاري المرّة[٦٠١] وبعضه لا يعرف أصلاً بمنزلة الآفات الحادثة في الأمعاء وفي مجاري البول، وكلّ واحد من هذه الأعضاء التي جرى ذكرها تحدث فيها أمراض[٦٠٢].

---

٥٩٠ M: + والمعتدل؛ M^1: المعتدل؛ M^1: والمعتدل | ٥٩١ AM: والالطا؛ D: أو اللاطي؛ Y: أو الأطي | ٥٩٢ ADMY: أو الكبير | ٥٩٣ F: والصّغير؛ AM: - أو المعتدل | ٥٩٤ A: - والمعتدل؛ DMY: والمعتدل؛ A^hM^h: حاشية له: ذكر المعتدل في مقادير الرّأس والصّدر مع تصدير الفصل بذكر الآفات ليس بجيّد، وليس بمثبت في بعض النسخ، وقوله بمنزلة السّاقيْنِ إذا كانتا على استقامة لا يشبه ذكر المعتدل لأنّ الاستقامة في السّاقيْنِ، أعني في عظميهما آفة واستقامة خلقتهما كما ذكر جالينوس وغيره أن يكون أعلاهما إلى الوحشيّ وأسفلهما مقبلاً إلى الإنسيّ، فلذلك ساغ ذكر الاستقامة مع التّقوّس إلى داخل أو خارج في أمثلة الآفات، ولا يسوغ ذكر الاعتدال في مقداري الرّأس والصّدر في أمثلة آفاتهما. | ٥٩٥ F: كانت؛ DSY: كانا | ٥٩٦ DSY: استقامة | ٥٩٧ F: أو إذا كانت؛ DSY: أو كانا | ٥٩٨ ADMY: يعرف | ٥٩٩ ADM: معرفة بمنزلة | ٦٠٠ D: + وفي المثانة وبعضه لا يعرف أصلاً بمنزلة الأفات والحادثة في الإمعاء؛ D^2: المعدة | ٦٠١ F^1: المرارة؛ S: المياه | ٦٠٢ F: + ما من الأمراض الآلميّة

(57) Three diseases occur in the stomach.[130] The first involves size, since we find that sometimes[131] it is smaller than it ought to be. The second involves structure, since we sometimes find that it is firm and round. Third, there is the position, since we find that sometimes and in some people it bulges outward. Two diseases occur in the bladder: The first involves size, for sometimes we find that it is smaller than it ought to be. The second involves position, for sometimes we find it bulging outward. Three diseases occur in the liver: The first involves the structure, for we sometimes find it restricting the blood vessels and ducts. The second involves the position, for we sometimes find it outside of its natural position. The third involves the size, for sometimes we find that it is too small.

130. B 19.3–7; K 1:353–54; G[a] 91–93: The specific diseases referred to in this paragraph are case studies that Galen gives in *The Small Art*, in which, for exceptional reasons, he could observe the internal organs directly or deduce their condition. Two MSS have the gloss: "A gloss of his: These diseases in the stomach, bladder, and liver with these diseases are those that Galen gave as particular examples that he had seen. He ought to have said that blockage of the urine occurs sometimes, not always."

131. Some MSS, here and in the following passages, read: "In some people."

(٥٧) وأمّا المعدة فتحدث فيها ثلثة أمراض، أحدها في المقدار، وذلك أنّا نجدها في بعض الأوقات[٦٠٣] أصغر ممّا ينبغي[٦٠٤]، والثاني[٦٠٥] في الخلقة، وذلك أنّا نجدها في بعض الأوقات[٦٠٦] مستحكمة[٦٠٧] الاستدارة، والثالث في الوضع، وذلك أنّا نجدها في بعض الأوقات وفي بعض[٦٠٨] النّاس ناتئة إلى خارج، وأمّا[٦٠٩] المثانة فيحدث فيها[٦١٠] مرضانِ، أحدهما[٦١١] في المقدار، وذلك أنّا نجدها في بعض النّاس أصغر ممّا ينبغي، والثاني[٦١٢] في الوضع، وذلك أنّا نجدها في بعض النّاس ناتئة إلى خارج. وأمّا الكبد فيحدث فيها ثلثة أمراض، أحدها في الخلقة، وذلك أنّا نجدها ضيّقة العروق والمجاري، والثاني[٦١٣] في الوضع، وذلك أنّا نجدها خارجة عن موضعها الطّبيعيّ، والثالث في المقدار، وذلك أنّا نجدها في بعض الأوقات[٦١٤] صغيرة.

---

٦٠٣ ADMY: النّاس | ٦٠٤ S: ما كانت عليه؛ DY: + عليه | ٦٠٥ AM: والآخر | ٦٠٦ ADMY: النّاس | ٦٠٧ AM: محكمة | ٦٠٨ ADMY: -الأوقات وفي بعض | ٦٠٩ A^hM^h: حاشية له: المعدة والمثانة والكبد بهذه الأمراض التي ذكرها هي في كلام جالينوس أمثلة جزئيّة رآها وتسديد البول إن يقال مكان يحدث في الجميع قد يحدث. | ٦١٠ F: - فيحدث فيها | ٦١١ M: أحدها | ٦١٢ AM: والآخر | ٦١٣ AM: والآخر | ٦١٤ ADMY: النّاس

[Chapter 20][132]

[Diagnosis of diseased states]

(58) If organs that have a particular disease are organs that are on the outside of the body, the disease can be diagnosed by change of complexion, by whether the organ is soft or hard to the touch, by whether it is hot or cold, and by size and number. If they are internal organs, their diseases can be diagnosed by the impairment of the functions specific to them, by what is excreted from the body, by the pain specific to that place, and by the place of the organ and the correspondence of the symptoms.

(59) If a function is in some way impaired, the impairment can be of three sorts.[133] First, its function can be lost completely, as occurs to vision in blindness. Second, its function can be weakened, as occurs

---

132. B 20.1; K 1:355; G[a] 94.

133. Two MSS include the gloss: "The variant 'kinds' [reading *anḥāʾ kh kh* for *ḥbʾḥ kh*] is found in another manuscript and is correct, but it has been altered in other manuscripts without good reason and is incorrect. That change is to "*ḥb*' [?]. [Galen] explains, 'There cannot be another genus of diseases [another MS reads "symptoms"] here—only the one genus that I am mentioning to you.' They are wrong to imagine that this change is necessary, since they think that his statement 'No other genus of diseases occurs in the body,' contradicts his statement later, 'There is another genus of diseases.' But it is not as they imagine, since what he actually said was, 'No other genus of disease occurs in the body—in either the tissues or the compound organs—that is peculiar + [?] to one of the two.' After that, he says, 'But there is another genus of diseases that I will mention to you afterwards, and that is common to all the organs that [the body] has.' [A paraphrase of B 27.1; K 1:379; G[a] 133: 'There remains one genus of disorders common to tissues and organs. This is dissolution of continuity.'] If he had said, 'There is no specific genus except these two; but, rather, a common genus occurs'—that is, not specific to them—'and is a third,' then the statement would be correct and would not need to be changed to something else.

"What is in the text of this manuscript is correct, and one could take as evidence this sign that is found in three places. The proof that the change made to the text is an error is that it amounts to saying, 'There is no third specific genus [of disease or symptom] in the body—only a third general genus.' So it is as though he had said, 'There is no third specific genus apart from these two, with the exception of one that is not specific but is common instead.' This is a blatant error."

This is certainly not entirely clear. The dingbat σ˥+ that appears at the beginning of the note before a variant presumably indicates a correction; but

[الفصل العشرون]

[ذكر التّعرّف بأحوال مرضيّة]

(٥٨) الأعضاء المريضة مرضاً خاصّاً، إن كانت من الأعضاء التي هي في ظاهر البدن فأمراضها[٦١٥] تعرف[٦١٦] من تغيّر اللّون ومن لين الملمس وصلابته وحرارته وبرودته، ومن المقدار ومن العدد، وإن كانت من الأعضاء الباطنة فأمراضها تعرف[٦١٧] من مضارّ الأفعال[٦١٨] ومن الأشياء التي تستفرغ من البدن ومن الوجع الخاصّ بالموضع[٦١٩] ومن موضع[٦٢٠] العضو ومن مناسبة الأعراض.

(٥٩) وكلّ فعل يناله ضرر[٦٢١] فمضرّته تكون على ثلثة وجوه[٦٢٢]، إمّا بأن يبطل فعله أصلاً بمنزلة ما يعرض للبصر عند العمي[٦٢٣]، وإمّا بأن يضعف فعله بمنزلة ما يعرض

---

٦١٥ AM: وأمراضها | ٦١٦ AM: تعرّف | ٦١٧ AM: تعرّف | ٦١٨ ADMY: + الخاصّة [M: + بالموضع] بها | ٦١٩ F^2: بالعضو | ٦٢٠ F: وضع؛ S: – موضع | ٦٢١ AM: مضرّة؛ DY: الضّرر | ٦٢٢ F: أنحاء؛ DSY: أوجه؛ A^hM^h: + +σ٦ حاخ خ، وهو الصّحيح، وقد غيّر في نسخ تغييراً لا حاجة إليه، وليس بصواب، وذلك التّغيير -+σ٦ حا، وبيّن ذلك أنّه ممّا لا يمكن أن يكون ها هنا جنس آخر من الأمراض [A^h: الأعراض] إلّا جنس واحد أنا ذاكره لك، وهذا التّغيير إنّما توهّموا الحاجة إليه توهّماً باطلاً لأنهم ظنّوا قوله، وليس تحدث في البدن جنس آخر من الأمراض يناقضه قوله بعد لكن ها هنا جنس آخر من الأمراض، وليس الأمر كما ظنّوا لأنّه إنّما قال، وليس يحدث في البدن جنس آخر من الأمراض في الأعضاء البسيطة المفردة ولا في الأعضاء المركّبة ممّا هو خاصّ + بواحد من الاثنيْنِ، ثمّ قال بعد لكن ها هنا جنس آخر من الأمراض، أنا ذاكره لك بعد مشترك لجميع الأعضاء له، فإذا قال ليس يوجد جنس خاصّ إلّا هذيْنِ، ولكن يحدث جنس مشترك، أي ليس بخاصّ هو ثالث كان الكلام مستقيماً لا يجوز تغييره بما غير.
صحّ ما في متن هذه النّسخة، وقد يشهد على صحّته بهذه العلامة في ثلثة مواضع، وأمّا الحجّة على أنّ العبارة التي عدل إليها عمّا في المتن خطأ، فهو [A^h: فهي] أن محصولها أن يقال، وليس يحدث في البدن جنس ثالث خاصّ إلاّ جنس ثالث عامّ فيكون كأنّه يقول فلا جنس خاصّ بعد هذيْنِ إلاّ واحداً ليس بخاصّ بل مشترك، وهذا خطأ فاحش. صح | ٦٢٣ ADMY: العماء

with vision when there is a film [over the eye]. Third, its function can operate badly and not in the way it ought to, as occurs with vision when the patient imagines that he sees a bug, a speck, or, in general, things that do not exist in reality.

(60) Some of the things that are excreted and expelled from the body are parts of the diseased organs.[134] These excretions may indicate their organs by the specific property of their substance, as the throat—which belongs to the windpipe and its parts—indicates that the disease is in the lungs. The excretions may indicate the organ by their quantity, such as the coarse crust that flakes off an ulcer and comes out with the feces and indicates that there is an ulcer in the large intestines, or the fine crust that indicates that there is an ulcer in the small intestines. They may indicate the organ by their position, such as the crusts that come out with the feces and indicate an ulcer in the intestines, and the crusts that come out with coughing and indicate that the disease is in the lungs. There are also the things contained in the organs, some of which the organs contain naturally and some of which, when contained by the organ, are unnatural. Some of the things that the organs contain naturally are naturally excreted from the body, but the excretion of them deviates from the natural mode—either in quality or in quantity—such as feces that are too much or too little, too soft or too hard, and urine

---

the fact that it appears again as σר before a possibly different form of the variant and that the variant itself—in both places—makes no sense might also mean that the scribes of **A** and **M** could not read their exemplar and, as was sometimes done, drew the form of what they saw in the hope that their reader could figure it out. If so, I have disappointed them. It is also not clear to me why the cross appears in the middle of the note. As for the variant, it is undotted. I have read it as *ḥbā kh kh*, with the *kh kh* being an abbreviation for *nusakh* (manuscripts), rather than *ḥbāḥ*, which would require me to explain why it appears at the end of the line as *ḥbā*. I have read it as a corruption of *anḥa*ʾ (kinds), which has the support of the sloppy but old and authoritative MS **F**; but *ajnās* (genera) is also a possibility. Finally, the quotations in the gloss do not, so far as I can tell, correspond exactly to anything in the epitome or in *The Small Art* itself.

The issue is the apparent contradiction between the interpretation of B 4.6; K 1:316; G[a] 27, given on pp. 72–73, paragraph 23, above, which identifies two genera of disease, and B 27.1; K 1:379; G[a] 133, which says that there is another genus of disease, loss of continuity—a problem also discussed in the gloss on p. 73, n. 60, above. The point of the variant would be that most of the scribes changed *anḥā*ʾ (kinds), which implies genera, to *wujūh* or *awjuh* (sorts), which does not.

134. B 20.10–11; K 1:357–58; G[a] 99: Galen mentions this distinction without elaboration, referring readers to his work *The Diagnosis of Diseases of the Internal Organs*, which, he boasts, is the first book to treat this subject systematically.

للبصر عند الغشاوة، وإمّا بأن يجري فعله مجرى رديًّا[٦٢٤] على غير ما ينبغي بمنزلة ما يعرض للبصر إذا[٦٢٥] يرى خيال البق أو خيال القذا، وبالجملة[٦٢٦] إذا رأى[٦٢٧] أشياء ليست موجودة[٦٢٨] في الطّبع.

(٦٠) والأشياء التي تستفرغ وتخرج من البدن، منها أشياء هي أجزاء من الأعضاء العليلة، وتدلّ على تلك الأعضاء إمّا بخصوصيّة[٦٢٩] جوهرها بمنزلة الحلق التي من[٦٣٠] قصبة الرّئة وأقسامها التي تدلّ على أنّ العلّة في الرّئة، وإمّا من مقدارها[٦٣١] بمنزلة القشرة الغليظة التي تنقشر من القرحة إذا خرجت مع البراز[٦٣٢]، فإنّها تدلّ على أنّ القرحة في الأمعاء الغلاظ، والرقيقة[٦٣٣] تدلّ على أنّ القرحة في الأمعاء الدّقاق، وإمّا[٦٣٤] وضعها[٦٣٥] بمنزلة ما تدلّ[٦٣٦] القشرة التي تخرج مع البراز[٦٣٧] على أنّ القرحة في الأمعاء والقشرة التي تقذف بالسّعال تدلّ على أنّ العلّة في الرّئة، ومنها أشياء هي ممّا[٦٣٨] يحتوي الأعضاء عليها، ومن هذه[٦٣٩] الأشياء ما[٦٤٠] احتواء[٦٤١] تلك الأعضاء عليها بالطّبع، ومنها أشياء احتواء[٦٤٢] تلك الأعضاء[٦٤٣] عليها خارج عن الطّبع، والأشياء التي احتواء الأعضاء عليها بالطّبع، إمّا أن يكون أشياء خروجها من البدن موجود في الطّبع إلاّ أنّها قد خرجت[٦٤٤] عمّا عليه مجراها بالطّبع، إمّا في كيفيّتها وإمّا في كمّيتها بمنزلة البراز[٦٤٥] إذا كثر[٦٤٦] أو قلّ، وإذا كان ليّنًا[٦٤٧] أو صلبًا والبول إذا

---

٦٢٤ F: رديّ ؛ DY: - مجري رديًّا | ٦٢٥ DFY: إن؛ ADMY: + كان | ٦٢٦ D: أو جملة؛ F: الهباء بالجملة؛ Y: أو بالجملة | ٦٢٧ F: + الشّيء على خلاف ما هو به أو؛ DY: يرى | ٦٢٨ AM[1]: بموجودة | ٦٢٩ F: لخصوص؛ DY: بخصوص | ٦٣٠ AM: في | ٦٣١ AM: بمقدارها | ٦٣٢ AM: الثّقل؛ DY: براز | ٦٣٣ ADMY: + منها | ٦٣٤ DSY: + من | ٦٣٥ F: بوضعها | ٦٣٦ DSY: - ما يدلّ | ٦٣٧ AM: الثقل؛ DSY: + فإنّها تدلّ | ٦٣٨ DS: ما | ٦٣٩ AM: وهذه | ٦٤٠ AM: منها أشياء؛ DY: - ما | ٦٤١ F: احتوى | ٦٤٢ F: احتوى | ٦٤٣ AM احتواءها | ٦٤٤ DFY: زالت | ٦٤٥ AM: الثّقل | ٦٤٦ DY: أكثر | ٦٤٧ AM: سيّالاً

that is either too much or too little or is black or white. Other things are excreted from the body in a way that is not natural, even though their quality is natural—such as blood, which is not naturally excreted from the body, even though the body contains it naturally. Some of the things that are not naturally contained in the organs are of the same genus as the things that exist naturally but are changed, such as blood,[135] while others are of an unnatural genus, such as worms and stones.

(61) Pain may arise from a sudden change of the temperament[136] with respect to either heat, coldness, dryness, or moisture. It may be caused by the dissolution of continuity, which may, in turn, be due to things that cut, things that stretch, or things that bruise the organ. It may also be due to position.

(62) The diseased organ is most often indicated by something swollen.[137] There are four kinds of swelling: those from blood, which are called phlegmona; those from the bilious humor, which are called erysipelas; those from phlegm, which are called edema; and those from the melancholic humor, which are called induration.[138]

---

135. Two MSS contain the gloss: "A gloss of his: This division is correct in putting blood in the two places where you know that it is—that is, blood, so long as it is unaltered, is contained naturally in the body, while, when it is altered, it is contained unnaturally, such as menstrual blood. It is sufficient for me to mention menstrual blood, relying on the two manuscripts, for prior to the time of menses, it is unaltered, and it is contained there naturally; but during menses, it alters and its being contained there is unnatural." There is a textual variant, probably also a gloss, that clarifies this point.

136. B 20.9; K 1:357; G[a] 98.

137. B 20.8; K 1:357; G[a] 98: *Wārim* being an unusual form, three MSS contain the gloss: "He means, 'an organ with a swelling on it.'"

138. Some MSS of Galen's text, including those from which the Arabic translation was made, omit erysipelas. These terms are also translated differently in G[a]. On the textual issue, see B, pp. 421–22. The terminological variants can be summarized as follows:

Greek text of *The Small Art*	Arabic translation and some MSS of *The Small Art*	Epitome of *The Small Art*	English translation in Epitome
φλεγμονή	*al-waram al-ḥārr* (inflammation)	*filaghmūnī; waram damawī* (hematoma)	phlegmona, from blood
ἐρυσαφέλη	——	*ḥumra*	erysipelas, from yellow bile
οἴδημα	*al-rakhw al-manfūkh*	*tahayyuj, waram rakhw*	edema, from phlegm
σκίρρος	*al-jāsī al-ṣalb*	*ṣalāba*	induration, from black bile

كثر أو قلّ[٦٤٨]، وإذا كان أسود أو أبيض، وإمّا أن يكون أشياء خروجها عن البدن على غير مجرى الطّبع، وكيفيّتها موجودة في الطّبع بمنزلة الدّم، فإنّ الدّم ليس له في الطّبع أن يجري[٦٤٩] من البدن لكن أن يوجد للبدن بالطّبع[٦٥٠]، وأمّا الأشياء التي احتواء الأعضاء عليها خارجاً عن الطّبع، فإمّا أن تكون من جنس الأشياء الموجودة في الطّبع إلاّ أنّها قد تغيّرت بمنزلة الدّم[٦٥١]، وإمّا أن يكون من جنس الأشياء التي جملتها خارجة عن الطّبع بمنزلة الدّود والحصى.

(٦١) وأمّا الوجع فإنّه يحدث إمّا بسبب تغيّر المزاج دفعةً، وذلك يكون إمّا من قِبَل الحرارة، وإمّا من قبل البرودة، وإمّا من قبل اليبوسة، وإمّا ن قبل الرّطوبة، وإمّا بسبب تفرّق الاتّصال، وذلك يكون إمّا من قبل الأشياء التي تقطع، وإمّا من قبل الأشياء التي تمدّد، وإمّا من قبل الأشياء التي ترضّ العضو، وإمّا الوضع.

(٦٢) فأكثر ما يدلّ على العضو العليل الوارم[٦٥٢]، وأصناف الورم أربعة، وذلك أنّه إمّا أن يكون من الدّم ويسمّى فلغمونيّ، وإمّا من الخلط المراريّ ويسمّى حمرة، وإمّا من البلغم ويسمّى تهيّجاً، وإمّا من الخلط السّوداويّ ويسمّى صلابة.

٦٤٨ DF: وقلّ | ٦٤٩ ADMY: يخرج | ٦٥٠ AFM^2: وأمّا [F: وإنّما] الموجودة [F: الموجود] في الطّبع طبيعة الدّم فقط؛ A^h: ـo من هنا ما بين العلامتين في الحاشية في نسخة | ٦٥١ A^hM^h: + حـ[شية] له: سلّم له هذه القسمة إدخال الدّم في الموضعيْنِ الذين علمت عليه فيهما، أي الدّم ما دام لم يتغيّر، فالاحتواء عليه بالطّبع، وإذا تغيّر فالاحتواء عليه خارجاً عن الطّبع بمنزلة دم الحيض، ويكفي قولي بمنزلة دم الحيض مثالاً على النّسختين جميعاً، فإنّه قبل وقت الأقراء يكون غير متغيّر، فيكون الاحتواء عليه في الطّبع وفي وقت الأقراء يكون قد تغيّر، فيكون الاحتواء عليه خارجاً عن الطّبع. | ٦٥٢ D^1F: بالورم؛ A^1DY: الورم؛ $A^hD^hM^hY^h$: + ومعناه، أي العضو الذي عليه [A^h: علّته] ورم.

(63) Some symptoms[139] become apparent in the impairments of functions; from them is derived the standard and method by which one is able to determine the affected organ by means of the harm done to its functions. Other symptoms become apparent through the things that are excreted from the body, and from them is derived the method of making determinations by means of what comes out of the body. Other symptoms become apparent in the states of the body. Some of these states are visible, such as jaundice; some can be heard, such as growling in the stomach; some can be smelled, such as fetid odors; some can be tasted, such as bitterness in the mouth; some are palpable, such as heat, coldness, moisture, dryness, hardness, softness, coarseness, fineness, roughness, and smoothness. Some of these symptoms that are apparent from the states of the body are apparent to the senses in the diseased organ itself; and so, from that, it is possible to derive the methods of diagnosis from the location of the organ and from the pain specific to the organ's place. Other symptoms are apparent to the senses in a different organ; and so, from that, it is possible to derive the method of deduction from the specific properties of the symptoms. For example, when the lung is diseased and its disease is inflammation, there is an accompanying redness in the cheeks, while, if the disease is a lesion, there is an accompanying bowing of the fingernails.

139. B 20.1, 8–10; K 1:355, 357; G[a] 94, 98–99: Two MSS have the gloss: "That is, the correspondence of symptoms."

(٦٣) وأمّا الأعراض[٦٥٣]، فمنها ما تظهر في مضارّ الأفعال[٦٥٤] ويستنبط[٦٥٥] منها القانون، والطّريق الذي منه[٦٥٦] يستدلّ به[٦٥٧] على الأعضاء بمضارّ[٦٥٨] أفعالها، ومنها ما يظهر في الأشياء التي تخرج من البدن ويستخرج منها الطّريق الذي يستدلّ به[٦٥٩] ما[٦٦٠] يبرز من البدن، ومنها ما يظهر في حالات الأبدان، وهذه الحالات منها مبصورة[٦٦١] بمنزلة اليرقان، ومنها مسموعة بمنزلة القراقر، ومنها المشمومة بمنزلة الرّوائح المنتنة، ومنها مذوقة بمنزلة مرارة الفم، ومنها ملموسة بمنزلة الحرارة والبرودة والرّطوبة واليبوسة والصّلابة واللّين والغلظ والرّقّة والخشونة والملاسة، وجميع هذه الأعراض التي تظهر في حالات الأبدان بعضها يدركه الحسّ[٦٦٢] في[٦٦٣] نفس العضو العليل، ويستخرج منه الطّريق الذي يستدلّ به من موضع[٦٦٤] العضو والطّريق الذي يستدلّ به من الوجع الخاصّ بالموضع، وبعضها يدركه الحسّ في عضو آخر، ويستخرج منها الطّريق الذي يستدلّ بها[٦٦٥] من[٦٦٦] خصوصيّات الأعراض بمنزلة ما يعرض إذا اعتلّت الرّئة لأنّه إذا كانت علّتها ورماً حارّاً عرض معه حمرة في الوجنتين، وإذا كانت علّتها قرحة عرض معها[٦٦٧] تقوّس[٦٦٨] الأظفار[٦٦٩].

---

٦٥٣ $A^{h}M^{h}$: + أي وأمّا من مناسبة الأعراض | ٦٥٤ AM: بمضارّ أفعالها | ٦٥٥ $A^{2}F^{2}M^{2}$؛ AM: ويستخرج؛ F: ويبسط | ٦٥٦ AM: – منه | ٦٥٧ DSY: – به | ٦٥٨ DSY: ومضارّ | ٦٥٩ F^{1}: منه على الأعضاء ومضارّ أفعالها ومنها ما يظهر في الأشياء التي تخرج من البدن وتستخرج منها التي يستدلّ به صحّ ما | ٦٦٠ ADMY: ممّا | ٦٦١ $A^{1}S$: مبصرة | ٦٦٢ AM: يدرك بالحسّ | ٦٦٣ DSY: من | ٦٦٤ DFY: على وضع | ٦٦٥ ADY: به | ٦٦٦ AM: على | ٦٦٧ AM: معه | ٦٦٨ $A^{2}M^{2}$: تعقف؛ F: + في | ٦٦٩ A^{1}: خ تعقّف؛ M^{2}: + تعقف الآطفاه [؟]

[Chapters 21 and 22][140]

[Signs]

(64) There are signs of health, of disease, or [indicating] neither health nor disease. Some of the signs [indicating] neither health nor disease indicate both health and disease simultaneously; some indicate health at one time and disease at another; and some do not indicate either complete health or complete disease. Each of these [three][141] species indicates either what is the case at the present moment, prognosticates what is to come, or is mnemonic of what has already been. Some of the signs prognosticating a disease that is to come are of the same genus as the things that exist naturally, except that they indicate a disease that will occur. This is because the things have changed from their natural state in their quality, quantity, or time. Others belong to the genus of things that have diverged from the natural state, but only by a slight amount. Some of the signs prognosticating health and some of the signs prognosticating disease are apparent in impairments of function. These are signs that are always indicative. Other such signs are apparent in the states of bodies, but these signs are not always and primarily indicative.[142] Other such signs are apparent in the things that are evacuated from the body. These signs are always indicative, but they are not primarily indicative; rather, they are indicative by means of coction[143] and its opposite.

140. B 21.1–22.4; K 1:358–65; G^a. 99–111. A summary of two chapters in which Galen treats these topics in more detail and with examples.

141. Added in some MSS.

142. Three MSS have the gloss: "A gloss of his: If, by saying 'primarily,' he meant 'without intermediary,' then he is right; but if, by saying 'primarily,' he meant 'first for the senses,' then he is not right. This is because it is possible to diagnose impairment of the function and to diagnose on the basis of it; yet, in both cases, the disease is diagnosed without an intermediary."

143. This old term, meaning "cooking," indicates the preparation of the waste products of digestion and disease for elimination from the body.

[الفصلان الحادي والعشرون والثّاني والعشرون]

[ذكر العلامات]

(٦٤) والعلامات منها صحيّة ومنها مرضيّة ومنها لا صحيّة ولا مرضيّة، وهذه التي ليست بصحيّة ولا مرضيّة منها ما يدلّ على الصّحّة وعلى المرض[٦٧٠] معاً، ومنها ما يدلّ مرّة على الصّحّة ومرّة على المرض، ومنها ما لا[٦٧١] يدلّ لا على صحّة تامّة ولا على مرض تامّ، وكلّ واحد من هذه[٦٧٢] الأنواع، إمّا[٦٧٣] أن يكون دالاًّ[٦٧٤] على ما هو حاضر، وإمّا منذر بما سيكون، وإمّا مذكّرة[٦٧٥] بما قد سلف[٦٧٦]، والعلامات المنذرة بالمرض الذي سيكون، منها ما هو من جنس الأشياء الموجودة في الطّبع إلاّ أنّها تدلّ على مرض سيكون لأنّها قد تغيّرت عن الحال الطّبيعيّة، إمّا في كيفيّتها، وإمّا في كمّيّتها، وإمّا في وقتها، ومنها ما هو من جنس الأشياء الخارجة عن الطّبيعة[٦٧٧] إلاّ أنّ مقدارها يسير، والعلامات المنذرة بالصّحّة والعلامات المنذرة بالمرض، منها ما يتبيّن في مضارّ الفعل، وهذه علامات تدلّ دائماً[٦٧٨]، ومنها ما يتبيّن في حالات الأبدان، وهذه علامات ليس تدلّ دلالة أوّليّة[٦٧٩] ولا دلالة دائمة[٦٨٠]، ومنها ما يتبيّن في الأشياء التي تستفرغ من البدن، وهذه علامات تدلّ دائماً إلاّ أنّ دلالتها ليست بدلالة أوّليّة، لكن[٦٨١] إنّما يدلّ بتوسّط النّضج وخلافه.

---

٦٧٠ DSY: والمرض | ٦٧١ AM: -لا | ٦٧٢ AM: + الثّلثة | ٦٧٣ DSY: فإمّا | ٦٧٤ F: دالّ؛ DY: يدلّ | ٦٧٥ ADMY: مذكّر | ٦٧٦ AM: كان | ٦٧٧ AM: الطّبع | ٦٧٨ AM: أوّلاً ودائماً | ٦٧٩ $A^hM^hY^h$: إذا [M^h: حاشية له إن] أراد بقوله أوّلاً أي بغير متوسّط، فهو صحيح وإن أراد أوّلاً أي أوّلاً عند الحسّ فليس بصحيح لأنّ ضرر الفعل قد يستدلّ [Y^h: ويستدلّ] عليه ثمّ يستدلّ به وفي [Y^h: في] الحالين [M^h: الحال] يدلّ هو على المرض بغير متوسّط. | ٦٨٠ AS: دائماً | ٦٨١ AM: بل

[Chapter 23][144]

[Causes]

(65) Some causes create health, some create disease, and some create neither health nor disease. Some healthful causes preserve health, and some bring about health. Some of the causes that create health preserve health, and some bring about health. Some of the causes that preserve health preserve the health of a body whose form is excellent and which has no flaw in its structure. Other such causes preserve the health of the body whose structure and form are inferior to the structure of the body that has an excellent form. Some of these causes preserve the body in its current state by means of things similar to it, while others—the transformative causes—change it from its current state by means of things opposite to it. Others transform the body and bring it to excellent form and structure. Still others transform its capacity and suitability to receive disease by eradicating the causes predisposing it to disease and thus keeping it in its original nature.

(66) Some of the causes that change bodies are things that change them necessarily;[145] there are six genera of these. Other causes are things that do not necessarily change the body, such as a vicious animal, a stone, swords, and the like. The six necessary causes are the air surrounding the body; the genus of things eaten and drunk; the genus of sleep and wakefulness; the genus of motion and rest, either in the

144. B 23.1–5; K 1:365–67; G[a] 111–13.
145. B 23.6–10; K 1:367–68; G[a] 113–16.

[الفصل الثّالث والعشرون]

[ذكر الأسباب]

(٦٥) الأسباب منها مصحّة[٦٨٢]، ومنها ممرّضة، ومنها لا مصحّة ولا ممرّضة، والأسباب المصحّة منها ما تحفظ الصّحّة، ومنها ما تفعل الصّحّة، والتي تحفظ الصّحّة منها ما تحفظ صحّة البدن الذي هيئته[٦٨٣] فاضلة، وبنيته لا نقص[٦٨٤] فيها، ومنها ما تحفظ صحّة الأبدان التي بنيتها وهيئتها دون بنية البدن الفاضل الهيئة، وهذه الأسباب منها ما تحفظ البدن على ما هو عليه بالأشياء المشابهة به[٦٨٥]، ومنها ما ينقله عمّا هو عليه بالأشياء المضادّة له، وهذه الأسباب النّاقلة، منها ما ينقل البدن ويبلغ[٦٨٦] به إلى الهيئة والبنية الفاضلة، ومنها ما ينقله عن الاستعداد والملاءمة[٦٨٧] لقبول المرض بقطع[٦٨٨] الأسباب المتهيّئة فيه ويحفظه على طبعه الأوّل.

(٦٦) الأسباب المغيّرة للأبدان منها أشياء تغيّر ضرورةً، وهي ستّة أجناس، ومنها أشياء ليست تغيّره[٦٨٩] ضرورةً، مثل الحيوان المفسد والحجارة والسّيوف وما أشبه ذلك، فأمّا السّتّة الأسباب الاضطراريّة، فهي الهواء المحيط بالبدن وجنس الأشياء التي تؤكل وتشرب وجنس النّوم واليقظة وجنس الحركة والسّكون، إمّا في جميع

---

٦٨٢ F: مصحّة | ٦٨٣ AM: + هيئة | ٦٨٤ ADMY: – نقص؛ AFM: عيب | ٦٨٥ AM: الشّبيهة به؛ DSY: له | ٦٨٦ AM: ويصير؛ F: + ويصير ويبلغ | ٦٨٧ AMS: والملاومة | ٦٨٨ A^2M^2؛ AM: بقلع | ٦٨٩ ADM: تغيّر؛ Y: تغيرها

entire body or in some organ to the exclusion of others; the genus of evacuation or retention of what can be evacuated from the body; and the genus of the accidents of the soul, which are joy, grief, anxiety, envy, anger, and fear.[146] These six genera can be causes of health when they preserve its proper quality and quantity to achieve moderation; but they can be causes of disease when they diverge from moderation toward one extreme or the other, either in quantity or in quality.

146. Necessary causes are the factors that people cannot avoid being exposed to—the famous "six non-naturals"; see pp. 18–19, n. 43, above.

البدن وإمّا في بعض الأعضاء دون بعض، وجنس استفراغ ما يستفرغ من البدن واحتباسه وجنس عوارض النّفس، وهي الفرح والحزن والغمّ[٦٩٠] والحسد والغضب والفزع، وهذه السّتّة الأجناس قد تكون أسباباً للصّحّة إذا هي حفظت كيفيّتها وكمّيّتها على ما ينبغي على الاعتدال، وتكون أسباباً للمرض إذا هي زالت عن الاعتدال إلى أحد الطّرفين[٦٩١] إمّا في كمّيّتها[٦٩٢] وإمّا في كيفيّتها[٦٩٣].

٦٩٠ AM: والغمّ والحزن | ٦٩١ AM: الجانبين | ٦٩٢ AM : كيفيّتها | ٦٩٣ AM : كمّيّتها

[Chapter 24][147]

[The causes of health]

(67) When the form and structure of the body are excellent, the body needs these six genera to be moderate in order to preserve health—which is to say that there must be moderation in the air surrounding the body, moderation in its food and drink, moderation in the accidents of its soul, moderation in its sleep and wakefulness, moderation in its motion and rest, and moderation in the evacuation of the superfluities that are evacuated.[148] There are three kinds of superfluity: feces, which are the superfluity remaining from the food that is assimilated by the stomach and belly; urine, which is the superfluity remaining from the food that reaches the liver;[149] and sweat, which is the superfluity remaining from the food that reaches the body as a whole.

147. B 24.2–10; K 1:370–72; G[a] 119–22.

148. One MS has the gloss: "And also moderation of foods and drinks."

149. Two MSS add: "and the blood vessels."

[الفصل الرّابع والعشرون]

[ذكر أسباب الصّحّة]

(٦٧) والذي يحتاج إليه في حفظ صحّة البدن الذي هيئته وبنيته[٦٩٤] فاضلة[٦٩٥] اعتدال هذه السّتّة الأجناس[٦٩٦]، أعني اعتدال الهواء المحيط بالبدن واعتدال مطعمه ومشربه واعتدال[٦٩٧] عوارض نفسه واعتدال نومه ويقظته واعتدال حركته وسكونه واعتدال استفراغ ما يستفرغ من فضوله[٦٩٨]، وهي ثلثة، البراز[٦٩٩] وهو فضل الطّعام الذي تستمريه المعدة والبطن، والبول، وهو فضل الغذاء الذي يصير إلى الكبد[٧٠٠]، والعرق، وهو فضل الغذاء الذي يصل[٧٠١] إلى جميع البدن[٧٠٢].

---

٦٩٤ AM: + الهيئة والبنية | ٦٩٥ AM: الفاضلة | ٦٩٦ M[1]؛ AM: أجناس | ٦٩٧ F[1]؛ FS: – مطعمه ومشربه اعتدال | ٦٩٨ S: فضله؛ Y: فضول | ٦٩٩ AM: الثّقل | ٧٠٠ AM: + والعروق | ٧٠١ DFY: يصير | ٧٠٢ F[h]: واعتدال المآكل والمشارب

[Chapter 25][150]

[The cure of diseases]

(68) In order to cure a disease, one needs excesses of these genera that exceed moderation in the direction opposite to that of the disease. One of two things is needed to preserve the health of bodies whose health is inferior to the health of the body with an excellent structure. If you wish to preserve these bodies as they naturally are, you must employ these genera so as to exceed moderation in the direction to which that body inclines, so as to be similar to it. However, if you wish to change such bodies to an excellent temperament, then you must employ these genera, during the time in which you are trying to change the temperament, in a way opposite to the aspects in which these bodies diverge from moderation. But if you are afraid to change the temperament, you must use moderation.

(69) The corruption of any organ that, in some way, falls short of the excellent form is either in the tissues or in the compound organs.[151] If the corruption is in the tissues, then either it is in all of them equally, or it is not in them equally—that is, in some of them but not in others. Each of these two kinds is also either simple or compound. Moreover, each of these two kinds must either be preserved by things similar to it,

150. B 25.6–11; K 1:374–76; G[a] 123–27: Galen observes that the natural temperament of the individual must also be taken into account, not just an ideal moderation, so that bodies that are naturally hotter, for example, must be treated with hotter regimens.

151. B 26.1–2; K 1:376; G[a] 127–28: Galen goes on to discuss the treatment of the various classes of defects in tissues and organs.

[الفصل الخامس والعشرون]

[ذكر شفاء الأمراض]

(٦٨) والذي يحتاج إليه في شفاء المرض[٧٠٣] هو إفراط هذه الأجناس ومجاوزتها[٧٠٤] الاعتدال[٧٠٥] إلى خلاف جهة المرض، والذي يحتاج إليه في حفظ صحّة الأبدان التي صحّتها دون صحّة البدن الفاضل البنية أحد أمرَيْنِ، وذلك أنّك إذا أردت أن تحفظ هذه الأبدان على طبائعها، فينبغي لك[٧٠٦] أن تستعمل فيها من هذه الأجناس ما هو مجاوز الاعتدال[٧٠٧] إلى الطّرف[٧٠٨] الذي ذلك البدن مائل إليه ليكون مشبهاً له، وإن أردتَ أن تنقله[٧٠٩] إلى المزاج الفاضل، فينبغي لك ما دمتَ في نقله[٧١٠] أن تستعمل من هذه الأجناس ما هو مخالف للوجه الذي[٧١١] قد[٧١٢] تجاوزت[٧١٣] تلك الأبدان الاعتدال إليه، وإذا فرغت من [٧١٤] نقلها استعملت الاعتدال.

(٦٩) لا يخلوا فساد الأعضاء النّاقصة عن الهيئة الفاضلة من أن يكون إمّا في الأعضاء المتشابهة الأجزاء وإمّا في الأعضاء المركّبة، فإن كان الفساد في الأعضاء المتشابهة الأجزاء، فليس يخلوا من أن يكون إمّا فساد في جميعها[٧١٥] بالسّواء وإمّا في بعضها دون بعض على غير تساوي[٧١٦]، وكلّ واحد من هذيْنِ أيضاً[٧١٧] إمّا أن يكون[٧١٨] بسيطاً وإمّا[٧١٩] مركّباً، وكلّ واحد من هذيْنِ أيضاً إمّا أن يحتاج إلى

---

٧٠٣ AM: المرضيّ؛ F: الأمراض | ٧٠٤ DSY: تجاوزها | ٧٠٥ AM: للاعتدال؛ DY: عن الاعتدال | ٧٠٦ DFY: – لك | ٧٠٧ AM: للاعتدال | ٧٠٨ F^1؛ AM: الجانب | ٧٠٩ AM: تنقلها | ٧١٠ AM: نقلها | ٧١١ AM: للوجوه التي | ٧١٢ DSY: – قد | ٧١٣ AM: جاوزت | ٧١٤ F: إلى ناحيته فإذا استتمّت؛ DSY: فرغت من > استممّت | ٧١٥ AM: جميعاً | ٧١٦ AM: مساواة؛ S: التّساوي | ٧١٧ AM: الفسادَيْن | ٧١٨ S: + كيفيّة واحدة أعني | ٧١٩ F: أو؛ S: أو في كيفيّتَيْنِ أعني

which is when a person employs the necessary entities,[152] or it must be corrected at the time of treatment by things opposite to it over a long period of time. <If the corruption is in the compound—which is to say, the instrumental—organs, then it is either simple or compound. If it is simple, then it is in either the structure, the number, or the position. The corruption that occurs in the structure is in either the shape, the cavity, the ducts, the roughness, or the smoothness. The corruption that occurs in the number is either excess or deficiency. The corruption that occurs in size is in being either too large or too small. The corruption that occurs in position is either in the organ being moved from its place or in its relation to the organs with which it is associated.>[153]

(70) Some of the necessary causes that affect the body dry, and some moisten.[154] Those that dry are exercise; hot, dry air; insomnia; evacuation; a scanty diet; and all the accidents of the soul. Those that moisten are rest; sleep; retention of what might be evacuated; a heavy diet; and cool, moist air. Some of the necessary causes that affect the body warm, and some cool. Those that warm are moderate exercise; a moderate diet,[155] especially if hot; hot air; retention of something hot, such as vapor; a preference for moderate food;[156] moderate waking; and moderate sleep. Among the accidents of the soul, anger always warms, worry usually does, and joy sometimes does. Those that cool are strenuous

152. That is, the six non-naturals discussed on pp. 111–12, paragraph 66, above.

153. This reads as though it could properly be part of the text, although three of the four oldest MSS omit it, perhaps because the topic is discussed again in paragraph 73 below.

154. B 25.3–5; K 1:373–74; G^a 123–25: An elaboration of a brief comment by Galen.

155. One MS has the gloss: "A gloss of his: That is, moderate in quantity."

156. Some MSS read: "a small quantity of moderate food."

حفظه بالأشياء المتشابهة[٧٢٠] له، وذلك عند ما يكون الإنسان متشاغلاً بأمور اضطراريّة، وإمّا أن يحتاج إلى إصلاحه في وقت الفراغ على طول المدّة بالأشياء المضادّة له، ‹وإن كان الفساد في الأعضاء المركّبة، وهي الآليّة، فليس يخلوا أن يكون إمّا بسيطاً وإمّا مركّباً، فإن كان بسيط، فإمّا أن يكون في الخلقة وإمّا في العدد وإمّا في الوضع. أمّا في الخلقة فيعرض فيها الفساد إمّا في الشّكل وإمّا في التّجويف وإمّا في المجاري وإمّا في الخشونة وإمّا في الملاسة ، وأمّا العدد فيعرض فيه الفساد إمّا بالزّيادة وإمّا بالنّقصان، وأمّا المقدار فيعرض فيه الفساد في العظم وأمّا في الصّغر وأمّا الوضع فيعرض فيه الفساد أمّا بنقل العضو عن موضعه وإمّا المشاركة لما يشاركه من الأعضاء ما[٧٢١].›

(٧٠) والأسباب[٧٢٢] التي تعرض للبدن باضطرار، منها ما يجفّف ومنها ما يرطب، أمّا التي تجفّف فالحركة والهواء الحارّ اليابس والسّهر والاستفراغ وقلّة الغذاء وجميع عوارض النّفس، وأمّا التي ترطّب فالسّكون والنّوم واحتباس ما يستفرغ وكثرة الغذاء والهواء الرّطب البارد، والأسباب التي تعرض للبدن باضطرار، بعضها يسخّن وبعضها يبرّد، أمّا التي تسخّن فالحركة المعتدلة والغذاء المعتدل[٧٢٣]، ولا سيّما إن كان حارّاً والهواء الحارّ واحتباس الشّيء الحارّ بمنزلة البخار وقصد في[٧٢٤] الغذاء المعتدل[٧٢٥] والسّهر المعتدل والنّوم المعتدل ومن عوارض النّفس والغضب[٧٢٦] يفعل ذلك دائماً، والهمّ في أكثر الأمر[٧٢٧] والفرح في بعض الأوقات، وأمّا[٧٢٨] التي تبرّد،

---

٧٢٠ DFY: المشابهة | ٧٢١ AMS: – وإن كان الفساد. . . الأعضاء ما؛ DY: – لما. . . ما | ٧٢٢ ADMY: الأسباب | ٧٢٣ M^h: حـ [اشية] له: أي في مقداره | ٧٢٤ F^1؛ $AFMY^2$: وقلّة: والقصد في؛ DY: والفضل في | ٧٢٥ AM: باعتدال؛ Y^h: وفي نسخة مكان المنعوظ؟ وقلّة الغذاء باعتدال | ٧٢٦ ADMY: الغضب | ٧٢٧ F: على الأكثر؛ DSY: الأوقات | ٧٢٨ DSY: فأمّا

exercise; an excessive diet, especially if the food is cold; air that is excessively hot or cold; delay in evacuation; excessive evacuation; an excessively scanty diet; excessive insomnia; and excessive sleep. Among the accidents of the soul, anxiety always cools, fear[157] usually does, and joy sometimes does, provided that it is excessive. Some foods and drinks warm the body, such as meat and foods seasoned with pepper and mustard, and spiced, honeyed wine.[158] Others cool, such as fruit, *rūsātaq* fish,[159] and cold water. Some occupations warm, such as wrestling and the craft of the blacksmiths, while others cool, such as seafaring and fulling.

(71) There are five genera of causes of heat:[160] first, exercise that is not excessive;[161] second, contact with things that warm when the contact is moderate; third, matter analogous to heat such as hot foods, drinks, and drugs; fourth, retardation of the dissolution of something hot, which is accomplished by means of compression; and, fifth, putrefaction. There are six genera of causes of coldness: first, very excessive exercise; second, rest; third, contact with things that heat excessively, which means the same thing as dissolution; fourth, contact with things

157. Some MSS read "leisure."

158. See p. 21, paragraph 20, n. 49, above.

159. Another word that the MSS have difficulty with, rendered variously as *rūsātaq*, *rūsāṭan*, and *dūstāṭan*. A gloss in several MSS reads: "This is a species of fish found along the shores of *khilāṭ* [or *akhlāṭ*] *zaᶜm* [?]."

160. B 23.9–10, 25.3–5; K 1:368, 373–74; G[a] 115–16, 124–25: A systematization of Galen's sketchy examples.

161. Some MSS read: "excessive exercise."

فالحركة العنيفة[٧٢٩] والاستكثار من الغذاء وخاصّة إن كان بارداً والهواء المفرط في الحرّ والبرد[٧٣٠] والإبطاء فيما[٧٣١] يستفرغ والاستفراغ المفرط وقليل من[٧٣٢] الغذاء جدّاً[٧٣٣] والسّهر المفرط والنّوم المفرط، ومن عوارض النّفس الغمّ يفعل ذلك[٧٣٤] دائماً[٧٣٥] والتّفزّع في أكثر الأمر[٧٣٦] والفرح في بعض الأوقات إذا[٧٣٧] أفرط. الأطعمة والأشربة منها ما يسخن بمنزلة اللّحم والأطعمة التي المتّخذة[٧٣٨] بالفلفل والخردل والخنديقون[٧٣٩]، ومنها ما يبرّد بمنزلة الفاكهة والرّوساطق[٧٤٠] والماء البارد، والصّناعات منها ما تسخن البدن[٧٤١] بمنزلة المصارعة وصناعة الحدّادين، ومنها ما يبرّد بمنزلة الملاّحة والقصارة.

(٧١) أجناس أسباب الحرارة خمسة، أحدها الحركة غير[٧٤٢] المفرطة[٧٤٣] والثّاني[٧٤٤] لقاء الأشياء المسخنة[٧٤٥] إذا كان لقاءها معتدلاً، والثّالث المادّة الموافقة للحرارة بمنزلة الأطعمة والأشربة والأدوية الحارّة، والرّابع إبطاء تحلّل الشّيء الحارّ، وذلك يكون بسبب التّكاثف، والخامس العفونة. وأجناس أسباب[٧٤٦] البرودة ستّة، أحدها الحركة المفرطة جدّاً[٧٤٧]، والثّاني[٧٤٨] السّكون[٧٤٩]، والثّالث ملاقاة الأشياء التي تسخن بإفراط، ومعنى هذا القول[٧٥٠] معنى التّحلّل[٧٥١]، والرّابع ملاقاة الأشياء التي تبرّد،

---

٧٢٩ ADMY: المفرطة | ٧٣٠ AM: و في البرد؛ DY: أوالبرد | ٧٣١ AM: والإفراط في احتباس ما | ٧٣٢ AM: وقلّة؛ S: والتّقليل من | ٧٣٣ AM: المفرط | ٧٣٤ DSY: – يفعل ذلك | ٧٣٥ DY: دائم | ٧٣٦ AM: والفزع في الأمر الأكثر | ٧٣٧ DSY: إن | ٧٣٨ AM: التي تتّخذ | ٧٣٩ A: والفنديقون؛ A^1Y^1: الخنديقون؛ SY: والقنديطون | ٧٤٠ DFS: الرّوساطن؛ Y: والدّوساطن؛ $A^hM^hY^h$: حـ[اشية؛ Y^h: في نسخة حاشية زعم أنّه] هذا [Y^h: – هذا] نوع من السّمك يكون في نواحي خلاط زعم [Y^h: خلاط زعم > أخلاط]. | ٧٤١ AM: – البدن | ٧٤٢ FSY: – غير | ٧٤٣ S: المعتدلة؛ F^h: بمنزلة الهواء الحارّ location uncertain | ٧٤٤ AM: والآخر | ٧٤٥ AM: التي يسخّن | ٧٤٦ F: وأسباب أجناس | ٧٤٧ DSY: – جدّاً | ٧٤٨ AM: والآخر | ٧٤٩ DFY: + المفرط | ٧٥٠ S^1؛ AMSY: + هو | ٧٥١ ADSY: التّخلخل؛ A^2: التّخلّل؛ F: التّحليل

that cool; fifth, matter compatible with coldness, such as cold foods and drinks; and, sixth, an excessively meager diet.

(72) If corruption occurs in the tissues and the defective temperament is uniform throughout all the organs, then a single species of treatment ought to be used for the entire body.[162] If the defective temperament is not uniform, then each of the organs ought to be treated with something specifically suited to it.

162. B 25.10–11; K 1:375–76; G[a] 127.

والخامس المادّة الملاءمة[٧٥٢] للبرودة بمنزلة الأطعمة والأشربة الباردة، والسّادس تقليل[٧٥٣] الغذاء المفرط[٧٥٤].

(٧٢) إذا حدث في الأعضاء المتشابهة الأجزاء فساد، فكان سوء[٧٥٥] المزاج الرّديّ متساويًا في الأعضاء كلّها، فينبغي أن يستعمل في مداواة البدن كلّه[٧٥٦] نوعًا واحدًا من المداواة، فإن[٧٥٧] كان سوء[٧٥٨] المزاج الرّديّ غير متساوي[٧٥٩]، فينبغي أن يداوى كلّ واحد من الأعضاء بالشّيء الذي هو موافق[٧٦٠] له خاصّةً.

٧٥٢ AM: الملاومة؛ F: الملامة؛ DSY: الملائمة | ٧٥٣ AM: قلّة | ٧٥٤ AM: بإفراط | ٧٥٥ AM: فإنّه إن كان؛ S : وكان سوء | ٧٥٦ DSY: بأسره | ٧٥٧ ADMY: وإن | ٧٥٨ AM: – سوء | ٧٥٩ AM: على غير مساواة | ٧٦٠ AM: الموافق

[Chapter 26][163]

[Classes of organic diseases and their treatment]

(73) There are four aspects of the functional organs in which the disorders of these organs occur.[164] These are the structure of the organ, its dimensions, its number, and its position. Each of these four exists in bodies in one of four ways. First, it may be in the excellent form, which is when it has the most excellent form and is most moderate. Second, it may deviate only slightly from the most excellent form, and for this reason it is accounted among the states of healthy bodies. Third, it may have deviated enough from the most excellent states so as to be not far from disease. A body in such a state is called "sickly." Fourth, it may have deviated greatly from the state of the healthy body. When a body is like this, it is called "diseased."

(74) Five kinds of defects and diseases occur in the structure of the organs: First, they may involve the shape, which happens whenever the shape of the organs changes—as when what is round becomes elongated. Second, the ducts, apertures, and orifices may contract, expand, or be

---

163. B 26.1–5; K 1:376–77; G[a] 127–29.

164. Galen expresses this idea in slightly different ways in his separate discussions relating to the structure, dimensions, number, and positions of the organ. This particular formulation is the one he used in referring to defects of position and is quoted almost verbatim here. Throughout this chapter, the epitomist adds examples of his own or draws examples from later in *The Small Art* to fill out Galen's chapter 26.

[الفصل السّادس والعشرون]

[ذكر أقسام الأمراض الآليّة وعلاجها]

(٧٣) الأشياء التّابعة[٧٦١] للأعضاء[٧٦٢] الآليّة التي فيها تحدث آفات هذه الأعضاء[٧٦٣] أربعة أشياء، وهي خلقة الأعضاء ومقاديرها وعددها و وضعها، وكلّ واحد من هذه الأربعة يوجد في الأبدان على أحد[٧٦٤] أربعة أصناف، أحدها[٧٦٥] الهيئة الفاضلة، وهو أن يكون على أفضل الهيآت وأشدّها اعتدالاً، والثّاني[٧٦٦] أن يكون دون[٧٦٧] الفاضلة[٧٦٨] بقليل، فيدخل بهذا السّبب في عداد حالات الأبدان الصّحيحة، والثّالث أن يكون قد بعد[٧٦٩] عن الهيئة التي هي[٧٧٠] أفضل الهيآت[٧٧١] حتّى يكون قد قارب[٧٧٢] المرض، ويقال إنّ حينئذ[٧٧٣] للبدن[٧٧٤] الذي حاله هذه الحال ممراض، والرّابع أن يكون قد بعد عن حال البدن الصّحيح بعداً كثيراً، ويقال له[٧٧٥] إذا كان كذلك[٧٧٦] مريضاً.

(٧٤) والذي يحدث في الخلقة من الآفات والأمراض خمسة أصناف، أحدها ما يحدث في الشّكل، وذلك عندما يتغيّر شكل[٧٧٧] العضو[٧٧٨] بمنزلة ما يصير المستدير مطاولاً، والثّاني[٧٧٩] ما يحدث في المجاري والثّقب والمنافذ عندما يضيق أو يتّسع

٧٦١ SY: تلزم | ٧٦٢ DSY: الأعضاء | ٧٦٣ DSY: التي. . . الأعضاء> إذا حدثت فيها الآفات | ٧٦٤ DFY: -أحد | ٧٦٥ AM: + على | ٧٦٦ AM : والآخر | ٧٦٧ AM: أبعد عن | ٧٦٨ AM : التي هي أفضل الهيآت | ٧٦٩ AM: قد بعد> أبعد | ٧٧٠ DSY: -الهيئة. . . هي | ٧٧١ AM: + بقليل | ٧٧٢ DSY: حتّى. . . قارب> وقارب | ٧٧٣ AM: إنّ | ٧٧٤ AMS: البدن | ٧٧٥ AM: للبدن | ٧٧٦ DSY: ذلك | ٧٧٧ AM: + أحد | ٧٧٨ AM: الأعضاء | ٧٧٩ AM: والآخر

blocked. Third, the cavities may shrink, tighten, expand, loosen, fill, or become blocked. Fourth, there is what happens when the roughness of an organ that is naturally rough becomes smooth. Fifth, there is what happens when the natural smoothness of the organ that is naturally smooth is roughened. There are two kinds of disorders and diseases occurring in the dimensions of the organs: first, when an organ becomes small that ought by nature to be large; and, second, when an organ is enlarged that ought by nature to be small. There are four kinds of disorders and diseases occurring with respect to the number of the organs, two occurring when the number is greater than is natural and two when the number is less than is natural. One of the two kinds of superfluity involves an excess of things that exist naturally, such as a sixth finger. The other involves things that do not exist naturally, such as worms growing in the belly and scrofula growing on the neck.[165] One of the two kinds of deficiency is when a part of an organ is lacking, and the other is when an organ is lacking entirely. There are two kinds of disorders occurring with respect to the position: First, the organ may be removed

165. Scrofula is tuberculosis of the lymphatic glands, characterized by swellings on the neck.

أو ينسدّ[٧٨٠]، والثّالث ما يحدث في التّجويف إذا صغر[٧٨١] أو ضاق[٧٨٢] أو كبر[٧٨٣] أو اتّسع[٧٨٤] أو امتلى[٧٨٥] أو انسدّ[٧٨٦]، والرّابع ما يحدث في الخشونة الطّبيعيّة عندما[٧٨٧] يملّس[٧٨٨] العضو الذي هو بالطّبع خشن، والخامس ما يحدث في الملاسة الطّبيعيّة[٧٨٩] إذا خشن عضواً[٧٩٠] هو بالطّبع أملس، والذي يحدث في مقادير الأعضاء من الآفات والأمراض صنفان، أحدهما أن يكون العضو الذي ينبغي أن يكون بالطّبع[٧٩١] كبيراً[٧٩٢] يصغر، والآخر[٧٩٣] أن يكون العضو للذي ينبغي أن يكون[٧٩٤] بالطّبع صغيراً يعظم، والذي يحدث في العدد من الآفات أربعة أصناف اثنانِ منها يحدث[٧٩٥] إذا كان العدد زايداً على ما في الطّبع، واثنانِ إذا كان العدد ناقصاً عمّا في الطّبع، أمّا الصّنفانِ الزّائدانِ، فأحدهما يكون من جنس الأشياء الموجودة في الطّبع بمنزلة الأصبع السّادسة، والثّاني[٧٩٦] من جنس الأشياء الخارجة عن الطّبع بمنزلة الدّود المتولّد في البطن والخنازير المتولّدة في الرّقبة، وأمّا الصّنفانِ النّاقصانِ فأحدهما يكون إذا نقص[٧٩٧] جزء من العضو، والثّاني إذا نقص[٧٩٨] عضو بأسره[٧٩٩]، والذي يحدث في الوضع[٨٠٠] من الآفات صنفانِ، أحدهما أن ينقل العضو[٨٠١] عن موضعه بمنزلة ما

٧٨٠ DF: إذا ضاقت أو اتّسعت أو اشتدّت؛ SY: إذا ضاقت واتّسعت [Y: أو اتّسعت] أو أنسدت | ٧٨١ AM: إذا صغر > عندما يصغر أو | ٧٨٢ AM: يضيق؛ F: ذاق | ٧٨٣ AM: يكبر؛ D: وكبر ؛ F: تكبّر | ٧٨٤ AM: يتّسع؛ DY: واتّسع | ٧٨٥ AFM: يمتلي؛ DY: وامتلى | ٧٨٦ AM: ينسدّ؛ DY: وانسدّ؛ F: ينسدّ | ٧٨٧ DSY: إذا | ٧٨٨ A: يتملّس؛ M: تماسّ | ٧٨٩ A^2M^2: الملوسة | ٧٩٠ DFY: عضو | ٧٩١ S: ينبغي. . . بالطّبع > هو بالطّبع سبيله؛ DY: > سبيله | ٧٩٢ F: كير؛ DSY: أكبر | ٧٩٣ S: الآخر | ٧٩٤ F: – للذي. . . يكون؛ DSY: للذي. . . يكون > سبيله | ٧٩٥ F: يحدثانِ؛ DSY: – يحدث | ٧٩٦ AM: والآخر | ٧٩٧ AM: عندما ينقص | ٧٩٨ AM: والآخر عندما ينقص | ٧٩٩ AM: كما هو جملة | ٨٠٠ F: الموضغ | ٨٠١ AM : يكون العضو ينتقل ويزول

entirely from its position, as occurs with the intestines when they move from their place in the coils and drop down near the testicles,[166] or with one of the joints when it is dislocated and leaves its normal position. Second, an organ that is normally near to or apart from another organ when it is in use is changed from the way that it usually is, as happens with the fingers, the lips, or the eyelids when they are close together and cannot separate or are separated and cannot come together.

(75) There are two means by which it is possible to treat the disorders occurring in the shape of organs: first, by straightening and correcting through surgery; and, second, by binding.[167] The causes with which it is possible to treat the disorders occurring in the cavities differ in accordance with the disorder. For example, if the cavity is enlarged and needs to be reduced in size, it is treated by binding and rest. If it is too small and needs to be enlarged, it is treated by exercising it or by that retention of the breath called, in Greek, κατάληψις, which is when someone holds his breath but nonetheless tries to force out the air of his breath. The means by which it is possible to treat the disorders arising in the dimensions of the organs also differ in accordance with the various kinds of these disorders. For example, if the disorder involves the organ being of excessive size, it is treated by rest and binding, while if it involves the organ's dimensions being insufficient, it is treated by exercise and massage. There are a variety of diseases involving

166. Some MSS read: "In a rupture."

167. B 26.6–10, 33.2–3; K 1:377–78, 391; G[a] 129–32, 153–54: The reader should remember that the original meaning of surgery was "the work of the hand" and that, in Galen's works, it refers to physical therapy as well as the use of the knife.

يعرض للأمعاء إن ينتقل ويزول عن موضعها في الفتل[٨٠٢] وينحدر[٨٠٣] إلى[٨٠٤] الانثيين أو يعرض لبعض[٨٠٥] المفاصل الانخلاع[٨٠٦] والخروج[٨٠٧] عن موضعه، والثّاني[٨٠٨] أن يكون العضو الذي[٨٠٩] شأنه أن يقرب أو يبعد من عضو آخر في أوقات الحاجة إلى ذلك يتغيّر عمّا كان عليه بمنزلة ما يعرض للأصابع والشّفتين أو للجفنين إن[٨١٠] يقرب الواحد من الآخر ولا يتباعد عنه، أو يتباعد ولا يدنوا منه.

(٧٥) والأسباب[٨١١] التي بها يكون[٨١٢] إصلاح[٨١٣] ما يحدث في الشّكل من الآفات اثنان، أحدهما التّقويم والإصلاح باليد، والآخر الرّباط[٨١٤]، وأمّا الأسباب التي يكون بها[٨١٥] إصلاح ما يحدث في التّجويف من الآفات فيختلف بحسب اختلاف الآفة، وذلك أنّه إن كان التّجويف قد عظم ويحتاج إلى أن يصغر فإصلاحه يكون بالرّباط والسّكون، وإن كان قد صغر ويحتاج إلى أن يكبر[٨١٦]، فإصلاحه يكون بتحريكه بالعمل وحصر[٨١٧] النّفس الذي يقال له باليونانيّة قاطالبسس[٨١٨]، وهو أن يحبس[٨١٩] الإنسان نفسه ويدفع مع ذلك[٨٢٠] هواء[٨٢١] التّنفّس[٨٢٢]، وأمّا الأسباب التي بها يكون إصلاح ما يحدث في مقادير الأعضاء، فهي أيضاً تختلف بحسب أصناف هذه الآفات، وذلك أنّه إن كانت الآفة إنّما تحدث[٨٢٣] من طريق أنّ مقدار العضو زاد فإصلاحه يكون بالسّكون والرّباط، وإن كانت إنّما حدثت من طريق أنّ المقدار نقص فإصلاحه يكون بالحركة والدّلك. شكل[٨٢٤] الأعضاء[٨٢٥] يفسد[٨٢٦]

---

٨٠٢ F: – إن ينتقل. . . القتل؛ DSY: إن يزول عن وضعها في الفتق | ٨٠٣ F: أو ينحدر؛ Y: وينجذب | ٨٠٤ F: + موضعها فمن الفتق | ٨٠٥ AM: لمفصل من | ٨٠٦ AM : أن ينخلع | ٨٠٧ AM : ويخرج | ٨٠٨ AM : والآخر | ٨٠٩ AM : + من | ٨١٠ F: أو؛ S: أنّه | ٨١١ DSY: الأسباب | ٨١٢ AM: يكون بها | ٨١٣ AM: صلاح | ٨١٤ DF: بالرّباط | ٨١٥ DFY: بها يكون | ٨١٦ AM: يعظم | ٨١٧ AM: وإمساك | ٨١٨ AM: قاطانيسيس؛ S: قاطاليسيس | ٨١٩ AM: يمسك | ٨٢٠ DSY: ويدفع. . . ذلك> ويدافع | ٨٢١ DY: هو؛ F: هذا | ٨٢٢ FM: النّفس | ٨٢٣ AM: حدثت | ٨٢٤ AM: وشكل | ٨٢٥ DSY: + إمّا أن | ٨٢٦ F: + في سائر البدن

distortion in the shape of an organ. It can happen that the head is long and narrow. A person's spine can have a hump. The legs can be bowed inwards or outwards. In other organs, there can be conditions such as when the thigh is twisted.

(76) A blockage occurs either primarily or accidentally.[168] The accidental blockage occurs, for example, from some kind of swelling.[169] The primary blockage may occur from coarse, viscid humors, in which case it is treated by things that cut it and by things that cleanse it, such as oxymel and honey-water. It can also occur from some other coarse superfluity, such as hard feces, in which case it is treated first by moistening and then by cutting that coarsening, using enemas having an excess of sharpness. Finally, they can occur from things whose genus is completely unnatural, such as stones. Things that are entirely contrary to nature are treated by removing them from the body completely. Things whose quantity alone is contrary to nature are to be treated by decreasing them. If there is something in the blood vessels whose quantity is excessive, or something whose quantity is becoming excessive, it is to be treated by bloodletting from the vein. If it is something in the stomach, it is to be treated by an emetic. If it is something in the chest, it is to be removed by coughing. If it is something in the liver, then it depends on which side it is on. If it is on the concave side of the liver, then it is to be removed through the intestines by purgation; but if it is on the convex side of it, it is to be removed by means of the urine. An accidental

168. B 33.4–15; K 1:391–93; G^a 154–58.
169. Some MSS omit this sentence.

إمّا في الرّأس بمنزلة ما يعرض له إذا كان مسفّطاً، وإمّا في الصّلب بمنزلة ما يعرض إذا صارت بالإنسان[٨٢٧] حدبة، وإمّا في السّاق بمنزلة ما يعرض إذا كان السّاق مقوساً[٨٢٨] إلى داخل أو إلى خارج، وإمّا في غير هذه الأعضاء فبمنزلة[٨٢٩] ما يعرض للفخذ إذا أعوجت[٨٣٠].

(٧٦) السّدّة تحدث إمّا حدوثاً أوّليّاً وإمّا حدوثاً عرضيّاً، فأمّا السّدّة العرضيّة فبمنزلة ما يعرض منها بسبب ورم من الأورام وأمّا السّدّة[٨٣١] الأوّليّة[٨٣٢]، فيكون إمّا من أخلاط غليظة[٨٣٣] لزجة[٨٣٤] ومداواتها[٨٣٥] بالأشياء التي تقطع، والأشياء التي تجلوا بمنزلة السّكنجبين وماء العسل، وإمّا من فضل آخر غليظ بمنزلة الرّجيع[٨٣٦] الصّلب ويداوى أوّلاً بالتّرطيب ثمّ[٨٣٧] بتقطيع ذلك الغلظ بالحقن التي لها فضل حدّة، وإمّا من شيء[٨٣٨] جملته من جنس ما هو خارج عن الطّبيعة بمنزلة الحصاة، ومداواة الأشياء التي جملتها من جنس ما هو خارج عن الطّبيعة يكون بإخراجها عن البدن أصلاً، فأمّا الأشياء التي مقدارها فقط خارج عن الطّبيعة[٨٣٩]، فمداواتها تكون بتنقيصها، وأمّا من شيء قد كثر مقداره والشّيء الذي يكثر مقداره إن كان[٨٤٠] في العروق، فينبغي أن يستفرغ بفصد العرق، وإن كان شئ في المعدة فبالقيء، وإن كان شيء في الصّدر فبالسّعال، وإن كان شيء في الكبد فبحسب النّاحية التي هو فيها، وذلك أنّه إن كان في الجانب المقعر من الكبد[٨٤١] فينبغي أن يستفرغ من الأمعاء بالإسهال، وإن كان في الجانب المحدب منها فينبغي أن يستفرغ بالبول، وأمّا السّدّة

٨٢٧ DFY: للإنسان | ٨٢٨ F: مقوّس؛ SY: مقوّسة | ٨٢٩ ADM: بمنزلة | ٨٣٠ AM: للفخذ إذا أعوجت > عند تعويج الفخذ | ٨٣١ AM: - العرضيّة. . . السّدّة | ٨٣٢ F: الآليّة | ٨٣٣ DFY: خلط غليظ | ٨٣٤ DFY: لزج | ٨٣٥ AM: وتداوي | ٨٣٦ ADMY: الثّقل | ٨٣٧ AM: ومن بعده | ٨٣٨ D: جنس؛ FY: جنس ما | ٨٣٩ AM: الطّبع | ٨٤٠ A: + شيئاً؛ M: + شيء | ٨٤١ DSY: منها

blockage may result from a hematoma, induration, edema, dryness, or distortion of the shape of the organ.[170]

(77) Roughness may occur in a bone, in which case that bone should be scraped until it is smooth.[171] If it is in the tongue, it should be smoothed by viscid, glutinous things such as gum and ispaghula. If it is in the windpipe, it is to be smoothed with gum tragacanth or licorice. If there is a smoothness in the bone, it should also be scraped until it is rough. If it is in the womb, then the humor that makes it smooth should be purged.

(78) Things whose number is in unnatural excess should be removed by either steel, fire, or caustic drugs.[172] Those that are too few in number should be treated in accordance with the source that they come from. For example, if the deficiency is in an organ created from blood, it might be possible to complete it, as is seen with flesh that is missing due to a deep ulcer. If the deficiency is in an organ that is generated from semen, it is not possible to make a complete replacement for it; but we can substitute other things for it, some of them acting as binding, as is done when something hard is fastened at the place where a bone is broken in order to bind it and connect it as flesh does. Some can be corrected by roughening the place, as is done with a lip that is shorter than it ought to

170. See p. 108, paragraph 62 and n. 138, on the kinds of swellings.

171. B 33.14–15; K 1:393; G[a] 158. Two MSS have the gloss: "A gloss of his: In certain ulcers, the bone may become so rough as to mar its surface, the damage making it difficult for the flesh to adhere to it. That is what must be smoothed until the damage is gone. In other ulcers, the bone may become so smooth due to an injurious moisture that it is also difficult for the flesh to adhere to it. In that case, the bone must be roughened until that moisture is gone, allowing the flesh to adhere to it."

172. B 26.9–10, 35.1–6; K 1:378, 401–2; G[a] 131–32, 169–72: Galen also mentions cosmetic surgery in the case of mutilation or deformity.

العرضيّة، فتكون إمّا من ورم دمويّ وإمّا من ورم صلب وإمّا من ورم رخو وإمّا من يبس وإمّا من فساد شكل العضو.

(٧٧) الخشونة[٨٤٢] تكون إمّا في عظم، وينبغي حينئذ أن يحكّ ذلك العظم حتّى يتملّس، وإمّا في اللّسان، فينبغي[٨٤٣] أن يملّس بالأشياء اللّزجة التي تغرّي مثل الصّمغ[٨٤٤] والبز رقطونا[٨٤٥]، وأمّا ماكان في قصبة الرّئة، فينبغي أن يملّس بالكثيراء وأصل السّوس، والملاسة[٨٤٦] أمّا[٨٤٧] أن تكون في عظم، وينبغي[٨٤٨] أيضاً أن يحكّ حتّى يخشن، وأمّا في الرّحم فينبغي أن يستفرغ ذلك الخلط الذي ملسها[٨٤٩].

(٧٨) الأشياء التي عددها خارج عن الطّبيعة ماكان منها من طريق الزّيادة، فينبغي أن يقطع إمّا بالحديد وإمّا بالنّار وإمّا بالأدوية المحرقة، وماكان منها من طريق النّقصان، فينبغي أن يعمل في أمره بحسب الأصل الذي منه كان، وذلك أنّه إن كان[٨٥٠] في عضو خلق[٨٥١] من الدّم، فقد[٨٥٢] يمكن أن يتمّم بمنزلة ما يرى ذلك في اللّحم إذا نقص في قرحة غائرة، وإن كان النّقصان في عضو كونه من المني، فليس يمكن أن يتمّم وتخلف[٨٥٣] بدلاً منه[٨٥٤]، ولكنّا نعوض منه بأشياء آخر، بعضها يقوم مقام الرّباط بمنزلة ما يفعل في العظم المكسور بإثباتنا على موضع الكسر شيئاً صلباً يربطه ويمكسه[٨٥٥] بمنزلة اللّحام، وبعضها يصلح لتخشين الموضع بمنزلة ما يفعل[٨٥٦] في الشّفة التي قد قصرت عمّا

٨٤٢ A^hM^h: [حاشية] له: يعرض في بعض القروح أن تخشن العظم خشونة تفسد سطحه، ويعسر لفساده ثبات اللّحم عليه، فذلك هو الذي يملّس حتّى يذهب الفساد، وقد يعرض في بعض القروح أن يملس العظم ملاسة برطوبة رديّة فيعسر أيضاً ثبات اللّحم عليه، فذلك هو الذي يخشن حتّى يذهب تلك الرّطوبة، فتثبت اللّحم عليه. | ٨٤٣ AM: وينبغي | ٨٤٤ AM: كالصّمغ | ٨٤٥ ADMY: وبز رقطونا؛ F: والبزقطونا؛ | ٨٤٦ F: والملائة | ٨٤٧ DSY: فأمّا | ٨٤٨ DY: فينبغي | ٨٤٩ AM: صار به أملساً | ٨٥٠ AM: + النّقصان | ٨٥١ F^2: + كونه | ٨٥٢ DSY: قد | ٨٥٣ DSY: + عليه؛ D^1Y^1: يخلق | ٨٥٤ AM: بدله | ٨٥٥ F: يمسكه | ٨٥٦ AM: + ذلك

be when we split it, lest it become limp and cover the teeth. Some of the organs are created from sperm. These include all the primary hard organs, such as the arteries and veins, the nerves, and the bones. Therefore, these organs are not likely to grow a replacement when something is cut from them, since the matter from which they are formed is not prepared and ready in the body. Other organs—the flesh and the fleshy organs—are created from the menstrual blood. Because the blood is continually prepared and made ready in the body, if something is cut from these organs, it might grow back. If the things whose number is unnatural[173] in the sense that they are too few are created from blood, it is possible for them to return to their original state at any age. However, if they are among the organs created from the semen, it is only possible for them to come into being in childhood, since it is only at this age that there is a remnant of the substance of the semen ready in the body. If it is unnatural in the sense of being too many, then it must either be removed completely by surgery, as in the case of scrofula, or be moved so as no longer to be in that place, as is the case with water flowing down into the eye.

(79) The diseases occurring due to the positions of organs are treated by returning the organs to their natural positions and keeping them there by cauterization and binding.[174] One of the diseases of position is dislocation, which is something that occurs by a forcible stretching. It is

173. Two MSS have the gloss: "A gloss of his: It can only truly be said that things are unnatural in number if they have, by nature, a particular number."

174. B 26.10, 35.7–8; K 1:378, 402–3; G[a] 132, 172–73.

ينبغي أنا[٨٥٧] نشقّها كيما تسترخي وتغطي الأسنان. الأعضاء منها ما خلق من المنى، وهي جميع الأعضاء الأصليّة الصّلبة بمنزلة العروق[٨٥٨] الضّوارب وغير الضّوارب[٨٥٩] والعظام، ولذلك صارت هذه الأعضاء متى انقطع منها شيء لم ينبت بدلًا[٨٦٠] منه لأنّ المادّة التي منها خلقت ليست معدّة مهيّأة[٨٦١] في البدن، ومنها ما خلق من دم الطّمث، وهي[٨٦٢] اللّحم والأعضاء اللّحميّة، ولأنّ الدّم لا يزال مهيّئًا معدًّا في البدن دائمًا صارت هذه الأعضاء إذا انقطع منها شيء عاد ونبت من الرّأس[٨٦٣] الأشياء التي هي خارجة عن الطّبع في عددها[٨٦٤] ما كان منها من طريق النّقصان[٨٦٥] إن كان ممّا خلق من الدّم، فهو يمكن أن يعود إلى حاله[٨٦٦] في كلّ سنّ من الأسنان، وإن كان ممّا خلق من المنى، فإنّما يكون ممكنًا[٨٦٧] أن يعود[٨٦٨] في سنّ الصّبيّ فقط، لأنّ في هذا السّنّ فقط[٨٦٩] بقيّة من جوهر المنى معدّة في البدن، وأمّا ما كان منها من طريق الزّيادة، فيحتاج إمّا إلى[٨٧٠] أن يستأصل كلّه[٨٧١] بمنزلة الخنازير، وإمّا إلى[٨٧٢] أن ينقل[٨٧٣] عن موضعه بمنزلة الماء النّازل في العين.

(٧٩) وأمّا[٨٧٤] الأمراض الحادثة في وضع الأعضاء، فإنّها تداوى بردّ الأعضاء إلى مواضعها الطّبيعيّة وحفظها فيها[٨٧٥] بالكيّ والرّباط، ومن الأمراض الحادثة في الوضع الخلع، وهو شيء يحدث بسبب المدّ القاسر، ويداوى بالمدّ المخالف وبالتّقويم

---

٨٥٧ AM: أن | ٨٥٨ DSY: الأعصاب والعروق | ٨٥٩ AM: + والأعصاب | ٨٦٠ F: بدل؛ DY: بدله | ٨٦١ AM: مهيّأة معدّة؛ Y: مؤدّة مهيّأة | ٨٦٢ ADMY: وهو | ٨٦٣ DSY: – من الرّأس | ٨٦٤ A^hM^h: حاشية له: إنّما يقال بالحقيقة خارجة عن الطّبع في عددها لما في الطّبع منه عددمّا [A: عددها]. | ٨٦٥ AM: + فإنّه | ٨٦٦ AM: يعود إلى حاله > يحدث | ٨٦٧ AM: يكون ممكنًا > يمكن | ٨٦٨ AM: يحدث؛ S: + إلى حاله | ٨٦٩ AM: وحده؛ DY: فيه | ٨٧٠ DSY: – إلى | ٨٧١ AM: + جملة | ٨٧٢ ADMY: – إلى | ٨٧٣ AM: + ويزال؛ F: نقل؛ S: ينتقل | ٨٧٤ DSY: فأمّا | ٨٧٥ DSY: – فيها

treated by stretching the joint in the opposite way, straightening it, putting it back in, and returning it to its place. Another such disease is the rupture—which is to say, the intestinal hernia. It is caused either by a tear occurring in the peritoneum because the latter has become hard, or by a distention in the peritoneum due to its softness.

(80) In connection with the compound diseases occurring in the interior organs, Galen gives the example of two patients whom he had seen. The stomach of one of them was cold, round, small, and protruding outward.[175] There were four diseases in his stomach, three of which were diseases of the compound organs, which are instruments, and one a disease of the tissues: a cold temperament. One of the three organic diseases was in the size of the organ, which was its small size. The second was in its structure, which was its roundness. The third was in its position, which was its protrusion outwards. The other patient's food was having difficulty rising from his stomach and belly to his liver.[176] He applied his intuition and realized that there were two diseases in the patient's liver: one in its structure, which was the narrowness of the blood vessels in it; and the other in its size, which was the small size of the liver itself.

175. B 19.2, 26.11; K 1: 353, 378–79; G^{a} 91, 132: Galen gives this as an example of one of the rare cases when it is possible to directly diagnose diseases of the internal organs—in this case, because the position, size, and shape of the stomach were plainly outlined under the skin. See p. 105, paragraph 57 and n. 130, above.

176. B 19.4, 26.12; K 1:353–54, 379; G^{a} 91–93, 133: Intuition was required in this case because the condition of the liver could not be observed directly. Galen does not actually mention the small size of the liver, so it is possible that the epitomist was conflating this case with the case of a small and misplaced bladder mentioned at B 19.3; K 1:353; G^{a} 91.

والإدخال والرّدّ، ومنها الفتق، وهي[٨٧٦] قيلة الأمعاء، وذلك يكون إمّا بسبب انخراق[٨٧٧] يحدث في الصّفاق لصلابته، وإمّا بسبب تمدّد في[٨٧٨] الصّفاق للينه.

(٨٠) إنّ جالينوس[٨٧٩] يتمثّل في الأمراض المركّبة الحادثة في الأعضاء الباطنة بإنسانيْن رآهما[٨٨٠]، أحدهما كانت معدته باردة مستديرة[٨٨١] صغيرة نائتة إلى خارج، وكان[٨٨٢] بهذه المعدة أربعة أمراض، ثلثة منها من أمراض الأعضاء المركّبة التي هي الآلات، وواحد من أمراض الأعضاء المتشابهة الأجزاء، وهو سوء المزاج البارد، وأمّا[٨٨٣] الثلثة الأمراض الآليّة، فواحد منها كان في مقدار العضو، وهو صغره، والثّاني[٨٨٤] في خلقته، وهي[٨٨٥] استدارته، والثّالث في وضعه[٨٨٦]، وهو نتوءه إلى خارج، وأمّا الإنسان الآخر، فكان غذاؤه[٨٨٧] لا يرتقي[٨٨٨] من معدته وبطنه إلى كبده الاّ بكدّ، فاستعمل[٨٨٩] فيه الحدس ووقف على أنّ في كبده مرضيْن[٨٩٠]، أحدهما في خلقتها، وهو ضيق العروق التي فيها، والآخر صغر[٨٩١] مقدارها، أعني[٨٩٢] الكبد نفسها.

---

٨٧٦ DFY: وهو | ٨٧٧ AM: خرق | ٨٧٨ AM: من | ٨٧٩ DFSY: إنّ جالينوس < وجالينوس | ٨٨٠ DSY: – رآهما | ٨٨١ AM: مدوّرة | ٨٨٢ AM: فكان؛ DY: وكانت | ٨٨٣ DFY: فأمّا | ٨٨٤ AM: والآخر | ٨٨٥ ADMY: وهو | ٨٨٦ D: موضعه؛ F: وضعها | ٨٨٧ AM: طعامه | ٨٨٨ AM: ينفذ؛ A^2M^2: يصعد | ٨٨٩ M: واستعمل؛ S: استعمل؛ ADY: + جالينوس | ٨٩٠ S: + آليْن | ٨٩١ AM: في | ٨٩٢ AM: وهو؛ ADMY: + صغر

[Chapter 27][177]

[Dissolution of continuity]

(81) Dissolution of continuity can occur in fleshy tissues, nervous tissues, or bony tissues. Those that occur in the fleshy tissues are called "lesions" and need to be treated in four steps—first, to bring together the parts that are separated; second, to protect them after they are brought together; third, to ensure that nothing gets in between those parts, either at the beginning or after a time; and, fourth, to consume a diet whose quality is coarse and viscid and whose quantity is moderate. Those that occur in the nervous organs are called σχίσμα in Greek, which means "break," and ἄμυγμα, which means "tearing." Those that occur in the bony organs are called "fractures." Knitting of the fracture corresponds to the knitting of the flesh in the sense that they both result from the functioning of nature and involve matter existing in the body that is one and the same. However, the knitting of a fracture is different in respect to hardness. That is because the thing by which symphysis occurs in the hard bone comes from the symphysis of the flesh, since it is close to the substance of bone.[178] The dissolution of continuity is compounded in three ways: It can occur with one of the causes of diseases, as happens with fractures when there is matter that gets into the fractured tissue. It can occur with one of the diseases, as happens with swellings, concavity,[179] and bad temperament. Finally, it can occur with symptoms, such as pus and ichor in the urine.

177. B 27.1–2, 29.1–31.3; K 1:379, 385–89; G[a] 133–34, 142–48.

178. An early MS adds, "And it is called the *z* [or *r*]-*sh-b-dh*." It is not clear what is meant by this. One old MS, followed by two late MSS, reads: "The thing by which the bone knits is harder than the knitting of flesh." It may relate to the appropriate diet for healing a fracture, which is what Galen is discussing.

179. Galen is discussing cases where a fracture cannot be set without leaving a gap.

[الفصل السّابع والعشرون]

[ذكر تفرّق الاتّصال]

(٨١) تفرّق الاتّصال يحدث إمّا في الأعضاء اللّحميّة[٨٩٣]، وإمّا في الأعضاء العصبيّة[٨٩٤]، وإمّا في الأعضاء العظميّة، والذي يكون في الأعضاء اللّحميّة تقال له قرحة، والقرحة تحتاج في مداواتها إلى أربعة أشياء، أحدها جمع الأجزاء التي تفرّقت، والثّاني[٨٩٥] حفظها بعد الجمع، الثّالث التّوقّي[٨٩٦] من وقوع شيء[٨٩٧] فيما[٨٩٨] بين تلك الأجزاء[٨٩٩] في مبدأ[٩٠٠] الأمر أو بعد زمان، والرّابع الغذاء الذي يكون كيفيّته غليظة لزجة[٩٠١] ومقداره معتدل[٩٠٢]، وأمّا الذي يكون في الأعضاء العصبيّة، ويقال[٩٠٣] له باليونانيّة سقاسما، وتفسيره الفسخ وأونغما[٩٠٤] وتفسيره الهتك، وأمّا الذي يكون في الأعضاء العظميّة، فيقال له الكسر، والتحام الكسر مشارك لالتحام اللّحم من طريق أنهما جميعاً يكونانِ من فعل الطّبيعة ومن المادّة الموجودة في البدن التي هي واحدة بعينها، وهو مخالف له من جهة[٩٠٥] الصّلابة، وذلك أنّ الشّيء[٩٠٦] الذي به يلتحم[٩٠٧] العظم الصّلب[٩٠٨] من لحام اللّحم لأنّه قريب من جوهر العظم[٩٠٩]. تفرّق الاتّصال[٩١٠] في اللّحام[٩١١] مركّب[٩١٢] على ثلثة أوجه[٩١٣]، إمّا مع سبب من أسباب الأمراض بمنزلة ما يعرض إذا كان مع الكسر مادّة تنصبّ إلى العضو المكسور، وإمّا مع مرض من الأمراض بمنزلة الورم والتّقوير وسوء المزاج، وإمّا مع عرض من الأعراض بمنزلة بول القيح[٩١٤] والصّديد.

---

٨٩٣ S^h: هذه شفي | ٨٩٤ S^h: هذه لا شفي | ٨٩٥ AM: والآخر | ٨٩٦ AM: المنع | ٨٩٧ AM: وقوع شيء > أن يقع | ٨٩٨ F: – فيما؛ S: فيها | ٨٩٩ AM: + شيء | ٩٠٠ A: ابتداء؛ M: ابتدئ | ٩٠١ DFY: غليظاً لزجاً؛ S: غلظ ولزوجة | ٩٠٢ AM: معتدلاً | ٩٠٣ AMY: فيقال | ٩٠٤ DY: اواونغما؛ S: واوفرما؟ | ٩٠٥ AM: في | ٩٠٦ AM: اللّحام | ٩٠٧ DSY: يلحم | ٩٠٨ S: – الصّلب؛ DSY: + أصلب | ٩٠٩ F: + ويسمّى الرشبذ | ٩١٠ DFY: + يكون | ٩١١ DFY: + في اللّحم؛ S: + في اللّحم | ٩١٢ AM: يتركّب | ٩١٣ AM: وجوه | ٩١٤ AM: القيح؛ A^2M^2: بول المدّة؛ DY: تولّد المدّة

[Chapter 28][180]

[Treatment, prophylaxis, and convalescence]

(82) A poor temperament may be well established and complete. It is cured by what is called a "treatment"[181]—that is, by things whose potency is opposite to it. It may be on the verge of coming to be and then is cured by a treatment compounded with prophylaxis. Finally, it may be tending to come to be and is prevented by what is called "prophylaxis combined with treatment." In general, this is called "prophylaxis." When we say "treatment," we mean, for example, using theriac to alter and destroy the putrefaction in a quartan fever or using a douche of cold water to quench and still the fever's heat in a tertian fever. When we say "prophylaxis," it is, for example, using hellebore to evacuate the melancholic humor and using scammony to evacuate the bilious humor in tertian fever, thereby preventing a recurrence. Evacuation may be performed by attraction through an organ located on the opposite side from the diseased organ but that corresponds to it. The matter will thus flow afterwards to the diseased organ. Evacuation may also be effected by transfer and extraction of the superfluity from that same diseased organ or from an organ close to it or that

180. B 28.1–8; K 1:380–85; G[a] 134–42.
181. Some MSS read: "simple treatment."

[الفصل الثّامن والعشرون]

[ذكر المعالجة والتّقدّمة بالحفظ والإنعاش]

(٨٢) سوء المزاج منه ما قد استحكم وانتهى[٩١٥]، وإصلاحه تقال له مداواة[٩١٦]، وتكون بالأشياء الضّادّة له في قوّتها، ومنها[٩١٧] ما هو في حدّ الكون[٩١٨]، وإصلاحه تقال له مداواة مركّبة مع التّقدّم بالحفظ، ومنه ما هو[٩١٩] يريد[٩٢٠] أن يكون ومنع هذا من أن يكون[٩٢١] يقال له الحفظ المركّب[٩٢٢] مع المداواة[٩٢٣]، ويقال له بالجملة التّقدّم بالحفظ. أمّا قولنا مداواة، فمثل تغيّر العفونة وإحالتها في حمّى الرّبع بالتّرياق[٩٢٤] وإطفاء[٩٢٥] حرارة الحمّى وتسكينها في[٩٢٦] الغبّ بإسقاء الماء البارد، وأمّا قولنا التّقدّم بالحفظ فمثل استفراغ الخلط السّوداويّ في حمّى الرّبع بالخربق الأسود واستفراغ الخلط المراريّ في حمّى الغبّ بالسّقمونيا ليمنع بذلك من عودة[٩٢٧] دور[٩٢٨]، والاستفراغ يكون إمّا على جهة الجذب[٩٢٩] من عضو موضعه في خلاف النّاحية الّتي فيها العضو العليل، وهو مع هذا مشارك له، وذلك عند ما تكون المادّة هو ذا تنصبّ بعد إلى العضو العليل، وإمّا على جهة النّقل والانتزاع[٩٣٠] من نفس العضو العليل أو من عضو

---

٩١٥ AM: وفرغ | ٩١٦ ADMY: + مفردة | ٩١٧ ADM: ومنه | ٩١٨ F: ما يتكوّن؛ SY: ما سيكون | ٩١٩ AM: – هو | ٩٢٠ $DF^{1}SY$: مزمع | ٩٢١ Y^{2}؛ DSY: – من أن يكون | ٩٢٢ AM: حفظ مركّب | ٩٢٣ ADM: مداواة | ٩٢٤ F: باليرقان؛ F^{2}: بالتّلطيف؛ S: بالتّبريد | ٩٢٥ DFSY: أو تطفية | ٩٢٦ AM: + حمّى | ٩٢٧ AMY^{2}: حدوث | ٩٢٨ F^{1}؛ F: الدّود؛ AMY^{2}: + آخر | ٩٢٩ AM: اجتذاب؛ F: الحدث؛ S: الجاذب | ٩٣٠ AM: والسّلّ

corresponds to it. That is done when the matter is fixed solidly in the organ and we wish to extract it and evacuate it from the organ. Both these two kinds of evacuation are performed by bloodletting, enemas, diuretics, and sweats.

(83) When a swelling occurs in one of the organs,[182] the matter causing it is to be evacuated either by its being returned to where it came from or by its being removed from that organ. Returning the matter to where it came from can be done by repelling it, by attracting it, or by sending it away.[183] Its removal from that same organ is either a removal that can be perceived by the senses, as when the matter that is in a swelling is removed by scarification, or a removal that can be perceived by the mind, such as the dissolution of the contents of the swelling by a hot poultice.

(84) It is possible to deduce the treatment needed for a diseased organ from the position of the organ, its temperament, its structure, or its faculty.[184] In the case of its position, the treatment is deduced from whether the organ is near to or far from or communicates with the place from which it is possible to evacuate what is in it. In the case of its structure, it is deduced with respect to whether or not it has a cavity.

---

182. Some MSS add: "and remains."

183. One MS has the gloss: "[Galen's] text says [paraphrasing B 34.4; K 1:395; G[a] 161], 'Either by repelling, attracting, or moving.' He himself explains it in the text and says later [B 34.7; K 1:396; G[a] 162], 'If you strengthen the blood vessels by astringent drugs, they will transport that superfluity from the diseased organ to someplace else.' It may be that the difference between that and expulsion is that expulsion is done by means of the action of the astringent on the matter without the mediacy of the power of the blood vessels. That which is named 'transporting' is the action of the astringent by means of the faculty of the blood vessels and their dealing with the matter [of the superfluity]."

184. B 34.8–19; K 1:397–400; G[a] 163–69: Galen is discussing cases where there is more than one cause of the disease.

قريب منه مشارك له، وذلك إذا كانت[٩٣١] المادّة قد تمكّنت واستقرّت[٩٣٢] في العضو، ونريد أن نخرجها[٩٣٣] ونستفرغها منه، وهذانِ النّوعانِ كلاهما من الاستفراغ يكونانِ بإخراج الدّم وبالحقنة وبالأدوية المدرّة للبول وبالعرق.

(٨٣) متى حدث في عضو من الأعضاء ورم[٩٣٤]، فالمادّة الفاعلة له يستفرغ إمّا بأن يرجع[٩٣٥] إلى ورائها وإمّا بأن يخرج من نفس العضو العليل[٩٣٦]، ورجوع المادّة إلى وراءها يكون إمّا بأن يدفع[٩٣٧] وإمّا بأن يجذب[٩٣٨] وإمّا[٩٣٩] بأن يرسل[٩٤٠]، وخروجها من نفس العضو يكون إمّا خروجاً يدركه الحسّ مثل خروج ما في الورم بالشّرط وإمّا خروج يدرك[٩٤١] بالعقل مثل[٩٤٢] تحلّل ما في[٩٤٣] الورم بالضّماد المسخّن.

(٨٤) قد يستدلّ على ما يحتاج اليه في مداواة العضو العليل من وضع العضو ومن مزاجه ومن خلقته ومن قوّته، أمّا من وضعه فبحسب موضعه إن كان قريباً أو بعيداً أو بحسب[٩٤٤] مشاركته للموضع[٩٤٥] الذي منه يمكن أن يستفرغ ما فيه، وأمّا من خلقته فبحسب ما هو عليه من أن له تجويفاً أم[٩٤٦] لا تجويف له، وأمّا من قوّته فبحسب

---

٩٣١ AM: عند ما تكون | ٩٣٢ F^1؛ FS: واستفرغت | ٩٣٣ AM: + عنه | ٩٣٤ ADM: + ودام | ٩٣٥ S: يردّ؛ DY: تردّها؛ Y^2: يرج [كذا] إلى وائها نسخ | ٩٣٦ A^1M^1: الوارم | ٩٣٧ DSY: بالدّفع | ٩٣٨ F: يحدث؛ DY: بالجذب؛ S: أو بالجذب | ٩٣٩ A^hM^h: حـ[اشية] في الفصّ: إمّا بأن يدفع وإمّا بأن يجذب وإمّا بأن يسيّر، وفسّرهو في نفسها في الفصّ ذلك فقال بعد والعروق إذا قوّيتها بالأدوية القابضة سيّرت ذلك الفضل من العضو العليل إلى ما وراءه له، فعسى أن الفرق يكون بين هذا وبين الدّفع أنّ الدّفع يكون من فعل القابض في المادّة من غير توسّط قوّة العروق، وهذا المسمّى التسيّر فعل القابض بتوسّط قوّة العروق به وتصريفها هي للمادّة. | ٩٤٠ DSY: بالسّلّ؛ F: بالانتزاع | ٩٤١ DSY: - خروج يدرك | ٩٤٢ DY: + ما؛ F: بمثل | ٩٤٣ DSY: - ما في | ٩٤٤ ADY: أو بحسب> وبحسب | ٩٤٥ F: الموضع؛ DY: مشاركة الموضع | ٩٤٦ F^1؛ AM: أو؛ F: إن

In the case of its faculty, the treatment is deduced from its condition in itself—whether it is the root or origin from which the faculty penetrates to other organs, such as the liver; or whether its function is general and benefits the entire body, such as the stomach; or whether it is a sense faculty, such as the eye.

(85) Prophylaxis tends to be concerned with the problem of the humors.[185] The humors vary in quantity when they increase and in quality when they are transformed. They can increase in two ways: first, by increasing entirely from a single cause; and, second, by having something generated from them by which the rest of what is there changes and is transformed, thereby increasing in that respect. They can change in quality in three ways: First, they can become more subtle and thus be finer, or they can become coarser. Second, their color can change so that they become yellow, bright red, or the color of fire. Third, their smell can change so that they come to have a bad smell. The humors can be returned to their natural state in one of two ways: by changing and transforming them by coction, or by evacuation. Evacuation is done by enemas, diuretics, and sweats.

(86) Convalescence is divided into three classes:[186] first, the regimen for the bodies of children; second, the regimen for the bodies of old people; and, third, the regimen for the bodies of people convalescing from disease. These all have bodies in which the blood is excellent but deficient in quantity, so that the body is dry and, for that reason, weak. Such

185. B 36.2–4; K 1:403–4; G[a] 173–75.

186. B 37.1–4; K 1:405–7; G[a] 178–80: The Arabic translation of Galen's text renders this as *al-taqwiyah wa-al-taghdhiyah* (strengthening and nutrition), for the Greek τὸ ἀναληπτικόν τε καὶ ἀναθρεπτικόν (recuperation and convalescence), which is closer to the epitome's translation. Galen mentions that these measures are also suitable for the old but does not mention children. Apparently, the characteristic state of all three groups is a lack of blood, which results in the body being cold and dry overall.

حاله في نفسه هل هو أصل ومعدن تنفذ[٩٤٧] منه قوّة إلى الأعضاء بمنزلة الكبد أو هل هو يفعل فعلاً عامّاً[٩٤٨] ينتفع[٩٤٩] به[٩٥٠] البدن كلّه بمنزلة المعدة أو هل هو قوى الحسّ بمنزلة العين.

(٨٥) التّقدّم[٩٥١] بالحفظ من شأنه العناية بأمر الأخلاط. والأخلاط تتغيّر إمّا في كمّيّتها إذا هي تزيّدت، وإمّا في كيفيّتها إذا هي استحالت. وتزيّدها يكون على ضربين، أحدهما أن يكون[٩٥٢] كلّها يتولّد[٩٥٣] من سبب واحد، والثّاني[٩٥٤] أن يكون قد تولّد منها شيء ويكون ذلك الذي تولّد تغيّر وتحيل سائر ما هناك. فيكثر من هذا الوجه. وتغيّرها في كيفيّتها يكون على ثلثة أوجه[٩٥٥]، أحدها أن[٩٥٦] يلطف فيرقّ[٩٥٧] وإمّا بأن[٩٥٨] يغلظ، والثّاني[٩٥٩] أن[٩٦٠] يتغيّر لونها، فيصير إمّا أصفر وإمّا[٩٦١] أحمر ناصع وإمّا بلون النّار، والثّالث أن يتغيّر رائحتها فيصير[٩٦٢] رديّة الرّائحة[٩٦٣]. وردّ الأخلاط إلى الحال الطّبيعيّة يكون بأحد[٩٦٤] وجهين، إمّا بالتّغيّر والإحالة بالنّضج وإمّا بالاستفراغ، والاستفراغ يكون بالحقن وبالأدوية المدرّة للبول وبالعرق.

(٨٦) الإنعاش[٩٦٥] ينقسم ثلثة أقسام، أحدها تدبير أبدان الصّبيان، والثّاني[٩٦٦] تدبير أبدان الشّيوخ، والثّالث تدبير أبدان النّاقهين من المرض[٩٦٧]، وهي الأبدان التي الدّم فيها جيّد إلاّ أنّه يسير[٩٦٨] المقدار[٩٦٩] والبدن يابس، فهو لذلك ضعيف،

---

٩٤٧ AM: تجري | ٩٤٨ ADMY: عاميّاً؛ F: عامّ؛ S: عامّا | ٩٤٩ DSY: ينفع | ٩٥٠ AM: + من؛ DSY: – به | ٩٥١ DFY: والتّقدّم | ٩٥٢ DSY: – يكون | ٩٥٣ DSY: تتولّد كلّها | ٩٥٤ AM: والآخر | ٩٥٥ AM: وجوه | ٩٥٦ F: + يتغيّر قوامها إمّا بأن؛ DSY: إمّا بأن | ٩٥٧ ADSY: ويرقّ | ٩٥٨ AM: أو | ٩٥٩ AM: والآخر | ٩٦٠ DSY: بأن | ٩٦١ DSY: أو | ٩٦٢ F: فيصير > أمّا ألى ما يكره ويصير | ٩٦٣ DSY: فيصير. . . الرّائحة > إلى ما يكره | ٩٦٤ MS: بإحدى | ٩٦٥ AM: والأنعاش | ٩٦٦ AM: والأخر | ٩٦٧ ADMY: الأمراض | ٩٦٨ AM: قليل | ٩٦٩ AM: + يسير

bodies ought to be treated with foods that are quickly digested and of moderate temperament, such as the meat of chickens, [other] poultry, and fish; appropriate drinks such as light, clear, aromatic young wines; and moderate exercise, such as moderate walking, bathing, and riding.[187]

This completes

the Alexandrians' epitome

of Galen's book known as

The Small Art of Medicine

translated by

Abū Zayd Ḥunayn ibn Isḥāq[188]

---

187. Galen concludes with an extended advertisement for some four dozen of his own books, with advice on the order in which they should be read. Many of them were mentioned earlier in the text.

188. The colophons vary in wording, with some adding blessings.

وينبغي أن يصلح هذه الأبدان، بالأطعمة[٩٧٠] السّريعة[٩٧١] الانهضام المعتدلة المزاج مثل[٩٧٢] لحم[٩٧٣] الدّجاج[٩٧٤] والفراريج والسّمك الرّضراضيّ وبالشّراب الموافق مثل الخمور[٩٧٥] اللّطيفة الرّقيقة[٩٧٦] الرّيحانيّة التي لم تبق[٩٧٧] وبالرّياضة المعتدلة مثل[٩٧٨] المشي المعتدل والحمام والرّكوب.

تمّت[٩٧٩] جوامع الإسكندرانيّين[٩٨٠]
لكتاب[٩٨١] جالينوس المعروف[٩٨٢]
بالصّناعة الصّغيرة[٩٨٣] الطّبيّة[٩٨٤]
نقل[٩٨٥] ابي زيد حنين ابن إسحق[٩٨٦]

---

٩٧٠ S^h: + لأجل ضعف قواها | ٩٧١ AM: المسهلة | ٩٧٢ AM: بمنزلة | ٩٧٣ AM: لحوم | ٩٧٤ FM^2: الدّجاج | ٩٧٥ S: الخمرة؛ DY: الخمر | ٩٧٦ DSY: – الرّقيقة | ٩٧٧ AM: مثل. . . تبق > بمنزلة الشّراب اللّطيف الرّقيق الرّيحانيّ الذي ليس بعقيق؛ DSY: يعتق | ٩٧٨ AM: بمنزلة | ٩٧٩ ADMY: تمّ | ٩٨٠ AM: – الإسكندرانيّين | ٩٨١ AM: كتاب | ٩٨٢ AM: – المعروف؛ S: المعروفة | ٩٨٣ AM: بالصّناعة الصّغيرة > في الصّناعة | ٩٨٤ AM: + للإسكندرانيّين؛ DSY: – الطّبيّة؛ D: + والحمد لله ربّ العالمين والصّلوة على سيّدها نبيّه محمّد وآله أجمعين وحسبنا الله ونعم الوكيل؛ Y: + بعون الله تعالى | ٩٨٥ AM: ترجمة | ٩٨٦ A: + والحمد لله وحده وحسبي الله ونعم الوكيل؛ DS: – نقل. . . إسحق؛ M: + والسّبح لله دائماً أبداً

THE ALEXANDRIAN EPITOMES OF GALEN

VOLUME 1

The Alexandrian Epitome of Galen's Book *On the Elements According to the Opinion of Hippocrates*

Part two of the epitome of Galen's books,

In which are *On the Elements, On the Temperament,*
On the Natural Faculties, and *On Anatomy*,

According to the opinion of the Alexandrians,

Translated by Ḥunayn ibn Isḥāq

Ex libris ʿAbd al-Wāḥid ibn Muḥammad, the physician;

Then it passed to Hibat Allāh ibn Haykal, the physician;

[then to] ʿAbd Allāh ibn al-Ḥusayn, the physician;

then to his son Ḥasan ibn ʿAbd Allāh ibn al-Ḥusayn, the physician,

on 14 Ṣafar 547/[21 May 1152];

then it came into the possession of this writer,
the humble Muḥammad,

son of the late Nāṣir al-Dīn ibn ʿAlī ibn Muḥammad
al-Bulyānī al-Shāfiʿī al-Azharī,

in the year 984/[1576]—may his end be good!*

---

* Text from the flyleaves and title pages of **R**.

الجزء الثّاني من جوامع كتب جالينوس

فيه «الإسطقسات» و«المزاج» و«القوى الطّبيعيّة» و«التّشريح»

على رأي الإسكندرانيّين

ترجمة حنين بن إسحق[١]

من كتب عبد الواحد بن محمّد الطّبيب،

ثمّ صار إلى هبة الله بن هيكل المتطبّب،

عبد الله بن الحسين المتطبّب كان،

ثمّ لولده حسن بن عبد الله بن الحسين الطّبيب، وذلك في الرّابع عشر من صفر، سنة سبع وأربعين وخمسمائة،

ثمّ هو في نوبة كاتبه الفقير محمد بن المرحوم ناصر الدّين بن علي بن محمد البُلْيَنيّ الشّافعيّ الأزهري

في سنة ٩٨٤، أحسن الله ختامه![١(آ)]

---

١ R | ١(آ) R^h

[The eight headings to the epitome of Galen's *On the Elements*][1]

If the customary laws are observed, your end and goal will be successfully attained; but if they are discarded in neglect, the end will be thwarted, and after much thought you will come to the opposite of your end. The laws of logic command that each art begin with the investigation of its subject—not with respect to its existence, but, rather, with respect to the questions that follow from what is to be investigated. These are what it is, which thing it is, and why it is.

The subject of this art of ours is the body of man. Our goal is the preservation of health existent in it, or the restoration of health that has been lost from it. Thus, we must first begin with the knowledge of the human body, for without that, we will not have the ability to affect it. Because the human body exists in two states—a natural state and an unnatural state—we must know it in both of the two states, so that we know the natural entities and the unnatural entities that diverge from the natural, in order that we might preserve the natural entity by its like and restore what has become unnatural by means of its opposite. Therefore, it is true to say that our art is divided into four parts: the knowledge of natural things, the knowledge of the unnatural things, the preservation of natural things, and the reversal of unnatural things. Because the knowledge of the natural is prior to the knowledge of the unnatural, it is necessary to begin with it first. You already know that the knowledge of the thing is completed by the knowledge of its causes and first principles. The principles of the body are the four elements, and its genesis from them is generated and completed by their mixture. When the elements are mixed, the four humors result from them; and from these result the tissues; from the tissues, the organs; and from the organs, the complete body. Each of the organs has a faculty, and the faculties have functions. Thus, the one who wishes to know the human body is led back to the necessity of knowing the things that we have enumerated.

---

1. One of the introductory glosses found before most treatises in **S**. It is in an unpleasant and not-always-legible, crabbed hand. The convention of introducing a book to students by explaining its "eight heads" apparently originated in late antiquity and was sometimes used by Islamic scholars, particularly in the scientific and philosophical tradition.

[الرّؤوس الثّمانية لجوامع كتاب جالينوس في العناصر]

عادة القانون إذا اتّبعت أنجحت وبلغت الغرض والغاية، وإذا أُهملت وأطحت أمحيت وأكثرت الفكر وتأتت إلى خلاف الغاية. والقانون المنطقيّ يأمر أن يفتتح كلّ صناعة بالنّظر في موضوعه لا في وجوده، بل في المطالب التّابعة لهذا المطلوب وهي ما هو وأي شيء هو ولِمَ هو.

وموضوع صناعتنا هذه هو بدن الإنسان، وغرضنا حفظ صحّة موجودة فيه أو ردّ صحّة قد فقدت عنه، فيجب أن نشرع أوّلاً في العلم ببدن الإنسان، فمن دون ذلك لا يكون لنا قدرة على الفعل فيه. ولأنّ بدن الإنسان وجد على حالين، حال طبيعيّة وحال خارجة عن الطّبيعة، فيجب أن نعلمه على حالتيه جميعاً، فنعلم الأمور الطّبيعيّة والأمور الخارجة عن الطّبيعة لنحفظ الأمر الطّبيعيّ بشبهه ونردّ الأمر الخارج عن الطّبيعة بضدّه. ولهذا صدق من قال إنّ صناعتنا تنقسم إلى أربعة أقسام، إلى علم الأشياء الطّبيعيّة وعلم الأشياء الخارجة عن الطّبيعة وإلى حفظ الأشياء الطّبيعيّة وردّ الأشياء الخارجة عن الطّبيعة. ولأنّ العلم الطّبيعيّ يتقدّم لعلم الخارج عن الطّبيعة ‹ما› يجب أن نشرع فيه أوّلاً. وأنتم فقد علمتم أنّ العلم بالشّيء يتمّ من العلم بأسبابه ومبادئه الأوَل، ومبادئ الجسم الإسطقسات الأربعة، وكونه منها يكوّن ويتمّم بامتزاجها، وإذا امتزجت حدثت عنها الأخلاط الأربعة، وهذه يحدث عنها الأعضاء المتشابهة الأجزاء، وعن المتشابهة الآليّة، وعن الآليّة جملة البدن. ولكلّ واحد من الأعضاء قوّة، وللقوى الأفعال. فتعود الضّرورة لمن أراد علم بدن الإنسان أن يكون عالماً بالأشياء التي عدّدناها.

Galen teaches us about the elements in his book entitled *On the Elements According to the Opinion of Hippocrates*, about the temperament in his book *On the Temperament*, and about the humors in his essay *On the Humors* and in the end of *On the Elements*, since the humors are the ultimate principles of the body, but are also basic. He teaches about the natural faculties in his book *On the Natural Faculties*, about the vital faculties in his book *On the Pulse*, about the psychic faculties in his book *On the Doctrines of Hippocrates and Plato*, about the organs in his book *On Anatomy*, and about the actions and their benefits in the book *On the Uses of the Parts*.

1. Our goal in this book. Out of all the things that we have enumerated, our goal is the investigation of the elements, because the beginning must occur in accordance with how nature begins. Nature is prior, so you take the elements and mix them, and from them you make the humors and the organs. This is also because the discussion of simple things is prior to the discussion of compound things.

2. The usefulness of this book. This book includes both theoretical and practical medicine. It includes theoretical medicine since the human body can be perfectly known only by knowing its principles and its principles can be known only by the four elements. The inclusion of the practical is due to the fact that the goal of the practical part of medicine is the preservation of existent health and the restoration of lost health. Health and disease are moderation and lack of moderation, and moderation and lack of moderation both exist only in the temperament of the elements.

3. Its title. The Book of the Elements According to the Opinion of Hippocrates. The book's title corresponds to its purpose. We say "according to the opinion of Hippocrates" because Galen follows the opinion of Hippocrates in this, for it was Hippocrates who was the first to make this great discovery concerning natural entities—that is, that there are four elements. In this he is followed by Plato and Aristotle, the greatest of the

وجلينوس يعلّمنا عن الإسطقسات في كتابه ترجمه ب«الإسطقسات على رأي بقراط» وفي المزاج في كتابه «في المزاج»، وفي الأخلاط في مقالته «في الأخلاط»، وفي آخر «كتاب الإسطقسات» لأنّ الأخلاط أقصى مبادئ البدن ولكن قريبة، وفي القوى الطّبيعيّة في كتابه «في القوى الطّبيعيّة»، وفي الحيوانيّة في كتابه «في النّبض»، وفي النّفسانيّة في كتابه «في آراء بقراط وفلاطن»، وفي الأعضاء في كتابه «في التّشريح»، وفي الأفعال ومنافعها في «كتاب منافع الأعضاء».

آ - فغرضنا في هذا الكتاب من جملة ما عدّدنا النّظر في الإسطقسات لأنّ المبدأ يجب أن يقع من حيث ابتدأت الطّبيعة، والطّبيعة ابتدأت فأخذت الإسطقسات فمزجتها وعملت منها الأخلاط والأعضاء ولأنّ الكلام في الأشياء البسيطة يتقدّم على الكلام في الأشياء المركّبة.

ب - فأمّا منفعة هذا الكتاب، فإنّها يشتمل قسميَ الطّبّ العلميّ والعمليّ، أمّا العلميّ فلأنّ العلم ببدن الإنسان إنّما يتمّ بعلم مبادئه ومبادئها به الإسطقسات الأربعة، وأمّا في العمل فمن قِبَل أنّ غرض الجزء العمليّ حفظ صحّة موجودة وردّ صحّة مفقودة، والصّحّة والمرض هما اعتدال ولا اعتدال، والاعتدال ولا اعتدال إنّما يوجدانِ في مزاج الإسطقسات.

ج - فأمّا سمته، ف«كتاب الإسطقسات على رأي إبقراط»، وهذه السّمة موافقة للغرض، وقولنا على رأي بقراط لاتباعه في هذا الرّأي إبقراط، فأوّل من وجد هذا الكبير العظيم في الأمور الطّبيعيّة إبقراط، أعني أنّ الإسطقسات أربعة،

philosophers. Some of the elements are remote and some proximate. The remote elements are fire, air, water, and earth, while the proximate elements are the four humors. These encompass all the rest of the compound entities that are beneath the sphere of the moon—though, to be sure, some animals exist that do not have these humors (for example, the worm, though there is something in it performing the same function).[2]

4. Its rank. You read the *Physics* before it. After it you read all the physical works, one after another.[3] This is because man is included within the subject of the physical books, which include the four elements and the nine temperaments. You then pass on and read Galen's books *On Diseases and Symptoms, The Kinds of Fevers,*[4] and *The Method of Healing.*

5. Its being by Galen. Its authorship is obvious from its style, its accuracy, the testimony of the commentators, and its conformity in content to the views of Hippocrates, Plato, and Aristotle.

6. To which of the two parts of the art [of medicine] does it belong. This book belongs to the theoretical part of medicine; and within the theoretical part, to the knowledge of natural things.[5]

7. Its divisions. It has two parts.[6] The first of them discusses the primary elements—that is, fire, air, water, and earth—while in the second it discusses the proximate elements—that is, the humors.

8. The method of the science. It follows the method of combination in that it begins with the elements and proceeds to the humors, but the method of analysis is the opposite, which begins with the humors and ends with the elements.

---

2. Worms were not thought to have blood; see p. liii above.

3. Presumably, Aristotle's *Physics, On Generation and Corruption*, and *On the Heavens*, which were part of the Alexandrian curriculum.

4. The ninth and fourteenth works of Galen in the Alexandrian curriculum; pp. xxvii–xxviii above.

5. See p. 3, paragraph 2, of the epitome to *The Medical Sects* above.

6. In accordance with a Greek tradition of dividing the work into two books; see p. lii above and L 10.1; K 1:492; G^a 117. The epitome, however, supports the Arabic tradition and some Greek sources in holding that *On the Elements* consists of one book, not two.

اتّبعه في ذلك عظماء الفلاسفة، أعني فلاطن وأرسطوطاليس، والإسطقسات منها بعيدة ومنها قريبة، والبعيدة النّار والهواء والماء والأرض، والقريبة الأخلاط الأربعة، وهذه شاملة لسائر الأمور المركّبة التي تحت فلك القمر، وإن وجد بعض الحيوان ليس فيه هذه الأخلاط كالدّود ففيه ما تنوب منابها.

د - فأمّا مرتبته، فأن يقرأ قبلَ العلم الطّبيعيّ، ومن بعده يجب أن يقرأ الكتب الطّبيعيّة بأسرها واحداً بعد واحد. هذا أن أخذ الإنسان بضميمة الكتب الطّبيعيّة وهو الإسطقسات أربعة، والأمزجة تسعة، وتنتقل فتقرأ «كتاب العلل والإعراض» و«فصول[٢] الحمّيات» و«حيلة البرء».

هـ - وأمّا أنّه لجالينوس، فظاهر من نمط كلامه وصحّة معانيه وشهادة المفسّرين له ولا تبّاعه فيها آراء بقراط وأفلاطن وأرسطوطاليس.

و - فأمّا إلى أي قسميَ الصّناعة ترتقي، فإلى الجزء النّظريّ ومن جملة الجزء النّظريّ إلى علم الأشياء الطّبيعيّة.

ز - وأقسامه جزآنِ في الأوّل منها يتكلّم في الإسطقسات الأوَل، أعني النّار والهواء والماء والأرض ‹ح -› وفي الثّاني يتكلّم في الإسطقساتت القريبة، أعني الأخلاط.

[ح] - ونحو العلم، وذلك أنّه يسلك فيه طريق التّركيب، وهذا بأن يبتدئ من الإسطقسات وتأتي إلى الأخلاط ونحو التّحليل بالعكس، وذاك إذا يبتدئ من الأخلاط وانتهى إلى الإسطقسات ـــهـ

٢ S^h: في فصول

In the Name of God, the Merciful, the Compassionate

The Epitome of Galen's Book

On the Elements[7] *According to the Opinion of Hippocrates*

Translated by Ḥunayn ibn Isḥāq[8]

[Chapter 1][9]

<The genera of the elements>

(1) There are three genera of elements. The remote elements—those common to all composite corporeal bodies[10] whatsoever—are fire, air, water, and earth. The proximate elements, those specific to the living bodies of animals that have blood, are the four humors—that is, blood, phlegm, yellow bile, and black bile. The most proximate of

7. The oldest MS uses the Arabic loanword from Greek, *isṭaqisāt*, following the title used in the Arabic translation of Galen's original. The word used for στοιχεῖον in the translation is *ʿunṣur*. Some manuscripts periodically gloss *ʿunṣur* with *isṭaqis*, which is one of the relatively few Greek loanwords to survive in ordinary use in philosophical Arabic. Ḥunayn avoids loanwords and transcriptions from Greek as a matter of principle, so there are very few Greek words in this text, apart from proper names and the names of a few drugs. However, Ḥunayn's Arabic translation of *On the Elements* uses *isṭaqis* in preference to *ʿunṣur*. I will normally translate both *ʿunṣur* and *isṭaqis* as "element," except when the text uses the latter for the Greek word.

8–10. See the following page.

بسم الله الرحمن الرحيم[٣]

جوامع[٤] كتاب جالينوس

في العناصر[٥] على[٦] رأي إبقراط[٧]

ترجمة[٨] حنين بن إسحاق[٩]

[الفصل الأوّل]

[أجناس العناصر]

(١) أجناس العناصر ثلثة، فمنها عناصر بعيدة تعمّ[١٠] الأجسام المركّبة كلّها، وهي النّار والهواء والماء والأرض، ومنها[١١] قريبة تخصّ أبدان الحيوان الذي له دم،

---

٣ A:+ توكّلت على الله؛ D: – بسم. . . الرّحيم؛ MY: + ربّ يسّر | ٤ DSY: مبدأ جوامع؛ F: + الإسكندرانيّين | ٥ G^aR: الإسطقسات | ٦ G^a؛ DSY: بحسب | ٧ M: بقراط؛ FU: – بحسب. . . إبقراط؛ R: + للإسكندرانيّين | ٨ M: + أبي زيد | ٩ AF: – ترجمة. . . إسحق؛ A: + وهي الإسطقسات للإسكندرانيّين ربِّ عونكّ؛ S: + العبّاديّ؛ U: + الحيري | ١٠ DMSY: + جميع | ١١ AR: + عناصر

elements are the organs specific to each individual species of animal—that is, the tissues such as fat, flesh, horn,[11] teeth, marrow, and nail.

(2) The element is a single, simple part of the thing of which it is an element.[12] The thing can be single and simple in two ways: either to the senses or in nature. The things that are simple and single to the senses are of the order of nerve and bone; but though they may be simple and single to the senses, in their nature they are composed of fire, air, water, and earth. That which is simple in nature—that is, what is truly simple—is of the order of fire, air, water, and earth; for our knowledge that they are elements is not sensible knowledge, but rational knowledge.

---

8. The title and ascription of authorship vary considerably in the manuscripts. The main variations are (1) the omission in some MSS of "*According to the Opinion of Hippocrates*"; (2) the addition of "by the Alexandrians" at one point or another; (3) additions to Ḥunayn's name of "Abī Zayd," "al-ʿIbādī," and "al-Ḥīrī"; and (4) the addition of pious formulae.

9. L 1.1; K 1:413; G[a] 9: An introduction to *On the Elements* 1 and, by extension, Hippocrates, *De natura hominis* (On the nature of man) 1–2. In this chapter, Galen defines "element" but neglects to give the specific senses in which "element" can be used in medicine. The explanation is necessary because "element" originally meant the indivisible part or fundamental principle of something; but, by Alexandrian times, students would certainly have thought mainly of the four elements. In medicine, the term would be used of either the four elements or the four humors. The epitomes characteristically present the accepted view in the form of a list, as is done here.

10. *Ajsām*. Galen's text uses two words that can be translated as "body": σῶμα, *jism*, "corporeal body"; and *badan*, "animal or human body," depending on the context. Where the context makes the meaning clear—which it usually does—I simply use "body." The text systematically maintains this distinction between material and living bodies.

11. One MS reads "blood vessels."

12. L 1.1–9; K 1:413–14; G[a] 11–12: The chapter in *On the Elements* begins with the less precise definition, "An element is the smallest part of that of which it is an element," and then uses the example of a mixture of finely ground powders to argue that reason, not the senses, must be used to determine what the elements of something actually are and that the true element is the simplest part by nature, not by sense. One MS adds: "which is to say, the *stoicheion*."

A gloss in one MS reads: "Aristotle defines the element as the first thing from which the composite is compounded, which exists in it potentially, not actually, and whose increase in it is potential in the sense that the element from which our bodies are compounded are compounded not by way of combination, but by way of temperament. Therefore, their forms are not preserved and are existent

وهي الأربعة الأخلاط[١٢]، أعني الدّم والبلغم والمرّة الصّفراء والمرّة السوداء، ومنها أقرب ما يكون، وهي الأعضاء[١٣] التي بدن كلّ نوع من أنواع الحيوان مخصوص منها بشيء، أعني الأعضاء[١٤] المتشابهة الأجزاء بمنزلة الشّحم واللّحم والقرون[١٥] والأنياب والمخّ[١٦] والمخاليب.

(٢) والعنصر[١٧] هو جزء مفرد بسيط للشّيء الذي هو عنصر له[١٨]، والشّيء المفرد البسيط على وجهين، أحدهما عند الحسّ[١٩] والآخر عند الطّبيعة. البسيط المفرد عند الحسّ بمنزلة العصبة والعظم، فإنّ هذه وإن كانت عند الحسّ بسيطة مفردة إلاّ أنّها في طبيعتها مركّبة من النّار والهواء والماء ولأرض، وأمّا[٢٠] البسيط المفرد عند الطّبيعة، وهو البسيط الحقّ، فهو[٢١] بمنزلة النّار والهواء والماء والأرض لأنّ هذه ليس علمنا بأنّها عناصر علم حسّيّ بل علم عقليّ[٢٢].

---

١٢ DMSY: الأخلاط الأربعة | ١٣ DMY: + المبشابهة الأجزاء | ١٤ FU: - الأعضاء | ١٥ R: والعروق؛ R¹: خ والقرون. | ١٦ ADFMS: والحمم؛ RY: والجمم؛ R¹: والخمات صح؛ U: - والأنياب والنجم | ١٧ RU: العنصر؛ A¹MR: + وهو الأسطقس | ١٨ Sʰ: أرسطو يحدّ الإسطقس أوّل شيء يتركّب منه المركّب، و[هو] موجود فيه بالقوّة لا بالفعل و زيادته فيه أنّه بالقوّة من قِبَل أن الإسطقسات التي منها يتركّب أجسامنا تركّبت منها لا على سبيل الاجتماع لكن على سبيل الامتزاج، ولهذا ما لا يكون محفوظة الصّورة، فهي موجودة بالقوّة لا موجودة بالفعل. | ١٩ Sʰ: قد طعن طاعن فيقول، كيف يقولون اعتبار الإسطقس ما دامت مفردة على أنّها نار وهواء لا على أنّها لا يكون بالحسّ، ونحن ندرك إسطقسات، وإنّ النار والماء والأرض والهواء بحواسّ لسنا نقول إنّ هذه ندركها الإسطقس هو كذلك بالقياس إلى ما هو له إسطقس. | ٢٠ DMSY: فأمّا | ٢١ AS: - فهو | ٢٢ DRUY: علماً حسّيّاً بل علماً عقليّاً

[Chapter 2][13]

<Their disagreement about the elements>

(3) People disagreed in their views about the elements. Some said that there was one element and others that there were more than one. Some of those who said that there was one element said that it did not move, while the others said that it did move.[14] Some of those who said that it did not move said that it was finite, as was the view of Parmenides, while others said that it was infinite, as was the view of Melissus.[15] Some of those who said that it moved said that the one element was water. Among them were Thales and Hippon. Others said that it was air, as was the view of Anaximenes and Diogenes [of Apollonia]. Still others said that it was fire, as was the view of Heraclitus and Hippasus. Finally, there were those who said that it was earth, among whom was Xenophanes.[16] Some of those who said that there was more than one element held that they were finite in number, while others held that they were infinite in number. Of those who thought that they were finite in

---

potentially, not existent actually." Cf. Aristotle, *On the Heavens*, 3.3 302a15–17: "An element . . . is a body into which other bodies may be analyzed, present in them potentially or in actuality (which of these, is still disputable), and not itself divisible into bodies different in form."

Another gloss in that MS reads: "A skeptic has objected, 'How can they say with respect to the element that so long as it is pure—even though it is just fire, earth, or air—we do not apprehend the element by sensation?' We do not say that the element is this fire, water, earth, and air that we do apprehend by the senses. In this respect the element is analogous to what it is an element of."

13. L 2.1–8; K 1:415–16; G^a 12–15: Galen, commenting on Hippocrates, *On the Nature of Man* 1, alludes to disagreement about the number and nature of the elements and, like Hippocrates, does not list the different opinions. The names were mostly unfamiliar to the scribes of the Islamic period, so there are many errors. This section refers to several passages in *On the Elements*, but it is mainly based on Aristotle, *Physics* 1.2; cf. Galen's citation of *Physics* 1 at L 4.21; K 1:448; G^a 61.

14. *Ghayr mutaḥarrik* (unmoving) and *mutaḥarrik* (moving) render ἀκίνητος and κινούμενος, which, in philosophical usage, refer both to motion and to change in general.

15. L 4.16–5.6; K 1.447–49; G^a 59–62; Hippocrates, *On the Nature of Man* 1. Parmenides of Elea, fl. ca. 475 BCE, held an extreme form of monism in which reality was one, unchanging, and finite. Melissus of Samos followed Parmenides but held that being was infinite. Aristotle was contemptuous of him (see *Physics* 1.3, 185a10, φορτικός [vulgar]); and Hippocrates, Aristotle, and Galen all

16. See the following page.

[الفصل الثّاني]

‹اختلافهم في العناصر[٢٣]›

(٣) وقد[٢٤] اختلف النّاس في الكلام في العناصر، فمنهم من قال إنّ العنصر واحد، ومنهم من قال إنّ العنصر أكثر من واحد، والذين قالوا إنّ العنصر واحد منهم[٢٥] من قال إنّه غير[٢٦] متحرّك ومنهم من قال إنّه[٢٧] متحرّك، والذين قالوا إنّه غير[٢٨] متحرّك منهم من قال إنّه متناهٍ بمنزلة قول بارمانيدس[٢٩]، ومنهم من قال إنّه غير متناهٍ بمنزلة قول مالسيس[٣٠]، وأمّا الذين قالوا إنّه[٣١] متحرّك فمنهم من قال إنّ هذا العنصر الواحد[٣٢] الماء، وممّن[٣٣] قال ذلك ثالس وايبُن[٣٤]، ومنهم من قال إنّه الهواء، وممّن قال ذلك أنكسيمانس[٣٥] وديوجانس، ومنهم من قال إنّه النّار، وممّن قال ذلك ايراقلطيس[٣٦] وإيباسس[٣٧]، ومنهم من قال إنّه الأرض، وممّن قال ذلك كسانوفانس[٣٨]، فأمّا[٣٩] الذين قالوا إنّ العناصر كثيرة، فمنهم من قال إنّها متناهية العدد، ومنهم من قال إنّه[٤٠] لا نهاية لعددها والذين قالوا إنّها متناهية العدد منهم[٤١]

---

٢٣ R^2 | ٢٤ ADMY: قد | ٢٥ AU: فمنهم | ٢٦ AR: – غير | ٢٧ AR: + غير | ٢٨ AFR: – غير | ٢٩ S؛ A: مارمينوس؛ DU: بارمسدس؛ M: بارميدس؛ R: فارميناس؛ Y: بارميندس. | ٣٠ MRY: مالسس؛ M^2: ثالسس؛ U: بالسس | ٣١ AR: + غير | ٣٢ AR: + هو | ٣٣ R: والذي؛ S: ومن | ٣٤ S: ثالس وابين؛ A: مالَسس وابن؛ F: مالَيْس واس؛ DMR: ثاسلس وأين؛ Y: ثاسلس ولبين. | ٣٥ A: ابريسيمانوس؛ M: أنكسيمانوس؛ R: أناكسمانس؛ U: كسيمانس | ٣٦ SY: ايراقلديس؛ AF: ابراقليطس؛ D: ايراقليدس؛ M: ابرفليطس؛ R: ابراقليدن؛ U: ايرافليصس | ٣٧ DY؛ A: ولنياس؛ F: انماسس؛ M: اساسيس؛ R: اساسيون؛ S: أثيناوس؛ U: ايباسلس | ٣٨ S: كسوفونانيس؛ DY: كسايرفييس؛ M: كسوفانس؛ A: مامیندس؛ F: بارامدس؛ R: ارمانيداس؛ U: بارامامیندس. | ٣٩ MRU: وأمّا | ٤٠ MR: إنّها | ٤١ DMSUY: فمنهم

number, some held that there were two elements, some that there were three, and some that there were four. Among those who said that there were two elements was Empedocles. Though he claimed that there were six elements, he also said that two of them, strife and love, were principles[17] and therefore not elements, while, of the four remaining, one of them[18] was hot and the other three were cold.[19] Therefore, we must combine the four elements, resulting in two. One of those who held that there are three elements was Ion of Chios, who claimed that the elements were earth, water, and fire.[20] Among those who claimed that there were four elements was Hippocrates. He held that the elements were fire, air, water, and earth. Some of those who said that the elements were infinite in number held that they were homoeomerous bodies, as was the view of Anaxagoras, while others held that they were indivisible bodies. Among the latter were Epicurus, who held that they could be neither cut nor divided; Asclepiades [of Bithynia], who held that they could neither touch nor be attached to each other; and Leucippus, who claimed that they were bodies without parts.[21]

assumed that if a theory could be reduced to that of Melissus, it was self-evidently absurd. The objection given by Aristotle and Galen was that denying change made physics—and, by extension, its subfield of medicine—moot, since the subject of physics is change, generation, and corruption. On the Greek philosophers mentioned here and below, see appendix 1, pp. 187–202.

16. L 4.6–8, K 1:444, G[a] 54–55: Those named were Ionian physicists of the sixth and fifth centuries BCE, all of whom held—or were alleged to have held—that one of the four material elements was primal and the rest were derived from it. The doxographers, needing an Ionian champion of earth as the primal element, assigned the role to Xenophanes; cf. DK 21A32–33, 36; 21B27–28; Aetius, [Pseudo-Plutarch], *Aetius Arabus*, 3.11. Better authorities—such as Sextus Empiricus, Simplicius, and Galen—said that he held that the first elements were earth and water; see KR, 176–77; Galen, *History of Philosophy* 5, K 19:243. Some manuscripts read "Parmenides" for "Xenophanes." Galen mentions only Thales, Anaximenes, and Heraclitus here.

17. Literally, "roots and heads," perhaps rendering the "roots" ῥιζώματα of Empedocles; cf. DK 31B6; KR, 323, no. 417, or the "real first principles" (τὰς κυριῶς ἀρχάς) mentioned by Simplicius; DK 31A28; KR, 329–30, no. 426; see n. 19 below.

18. Some MSS add: "and that is fire."

19. Empedocles said that the four elements of earth, air, fire, and water were activated by love and strife. He thought that the four material elements were unchanging, being subject only to combination and separation, for which he is criticized below. Three marginal notes in one MS explain that it is the function of love and strife to combine and separate, making them the active causes,

20–21. See the following page.

من قال إنها اثنانِ، ومنهم من قال إنّها ثلثة، ومنهم من قال إنها أربعة، والذين قالوا إنّها اثنانِ، منهم إمبذقلس[٤٢]، فإنّ هذا، وإن كان يزعم[٤٣] العناصر[٤٤] ستّة، فإنّه يقول إنّ اثنين منها إنّما[٤٥] هي[٤٦] أصول و رؤوس وليست بعناصر، أعني الغلبة[٤٧] والمحبّة[٤٨]، ويقول في الأربعة الباقية إنّ واحداً[٤٩] منها[٥٠]، حارّ وثلثة باردة، فيجب من ذلك أن يكون بجمع العناصر الأربعة وتحصّلها[٥١] في اثنين، وأمّا الذين قالوا إنّها ثلثة، فمنهم إيون المنسوب إلى كيوس، فإنّ هذا زعم أنّ العناصر[٥٢] هي الأرض والماء والنار، وأمّا الذين قالوا إنّها أربعة، فمنهم إبّقراط[٥٣]، فإنّ هذا قال إنّ العناصر[٥٤] هي النّار والهواء والماء والأرض، وأمّا الذين قالوا إنّ العناصر غير متناهية العدد، فمنهم من قال إنّها أجسام متشابهة الأجزاء بمنزلة قول أنكساغورس[٥٥]، ومنهم من قال إنها أجسام لا تتجزّأ، ومن هؤلاء[٥٦] إبيقورس[٥٧]، وهو يزعم أنّها لا تنقطع[٥٨] ولا تنقسم، ومنهم أسقلبياذس[٥٩]، وهو يزعم أنها[٦٠] لا تتّصل ولا تلتحم[٦١]، ومنهم لوقيفس[٦٢]، وهو يزعم أنها أجسام ليس لها أجزاء[٦٣].

---

٤٢ AU: اسدفليس؛ DFY: امبدقليس؛ M: إنباذقلس؛ R: أفيدقلس؛ S: امتذقلس؛ S^{h}: هذا يعتقد أنّ الغلبة والمحبّة سببانِ فاعلانِ والأربعة أسباب هيولانيّة. | ٤٣ RU: + أنّ | ٤٤ R: الأسطقسات | ٤٥ DRU: - إنّما | ٤٦ R: - هي | ٤٧ S^{h}: شأنها أن يفرّق على رأيه | ٤٨ S^{h}: شأنها أن يجمع على رأيه | ٤٩ AM: الواحد | ٥٠ DFMSY: + وهو النّار | ٥١ FR: وتخلّصها؛ M: وتخصّصها؛: فتخلّصها صح | ٥٢ R: الأسطقسات | ٥٣ FMU: بقراط | ٥٤ R: الأسطقسات | ٥٥ A: فركساغورس؛ DY: ان كساغورس؛ F: اناكساغورس؛ U: اناكسوفورس | ٥٦ A: من هؤلائ؛ FRU: فمن هؤلاء؛ DY: ومن ذلك؛ D^{1}: هو آراء | ٥٧ DSY: إفيقورس؛ A: افيغورس؛ F: أبيغورس؛ M: فيقورس؛ R: أبيقورس؛ U: بيفورس | ٥٨ R: + ولا يجزّأ؛ S: يتّصل | ٥٩ D: اسقلادس؛ F: ابقليبادس؛ R: أسقليبياذس؛ S: اسمبياذس؛ A^{2}: خ لوقس | ٦٠ A: + أجسام ليس لها أجزاء، ومنهم لوقس، وهو يزعم أنّها | ٦١ R: وتلحم | ٦٢ DY؛ F: لوفكس؛ M: لوقيس؛ R: فبلينس؟؛ S: لوقيس؛ A^{2}: خ اسقليسادس | ٦٣ A: - ومنهم لوقيفس. . . أجزاء

<The difference between the element and the principle>[22]

(4) The difference between the element and the principle—which is to say, the source—is that the element is one of the simple parts of the thing of which it is an element. Thus, the element must necessarily be potentially existent in that of which it is an element, while the principle—which is to say, the source or the basis—need not necessarily be existent in that of which it is a principle. Thus, smoke is not existent in the wood, and the principle is necessarily not a part of that of which it is a principle, even though the stability and continuance of the thing may depend on the principle. This is the case with prime matter and form, for they are not two parts of the body, since they are not a body and it would be false and absurd to claim that what is not a body is part of a body.

<The principles of things>[23]

(5) Some of those who talked about the principles and sources of things said that there was one principle, while others said that they were multiple. Some of those who held that there was one principle said that it was one in species, as was the view of Democritus and Leucippus.[24] Others held that it was numerically one, as Parmenides and Melissus said. Some of those who said that the principles[25] were multiple claimed that there were two: God, may He be blessed and exalted, and prime

while the other four are passive. Both Aristotle (*Physics* 1.5, 188b34) and Simplicius (DK 31A28; KR, 329–30, no. 426) said that Empedocles actually posited two elements, but they are both referring to love and strife.

20. On Ion's three elements, see Philoponus, *On Generation and Corruption* 227.14, though at 207.19–20 he seems to give fire, earth, and air; DK 36A6, 36B1.

21. These were Atomists of various sorts. Anaxagoras held an eccentric theory of "homoeomerous seeds," in which all substances originally existed in scattered particles that Mind collected to make things of varying kinds. Leucippus and Democritus developed the classical atomic theory, in which they were followed by Epicurus. Asclepiades, a founder of Methodism, held a corpuscular theory of matter and was cordially detested by Galen.

22. Cf. L 6.39; K 1.470; G[a] 39: "For element [στοιχεῖον; *isṭaqis*] differs from first principle [ἀρχή, *mabdaʾ*] in this, that first principles are not necessarily homogeneous with the things whose first principles they are, but elements are entirely homogenous." The comment occurs in a passage where Galen criticizes Athenaeus for not understanding the distinction between the two. The difference

23–25. See the following page.

‹الفرق بين العنصر والرّأس[٦٤]›

(٤) الفرق بين العنصر والرّأس الذي هو الأصل أنّ العنصر جزء بسيط من أجزاء الشّيء الذي هو عنصره، والعنصر لا محالة موجود فيما هو عنصره بالقوّة، والرّأس، أعني الأصل والمبدأ، ليس يجب أن يكون لا محالة موجوداً[٦٥] في الشّيء الذي هو رأسه، من ذلك أنّ[٦٦] البخار ليس هو موجود في الخشب، والرّأس لا محالة ليس هو جزء[٦٧] ممّا هو رأس له[٦٨] ولو أنّه كان به ثباته وقوامه[٦٩] بمنزلة الهيولى والنّوع، فإنّ هذيّن ليس هما جزآن[٧٠] من الجسم، وذلك[٧١] أنّهما ليسا بجسم، وإن[٧٢] يكون شيء ليس هو جسم[٧٣] جزءاً من الجسم، فإنّ ذلك[٧٤] منكرشَنِعٌ.

‹رؤوس الأشياء[٧٥]›

(٥) الذين تكلموا في رؤوس الأشياء وأصولها منهم من قال إنّ الرّأس واحد، ومنهم من قال إنّها كثيرة، والذين قالوا إنّ الرّأس واحد منهم من قال إنّه واحد في النّوع على ما قال ديمقريطس[٧٦] ولوقبس[٧٧]، ومنهم من قال إنّه واحد في العدد على ما قال بارمانيدس[٧٨] ومالسس[٧٩]. فأمّا الذين قالوا إنّ الرّؤوس كثيرة، فمنهم[٨٠] من قال إنّها اثنان، الله تبارك وتعالى والهيولى[٨١]، ومنهم من قال إنّها ثلثة،

٦٤ R^2: الفرق بينهما | ٦٥ AM: موجود | ٦٦ Y: إلى؛ D: أنّي | ٦٧ MSY: جزءاً | ٦٨ DMSY: له رأس | ٦٩ DMSY: قوامه وثباته | ٧٠ ADMSY: جزيّن | ٧١ DMY: وذاك | ٧٢ ADFY: فإن | ٧٣ DMSUY: جسماً | ٧٤ M: ذاك؛ RU: فإنّ ذلك > وذلك | ٧٥ R^2 | ٧٦ ADFSY: ديمقراطيس؛ U: ديمقريصس | ٧٧ AD: ولوقس؛ R: ولوقلس؛ U: ووفيس | ٧٨ DSY: فارمانيدس؛ A: برميندس؛ FU: بارمانيدس؛ M: مارميندس؛ R: بارمنيدس. | ٧٩ ARSY؛ D: مالبس؛ F: ماليسس؛ MU: ثالسس | ٨٠ RU: منهم؛ R^2: كفرنعم؟ | ٨١ DMS: + بمنزلة أصحاب الرّواق؛ Y: + بمنزلة أصحاب [كذ]

matter. Others of them said that there were three principles: form, prime matter, and nonbeing[26]—which was what Socrates said. Some of them said that there were four: God, form, matter, and nonbeing—which was what Aristotle said. Some said that there were six principles: the four elements, strife, and love—which was what Empedocles said. Some said that there were ten principles, which was the view of the Pythagoreans. Some of those who, like Democritus, held that the elements were multiple believed that they were numerically multiple. Those who held this theory claimed that the bodies generated from these elements became different only by virtue of the differing shapes, positions, and oppositions of these elements; for these indivisible elements differed in that some of them were round and some oblong in shape, some erect and some horizontal in position, and some prior and some posterior in their oppositions. Others held that they were multiple in species, as was the view of Hippocrates. Those who held this view said that the differences among bodies compounded from these elements were due to different temperaments of these elements. According to Epicurus's theory, the diseases that occur in the tissues are of two kinds: one that is condensation and the other that is rarefaction.

is that elements are constituents of physical reality, while principles are their ultimate causes of whatever sort. See, in general, Aristotle, *Metaphysics* 1.4 and *Physics* 1.1. In the latter, Isḥāq ibn Ḥunayn translates ἀρχαὶ ἢ αἰτία ἢ στοιχεῖον (principles, causes, and elements) as *al-mabādiʾ aw al-asbāb aw al-isṭaqisāt*.

23. L 2.1–17; K 1:415–19; G[a] 12–18: Galen's criticism of the atomic theory of Democritus.

24. Helmreich's edition of Galen's text mentions only Leucippus here. De Lacy in L reads "Diodorus and Leucippus" on the strength of Sālim's edition of the Arabic translation, concluding that it refers to Diodorus Cronus, also an Atomist. However, the one MS that I am able to consult of the Arabic *On the Elements*, Aya Sofya 3593, f. 2a13, clearly reads "Democritus and Leucippus," as do the parallel passages in Aristotle's *Metaphysics* and this epitome.

25. Some MSS add: "such as the Stoics."

26. **S** alone reads "number," with an interlineal correction to *wa-al-ʿadam* (and nonbeing), the reading supported by the other manuscripts. *Aetius Arabus* 1.3.21 says that Socrates (and Plato: 1.7.31) believed in three principles: God, the element, and the form. Possibly the text refers to the Platonic doctrine of three classes of entities: the Ideas, the mathematicals, and the material objects; cf. Aristotle, *Metaphysics* 1.6, 987b15–18. If, however, "nonbeing" is correct, it must refer to Becoming (γέννησις).

النّوع والهيولى والعدم[٨٢] على ما قال سوقراطيس[٨٣]، ومنهم من قال إنّها أربعة، الله تبارك وتعالى والنّوع والهيولى والعدم، على ما قال أرسطوطاليس[٨٤]، ومنهم من قال إنّها ستّة، الأربعة العناصر والغلبة والمحبّة، على ما قال إمباذقليس[٨٥]، ومنهم من قال إنّها عشرة على ما قال أصحاب فوثاغورس[٨٦]. الذين قالوا إنّ العناصر كثيرة منهم من قال إنّها كثيرة في العدد بمنزلة قول ديمقريطس[٨٧]، وزعم أهل هذا القول[٨٨] أنّ الأجسام المكوّنة[٨٩] من هذه العناصر إنّما صارت مختلفة من قِبَل اختلاف هذه العناصر في أشكالها ووضعها وتضادّها، أمّا في الشّكل فإنّ هذه العناصر التي لا تتجزّأ تختلف لأنّ بعضها مدوّر وبعضها مطاول، وأمّا في الوضع فلأنّ بعضها منتصب وبعضها مبطوح[٩٠]، وأمّا في التّضادّ فلأنّ بعضها متقدّم وبعضها متأخّر، ومنهم من قال إنّها كثيرة في النّوع بمنزلة قول إبقراط[٩١]، وأهل هذا القول يقولون إنّ اختلاف[٩٢] الأجسام التي هي مؤلّفة من هذه العناصر إنّما جاء من قِبَل اختلاف مزاج العناصر. الأمراض الحادثة في الأعضاء المتشابهة الأجزاء على رأي إبيقورس[٩٣] ضربانِ[٩٤]، أحدهما التّكاثف والآخر التّخلخل.

---

٨٢ S²؛ S: والعدد | ٨٣ RS: سقراطس؛ M: سقراطيس. | ٨٤ FRU: أرسطاطاليس. | ٨٥ A: امبرقلس؛ DMSY: إمباذقلس؛ F: انبذقليس؛ RU: امبدوقلس. | ٨٦ D: فوثاغوروس؛ S: - أصحاب؛ U: فواغورس | ٨٧ AR: ديمقراطس؛ F: ديماقرطيس.؛ U: ديمقراطيس | ٨٨ ADFSY: الرّأي | ٨٩ DFMSY: المتكوّنة؛ U: المتكوّنين | ٩٠ F: مسطوح؛ M: مطبوح | ٩١ DMSUY: بقراط. | ٩٢ DMSY: + هذه | ٩٣ DMSUY: إفيقورس | ٩٤ DMSY: مرضان

[Chapter 3]

[Whether the elements sense and suffer][27]

(6) Some of those who said that the elements were more than one said that they could neither sense nor suffer. This was the view of Democritus. Others said that they could both sense and suffer, as Asclepiades held. Some, such as Anaxagoras, held that they could not suffer, but they could have sensation. Finally, there were those, such as Hippocrates, who held that, while they could suffer, they could not have sensation. "To suffer" here means the reception of influences. If suffering, sensation, not suffering, and lack of sensation are each matched with the other three, each pair will produce a combination, some of which are compatible and others not, in this pattern:

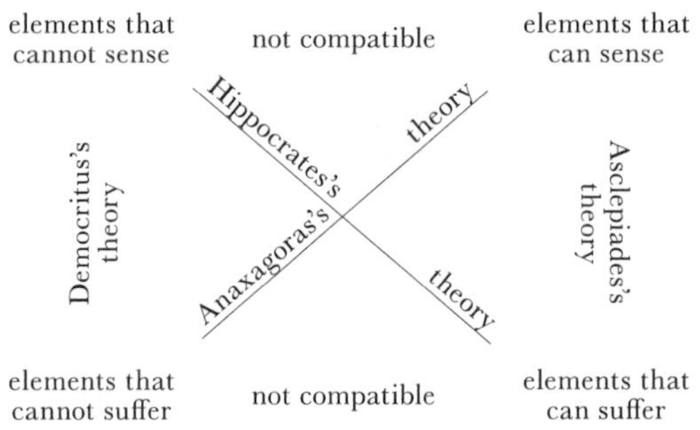

27. L 2.18–3.8, K 1.419–28; G[a] 18–29: The context is a discussion of whether a body can have sensation and sentience if its elements do not. It is unfortunate that the translator renders πάσχω (to suffer) by the verb *alima* (to feel pain) and its derivatives, which are thus confusingly used here in the philosophical sense of "to be affected," rather than in the medical and ordinary sense of "to suffer" something unpleasant. Toward the end of this paragraph of the epitome, the translator feels obliged to explain that the verb *alima* is not used in its usual sense of "to feel pain." The original passage from Hippocrates that Galen is quoting does say ἀλγεῖν (to suffer pain), but Galen switches to using πάσχειν (to suffer, in both senses) a few lines later.

[الفصل الثّالث]

[اختلافهم في أنّ للعناصر حسّ وألم]

(٦) الذين[٩٥] قالوا إنّ العناصر كثيرة منهم من قال إنّها لا تحسّ ولا تألم[٩٦] بمنزلة قول ديمقريطس[٩٧]، ومنهم من قال إنّها[٩٨] تحسّ وتألم[٩٩] بمنزلة قول[١٠٠] أسقليبياذس[١٠١]، ومنهم من قال إنّها لا تألم لكنّها تحسّ بمنزلة قول أنكساغورس[١٠٢]، ومنهم من قال إنها لا تحسّ لكنّها تألم بمنزلة قول إبقراط[١٠٣]، ومعنى[١٠٤] الألم هاهنا قبول الأحداث، وإذا جمع كلّ واحد من الألم والحسّ وعدم الألم وعدم الحسّ مع كلّ[١٠٥] واحد من الثلثة الآخر[١٠٦] تركّب من كلّ اثنين منهما[١٠٧] تركيباً، وبعض هذه[١٠٨] التّركيبات يلتأم وبعضها لا يلتأم على هذا المثال:

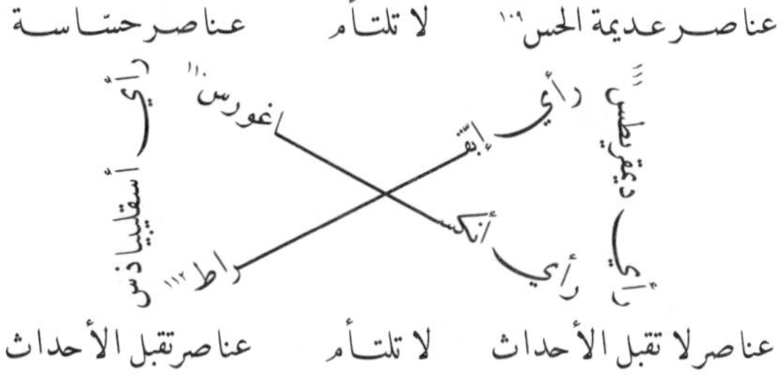

---

٩٥ AU: والذين | ٩٦ R: لا . . . بألم > تحسّ وتألم | ٩٧ AU: ديمقراطيس؛ F: ديمقرطس؛ R: أسقليبيادس | ٩٨ RU: + لا | ٩٩ Mʰ: حاشية له: لا تحسّ ولا تنفعل؛ R: ولا تألم | ١٠٠ DSY – قول | ١٠١ R: + ومنهم من قال إنّها لا يحسّ ولا يألم بمنزلة قول ديمقراطس | ١٠٢ A: بركساغورس؛ F: اناكسوغورس؛ U: أناكساغوارس | ١٠٣ MS: بقراط | ١٠٤ FY: معنى | ١٠٥ RU: – كلّ | ١٠٦ AU: الأجزاء | ١٠٧ AFS: منها | ١٠٨ F: وكبعض هذه | ١٠٩ RU: للحسّ | ١١٠ F: إبقراط | ١١١ AU: ديمقراطيس؛ R: ديمقراطس | ١١٢ S: بقراط؛ F: انكساغورس

(7) Some of those who insisted that the elements had no parts said that they were indivisible due to their hardness, and some said that they were indivisible due to their small size.[28] The one who held that they were indivisible due to their hardness was forced to two unacceptable absurdities. First, they have qualities and accept effects. That is because, if they have hardness, they must also have softness, and softness is something that implies ease of accepting effects. Moreover, hardness is itself a quality resulting from coldness or dryness. The second conclusion is that, if the cause of its resistance to accepting effects is its hardness, then what is it about them that makes them small rather than large? On the other hand, the one who says that the elements are indivisible because they are small is compelled to admit that, though these elements might be actually indivisible, they are nonetheless potentially divisible, since their nature is such as to accept division and distinction of parts, though their smallness prevents it.

<*The occurrence of pain*>[29]

(8) The occurrence of pain requires two things: first, to receive what it suffers and, second, to sense the suffering that occurs in it. Receptivity to what it suffers is required because, if a creature possessing sensation does not suffer, it will neither sense nor acquire pain. Sensation of what it suffers is required because, if that which suffers has no sensation of it, it will not experience pain from what it suffers.

<*Compounds*>[30]

(9) Some things that are compounded with each other are only compounded by contiguity and contact, as is the case of the things from which a house is built or the cases in which seeds[31] are mixed. When

28. L 2.17; K 1:418–19; G^a 17: Explaining that according to the Atomists, the qualities of the atoms do not change.

29. L 2.18–21; K 1:419–20; G^a 18–19: Cf. Hippocrates, *On the Nature of Man* 2.10: ". . . but I hold that if man were a unity, he would never suffer pain."

30. L 3.8–18; K 1:428–31; G^a 29–33: Galen is explaining how a compound in which the parts retain their individual qualities differs from a true temperament, in which new qualities occur. The examples of compounds that he gives are the construction of a house and the mixing of different kinds of seeds in a basket. This occurs in the course of his argument that the Atomists cannot give a satisfactory account of living bodies, since the atoms are neither sentient nor able to be affected.

31. Some MSS read "powder" for "seeds," following L 1.2–3; K 1:413–14; G^a

(٧) الذين أوجبوا عناصر لا أجزاء[١١٣] لها منهم من قال إنّها لا تتجزّأ لصلابتها ومنهم من قال إنّها لا تتجزّأ لصغرها، ويلزم من قال إنها لا تتجزّأ لصلابتها أمرانِ منكرانِ شنعانِ، أحدهما أنها[١١٤] تكون مكيّفةً[١١٥] قابلةً للأحداث، وذلك[١١٦] أنه إن كانت لها صلابة، فلها أيضاً لين، واللّين شيء يوجب سهولة قبول الأحداث مع أنّ الصّلابة نفسها هي أيضاً[١١٧] كيفيّة تابعة للبرودة أو اليبوسة[١١٨]، والأمر الآخر أنّه إن كان إنّما السّبب في بعدها عن قبول الأحداث صلابتها، فلِمَ صارت على ما يوحون[١١٩] به من أمرها يكون صغاراً ولا تكون كباراً؟ وأمّا من قال إنّها لا تتجزّأ لصغرها فيلزمه أن يكون هذه العناصر، وإن كانت لا تتجزّأ بالفعل، فهي تتجزّأ بالقوّة لأنّ طبعها من شأنه أن يقبل القسمة والتّجزئة، لكنّ صغرها يمنع من ذلك.

‹حدوث الوجع[١٢٠]›

(٨) حدوث الوجع يحتاج فيه إلى أمرين، أحدهما قبول الألم والآخر أن يحسّ ما يناله من الألم، أمّا قبول الألم فلأنّ ما له حسّ إن هو[١٢١] لم يألم لم يحسّ ولم يتّجع، وأمّا حسّ الألم فلأنّ الذي يألم إن لم يكن له حسّ لم يوجعه ذلك الألم.

‹تركيب الأشياء[١٢٢]›

(٩) الأشياء التي يتركّب بعضها مع بعض منها ما يكون تركيبها على طريق المجاورة والملامسة فقط بمنزلة الأشياء التي يُبنى بها بيت[١٢٣] والأشياء التي تخلط منها[١٢٤]

---

١١٣ F: الأجزاء؛ UY: إلاّ أجزاء | ١١٤ SUY: إنّه | ١١٥ AFU: مكتفية؛ M: متكفية؛ S: كيفيّة | ١١٦ SU: وذاك | ١١٧ AF: أيضاً هي | ١١٨ DRU: لليبوسة | ١١٩ D: تلوضون؟؛ F: يتوخون؛ MSY: يوجبون | ١٢٠ R^2 | ١٢١ AD: – إن هو | ١٢٢ R^2 | ١٢٣ AS: بيتاً؛ D: ثلث؛ F: بينه | ١٢٤ D^1M: فيها

things are compounded in this way, the whole of the compound thing contains nothing save that which is already contained in the simple parts from which it was compounded. Other things are compounded by way of composition and the tempering of one with the other, as is the case with vinegar and honey, which are compounded to form oxymel. If something is compounded in this way, something else results from combining its parts that it did not have while the parts were still simple, before they were combined. According to those who hold that the elements are numerically many—that is, the followers of Democritus—bodies can be compound only by way of contiguity, since these elements do not have an effect in each other, nor are they affected by each other in such a way that a temperament could result from them. Therefore, it is impossible for the adherents of this view to hold that they can generate bodies that are able to be affected and have sensation by compounding elements that neither have sensation nor accept changes. However, according to the view of those who hold that elements are of more than one species—that is, the followers of Hippocrates—bodies are compound in their theory only in the sense of temperament, because they say that these elements are able to be affected. Therefore, even though they say that the elements do not have sensation, they are able to deduce that bodies capable of sensation are generated from this temperament.

10–11, which gives the example of a mixture of four substances ground to a fine powder. If the reading "seeds" is correct, it comes from L 3.18, which criticizes the Atomists by comparing their theory of compounds to mixing different kinds of grain. In either case, the point is that the mixture would appear homogeneous to the eye, but its parts would nonetheless still be distinct substances unaffected by their being mixed together—unlike a true compound, in which a new substance comes into being. A gloss in one of the manuscripts reads, "That is, like a basket. For example, kinds of seed are mixed so that they are completely blended, even if that thing has a specific arrangement, such as a building—or, indeed, has no order at all. Galen's text [i.e., L 3.18, K.1.431] reads, 'Those who hold that this happens [i.e., that a body becomes alive] with [the parts] remaining as they were except for being mixed together—like wheat, barley, chickpeas, and beans in a heap—are making an absurd claim.'" Another gloss gives a shorter version and mentions that Galen's text refers to a "heap" of mixed seeds instead of a basket.

البزور[١٢٥]، وما كان تركيبه على هذه الجهة فليس يكون لجملة الشّيء المركّب شيء ليس هو للأجزاء[١٢٦] البسيطة التي منها تركّب[١٢٧]، ومنها ما يكون تركيبها[١٢٨] على طريق المخالطة والممازجة بعضها لبعض بمنزلة الخلّ والعسل اللذيْنِ تركّب[١٢٩] منهما السّكنجبين، وما كان تركيبه على هذا الوجه فقد يتولّد من الأجزاء إذا ركّبت شيء آخر لم يكن لها في وقت ما كانت بسيطةً قبل أن تتركّب، وكون[١٣٠] الأجسام المركّبة بحسب رأي القوم الذين ينتحلون عناصر كثيرة العدد، أعني أصحاب ديمقريطس[١٣١]، إنّما يكون على جهة المجاورة إذ كانت هذه العناصر لا يفعل بعضها في بعض ولا ينفعل بعضها من بعض ليكون منها مزاج، ولذلك ليس يمكن أهل هذا الرّأي أن ينتحلوا[١٣٢] ويولّدوا من تركيب عناصر لا تحسّ ولا تقبل الأحداث أجساماً قابلة[١٣٣] للأحداث حسّاسة. فأمّا[١٣٤] بحسب رأي القوم الذين ينتحلون عناصر كثيرة في النوع، أعني أصحاب إبقراط، إنّما[١٣٥] يكون الأجسام المركّبة عندهم[١٣٦] على جهة المزاج لأنّهم يقولون إنّ هذه العناصر تقبل الأحداث، ولذلك قد يمكنهم، وإن كانوا يقولون إنّ العناصر لا يحسّ أن ينتجوا[١٣٧] ويولّدوا[١٣٨] من[١٣٩] مزاجها أجساماً تحسّ.

---

١٢٥ AFS: الذّرور؛ cf. G 1.2؛ R^1: + حـ[اشية له] أي كحيل مثلاً خلط فيه عدّة بزور فتتجاوز وتتلاقا، وإن كان ذلك الشّيء على نظام مخصوص كأمر البناء بل على غير نظام. في الفصّ فأمّا من قال إنّ ذلك يحدث عنها وهي باقية على حالها وإن [G^a: بأن] خالطت بعضها بعضاً [G^a: + فقط] كمخالطة الحنطة للشّعير والحمص والباقلاء في صبة واحدة فدعواه محال. | A^2M^2: حاشية له: أي مثلاً مكيل يخلط فيه [M^2: + عدّة] أنواع من بزور النّبات وفي نصّ جالينوس مكان مكيل صبّه. | ١٢٦ AU: للآخر | ١٢٧ ARU: رُكّب | ١٢٨ AMS: تركيبه | ١٢٩ AMU: يتركّب | ١٣٠ DMUY: كون | ١٣١ AUY: ديمقراطيس؛ D: ديمقراطس | ١٣٢ D^1RY^1: ينتجوا | ١٣٣ DMS: قبوله؛ F: تقبل؛ Y: قبول | ١٣٤ AMSY: وأمّا | ١٣٥ DFMSUY: بقراط إنّما | ١٣٦ ADFMSUY: - إنّما يكون | ١٣٧ S: ينتحلون | ١٣٨ DY: ويولّدون | ١٣٩ DMSY: + قِبَل

<The tools of inference>[32]

(10) There are two devices and tools by which the knowledge of entities is acquired. One of them is inference and the other experience. The fact that the elements are not of a single species, as Democritus thought, may be known both by inference and by experience. It may be known by inference through the argument of Hippocrates when he says that if man were unitary, he would not feel pain; but since we find that he does feel pain, he is therefore not unitary—that is, he is not compounded from a single element. This is because whatever is subject to pain must necessarily change, and that which changes can only do so by moving and changing from one thing to another. It can also be shown by experience, for when we prick our body with a pin, it gives it pain. If Democritus were to say that that pin had merely entered the void between the atoms, then surely no pain would occur when the pin entered it, for the void and the space that are between these parts are nothing. If he were to say that the pin enters the atoms themselves, then that which he had posited to be indivisible would have been divided and that which he had posited as unable to be affected would be affected, just as Hippocrates had deduced when he said that the occurrence of pain implies that the element is not unitary. We can draw the very

32. L 2.9–43, 3.23–30; K 1:416–24, 432–34; G^a 15–24, 33–41; Hippocrates, *On the Nature of Man* 2: This passage summarizes and clarifies a passage where Galen argues against the Atomists on the grounds that they cannot explain the experience of pain. He first argues on the basis of reason; then he stops and points out that arguments can be made on the basis of both experience (ἐμπειρία, *tajribah*) and reason (λόγος, *qiyās*).

‹آلات الاستخراج[١٤٠]›

(١٠) الأدوات والآلات التي بها يستخرج معرفة الأمور اثنانِ[١٤١]، أحدهما القياس والآخر التجارب، وقد يُعلم أنّ العنصر[١٤٢] ليس هو واحد في النّوع على ما ظنّ ديمقريطس[١٤٣] من القياس[١٤٤] ومن التّجارب، أمّا من القياس فمن قول إبّقراط[١٤٥] حيث يقول إنّه لوكان الانسان[١٤٦] واحداً[١٤٧] لم يتّجع، وقد نجده يتّجع، فليس هو إذاً واحداً[١٤٨]، أي ليس هو مركّبا[١٤٩] من عنصر واحد، لأنّ الذي يناله الوجع يحتاج أن تتغيّر، والمتغيّر إنّما ينتقل ويتغيّر[١٥٠] من شيء إلى شيء، وأمّا من التّجارب، فمن أنّا إذا غرزنا البدن بإبرةٍ فأوجعته. إن قال ديمقريطس[١٥١] إنّ تلك الإبرة إنّما دخلت في الخلاء[١٥٢] الذي بين الأجزاء التي لا تتجزّأ، فقدكان ينبغي أن لا يحدث بدخولها وجعاً[١٥٣] لأنّ الخلاء والفضاء الذي[١٥٤] فيما بين هذه الأجزاء ليس هو[١٥٥] شيئاً، وإن قال إنّ الإبرة دخلت في نفس الأجزاء التي لا تتجزّأ، فقد تجزّأت التي هي عنده غير متجزّئة وقبلت[١٥٦] الأحداث التي هي عنده غير قابلة للأحداث[١٥٧] كما أنتج إبّقراط في[١٥٨] حدوث الوجع أنّ العنصر ليس بواحد. كذلك قد يمكنّا[١٥٩] أن ننتج من جميع التّغييرات

---

١٤٠ R^2 | ١٤١ ADMSY: اثنانِ | ١٤٢ SY: العناصر | ١٤٣ AUY: ديمقراطيس؛ DR: ديمقراطس؛ M: ديمقرطس | ١٤٤ AFY: القياسات | ١٤٥ DMY: بقراط | ١٤٦ DMR^1SY: + شيئاً | ١٤٧ FM: واحد | ١٤٨ S: واحد؛ F: بواحد | ١٤٩ DF: مركّب | ١٥٠ AFU: يتغيّر وينتقل | ١٥١ AFU: ديمقراطيس؛D: ديمقرطاس؛ MR: ديمقراطس | ١٥٢ A: التخلخل | ١٥٣ DMSUY: وجع | ١٥٤ R: الذين | ١٥٥ R: ليسا هما | ١٥٦ A: وقبله؛ DY: وقلت | ١٥٧ F^2؛ FS: – غير قابلة للأحداث؛ R: الأحداث | ١٥٨ DFSUY: من | ١٥٩ MSU: يمكننا

same inference from all other alterations. By other alterations I mean pleasure and grief, heat and coldness, and the other species. The argument is that, if man were compounded from a single element, he would experience neither pleasure nor grief, heat nor coldness, nor would any other changes affect him, since there could not be something that could hurt him—nor could that thing both be painful in itself and accept any of these effects [such as pain, pleasure, or grief], since it would be active and passive in a single respect in the same way.

<The refutation of the others' arguments>[33]

(11) The refutation of Democritus's argument serves also to refute the views of the rest of those who claim that the elements are unchangeable, such as Empedocles and Anaxagoras, for Empedocles also held that the elements were unchangeable and unalterable and claimed that composite bodies resulted only from their being compounded by way of contiguity, not by way of temperament. Anaxagoras claimed that the elements were indivisible atoms, that they were unchangeable, and that generation and corruption resulted only from their combination and separation. Empedocles and Anaxagoras were actually in agreement in that they both held that the elements were unchangeable and that the composite bodies were generated and corrupted only by the

33. L 9.11; K 1:483–84; G[a] 105–6: Their views are being contrasted to Hippocrates's belief in the transmutation of elements; cf. Aristotle, *Physics* 1.4, 187a21–187b8, and 3.4, 203a16–203b2, where he talks about their view that substantial change is blending (μῖξις) rather than temperament (κρᾶσις), as the Hippocratics and Galen would have it; cf. pp. 147–58, paragraph 13 and n. 37, below.

الأُخر. هذه النّتيجة بعينها، أعني بالتّغييرات الأُخر اللّذّة والغمّ والحرارة والبرودة وسائر الأنواع الأُخر، وذاك[١٦٠] أنّه لو كان الإنسان مركّبًا من عنصر واحد لكان لا يلذّ[١٦١] ولا يغتمّ ولا يسخن ولا يبرد ولا يناله شيء[١٦٢] من الأحداث الأُخر، إذ كان ليس هاهنا شيء يؤلمه ولا يمكن أن يكون الشّيء هو المؤلم لنفسه والقابل للفعل منها حتّى يكون هو الفاعل والمنفعل من جهةٍ واحدةٍ بعينه.

‹انفساخ أقوالهم[١٦٣]›

(١١) وبانفساخ قول ديمقريطس[١٦٤] قد انفسخت أقوال[١٦٥] سائر من زعم[١٦٦] أنّ العناصر لا تقبل الأحداث بمنزلة إمباذقليس[١٦٧] وأنكساغورس[١٦٨]، فإنّ إمباذقليس[١٦٩] ينتحل أيضاً[١٧٠] أنّ العناصر لا تقبل الأحداث[١٧١] ولا تتغيّر، ويزعم أنّ الأجسام المركّبة إنّما تكون من العناصر بتركيبها على طريق المجاورة لا على طريق المزاج، وأنكساغورس يزعم أنّ العناصر أجزاء لا تتجزّأ وأنّها غير قابلة[١٧٢] للأحداث وأنّ الكون والفساد إنّما يكونان[١٧٣] باجتماعهما وتفرّقهما[١٧٤]، وإمباذقليس[١٧٥] وأنكساغورس[١٧٦] يشتركانِ ويتّفقانِ في أنهما جميعاً ينتحلانِ أنّ العناصر لا يقبل الأحداث وأنّ الأجسام المركّبة

---

١٦٠ DFMSY: وذلك | ١٦١ ^{1}DMSUY: يلتذّ | ١٦٢ AM: شيئًا؛ A^1M^1: خ ولا ينال شيئًا | ١٦٣ R^2 | ١٦٤ AFU: ديمقراطيس؛ R: ديمقراطس | ١٦٥ DMSUY: أقاويل | ١٦٦ DMSY: قال وزعم | ١٦٧ DY: إمياذقليس؛ FU: إمبادوقليس؛ M: إنباذقلس؛ R: إنبذقلس | ١٦٨ A: وأبركساغورس؛ R: أناطاغوراس | ١٦٩ A: أمير وقليس؛ D: إمياذقليس؛ FU: إمبادوقليس؛ M: إنباذقلس؛ R: إنبدقلس | ١٧٠ R: ينتحل أيضاً > ممّن زعم | ١٧١ R: التّأثير والأحداث | ١٧٢ DSY: قبولة | ١٧٣ A: يكونا؛ RU: يكون؛ Y: +أن | ١٧٤ MY: باجتماعها وتفرّقها؛ S: باجتماعها وتفريقها | ١٧٥ D: إمباذقليس؛ F: إمبوقليس؛ M: إنباذقلس؛ R: أنبدقلس؛ S: وإمبادقلس؛ U: إماذوقليس | ١٧٦ A: وأبركساغورس؛ DY: وأناكساغورس

combination and separation of the elements. They disagreed in that Empedocles held that the elements were fire, air, water, and earth and that homoeomerous bodies were compounded from them, while Anaxagoras held the contrary view, saying that the homoeomerous bodies were the simple elements and that fire, air, water, and earth were merely compounded from them.

<*Their views on the elements*>[34]

(12) The Ancients held four opinions concerning the elements. The first opinion was that of those who said that the elements do not have sensation and cannot be affected. Their view was false and clearly absurd, but it was not incoherent. The second opinion was that of those who said that the elements have sensation and cannot be affected. Their opinion is false, clearly absurd, and incoherent. That is because it is impossible for a human being to understand how something could have sensation yet be entirely unable to be affected. The third opinion is that of those who said that the elements had sensation and could be affected. Their view is plausible but untrue. It is plausible because living bodies that can be affected and have sensation could result from elements that can be affected and have sensation and because it is also necessary that

---

34. L 3.1–8; K 1:427–86; G^a 26–29: An elaboration of an argument directed mainly against the Atomists, claiming that the fact that living bodies are sentient can only be explained by a system of changeable, nonsentient elements.

إنّما يكوّن ويفسد باجتماع هذه العناصر وتفرّقها فقط، ويختلفانِ في أنّ إمباذقليس[١٧٧] ينتحل أنّ العناصر هي النار والهواء والماء والأرض، وأنّ المتشابهة الأجزاء إنّما هي أجسام مركّبة من هذه[١٧٨]، وأنكساغورس يزعم خلاف هذا، وذاك[١٧٩] أنّه يقول إنّ الأجسام المتشابهة الأجزاء هي العناصر البسيطة، والنار[١٨٠] والهواء والماء والأرض، إنّما هي مركّبة من هذه.

‹انتحالهم في العناصر[١٨١]›

(١٢) الآراء التي انتحلوها[١٨٢] القدماء في العناصر أربعة، وأهل الرّأي الأوّل، وهم الذين قالوا إنّ العناصر لا تحسّ ولا تقبل الأحداث، قولهم[١٨٣] قول كذب صراح مُحال إلاّ أنّه ليس هو قولٌ لا يُفهَم، فأمّا أهل الرّأي الثاني، وهم الذين قالوا إنّ العناصر تحسّ ولا تقبل الأحداث، فقولهم[١٨٤] كذب صراح[١٨٥] مُحال لا يُفهَم، وذلك أنّه ليس يمكن إنسان[١٨٦] أن يفهم كيف يكون شيء يحسّ من غير أن يناله حدث من الأحداث، وأمّا قول[١٨٧] أهل الرّأي الثالث، وهم الذين قالوا إنّ العناصر تحسّ وتقبل الأحداث، وقولهم[١٨٨] قول ممكن إلاّ أنّه ليس بحقّ، أمّا[١٨٩] إمكانه فمن طريق أنّه يجوز[١٩٠] أن يكون من عناصر تقبل الأحداث وبحسّ أبدان تقبل الأحداث وتحسّ

---

١٧٧ AS: إميادقلس؛ A¹: أميرقلس؛ D: إمياذقليس؛ M: إنباذقلس؛ R: إنبدقلس؛ U: إماذوقليس | ١٧٨ DMY: هي النّار. . . من هذه> أجزاء لا تتجزّأ وأنّها غير قابلة للأحداث وأنّ الكون والفساد إنّما يكونانِ باجتماعها وتفرّقها | ١٧٩ ADMSY: وذلك | ١٨٠ SY: من النّار | ١٨١ R^2 | ١٨٢ RU: انتحلها | ١٨٣ AMS: فقولهم؛ MR: + هذا | ١٨٤ A: فقولهم> هؤلاء أيضاً؛ DMSU: + قول؛ Y: + قول قول؛ | ١٨٥ R: – صراح؛ S: صريح؛ DY: صرح | ١٨٦ DRUY: إنسانًا | ١٨٧ ADMSY: – قول | ١٨٨ DMFSY: فقولهم | ١٨٩ AS: وأمّا | ١٩٠ A¹FRU؛ A: يحرم D: يمكن؛ MSY: ممكن

every part of something having sensation has sensation itself. It is false because, if it were the case that the elements had sensation, then all bodies formed from them would have sensation, whereas we know that plants and many parts of animals are without sensation. The fourth opinion is that of those who say that the elements can be affected but do not have sensation—a view that is both plausible and true.[35] That is because we clearly find that some bodies have sensation and some do not have sensation. From this it is obvious to us that the elements do not have sensation and that things having sensation are generated from their temperament. That is because it is possible for something having sensation to be generated from that which does not have sensation, for nature is always such that it progresses toward what is more excellent,[36] extracting it from nonexistence into existence. However, for that which does not have sensation to be generated from that which does have sensation is not in the realm of the possible.

<*Their disagreement about the temperament*>[37]

(13) Men have disagreed about the question of the temperament. Some have said that it is the substance of sensation and the substance of all the psychic and natural faculties, which is what most of the physicians say. According to their opinion, sensation must be generated from the temperament. Some of them say that it is the first among the tools

35. A MS contains the gloss: "These have no parts. That which has sensation results from them by the effect that exists in them. If they were to remain as they were without being affected, nothing could result from them that had an essence different from their essence."

36. A gloss in one MS reads: "Someone might object that, if it were possible for there to exist something that is without sensation and something else that is affected and has sensation, then, since it has existence from things having sensation, what is baser would belong to it. We reply that nature is such that it transforms the baser thing into the more perfect, not the more perfect into the baser, and that which has sensation is nobler than that which does not have sensation." This is not clear to me, but perhaps the point is that if what has sensation is affected by what does not, then what has sensation is of a lower rank than that which is insensible.

37. L 3.26–30; K 1:433–34; G^a 36–38: Galen declines to explain the exact relation of the temperament to the soul and the psychic faculties, since, though

ومن أنّه يجب أيضاً أن يكون كلّ جزء من الشيء الحسّاس يحس، وأمّا كذبه فمن طريق أنّه لو كانت العناصر تحسّ لكان جميع ما هو منها من الأجسام تحسّ، ونحن نرى أنّه لا النّبات ولا كثيراً[١٩١] من أجزاء الحيوان له حسّ[١٩٢]. وأمّا أهل الرّأي الرّابع، وهم الذين قالوا إنّ العناصر تقبل الأحداث وليست بحسّاسة، فهو قول ممكن حق[١٩٣]، وذلك[١٩٤] إنّا[١٩٥] إذا كنّا نجد عياناً بعض الأجسام تحسّ وبعضها لا يحسّ[١٩٦]، فقد تبيّن[١٩٧] لنا من هذا أنّ العناصر لا حسّ لها، وأنّ الأشياء الحسّاسة إنّما تتولّد من المزاج، وذاك[١٩٨] أنّ كون ما له حسّ ممّا لا حسّ له أمر يمكن[١٩٩] لأنّ الطّبيعة أبداً إنّما شأنها[٢٠٠] المصير إلى الأمر الأفضل[٢٠١] والنّزوع من العدم ألى الوجود، فأمّا إن يكون ممّا له حسّ أشياء لا حسّ لها، فليس ذلك ممّا يمكن.

‹اختلافهم في المزاج[٢٠٢]›

(١٣) قد اختلف النّاس في أمر[٢٠٣] المزاج، فمنهم من قال إنّه جوهر الحسّ وجوهر جميع القوى النّفسانيّة والطّبيعيّة بمنزلة ما قال جُلّ الأطبّاء، وبحسب رأي هؤلاء يجب أن يكون الحسّ إنّما يتولّد من المزاج. ومنهم من قال إنّه الآلة

١٩١ DMUY: كثير | ١٩٢ S^h: + كالعظام والغضاريف | ١٩٣ S^h: هذه لا جزء لها وإنّما يحدث عنها ما له حسّ بالانفعال الذي هو موجود فيها، فلو بقيت على حالها من غير أن ينفعل لما حدث عنها شيء من الأشياء ذاته مخالف ذاتها. | ١٩٤ AMSU: وذاك | ١٩٥ S: وإنّا؛ AU: أن | ١٩٦ AFS: حسّ لها؛ D: أحسّ له؛ Y: حسّ له | ١٩٧ A: بيّن؛ DY: يتبيّن | ١٩٨ AY: وذلك | ١٩٩ SU: ممكن | ٢٠٠ DFSY: من شأنها أبداً؛ M: إنّما شأنها | ٢٠١ S^h: لقائل أن يقول إن جاز أن يوجد ما هو غير حسّاس ومنفعل حسّاس لأجل أن يوجد من الأشياء الحسّاسة فالأخسّ له، فنقول إنّ الطّبيعة شأنها أن ينقل الشّيء الأخسّ إلى الأكمل لا الأكمل إلى الأخسّ وما له حسّ أفضل ممّا لا حسّ له. | ٢٠٢ R^2 | ٢٠٣ DSY: – أمر

of the faculties, not their substance—as the deepest of the philosophers, Aristotle, and his disciples have said. According to their opinion, it is the suitability for the reception of sensation that is generated from the temperament.

<The genera of qualities>[38]

(14) There are four genera of qualities, in three of which the thing and its opposite exist, while in the fourth it does not exist. The three that include opposites are existence and nonbeing, possibility and impossibility, and action and passion. Thus, in these genera there can be change from the thing to its opposite. The fourth genus—the one in which the thing and its opposite do not both exist—is shape.[39] That is because no shape is the opposite of another shape, just as no size is the opposite of another size. Therefore, in shape there can be no transformation and change, because every change and every transformation is from something opposite to something that is its opposite in that respect. When a square is constructed from two triangles, we do not say that they have changed or been transformed into the square, since the two triangles continue to exist in the square, unaltered and without transformation. In just this way, something small is not transformed when it becomes large, since the small thing is subsumed in the large thing.

---

they are in some sense dependent on the temperament, his purpose here is only to refute those opposed to Hippocrates. "Temperament" (*mizāj*) is a mixture analogous to the result of a chemical reaction in which a new substance is formed, not the sort of mixture resulting from stirring together different kinds of seeds.

38. L 3.14–17; K 1:429–30; G^a 31–32: Galen gives the examples of combining triangles to make a square and building a house (see pp. 142, 149, paragraphs 9, 15) as examples of compounds where the constituent parts remain unchanged. The point may be that atoms of different shapes are not sufficient to explain the changes brought about by temperament.

39. This is a strange list. Shape is the only item normally considered a quality. Action and passion are categories of their own in Aristotle's list of categories. Perhaps this is a list of categories that was somehow garbled in transmission.

الأولى من آلات القوى، وليس هو جوهرها بمنزلة ما قال حُذّاق الفلاسفة[٢٠٤]، وهم أرسطوطاليس[٢٠٥] وأصحابه، وبحسب رأي هؤلاء يجب أن يكون الموافقة لقبول الحسّ إنّما يتولّد من المزاج.

‹أجناس الكيفيّات[٢٠٦]›

(١٤) أجناس الكيفيّات أربعة، ثلثة منها يوجد فيها الشّيء وخلافه والرّابع ليس يوجد ذلك فيه. أمّا الثّلثة الجامعة للمتخالفات[٢٠٧]، فهي الوجود والعدم والإمكان والامتناع والفعل والانفعال، ولذلك قد يكون التّغيّر[٢٠٨] في هذه الأجناس من الشّيء إلى خلافه، وأمّا الجنس الرّابع الذي لا[٢٠٩] يوجد فيه الشّيء وخلافه، فهو الشكل، وذاك[٢١٠] أنّه ليس من شكل هو مخالف لشكلٍ آخر كما لا يكون عظماً مخالفاً[٢١١] لعظمٍ آخر، ولذلك لا يكون في الشّكل استحالة وتغيير، وذلك[٢١٢] أنّ كلّ تغيّرٍ[٢١٣] وكلّ استحالة إنّما يكون من الشّيء المخالف إلى الشيء الذي هو خلافه من[٢١٤] أجل[٢١٥] ذلك، متى صار من مثلّثين مربّع لا نقول إنّ[٢١٦] المثلّثين قد تغيّرا أو استحالا[٢١٧] إلى المربّع لأنّ المثلّثين قائمانِ موجودان في المربّع لم يتغيّرا ولم يستحيلا كما لا يستحيل الصّغير إذا صار منه[٢١٨] عظيماً، وذاك أنّ الصّغير داخلٌ في الكبير.

٢٠٤ R: الأطبّاء والفلاسفة | ٢٠٥ DFRY: أرسطاطاليس؛ S: أرسطو | ٢٠٦ R^2 | ٢٠٧ R^2؛ DUY: للمختلفات؛ F: للمخالفات؛ R: المتخالفات [؟] | ٢٠٨ AMS: التّغيير | ٢٠٩ DSY: ليس | ٢١٠ RU: وذلك | ٢١١ ADMSY: لعظم مخالف | ٢١٢ FSU: وذاك | ٢١٣ AF: تغيير | ٢١٤ DMY: ومن | ٢١٥ FSU: ـأجل | ٢١٦ AF: ـ إنّ | ٢١٧ ADMSY: واستحالا | ٢١٨ DSY: ـ منه

<The compound by contiguity>[40]

(15) We have described the two species of compound, one being temperament and the other contiguity. The compound thing that is a compound by way of temperament may have qualities contained in it that the simple things from which it is compounded do not have—qualities that belong to the three genera of qualities containing both the thing and its opposite. However, the compound thing that is a compound by way of contiguity has no qualities whatever apart from those that the simple things from which it is compounded also have, nor is it in any of the three genera of qualities containing opposites, but only shape and size. This is because, if it gains anything from the compounding, it gains only these two. An example of that is a house, for when it is built, the color of the stone from which it is built remains in it, as do the weight and hardness of the stone, which remain as they were without alteration. Design and construction give something to the house that belongs specifically to it and does not belong to the stone: the shape in which the house is built and the exact size in which it is made.

<Absurdities>[41]

(16) The absurdities forced upon the one who says that the elements are unchangeable are of this sort: First, if a person were compounded from a single element, he would not have sensation, since sensation can occur only when the one who senses receives a change occurring in him from the thing that he senses. If the person did not have sensation, he

40. L 3.8–13; K 1:429; G^a 29–31: Galen gives the example of building a house to show why atomism cannot explain temperament.

41. L 3.27–29; K 1:433; G^a 36–37: An elaboration of the conclusion of Galen's long argument against the Atomists and others who would deny the temperament of the elements.

‹تركيب المجاورة[٢١٩]›

(١٥) ولمّا كانت أنواع التّركيب على ما وصفنا نوعيْن[٢٢٠]، أحدهما المزاج والآخر المجاورة، فقد يجب أن يكون الشّيء المركّب بتركيب المزاج تجتمع له كيفيّات لم تكن للأشياء[٢٢١] البسيطة التي منها رُكّب في الثّلثة الأجناس من أجناس الكيفيّات الجامعة للشّيء وخلافه، وأمّا الشّيء المركّب بتركيب المجاورة، فليس له من الكيفيّات التي لم يكن للأشياء البسيطة التي منها رُكّب شيء أصلاً، ولا في واحد من الثّلثة الأجناس الكيفيّات الجامعة للمتخالفات[٢٢٢] خلا الشّكل والعظم، وهذا إن اكتسب من التركيب شيئاً[٢٢٣]، فإنّما يكتسب هذيْن[٢٢٤] فقط. ومثال ذلك البيت، فإنّه إذا بُني بقي فيه لون الحجارة التي يُبنى بها وثقلها وصلابتها على حاله[٢٢٥] لا يتغيّر، وأكسبه التّأليف والبناء شيئاً هو للبيت[٢٢٦] خاصّةً، وليس هو لتلك الحجارة، وهو[٢٢٧] الشّكل[٢٢٨] الذي يُبنى عليه[٢٢٩] والمقدار الذي يُعمل له[٢٣٠] من العظم[٢٣١].

‹الشّناعات[٢٣٢]›

(١٦) الشّناعات التي تلزم من قال إنّ العناصر لا يقبل الأحداث هي على هذا النّحو، أوّلها أنّه لو كان الإنسان مركّباً من عنصرٍ واحدٍ لكان لا يحسّ، إذ كان الحسّ إنّما يكون بقبول الحاسّ للحدث الواقع به[٢٣٣] من الشّيء الذي يحسّه، ولو كان الإنسان

---

٢١٩ R^2 | ٢٢٠ AFS: نوعان | ٢٢١ AU: الأشياء | ٢٢٢ A: المخالفات؛ DR: للمخالفات؛ M: المتخالفات | ٢٢٣ RU: شيء | ٢٢٤ A: هاتيْن؛ R: هذانِ | ٢٢٥ DM: حالها؛ R: حال | ٢٢٦ ASU: البيت | ٢٢٧ FSU: –وهو؛ R^3: وهوصحّ | ٢٢٨ AR: وهو الشّكل < والشّكل | ٢٢٩ DMR^3SY: + البيت | ٢٣٠ A: –له؛ DY: عليه؛ FSU: به | ٢٣١ F: العظمة؛ S: –من العظم | ٢٣٢ R^2: + الشّناعات | ٢٣٣ F: منه به؛ S: منه

could not desire anything or reject something else. That is because desire must come to be either from a yearning for something helpful and beneficial, which is to be acquired, or from an aversion to something harmful, which is to be avoided. If something does not have sensation, it will not know the beneficial so as to acquire it, nor the harmful so as to avoid it. If a person had no desire, he would not have voluntary motion, since every voluntary motion by which a person moves is a motion due to his desire for something. Second, if a person did not have sensation, he also could not have imagination. That is because, if the images and forms of things were not taken up from sensation to imagination, a person could not imagine anything. If someone did not imagine, he could also not think, since there would be nothing for thought to exercise judgment upon if imagination had no forms or images taken from sensation.[42] Third, if these three—voluntary motion, sensation, and deliberative actions—were not in him, he would also not have a soul. If he had no soul, he would not breathe—that is, be one of those things that have breath—and, if he did not have breath, he would not be animate.

42. Some MSS add: "Thus, he also would not remember, since if he had not acquired forms and images from imagination and thought, he would remember nothing."

لا يحسّ لكان لا يشتهي شيئاً ويعاف[٢٣٤] آخر، وذاك أنّ الشّهوة إنّما تكون إمّا[٢٣٥] بالتوقان إلى الشّيء المرفد[٢٣٦] النافع[٢٣٧]، فيجتلب[٢٣٨]، وإمّا بالعيافة[٢٣٩] للشّيء[٢٤٠] الضّارّ، فيجتنب[٢٤١]، وإن لم يكن للشّيء حسّ لم يعرف النّافع فيجتلبه ولا الضّارّ فيجتنبه، ولوكان الإنسان لا شهوة له لكان أيضاً لا يكون له حركة إراديّة، إذكانت كلّ حركة إراديّة يحرّكها الإنسان، فإنّما يحرّكها بالشهوة منه للشّيء، والثانية أنّه لولم يكن للإنسان حسّ لكان يستبطل تخيُّله أيضاً، وذاك أنّه إن لم يتأدّي إليه مثالات الأشياء وصورها من الحسّ إلى التّخيّل لم يتخيّل للإنسان[٢٤٢] شيء[٢٤٣]، ولولم يكن الإنسان يتخيّل لكان أيضاً لا يتفكّر، وذاك أنّه ليس يكون للفكر[٢٤٤] شيء يحكم[٢٤٥] عليه، متى لم يكن للتّخيّل صور ومثالات أخذها عن[٢٤٦] الحسّ[٢٤٧]. والثّالثة[٢٤٨] أنّه لولم تكن في الإنسان هذه[٢٤٩] الثّلثة[٢٥٠]، أعني الحركة الإراديّة والحسّ والأفعال السّياسيّة، لم يكن له أيضاً نفس، ولولم يكن له نفس لم يكن متنفّساً، أي من ذوات الأنفس، ولولم يكن من ذوات الأنفس[٢٥١] لم يكن حيواناً.

---

٢٣٤ ADSY: + شيئاً | ٢٣٥ FSY: – إمّا؛ U: الماء | ٢٣٦ A: اللّذيذ؛ MR: – المرفد؛ S: الموافق؛ U: المرور | ٢٣٧ R: + المراد | ٢٣٨ DFUY: فيختلف؛ R: فتجتلبه | ٢٣٩ A: بالتّعاف؛ F: بالمعافاة؛ R: بالمعافة | ٢٤٠ AU: الشّيء | ٢٤١ AR: + وينفرمنه؛ FU: فيجتنب > قبيحا ويعرف؛ R: فتتجنّبه | ٢٤٢ DMY: الإنسان | ٢٤٣ DMRY: شيئاً | ٢٤٤ FM: الفكر؛ R: للفكرة؛ U: الفكرة | ٢٤٥ M: يتحكّم؛ R: + به | ٢٤٦ AS: على | ٢٤٧ S: – للتّخيّل. . . الحسّ؛ DMSY + وأنّه [Y: إن] لم يكن [S: – وأنّه لم يكن] الإنسان [D: إن] يتفكّر ولا يتخيّل، فهو أيضاً لا يذكر، وذلك أنّه إن لم يتحصّل [S: يتخيّل] عند التّخيّل والتّفكّر [DS: والفكر] الصّور [DY: الصّورة؛ S: الصّورة] والمثل لم يكن يتفكّر شيئاً. | ٢٤٨ FU: والثّانية | ٢٤٩ DY: + انخلال؛ S: + الخلال | ٢٥٠ DFMSY: الثّلث | ٢٥١ FMU: النّفس

<A syllogism, premises, and conclusion>[43]

(17) Hippocrates gave two formal syllogisms to show that man is not composed from a single element. First, if man were from a single element, he would not feel pain, but we find that he does feel pain. From this, it follows that man[44] is not compounded from a single element. In his second syllogism, he said that, if man did feel pain, even though he was unitary, he would always be treated medically[45] in the same way. But we find that he is not always to be treated in the same way. Therefore, it must not be the case that man both feels pain and is a unitary thing. The point of this argument is that, if man is compounded from a single element, the syllogism necessarily implies that he cannot feel pain, since there would be nothing that would make him feel pain. Were we to overlook the difficulty faced by those who hold this view and grant them that he can feel pain, then it would be obvious that he would feel pain only from himself. Then, since his self was a unitary thing, it would follow that his pain would be a single kind of pain. Because his pain would be the same, it would follow that he would always be treated in the same manner, though we can plainly see that he must be treated in various ways. The basis of this syllogism is sensation and is a matter of consensus. Its premises follow from that basis by means of demonstration, and its conclusion follows from it by demonstration. That first principle, based on a conviction founded on sensation, is that there are different means of treatment. That is because some treatments employ things

---

43. L 3.43–47; K 1:436–38; G^a 41–43: Hippocrates, *On the Nature of Man* 2, trans. De Lacy, cited by Galen at L 3.32; K 1:434; G^a 38–39, had said, "I say that if man were one thing, he would never feel pain, for there would be nothing that would cause him pain if he were one." Galen expounded this argument at length in L 2.23–3.42; K 1:421–36; G^a 19–41. Galen gives brief examples and mentions the structure of the hypothetical syllogisms underlying the extended argument. It is an example of Galen's doctrine that the good physician must be trained in logic. Various aspects of this argument have been discussed in the previous paragraphs.

44. One MS adds: "if he does feel pain."

45. This refers mainly to treatment by drugs, which work by changing the temperament.

‹قياس ومقدّمات ونتيجة[٢٥٢]›

(١٧) قد بيّن إبقراط[٢٥٣] أنّ الإنسان ليس هو من عنصر واحد بقياسيْن من القياسات الوضعيّة[٢٥٤]، الأوّل منهما[٢٥٥] أنّه قال لوكان الإنسان من عنصر واحد لكان لا يناله الوجع، ولكنّا قد نجده يناله الوجع، فيجب من ذلك أن لا يكون الإنسان[٢٥٦] مركّبًا من عنصر واحد، والقياس الثّاني أنّه قال لوكان الإنسان[٢٥٧] يناله الوجع، وهو شيء واحد، لكانت مداواته أيضاً نحواً واحداً[٢٥٨]، ونحن نجد مداواته ليست[٢٥٩] نحواً واحداً[٢٦٠]، فيجب من ذلك أن يكون الإنسان ليس يناله الوجع، وهو شيء واحد، ومعنى هذا القول أنّه إن كان الإنسان مركّبًا من عنصر واحد، فالقياس يوجب ضرورةً أنّه لا ينبغي أن يناله الوجع إذ كان لا يوجد شيء يوجعه[٢٦١]، وإن سامحنا أهل هذه المقالة في هذا وأعطيناهم أنّه قد يناله الوجع، فالأمر في ذلك بيّن أنّه إنّما يناله الوجع من ذاته، وإن[٢٦٢] كانت ذاته شيئاً واحداً، فقد يجب أن يكون وجعه وجعاً واحداً، ولأنّ[٢٦٣] وجعه واحداً، فقد يجب أن تكون مداواته نحواً واحداً، ونحن نجد عياناً أنّ مداواته تكون بأنحاء شتّى، ولهذا القياس ابتداء من الحسّ داخل في باب الإجماع[٢٦٤]، ومقدّمات تابعة لذلك الابتداء على طريق البرهان ونتيجة[٢٦٥] تحصل من ذلك برهانيّةً. أمّا الابتداء الداخل في باب الإقرار الموجود حسّاً، فهو أنّ أنحاء المداواة مختلفة، وذاك أنّ منها ما تكون بالأشياء التي تسخن، ومنها ما

---

٢٥٢ R^2 | ٢٥٣ DMUY: بقراط | ٢٥٤ FM: الوضيعة | ٢٥٥ RU: منها | ٢٥٦ S: + يناله الوجع وهو | ٢٥٧ DMSY: -الإنسان | ٢٥٨ AFSY: نحو واحد | ٢٥٩ DMSY: + من | ٢٦٠ AFMSY: نحو واحد | ٢٦١ FSUY: موجعه | ٢٦٢ DM: وإذ؛ SY: وإذا | ٢٦٣ R: وإن كان؛ S :وإذا كان | ٢٦٤ ADY: الاجتماع | ٢٦٥ R: ونتيجته

that warm, some things that cool, some things that dry, some things that moisten, some things that restrain and prevent, some things that release and loosen, some things that contract and block, and some things that expand and rarefy. There are two demonstrative premises that follow from this starting point. First, if the means of treatment differ, then the ways in which pains occur are also different. Second, if the ways in which pains occur are different, then the things from which pains occur are also different. The conclusion resulting from these two premises is that a human being must necessarily not be composed from just one thing.

تكون بالأشياء التي تبرّد، ومنها ما تكون بالأشياء التي تجفّف[٢٦٦]، ومنها ما تكون بالأشياء التي ترطب[٢٦٧]، ومنها[٢٦٨] بالأشياء التي تحبس وتمنع، ومنها[٢٦٩] بالأشياء التي تطلق وتحلّ، ومنها[٢٧٠] بالأشياء التي تقبض وتسدّ[٢٧١] ومنها بالأشياء التي توسع وتخلخل. وأمّا المقدّمات البرهانيّة التّابعة لهذا الأصل، فمقدّمتانِ، إحداهما[٢٧٢] أنّه إن كانت أنحاء المداواة مختلفةً، فقد يجب أن تكون أنحاء الأوجاع أيضاً مختلفة، والثّانية أنّه إن كانت أنحاء الأوجاع مختلفة، فالأشياء التي عنها يحدث الأوجاع مختلفة أيضاً. وأمّا النّتيجة الحاصلة عن هاتين المقدّمتين، فهي أنّه يجب[٢٧٣] من هذا أنّ الإنسان ليس هو من شيء واحد.

٢٦٦ ADSY: ترطب | ٢٦٧ ADSY: تجفّف | ٢٦٨ ADMSY: + ما تكون | ٢٦٩ ADSY: + ما تكون | ٢٧٠ A¹DY: + ما تكون | ٢٧١ R: تمسك | ٢٧٢ M: أحدهنّ؛ S: أحدهما | ٢٧٣ A¹R: + أن يكون؛ S: أن يجب

[Chapter 4]

[That the element is not numerically one]

(18) Some of those who, like the followers of Democritus, held that there was only one element said that it was one in species; but their view has been refuted by a clear proof.[46] Others of them said that it was one in number, but those who held this view differed among themselves. Some of them said that this element that was one in number did not alter, change, or move, as was the view of the followers of Melissus.[47] Others said that it moved, altered, and changed. Those who held the first view rejected sensation and cast doubt on that which is apprehended directly. We are content to leave them to their complacency, since sensation itself reproves them and refutes their theory. However, we must examine the views of those who held the second theory and then refute it.[48] Each member of this camp held his own theory that differed from the theory of every other member. One of them, Heraclitus, believed that the one element was fire. His demonstration of this was that, if fire condensed a little, it became air. If it condensed some more, it became water; and if it condensed even more than that, it became earth. Two others, Diogenes and Anaximenes, claimed that the element was air, their demonstration of this being that if air condensed a little, it became water, and if it condensed even more than that, it became

46. That is, atomism, which has been refuted by the arguments from the reality of pain and the diversity of drug therapy discussed on pp. 149–52, paragraphs 16–17.

47. See pp. 137–38, paragraph 3, n. 15, above.

48. L 4.1–2; K 1:442–43; G[a] 52–53, commenting on Hippocrates, *On the Nature of Man* 1; cf. pp. 137–38, paragraph 3, above: Refuting the views of the Ionian physicists, particularly their account of transmutation.

[الفصل الرّابع]

[ذكر في أنّ العنصر ليس بواحد العدد]

(١٨) الذين قالوا إنّ العنصر واحد منهم من قال إنّه واحد في النّوع بمنزلة أصحاب ديمقريطس[٢٧٤]، وقد انفسخ قول هؤلاء وظهرت عليهم الحجّة، ومنهم من قال إنّه واحد في العدد وأصحاب هذه المقالة يختلفون، فمنهم من يقول إنّ هذا الواحد في العدد لا يتغيّر ولا يقبل الأحداث ولا يتحرّك بمنزلة أصحاب مالسيس[٢٧٥]، ومنهم من يقول إنّه يتحرّك ويتغيّر ويقبل الأحداث، وأهل[٢٧٦] المقالة الأولى قوم يدفعون الحسّ ويجحدون ما يدركه العيان، فنحن من هذا الطّريق يدعهم يفرحون بما هم فيه[٢٧٧] إذ كان الحسّ هو الموبّخ[٢٧٨] لهم والفاسخ لقولهم، وأمّا أهل المقالة الثّانية، فنحن ننظر في قولهم ونفسخه، وهم جماعة كلّ واحد منهم ينتحل في هذا الواحد[٢٧٩] غير ما ينتحله الآخر، منهم إيراقليطس[٢٨٠]، وهو ينتحل[٢٨١] أنّ النّار هي العنصر، وبرهانه على ذلك أنّها إذا تكاثفت قليلاً صارت هواءً، وإذا زاد[٢٨٢] تكاثفها صارت ماءً، وإذا تكاثفت أكثر من هذا[٢٨٣] صارت أرضاً، ومنهم ديوجانس وأنكسيمانس[٢٨٤]، وهما يزعمانِ أنّ العنصر هو الهواء، وبرهانهما على ذلك أنّ الهواء إذا تكاثف قليلاً صار ماءً، وإذا زاد[٢٨٥]

---

٢٧٤ A: ديمقراطيس؛ F: ديمقرايطيس؛ R: ديمقراطس؛ U: ديمقريطيس | ٢٧٥ MU: ثالسس؛ R: مالسس | ٢٧٦ DSY: + هذه | ٢٧٧ R: عليه وفيه | ٢٧٨ FU: المربح؛ M: المزنح؛ Y: المبخّ | ٢٧٩ A¹؛ AR: الوجه | ٢٨٠ FR؛ A: أبوفليطس؛ DY: إيراقليدس؛ M: إيرقليطس؛ S: إيروقليدس؛ U: إبراولطس | ٢٨١ DR: وهو ينتحل > وينتحل | ٢٨٢ F: إذ إن زادت؛ MSU: ازدادت | ٢٨٣ ADFMY: + أيضاً | ٢٨٤ S؛ A: أنكسيماس؛ D: أنالسيماس؛ FM: أناكسيمانس؛ R: أراكسمانس؛ U: أبالسممانس؛ Y: أناكسيناس | ٢٨٥ FMS: ازداد

earth, while if it rarefied, it became fire. Two more were Thales and Hippon, who thought that the element was water. The demonstration they gave for this was that if water condensed, it became earth, while if it rarefied, it became air, and if it rarefied further, it became fire. Another one of them was Xenophanes of Colophon, who claimed that the element was earth, his demonstration for this being that if it rarefied a little, it became water. If it rarefied more, it became air, and if it rarefied still more, it became fire.

(19) Galen offered five proofs, refuting all of them collectively.[49] First, these people had intended to justify their beliefs concerning the elements, but they unwittingly did something else. What they actually did was to explain the transmutations of the elements and the transformations of one into another. Second, on the basis of this transmutation and transformation that they affirmed of the elements, they drew an improper conclusion, one that was not the conclusion of a valid syllogism. They described the transmutation and transformation of the elements into each other and explained that this transmutation and transformation was in a single thing. From this statement, they ought to have drawn the conclusion that the thing that was the substratum of the four elements was a single thing, having neither form nor species peculiar to it—which is to say, prime matter. However, they overlooked this and instead concluded that one of the four was the element. Third, they posited that the element and the principle that was the root of all things must surely have been one, and then they claimed that it changed

49. L 4.1–21; K 1:443–48; G[a] 53–61: The epitome clarifies Galen's refutations of the Ionian physicists.

تكاثفه صار أرضاً، وإذا تخلخل صار ناراً، ومنهم[٢٨٦] ثالس وإيبون[٢٨٧]، وهما يريانِ[٢٨٨] أنّ العنصر هو الماء، وبرهانهما على ذلك أنّ الماء إذا تكاثف صار أرضاً، وإذا تخلخل صار هواءً، وإذا ازداد[٢٨٩] تخلخله صار ناراً، ومنهم كسانوفانس[٢٩٠] القلوفويّ[٢٩١]، وهو يزعم أنّ العنصر هو الأرض، وبرهانه على ذلك أنّها إذا تخلخلت قليلاً صارت ماءً، وإذا زاد تخلخلها صارت هواءً، وإذا تخلخلت أكثر من ذلك صارت ناراً.

(١٩) وجالينوس يحتجّ[٢٩٢] على هؤلاء كلّهم عامةً بخمسة[٢٩٣] حجج، أوّلها أنّهم قوم أرادوا أن يثبتوا[٢٩٤] اعتقادهم في أمر العناصر، فتركوا ذلك وهم لا يشعرون وأخبروا باستحالات العناصر وانقلاب بعضها إلى بعض، والثّانية أنّهم أنتجوا[٢٩٥] من هذه الاستحالة والانقلاب اللذين أوجبوهما للعناصر شيئاً لا يشاكل ولا يجري على طريق النّتائج القياسيّة، وذلك[٢٩٦] أنّهم لمّا وصفوا استحالة العناصر وانقلابها بعض[٢٩٧] إلى بعض، وبيّنوا أنّه ينبغي أن يكون هذا الانقلاب وهذه الاستحالة في شيء واحد، وكان يتّبع هذا القول أن ينتجوا منه أنّ[٢٩٨] الشّيء الموضوع للعناصر الأربعة هو شيء واحد لا صورة له ولا نوع يخصّه، أعني الهيولى، فتركوا ذلك وأنتجوا أنّ واحداً من الأربعة هو العنصر، والثّالثة أنّهم وضعوا أنّ العنصر والرّأس الذي هو أصل الأشياء إنّما هو واحد، ثمّ زعموا أنّه يتغيّر ويستحيل، وإن[٢٩٩] كان يستحيل فقد بطل

---

٢٨٦ DSY: + من يقول وهو | ٢٨٧ A: مالسيس وأيثون؛ DR³Y: باسلس وأينون؛ F: تالس وأينون؛ M: ثالس وأيان؛ R: مالس واينون؛ S: ثاسلس وأيون؛ U: تالس وابيرور | ٢٨٨ MS: يزعمانِ | ٢٨٩ DRUY: زاد | ٢٩٠ AMS: كسانوفس؛ D: كسانولس؛ F: كسانومس؛ R: كينوفس؛ U: كساموس؛ Y: كسانونس | ٢٩١ A: العلقري؛ D: الغلوقوي FM: العلوقري؛ R: القلوذوي؛ SUY: العلوقيّ | ٢٩٢ R: يردّ | ٢٩٣ R: ويحتجّ بخمس | ٢٩٤ S: يبيّنوا | ٢٩٥ A: نتجوا؛ M: ينتجوا | ٢٩٦ DFMUY: وذاك | ٢٩٧ DFMSY: بعضها | ٢٩٨ AF: إلى | ٢٩٩ DMSY: فإن؛ وإن؛ F: وإذ

and was transmuted. However, if it were transmuted, it would be nullified and removed. Surely, the principle and root ought not to be removed and nullified but, rather, ought to continue and endure. If they replied that it did endure, how could they then explain how fire remained fire yet became water? Fourth, they all held opposing theories, yet each one supported his theory with one and the same demonstration: that, when this one element condensed and rarefied, the other elements were generated from it.[50] Fifth, if they were to claim that this one element condensed and rarefied, then it was clearly the case that it was not one; for it is unquestionably necessary that its alteration from one state to another must be from something else that, at one time, alters it toward rarefaction and, at another time, toward condensation. Because of this, they must be two.

50. Hippocrates, *On the Nature of Man* 1, pointed out that they had used the same arguments to reach differing conclusions.

وارتفع، والرّأس والأصل ليس ينبغي أن يرتفع ويبطل، بل ينبغي أن يلبث ويبقى، فإن قالوا إنّه يبقى، فكيف يجوز لهم أنّ يقولوا[٣٠٠] النّار تبقى نارًا وتصير ماءً؟ والرّابعة أنّ جميعهم يقولون[٣٠١] أقوالًا أضدادًا، وكلّهم يأتون ببرهانٍ واحدٍ بعينه، وهو أنّ هذا العنصر الواحد إذا تكاثف وتخلخل تولّدت منه العناصر الأُخر، والخامسة أنّهم إن كانوا يزعمون أنّ هذا العنصر الواحد يتخلخل ويتكاثف، فالأمر فيه بيّنٌ أنّه ليس بواحد إذ كان يجب لا محالة أن يكون تغيّره من حالٍ إلى حال إنّما يكون من شيء آخر يغيّره مرّةً إلى التّخلخل ومرّةً إلى التّكاثف، فيجب من ذلك أن يكونا اثنين.

٣٠٠ D: + إنّما؛ RU: + إنّ؛ F: يقولون | ٣٠١ FY: يقولوا

[Chapter 5]

[That human bodies do not come to be from a single humor]

(20) Some of those who held that there was one element were a group of practitioners of the science of physics, while others were a group of physicians. Those of them who were practitioners of the science of physics concluded that all bodies came to be from a single element, but they disagreed about what it was. Thales said that it was water; Heraclitus, that it was fire; Anaximenes, that it was air; and Xenophanes, that it was earth. Those of them who were physicians held that man comes to be from a single element, but they also disagreed.[51] Some of them say that man was from blood; some, that he was from phlegm; some, that he was from yellow bile; and some, that he was from black bile. However, the fact that man is not from a single element can be known from three things: first, generation; second, diversity of species; and, third, diversity of faculties. Generation is incompatible with a unitary thing because two things are unquestionably needed in the generation of simple bodies such as fire and water, one being active and the other passive. The generation of compound bodies from these can only

51. L 5.1–14, 14.1; K 1:448–51, 506–7; G[a] 61–65, 136–37; Hippocrates, *On the Nature of Man* 2–3. Hippocrates referred to physicians who held that the body is composed of, or generated from, only one of the four humors; but blood is the only alternative specifically considered either by him or by Galen, perhaps because none of the other humors had credible defenders. Galen pointed out that Aristotle and Hippocrates used identical arguments to refute those who held respectively that there was only one material element and that the human body was generated from only one humor.

[الفصل الخامس]

[ذكر في أنّ بدن الإنسان ليس من عنصر واحد]

(٢٠) والذين[٣٠٢] قالوا إنّ العنصر واحد منهم قوم من أصحاب علم الطّبيعيّات[٣٠٣] ومنهم قوم من الأطبّاء، ومن كان منهم من أصحاب علم الطّبيعيّات[٣٠٤] قضى بأنّ الأجسام كلّها من عنصر واحد واختلفوا فيه، فقال ثالس[٣٠٥] إنّه الماء، وقال إيراقليطس[٣٠٦] إنّه النّار، وقال أنكسيمانس[٣٠٧] إنّه الهواء، وقال كسونوفانس[٣٠٨] إنّه الأرض، ومن كان من الأطبّاء قضى بأنّ الإنسان من عنصر واحد، واختلفوا[٣٠٩] أيضاً، فقال بعضهم إنّ الإنسان من الدّم، وبعض[٣١٠] قال[٣١١] إنّه من البلغم، وبعض قال إنّه من المرّة الصّفراء، وبعض قال إنّه من المرّة السوداء، وقد يعلم أنّ الإنسان ليس هو من عنصر واحد[٣١٢] من ثلثة أشياء، أحدها الكون، الثّاني اختلاف الأنواع، الثّالث اختلاف القوى. أمّا الكون، فإنّه أبداً لا يلتأم من واحد، وذلك لأنّ كون الأجسام البسيطة بمنزلة النّار والماء يحتاج فيه لا محالة إلى اثنين، أحدهما فاعل والآخر منفعل، وكون الأجسام المركّبة من هذه إنّما يكون بالمزاج، والأمر في المزاج

٣٠٢ DFMSY: الذين | ٣٠٣ M: الطّبائع | ٣٠٤ M: الطّبائع | ٣٠٥ D: باسلس؛ U: ثالسس؛ Y: ساسلس | ٣٠٦ AF: إيراقليطس؛ DY: إيراقليدس؛ R: إبراقلس؛ U: إيراوليطس | ٣٠٧ D: باكسماس؛ M: أنكسامانوس؛ R: أناكسمانس؛ S: باكسمانس؛ U: أنكيمانس؛ Y: باسيماس | ٣٠٨ DRY؛ A: كسنوفاس؛ A¹: كسانوفانس؛ F: اكسافوفانس؛ M: كسوفانس؛ S: كسونوفانس؛ U: كسابونانس | ٣٠٩ DFMSY: وقد اختلفوا | ٣١٠ AF: وبعضهم؛ | ٣١١ A: يقول؛ FU: – قال؛ M: قالوا | ٣١٢ A: + لكن؛ R: + ولكن؛ U: + بل هو

occur by temperament; and it is quite obvious that temperament cannot result from one entity but, rather, must be from more than one. In the case of diversity of forms and species, they could not differ were they not formed from differing things. The diversity of faculties also indicates a diversity of sources.

(21) A person is most red in childhood and in springtime.[52] He is most yellow in youth and in summertime. In middle age and in autumn, he is most black, and in old age and wintertime, most white.

(22) There are two ways that we might know that man is composed of the four elements. First, there are things in him that so nearly resemble the four elements that they might almost be the elements themselves.[53] Thus, when we see something like bone in the body—something that is heavy, coarse, cold, and dry—we say that there is earth in man. When we see something formless, flowing, moist, and cold, like phlegm, we unhesitatingly say that there is water in man. When we place our hands on the body, we feel heat coming from it that is stronger than the heat of the air; and from this we judge that there is fire in the body. We also see winds in the body and judge that there is air in it.

(23) Second, the body is nourished only by the four elements, and the thing from which something is nourished and from which it grows also gives it its structure.[54] That is because the thing that increases in some measure, when it is receptive to increase and growth, is something similar to that on the basis of which the increase occurs.

52. L 11.14–16; K 1:494–95; G[a] 122–23: Galen mentions that blood—and, with it, the whole body—is, at various times, black, yellow, red, or white, giving these facts as evidence that blood is also different in different people. This is further evidence that the body is not composed of a single substance.

53. L 5.15–25, 11.3–5; K 1:452–54, 494–95; G[a] 66–68, 120: Galen is actually arguing almost the opposite: that, since the different tissues are never completely cold and hard, hot and soft, or the like, none can come to be from a single humor.

54. L 5.25–31; K 1:454–56; G[a] 68–70: An elaboration of Galen's reference to the growth of seeds into plants and trees as evidence that the elements constitute the body. Another allusion to the food chain appears at L 14.1–3; K 1:506–7; G[a] 136–38, quoting Hippocrates, *On the Nature of Man* 6, who is mainly interested in its relevance to purgation. Galen points out its relevance for understanding nutrition, and the epitome develops and systematizes the argument.

بيّنٌ أنّه ليس هو من واحد، بل أكثر من واحد[٣١٣]، وأمّا اختلاف الصّور والأنواع، فإنّه ما كانت[٣١٤] الأنواع والصّور[٣١٥] لتختلف[٣١٦] لولا إنّ بنائها[٣١٧] من أشياء مختلفة، وأمّا اختلاف القوى، فهو يدلّ على اختلاف الأصول.

(٢١) الإنسان يكون في سنّ الصّبى وفي[٣١٨] وقت الرّبيع أشدّ حمرةً، وفي سنّ الشّباب وفي وقت[٣١٩] الصّيف أشدّ صفرةً، وفي سنّ الكهول ووقت الخريف أشدّ سواداً، وفي سّن الشّيوخ ووقت الشّتاء أشدّ بياضاً.

(٢٢) وقد تعلم أنّ الإنسان مؤلّف من الأربعة العناصر من وجهين، أحدهما أنّ فيه أشياء نظائر الأربعة العناصر تكاد أن تكون هي العناصر بأعيانها، وذلك أنّا نرى في البدن شيئاً ثقيلاً كثيفاً بارداً يابساً بمنزلة العظم، فنقول إنّ في الإنسان أرضاً، ونرى في البدن شيئاً منحلاً[٣٢٠] سيّالاً رطباً بارداً[٣٢١] بمنزلة البلغم، فنقول من هذا قول بتّات[٣٢٢] إنّ[٣٢٣] في الإنسان ماءً، وإذا وضعنا إيدينا على البدن[٣٢٤] أحسسنا منه بحرارة تقهر حرارة الهواء، فنقضي عليه من هذا أنّ فيه ناراً[٣٢٥]، ونرى في البدن أيضاً رياحاً، فنقضي بأنّ فيه هواء.

(٢٣) والوجه الآخر من أنّ البدن إنّما يغتذي من الأربعة العناصر والشّيء الذي منه يغتذي الشّيء[٣٢٦] فبنيته أيضاً منه، وذلك[٣٢٧] أنّه إن كان الشّيء الذي يزيد[٣٢٨] في الشّيء[٣٢٩] عند قبوله للتّزيّد والنّماء هو شيء شبيه بالشّيء الذي صار[٣٣٠] هذا

---

٣١٣ DSY: – بل. . . واحد | ٣١٤ A: إن | ٣١٥ F: – والصّور؛ R: الصّور والأنواع | ٣١٦ ADSY: تختلف؛ F: مختلف | ٣١٧ A: بناتها؛ S: نفاها؛ S¹: وجودها | ٣١٨ DFU: في | ٣١٩ DFSY: ووقت | ٣٢٠ R: متخلخلاً | ٣٢١ DSY: – بارداً | ٣٢٢ A¹FR؛ ADMSUY: – قول بتّات | ٣٢٣ DSY: بأن | ٣٢٤ DSY: أبداننا | ٣٢٥ FS: نار | ٣٢٦ DMR¹SY: + ويبني منه | ٣٢٧ FSY: وذاك | ٣٢٨ DR¹Y: يوجد | ٣٢٩ A: المنى؛ A¹: خ الشّيء | ٣٣٠ A: + منه؛ DY: + فيه

Therefore, it is clear that the structure and subsistence of this thing that is receptive to increase can come only from something having the subsistence and structure of what increases in this thing. Take, for example, a tree whose diameter is one cubit and which then increases and grows, by means of earth and water, until it reaches a diameter of two cubits. I say that the structure and subsistence of the first cubit is also from earth and water because, if the second cubit is no different in any way from the first cubit and if the second cubit is from earth and water only, then the first must also be from earth and water. We may also learn that bodies are nourished by the four elements in the following way. Plants subsist in ways corresponding to the four elements, since they cannot subsist by earth without water, nor by water without earth. It therefore follows that plants come from earth and water. However, combining earth with water results only in clay, so it therefore follows that there must be something in plants other than earth and water. The only simple bodies other than earth and water are fire and air. Therefore, the subsistence of plants also specifically requires these two elements, in addition to earth and water, since we have seen clearly that plants cannot survive without contact with air and the heat of the sun in such a way as to demonstrate that they require nourishment from these two also. Therefore, we have thus proven that plants are nourished by the four elements. Animals are nourished by plants, and man is nourished by animals and plants. It thus follows that man is also composed of the four elements.

زيادةً عليه، فالأمر واضح أن بنية هذا الشّيء القابل للزّيادة وقوامه إنّما هو من الشّيء الذي منه قوام ذلك الشّيء الزّايد في هذا وبنيته. مثال[٣٣١] ذلك أنّ شجرةً كانت مقدار ذراع واحد، فتزيّدت ونمتْ بالأرض والماء[٣٣٢] حتّى صارت ذراعين. أقول إنّ بنية الذّراع الأوّل[٣٣٣] وقوامها إنّما كان أيضاً من الأرض والماء، وذلك أنّه إن كانت الذّراع الثّانية غير مخالفة للذّراع الأولى[٣٣٤] في شيء والذّراع الثانية إنّما هي من الأرض والماء، فالأولى[٣٣٥] أيضاً إنّما هي من الأرض والماء، وقد نعلم أنّ الأبدان إنّما تغتذي من الأربعة العناصر من هذا الوجه، وهو أنّ النّبات إنّما قوامه بالأربعة العناصر[٣٣٦]، إذ كان لا قوام له بالأرض دون الماء ولا بالماء دون الأرض، فيجب من ذلك أن يكون النّبات من الأرض والماء، ولكنّ مخالطة الأرض للماء إنّما تحدث عنها[٣٣٧] طين فقط، فيجب من ذلك أن يكون في النّبات شيء آخر غير الأرض والماء، ولم يبق بعد الأرض والماء من الأجسام البسيطة شيء خلا النّار والهواء، فيجب أن يكون قوام النّبات إنّما هو بهذينِ بعد الأرض والماء خاصّةً إذ كنّا قد نرى عياناً أنّ النّبات لا ثبات له دون لقاء الهواء وحرارة الشّمس من طريق أنّه يحتاج إلى الاغتذاء من هذينِ أيضاً، فالنّبات على ما بيّنّا يغتذي من الأربعة العناصر، والحيوان يغتذي من النّبات، والإنسان يغتذي من الحيوان ومن النّبات، فتحصل من ذلك أنّ الإنسان أيضاً[٣٣٨] مؤلّف من الأربعة العناصر.

---

٣٣١ R^2: + تمثيل | ٣٣٢ DSY: بالماء والأرض | ٣٣٣ DFUY: الأولى | ٣٣٤ ADMSY: الأول | ٣٣٥ DY: والأولى؛ R: فالأولى> فالذّراع الأولى | ٣٣٦ S: + فنعلم هذا من أنّه يغتذي منه | ٣٣٧ M^1S: عنهما | ٣٣٨ A^1؛ ADY: - أيضاً

[Chapter 6][55]

<*Hot and cold*>

(24) The names "hot" and "cold" carry two meanings, one being the quality of heat and coldness in the sense that we say that this body's state is a hot state or a cold state, and the other being the body in which this quality is. However, that quality by which the body is described is either alone in this body, or else it is combined with its opposite. An example of the quality alone is the heat without coldness that is in fire, or the coldness without heat that is in water. Any bodies that are of this sort are said to be in an extreme of heat or coldness, since anything that is completely pure and not combined with its opposite is in an extreme state. Qualities that are mixed with their contraries are, for example, those in all composite bodies. Each of these is described by heat or coldness in one of two senses: either by that which is preponderant in it, or on balance between it and another. An example of preponderance is the body in which there is more heat than coldness; so it is said, in this sense, to be hot. If there is more cold than hot in it, it is said, in this sense, to be cold. As for "on balance," if a body is compared to another and its heat is found to be more intense than the other's, it is said in this sense to be hot. If its cold is found to be more intense than the other's, it is said to be cold.

---

55. L 6.17–24; K 1:461–64; G^{a} 75–79: The passage occurs in the midst of a lengthy discussion of the distinction between the four elements and the four primary qualities, presented in the form of a debate between a nineteen-year-old Galen and a disciple of Athenaeus of Attaleia (ca. 1st century BCE), the founder of the Pneumatic school of medicine. Galen mentions him at this point because, under Stoic influence, Athenaeus had rejected the elements in favor of a system in which matter is acted on by the four qualities and πνεῦμα (spirit). Galen criticizes this theory for identifying sensible qualities as elements. The distinction made in this first sentence is more necessary in Greek than in Arabic. In Greek, including in the corresponding chapter of Galen's text, the noun "heat," a quality, is rendered as τὸ θερμόν, and the adjective "hot" as θερμός, thus blurring—at least for the young students in Alexandria—the distinction between the quality and the thing having the quality. In Arabic, the distinction is clear between *ḥarārah*, "heat," and *ḥārr*, "hot."

[الفصل السّادس]

‹الحارّ والبارد[٣٣٩]›

(٢٤) اسم الحارّ واسم البارد يجريان على معنيين، أحدهما كيفيّة الحرارة والبرودة بمنزلة ما نقول إنّ هذا الجسم حاله حال حارّة أو حال باردة، والآخر الجسم الذي فيه تلك الكيفيّة، إلاّ أنّ هذا الجسم إمّا أن تكون فيه تلك الكيفيّة التي يوصف بها وحدها، وإمّا أنّ يكون يخالطها فيه ضدّها، أمّا وحدها فبمنزلة الحرارة[٣٤٠] في النّار التي ليس معها برودة وبمنزلة البرودة في الماء التي ليس معها حرارة، وما كان من الأجسام على هذا، فهو يقال إنّه في غاية الحرارة والبرودة[٣٤١]، وذلك[٣٤٢] أنّ كلّ شيء خالص محض لا يخالطه ضدّه، فهو في الغاية، وأمّا الكيفيّة التي يخالطها ضدّها، فبمنزلة ما في جميع الأجسام المركّبة، وكلّ واحد من هذه توصف بالحرارة والبرودة على أحد وجهين ،إمّا بالأغلب عليه، وإمّا بالمقايسة بينه وبين[٣٤٣] آخر، وأمّا[٣٤٤] من جهة الأغلب، فإذا كانت الحرارة فيه أكثر من البرودة، فيقال من هذا الوجه أنّه حارّ، وإذا كانت البرودة فيه أكثر من الحرارة قيل[٣٤٥] من هذا الوجه إنّه بارد، وأمّا من جهة المقايسة، فإذا قيس بجسم آخر، فوجد أشدّ حرارةً منه، فقيل من هذا الوجه أنّه حارّ، أو وجد أشدّ برودةً منه، فقيل إنّه بارد.

---

٣٣٩ R^2: + [الحارّ و] البارد | ٣٤٠ AM: + التي | ٣٤١ FSY: أو البرودة | ٣٤٢ AF: وذاك | ٣٤٣ DMSY: + جسم؛ F: أجسام | ٣٤٤ DUY: أمّا | ٣٤٥ DFMSY: عليه أغلب ويقال؛ A^1: + عليه

(25) When Hippocrates says that living bodies are compounded from the hot and the cold, we need not understand him to mean by that these qualities; for the qualities are not bodies, and the element must be of the same genus as the thing of which it is an element.[56] From this it follows that, since the living bodies are corporeal bodies, their element is not merely a quality without matter. Instead, the quality is simply the principle and form[57] of the sensible corporeal body. When he says "this body," he should also not be understood to be referring to the one that is said to be hot or cold preponderantly or on balance, because the corporeal bodies to which this attribute applies are infinite in number, while the elements must necessarily be finite. The reason is that, if they were infinite, then generation by means of them could never be completed, since the infinite can neither be cut off nor delineated. Instead, we must understand him to mean the corporeal bodies that are in the extreme of heat, coldness, moisture, and dryness. These are four corporeal bodies: fire, air, water, and earth.

56. L 8.1–10; K 1:476–79; G[a] 96–100; Hippocrates, *On the Nature of Man* 3: Since Hippocrates often refers to qualities rather than elements, Galen claims that he is referring to the dominance of a particular quality and is not confusing sensible qualities with material elements.

57. *Nawᶜ*, which in later philosophical Arabic invariably means "species," must obviously be translated here as "form." The more usual philosophical Arabic term for the material form would be *ṣūrah*.

(٢٥) فإذا قال إبقراط[٣٤٦] إنّ الأبدان مركّبة من الحارّ والبارد، فليس ينبغي أن نفهم عنه أنّه يريد بذلك الكيفيّات لأنّ الكيفيّات ليست أجساماً[٣٤٧]، والعنصر مجانس للشّيء الذي هو له عنصر[٣٤٨]، فيجب من ذلك إذ كانت الأبدان أجساماً[٣٤٩] أن لا[٣٥٠] يكون عنصرها كيفيّةً ليس لها هيولى، بل إنّما الكيفية رأسٌ ونوع للجسم[٣٥١] المحسوس، ولا ينبغي أيضاً أن نفهم عنه أنّه يريد بقوله هذا الجسم الذي يقال إنّه كذلك على طريق الأغلب أو على طريق المقايسة، لأنّ الأجسام التي هي على هذه الصّفة لا نهاية لها، والعناصر ينبغي أن تكون متناهية، وذاك[٣٥٢] أنّها إن كانت غير متناهية لم يتمّ بها كون، لأنّ ما لا نهاية له لا يقطع ولا يحاز[٣٥٣]، لكن ينبغي أن نفهم عنه أنّه يريد بهذه الأجسام التي هي في الغاية من الحرارة والبرودة والرّطوبة واليبوسة، وهي أربعة أجسام، النّار والهواء والماء والأرض.

---

٣٤٦ DMSUY: بقراط | ٣٤٧ AF: أجسام | ٣٤٨ DFMSY: له عنصر > عنصره | ٣٤٩ F: أجسام | ٣٥٠ DFS: ألاّ | ٣٥١ FMS: الجسم | ٣٥٢ AMS: وذلك | ٣٥٣ DY: إيجاز؛ F: يخاز [كذا]

[Chapter 7][58]

<The states of bodies>

(26) The states of bodies alter either with respect to themselves or with respect to something external. Their alteration with respect to something external is, for example, moving from place to place. Their alteration with respect to themselves is of two kinds: one in which its substance remains in its current state and the other in which its substance does not remain in its current state. The alteration that does not allow the substance of the living bodies to remain falls outside the scope of medicine; and only the alteration that does allow the substance of living bodies to remain in its state falls within the scope of living bodies as they are of concern to medicine.

<Alteration of quality and quantity>[59]

(27) This alteration is of two kinds: one in quantity and the other in quality. Alteration in quantity is of two kinds, one in which something in the body diminishes, and the other in which something in it increases. If something has been diminished in the body, the physician must increase something else there that is similar to what was diminished in order to replace it. Because the body is compounded from the four elements, the food by which its increase occurs might well also need to be compounded from the four elements. For that reason, the body is not nourished by elements, which are simple—neither water, nor air, nor

58. L 9.6–7; K 1:482; G[a] 104: Galen is distinguishing substantial change from other sorts of bodily change. This passage provides a clearer analysis of the concepts.

59. L 7.1–11; K 1:473–76; G[a] 91–95: Galen observes that bodies change in two ways: by alteration of the quality of their substance and by loss of substance. Medical treatment—usually meaning treatment by drugs—reduces excesses of particular qualities and replaces qualities that were lost. Food replaces the substance that was lost. Drugs and food differ only in that food is intended to replace the same substance that was lost, whereas drugs are intended to change a quality. Thus, in some circumstances, something can be both a drug and a food.

[الفصل السّابع]

‹حالات الأبدان[٣٥٤]›

(٢٦) وحالات الأبدان تتغيّر إمّا من قِبَل أنفسها وإمّا من قبل شيء من خارج، وتغيُّرها من قبل شيء من خارج يكون بمنزلة الانتقال من موضع إلى موضع، وأمّا تغيُّرها من قبل أنفسها، فيكون على وجهيْنِ، أحدهما وجوهرها باقٍ على حاله، والآخر وجوهرها ليس بباقٍ على حاله، والتّغيير[٣٥٥] الذي لا يبقى معه جوهر الأبدان[٣٥٦] ليس يدخل فيما يعنى به الطّبّ من أمرها، فأمّا[٣٥٧] التّغيير[٣٥٨]، فهو الذي يبقى معه جوهر الأبدان على حاله، فهو الذي يدخل فيما يعنى به الطّبّ من أمر الأبدان فقط.

‹تغيّر في الكيفيّة والكمّيّة[٣٥٩]›

(٢٧) وهذا التّغيير[٣٦٠] صنفانِ، أحدهما في الكمّيّة والآخر في الكيفيّة. أمّا التّغيّر في الكمّيّة، فعلى ضربيْنِ، أحدهما أن ينقص من البدن شيء، والآخر أن يزيد فيه شيء، وإذا نقص من البدن شيء، فينبغي للطّبيب أن يزيد فيه مكان ما نقص منه من[٣٦١] شيء شبيه به، ولأنّ البدن مركّب من الأربعة العناصر قد[٣٦٢] ينبغي أن يكون الغذاء الذي به تكون الزّيادة فيه مركّباً[٣٦٣] أيضاً من الأربعة العناصر، ولذلك صار البدن لا يغتذي من العناصر، وهي بسيطة، لا من الماء ولا من الهواء ولا من الأرض

٣٥٤ R^2 | ٣٥٥ DRUY: والتّغيّر | ٣٥٦ $ADMR^1SY$: + على حاله | ٣٥٧ MR: وأمّا | ٣٥٨ D^1FUY: التّغيّر | ٣٥٩ R^2 | ٣٦٠ DFY: التّغيّر؛ S: + على الإطلاق | ٣٦١ ADY: – من | ٣٦٢ FSU: فقد؛ MY: – قد | ٣٦٣ AFS: مركّبةً

earth, nor fire. On the other hand, if something increases in the body, the physician must[60] find a way to decrease it by evacuation. Alteration of quality is, for example, what happens to the body when it becomes hot or cold. When that happens to it, the physician must consider what to do. If there is no great alteration, the patient is to be treated with a drug that will convert what has been altered and bring it toward its opposite—though not to the extreme of the quality that it has, but only inclining the body moderately away from its usual nature. Thus, if the body has inclined away from its usual state toward cold, it is to be treated with pepper or pellitory. If the body has inclined toward heat, it is to be treated with lettuce or barley water. If, however, the body is greatly altered, it must be treated with a remedy that changes and transforms it to the extreme opposite of what it was. Thus, if it has become cold, we heat it with fire; and if it has become hot, we cool it with water.

<*Instruction*>

(28) There are two courses of instruction based on the *stoicheia*—which is to say, the elements.[61] The first method starts with the completion and end of the thing in imagination and goes back until it reaches its source and beginning. The second method begins from the source and beginning of the thing and goes until it reaches its completion and end. The first of these two methods of instruction is called conversion

60. Some MSS read: "might need to."

61. Though both of these methods are used in *On the Elements*, they are not discussed theoretically; but see *The Small Art*, intro.; K 1:305–7; G^a 3–7, and its epitome, pp. 50–56, paragraphs 1–9, above. The epitome of *The Small Art*, p. 56, paragraph 9, says that *On the Elements* uses the method of synthesis. The discussion here may be occasioned by Galen's criticism of Athenaeus's failure to observe rigorous logical method; see L 9.20; K 1:486; G^a 108–9.

ولا من النّار، فأمّا إذا زاد في البدن شيء[٣٦٤]، ينبغي للطّبيب أن يحتال في تنقيصه[٣٦٥] بالاستفراغ، وأمّا التّغيّر[٣٦٦] في الكيفيّة، فيكون بمنزلة ما يعرض للبدن[٣٦٧] إذا هو سخن أو برد، ومتى[٣٦٨] عرض له ذلك، فينبغي أن ينظر، فإن كان لم يتغيّر تغيّرًا شديدًا، فينبغي أن يداوى بدواء يقلّبه ويحيله[٣٦٩] إلى خلاف ذلك من غير أن يكون في الغاية من الكيفيّة التي هي له، لكن يكون منها في حدّ معادل لمقدار[٣٧٠] ميل البدن عن طبيعته، فإن كان البدن قد مال عن حاله[٣٧١] إلى البرودة كانت مداواته بالفلفل أو بالعاقرقرحا[٣٧٢]، وإن كان قد[٣٧٣] مال البدن[٣٧٤] إلى الحرارة، كانت مداواته بالخسّ وبكشك[٣٧٥] الشّعير، وإن[٣٧٦] كان البدن قد تغيّر تغيّرًا شديدًا، فينبغي أن يداوى بدواء يحيله ويقلبه إلى خلاف ما هو عليه ممّا هو في الغاية، فإن كان قد برد[٣٧٧] أسخناه[٣٧٨] بالنّار، وإن كان قد سخن برّدناه بالماء

‹التّعليم[٣٧٩]›

(٢٨) من[٣٨٠] التّعليم المبني على[٣٨١] أمر الأسطقسات، وهي العناصر له سبيلانِ، أحدهما المسلك الذي يبتدئ[٣٨٢] من منتهى الشيء وتمامه في الوهم ويرجع حتّى ينتهي إلى أصله ومبدأه، والثّاني المسلك الذي يبتدئ من أصل الشيء ومبدئه وينتهي[٣٨٣] إلى تمامه وغايته، والطّريق الأوّل من هذينِ التعليمَين يقال له النّقض[٣٨٤] والتّحليل بمنزلة

---

٣٦٤ FMU: + فقد | ٣٦٥ R: تنقّصه | ٣٦٦ AMS: التّغيير | ٣٦٧ A: + مرض | ٣٦٨ DY: متى؛ R: فمتى | ٣٦٩ FS: يحيله ويقلّبه | ٣٧٠ DMY: بمقدار؛ FSU: المقدار | ٣٧١ DMSY: + عن حاله | ٣٧٢ A: بالعاقرقرحا أو الفلفل؛ DFM: بالفلفل وبالعاقرقرحا؛ Y: بالعاقرحا | ٣٧٣ ADFSY: – قد | ٣٧٤ DFMSUY: – البدن؛ M: + عن حاله | ٣٧٥ DMFSY: أو بكشك | ٣٧٦ ADY: فإن | ٣٧٧ S: + كوقت الشّتاء | ٣٧٨ DY: إسخانه؛ R: أسخنه | ٣٧٩ R^2 | ٣٨٠ DMRUY: – من | ٣٨١ $ADFR^3Y$: عن | ٣٨٢ DSY: + به | ٣٨٣ MSU: حتّى ينتهي | ٣٨٤ DSY: العكس

and analysis—as, for example, when we say that tissues are compounded from humors, humors from nutriments, nutriments from plants, and plants from the four principles, by which I mean the elements. The second method is called synthesis, as when we say that plants are from the four elements, nutriments are from plants, humors from nutriments, tissues from humors, organs from tissues, and the entire body from organs.

ما نقول إنّ الأعضاء المتشابهة الأجزاء مركّبة من الأخلاط، والأخلاط من الأغذية، والأغذية من النبات، والنبّات من الأربعة الأركان، أعني العناصر. والطّريق الثّاني يقال له التّركيب بمنزلة ما نقول إنّ النبّات يكون من الأربعة العناصر والأغذية من النبات والأخلاط من الأغذية والأعضاء المتشابهة الأجزاء من الأخلاط والأعضاء المركّبة من الأعضاء[٣٨٥] المتشابهة الأجزاء، وجملة البدن من الأعضاء المركّبة.

---

٣٨٥ DMFSY: -الأعضاء

[Chapter 8][62]

[Compounds]

(29) All the Ancients agreed in affirming that the compound corporeal bodies result from compounding simple elements. Democritus, however, said that the elements remained in the same state unchanged, and compound bodies resulted from them only by the alteration of the states of position, opposition, and shape. Anaxagoras said that there were parts of each of the organs among the elements. Thus, if the parts of bone were separated from among the elements that they were in and then they were rejoined and combined with each other, what would result from them would be bone; and if the parts of flesh were separated from the elements that they were in and then the parts of flesh were joined and combined with each other, what would result from them would be flesh. If these parts were scattered and then combined with the elements, fire, water, earth, and air resulted. This is because Anaxagoras believed that the elements were a mixture of homoeomerous bodies.[63] Empedocles said that the four elements were neither transmuted nor altered and that differing compound bodies resulted from them due to the difference of their composition, not due to alteration or transmutation. Hippocrates, Thales, Heraclitus, and Diogenes said that the compound bodies could only result from the alteration and transmutation of the elements, though Thales and his followers said

62. L 9.6–19; K 1:482–86; G[a] 104–8: A clarification of Galen's argument that the four primary qualities mediate the transmutation of the elements into each other. This stands in contrast to the theories of the Atomists and the various Ionian physicists, who all held, in one way or another, that transmutation involves combining or separating unchanging parts.

63. Anaxagoras thought that the different kinds of things were constituted from tiny fragments of each kind of substance that already existed in the form of scattered particles. See p. 139, n. 21, above.

[الفصل الثّامن]

[المركّبات]

(٢٩) جميع القدماء قد أقرّوا وأجمعوا أنّ الأجسام المركّبة إنّما تكون من تركيب العناصر البسيطة، إلاّ أنّ ديمقريطس يقول إنّ العناصر تبقى على حالها لا تتغيّر وتحدث عنها الأجسام المركّبة بتغيّر حالاتها في الوضع[٣٨٦] والتّضادّ والشّكل، وأنكساغورس[٣٨٧] يقول إنّ في العناصر أجزاء من كلّ واحد من الأعضاء، فإذا فارقت أجزاء العظم العناصر التي هي فيها واجتمعت والتأم بعضها إلى بعض[٣٨٨] صار منها عظم[٣٨٩]، وإذا فارقت أجزاء اللّحم العناصر التي هي فيها واجتمعت والتأم بعضها ببعض صار[٣٩٠] منها لحم، وإذا تفرّقت هذه الأجزاء واختلطت بالعناصر صارت ناراً وماءً وهواء وأرضاً[٣٩١]، وذلك لأنّ أنكساغورس يعتقد أنّ العناصر إنّما هي خلط من الأجرام[٣٩٢] المتشابهة الأجزاء، وإمباذقلس[٣٩٣] يقول إنّ العناصر الأربعة غير مستحيلة ولا متغيّرة، وإنّما تحدث عنها أجسامٌ مركّبة مختلفة من قِبَل اختلاف تركيبها عن غير تغيّر ولا استحالة، وإبّقراط[٣٩٤] وثالس وإيراقليطس[٣٩٥] وديوجانس يقولون إنّ الأجسام المركّبة إنّما تحدث عن تغيّر العناصر واستحالتها، إلاّ أنّ ثالس[٣٩٦] وأصحابه

٣٨٦ AF: الموضع | ٣٨٧ A: وأمّا أنكساغورس؛ D: وأناكساعوراس؛ U: وآنماكساغورس؛ Y: وأنالساغوراس | ٣٨٨ DSY: ببعض | ٣٨٩ F^2: خ أو لحم؛ U: لحم | ٣٩٠ DSY: كان | ٣٩١ ADFSY: وأرضاً وهواءً | ٣٩٢ S: الأجزاء | ٣٩٣ AF: امياذوقيلس؛ D: وأمّا دقيلس؛ M: إنباذقلس؛ R: ودميار وقلس [؟]؛ U: إمتادوقلس؛ Y: إمادقليس | ٣٩٤ DMSUY: بقراط | ٣٩٥ AMR: وثالس وإيراقليطس؛ DY: وباسلس وإيراقليدس؛ F: وبالينورايراقليطس؛ S: ثايلس وإيراقليطس؛ U: ثلسيس وإيرقليطيس | ٣٩٦ S: ثايلس؛ DY: باسلس

that the alteration of the elements occurred only by connection and separation. Hippocrates did not say that the alteration of the *stoicheia* occurred by their connection and separation, but, rather, by heat, coldness, moisture, and dryness. Therefore, he posited that it is these qualities that gave reality to the elements. For this reason, the elements were named hot, cold, moist, and dry.

يقولون إنّ تغيّر العناصر إنّما يكون باجتماعها[٣٩٧] وتفرّقها، فأمّا إبّقراط[٣٩٨] فليس يقول إنّ تغيّر[٣٩٩] الإسطقسات يكون باجتماعها وتفرّقها لكنّ بالحرارة والبرودة والرّطوبة واليبوسة، ولذلك جعل المحقّق لأنواع العناصر والمُثبِت لها هذه الكيفيّات، ولذلك سُمّي العناصر الحارّ والبارد والرّطب واليابس.

٣٩٧ S: + تخلخلها | ٣٩٨ DMSUY: بقراط | ٣٩٩ A: + العناصر يعني

[Chapter 9][64]

<The qualities>

(30) There are different genera of qualities. Some are apprehended by vision; these are the different colors. Some are apprehended by hearing; these are the different sounds. Some are apprehended by smell; these are the different scents. Some are apprehended by taste; these are the different flavors. Some are apprehended by touch; these are a variety of qualities: heat and coldness, moisture and dryness, hardness and softness, lightness and heaviness, density and thinness,[65] roughness and smoothness, solidity and brittleness, and flabbiness and firmness.[66] Some of these qualities do not alter the entire substance of the body, and some have their effect in the entire body. The qualities that do not alter the entire substance of the body are qualities that are seen, that are heard, that are smelled, and that are tasted. Those that are seen are like colors, for they do not alter the whole body, but only the eyes; nor do they alter the entirety of the two eyes, but only the optic spirit. Black, more than any other color, concentrates the optic spirit, while white disperses it. Qualities that are heard have their effect only in the two ears. Those that are smelled have their effect in the two front ventricles of the brain, which they alter to some degree. Those qualities that are tasted mostly affect only the tongue, while in the rest of the body they have either no effect at all or else only a small effect. The

64. L 9.13–17; K 1:484–85; G^a 106–8: Galen argues that only the four primary qualities—heat, cold, moisture, and dryness—have the capacity to alter substance and other qualities, in contrast to color, sound, scent, taste, or many tactile qualities. The categorization of qualities in these two paragraphs is much more detailed than the one in Galen's text.

65. That is, thinness in the sense of a thin soup.

66. On Galen's lists of tactile qualities in other works and their possible sources, see L, p. 198.

[الفصل التّاسع]

‹الكيفيّات[٤٠٠]›

(٣٠) أجناس الكيفيّات مختلفة، فمنها ما يدركه البصر، وهي الألوان المختلفة، ومنها ما يدركه السّمع، وهي الأصوات المختلفة، ومنها ما يدركه الشّمّ، وهي الرّوائح المختلفة، ومنها ما يدركه المذاق، وهي الطّعوم المختلفة، ومنها[٤٠١] ما يدركه اللمس، وهي كيفيّات شتّى، أعني الحرارة والبرودة والرّطوبة واليبوسة والصّلابة واللّين والخفّة والثّقل والكثافة والسّخافة والخشونة والملاسة[٤٠٢] والمتانة[٤٠٣] والهشاشة والرّخاوة والاكتناز، وبعض هذه الكيفيّات لا تغيّر جملة جوهر البدن[٤٠٤]، وبعضها يفعل فعله في جميع البدن. أمّا الكيفيّات التي لا تتغيّر بها جملة جوهر البدن، فهي الكيفيّات التي تبصر والتي تسمع[٤٠٥] والتي تشمّ والتي تذاق، وأمّا التي تبصر بمنزلة[٤٠٦] الألوان، فإنّها ليس تغيّر البدن كلّه لكنّ العينيْنِ فقط، وليس تغيّر أيضاً جملة كلّ واحدة[٤٠٧] من العينيْنِ، لكنّ تغيّر منها الرّوح الباصر فقط، فالأسود من جميع[٤٠٨] الألوان يجمع الرّوح الباصر والأبيض يفرّقه، وأمّا التي[٤٠٩] تسمع فإنّها تفعل ما تفعله في الأُذنيْنِ فقط، وأمّا التي يشمّ فإنّها تفعل فعلها في البطنيْنِ المقدّميْنِ من بطون الدّماغ وتغيّرهما تغيّرًا له قدر، وأمّا التي تذاق فإنها إنّما[٤١٠] تفعل أكثر ما تفعله في اللّسان فقط، وأمّا في سائر البدن، فإنّها إمّا أن تكون لا تفعل فيه شيئاً، وإمّا أن تفعل فيه فعلاً يسيرًا، والدّليل على ذلك أنّ

---

٤٠٠ R^2 | ٤٠١ A: المذاق. . . منها: مضاف في الهامش؛ R: –ما يدرك. . . ومنها | ٤٠٢ F: والملوسة؛ M: والملاسة والخشونة؛ U: واللّين | ٤٠٣ AD: –المتانة؛ S: واللّزوجة والعّل | ٤٠٤ A: –وإذا تفرّقت هذه الأجزاء. . . جوهر البدن | ٤٠٥ AFR: –والتي تسمع | ٤٠٦ AM: فبمنزلة | ٤٠٧ AS: واحد | ٤٠٨ A^1: –من جميع؛ FRU: –جميع؛ S: في جميع | ٤٠٩ DSY: الذي | ٤١٠ ADSY: –إنّما

proof of that is that the body can sometimes sense the alteration that occurs from it. Of the qualities that exercise their effect in the entire body, some have that effect only on the exterior of the body, without the effect reaching the interior of the body—for example, roughness, smoothness, hardness, and softness. Others affect both the exterior and the interior of the body—for example, heat, coldness, moisture, and dryness. We also say that some of the tactile qualities do no more than move the thing, as is the case with heaviness and lightness, while others have some sort of effect in the thing that exercises some influence on it, such as heat and coldness.

(31) This class may be divided in a different way, for it is said that some qualities are primary qualities,[67] the fundamental qualities from which the other qualities are generated. These are four: heat, coldness, moisture, and dryness. The others are secondary qualities generated from the primary qualities. These are the rest of the qualities that are felt, seen, heard, tasted, and smelled. Among the other tactile qualities are density, which is generated from coldness; thinness, which is generated from heat; softness, which is generated from moisture; hardness, which is generated from dryness; solidity, which is generated from moisture; and brittleness, which is generated from dryness. The visible qualities, which are the colors, result from density and thinness; and these two, in turn, result from heat and coldness. That is because density generates blackness, since blackness is the negation of the effect of vision, and its occurrence is from the concentration of the parts of the visible thing. Thus, when a person sees nothing, he thinks that he is

67. L 9.13–19; K 1:484–85; G[a] 106–8: The next three paragraphs are a restatement and elaboration of Galen's argument that only the four primary qualities constitute substances. Galen gives only a few specific examples of secondary qualities generated from the primary qualities.

البدن قد يحسّ بما يحدث عنها من التّغيير، وأمّا[٤١١] الكيفيّات التي تفعل ما تفعله في جميع البدن، فبعضها تفعل ذلك في ظاهره فقط، ولا يبلغ فعلها إلى باطنه بمنزلة الخشونة والملاسة[٤١٢] والصّلابة واللّين، وبعضها تفعل ذلك في ظاهر البدن وباطنه معاً بمنزلة الحرارة والبرودة والرّطوبة واليبوسة[٤١٣]، ونقول أيضاً إنّ الكيفيّات الملموسة منها ما هو محرّك للشّيء فقط بمنزلة الثّقل والخفّة[٤١٤]، ومنها ما يفعل في الشّيء فعلاً يؤثّر فيه بمنزلة الحرارة والبرودة[٤١٥].

(٣١) وقد يقسم هذا المعنى أيضاً بضرب آخر من القسمة، فيقال إنّ الكيفيّات منها كيفيّات أُوَل، وهي الأمّهات التي عنها تتولّد الكيفيّات الأُخر، وهي أربع كيفيّات، الحرارة والبرودة والرّطوبة واليبوسة، ومنها كيفيّات ثوانٍ متولّدة عن تلك[٤١٦] الأُوَل، وهي سائر الكيفيّات الملموسة والكيفيّات المبصورة[٤١٧] والمسموعة والمذوقة والمشمومة[٤١٨]، وأمّا[٤١٩] سائر[٤٢٠] الملموسة، فمنها الكثافة، وتولّدها عن البرودة، ومنها السّخافة، وتولّدها عن الحرارة، ومنها اللّين، وتولّده عن الرّطوبة، ومنها الصّلابة، وتولّدها عن اليبوسة، ومنها المتانة، وتولّدها عن الرّطوبة، ومنها الهشاشة، وتولّدها عن اليبوسة[٤٢١]، وأمّا الكيفيّات المبصورة[٤٢٢]، وهي الألوان، فإنّها تابعة للكثافة والسّخافة، وهاتانِ تابعتانِ للحرارة[٤٢٣] والبرودة، وذلك أنّ الكثافة تولّد السّواد، إذ كان السّواد إنّما هو سلب فعل البصر وحدوثه يكون عن اجتماع أجزاء الشّيء المبصور، ولذلك متى لم ير[٤٢٤] الإنسان شيئاً ظنّ أنّه ينظر إلى سواد،

---

٤١١ DSY: فأمّا | ٤١٢ D: والملموسة؛ F: الملوسة؛ Y: الملوسة | ٤١٣ A^1DFY: واليبوسة والرّطوبة | ٤١٤ DMSY: الخفّة والثّقل | ٤١٥ A^1 ends here | ٤١٦ S: + الأربع؛ F: الأربعة؛ U: + للأربعة | ٤١٧ FR: – والكيفيّات المبصورة؛ F^1R^1: وسائر الكيفيّات المنظورة | ٤١٨ F^1R: والمشمومة والمذوقة | ٤١٩ DY: أمّا؛ R: فأمّا | ٤٢٠ RU: + الكيفيّات | ٤٢١ R: + المفرطة | ٤٢٢ R: المنظورة | ٤٢٣ R: هاتان. . . للبرودة > تولّدها عن الحرارة | ٤٢٤ AFM: يرى

seeing blackness. Thinness generates whiteness because whiteness is the necessitation of the act of vision and its falling upon the visible thing. This results from the opening and spreading apart of the parts of the visible thing so that vision can meet and touch them. The rest of the colors are an intermediate between these two and are mixed from them, since those two are the extremes. The audible qualities are the kinds and species of sound and result from the hardness, softness, largeness, and smallness of bodies. Hardness and softness follow from moisture and dryness. The gustatory qualities are the species of flavors and result from the various species of temperament.[68] The olfactory qualities are the kinds of odors. The odors result from the kinds of temperament and the subtlety and coarseness of the substance.[69]

(32) The four fundamental qualities can be combined as follows: Heat and coldness do not combine. Moisture and dryness do not combine. Heat and dryness combine as fire. Heat and moisture combine as air. Coldness and dryness combine as earth. Coldness and moisture combine as water. A diagram of that:

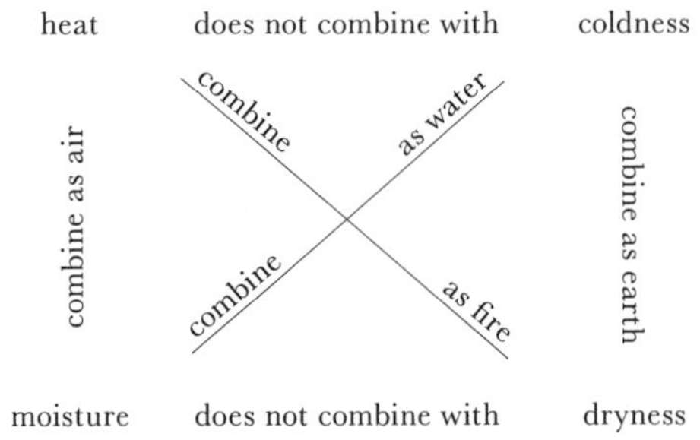

68. Some MSS add: "and the fineness and coarseness of the substance."
69. One MS reads: "of the air."

والسّخافة تولّد البياض لأنّ البياض إنّما هو إيجاب[٤٢٥] فعل البصر ووقوعه على الشّيء المبصور، وهذا تابع لانفتاح أجزاء الشّيء المبصور وانبساطها حتّى يلقاها ويماسّها البصر، فأمّا[٤٢٦] سائر الألوان، فهي وسط فيما بين هذين اللّونين مخلوطة منهما إذ كانا هما طرفانِ، وأمّا الكيفيّات المسموعة، فإنّها أصناف الأصوات وأنواعها، وهذه تابعة لصلابة الأجسام ولينها وعظمها وصغرها، والصّلابة واللّين تابعانِ[٤٢٧] للرّطوبة واليبوسة، وأمّا[٤٢٨] الكيفيّات التي يذاق، فهي[٤٢٩] أنواع الطّعوم، وهذه تابعة لأنواع المزاج[٤٣٠]، وأمّا الكيفيّات المشمومة، فهي أصناف[٤٣١] الرّوائح، والرّوائح تابعة لأصناف المزاج ورقّة الجوهر[٤٣٢] وغلظه.

(٣٢) وتأليف الأربع الكيفيّات[٤٣٣] الأمّهات يكون على هذا المثال، الحرارة والبرودة لا تأتلفانِ، والرّطوبة واليبوسة لا تأتلفانِ، والحرارة واليبوسة تأتلف منهما[٤٣٤] النّار، والحرارة والرّطوبة يأتلف منهما الهواء، والبرودة واليبوسة يأتلف منهما الأرض، والبرودة والرّطوبة يأتلف منهما الماء[٤٣٥]. مثال ذلك[٤٣٦]:

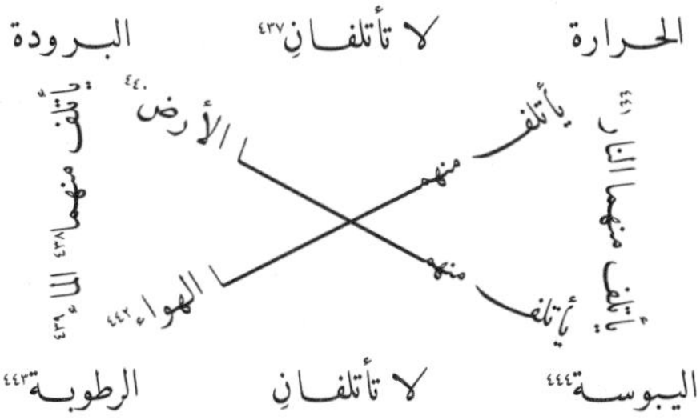

---

٤٢٥ A: أجاب؛ R: انجياز [؟]؛ DY: يجاب | ٤٢٦ DMY: وأمّا | ٤٢٧ DMY: تابعينَ؛ S: تابعتينَ | ٤٢٨ DMSY: فأمّا | ٤٢٩ FS: وهي | ٤٣٠ DMR^3SY + ورقّة الجوهر وغلظه | ٤٣١ A: تابعة لأصناف | ٤٣٢ A: الهواء | ٤٣٣ DMUY: كيفيّات | ٤٣٤ FSU: منها | ٤٣٥ DSY: + وهذا | ٤٣٦ FU: الماء. . . ذلك > النّار مثال ذاك وصورته؛ R: – مثال ذلك | ٤٣٧ AF: تلقيان | ٤٣٨ FM: منها | ٤٣٩ A: الهواء؛ DMSY: الأرض | ٤٤٠ DMSY: الماء | ٤٤١ DMSY: الهواء | ٤٤٢ DMSY: النّار | ٤٤٣ DMSY: اليبوسة | ٤٤٤ DMSY: الرّطوبة

(33) Qualities can be divided in a way more complete and superior to the previous way.[70] Some qualities are fundamental and generative; these are the tactile qualities. Others are generated from the tactile qualities. Some tactile qualities are themselves primary—those qualities from which other qualities are generated but which are not themselves generated from other qualities. There are four of these—which are to say, heat, coldness, moisture, and dryness. Others are secondary and are themselves of two kinds: some generate and are generated, while others are generated but do not generate. Those that are generated but do not generate are qualities such as heaviness and lightness. Of those that generate and are generated, some generate qualities like themselves, and some generate qualities not like themselves. Those that generate qualities like themselves are qualities such as roughness and smoothness; for from smoothness is generated smoothness like it, and roughness, by its nature, generates roughness like itself. Very often, roughness generates smoothness accidentally, in the way that something that chills can do. Qualities that generate qualities unlike themselves are, for example, compression and rarefaction; for blackness, heaviness, and hardness are generated from compression, and whiteness, lightness, and softness are generated from rarefaction. Some of the qualities generated from the tactile qualities are visible, some audible, some olfactory, and some gustatory. One of the visible qualities is whiteness, which is generated from the rarefaction of the parts of the body. Another is blackness, which is generated from the compression of the parts of the body. The rest of the colors that are between these two colors are generated from states that are between these two states. The audible states follow from how hard and soft the bodies are and how large and small their sizes are. The olfactory and gustatory qualities follow from the temperament, in the way that sharpness, pungency, bitterness, and saltiness follow from heat, while constriction, astringency, and sourness follow from coldness. From the balance between the two come sweetness and greasiness.

---

70. L 9.13–17; K 1:484–85; G[a] 106–9: Galen is arguing that only the four primary qualities of heat, coldness, moisture, and dryness are primary in the sense that they can directly affect the body as a whole.

(٣٣) وللكيفيّات قسمة أتمّ[٤٤٥] وأكمل من القسمة المتقدّمة، وهي أنّ الكيفيّات منها أمّهات مولّدة، وهي الكيفيّات الملموسة، ومنها متولّدة من الملموسات، فأمّا الكيفيّات الملموسة، فمنها كيفيّات أُوَل، وهي الكيفيّات التي يتولّد عنها غيرها ولا تتولّد هي عن غيرها[٤٤٦]، وهي أربع، أعني الحرارة والبرودة والرّطوبة واليبوسة، ومنها كيفيّات ثوانٍ، والثواني صنفانِ، فمنها ما تولّد وتتولّد، ومنها ما تتولّد ولا تولّد. أمّا التي تتولّد ولا تولّد، فبمنزلة الثّقل والخفّة، وأمّا التي تولّد وتتولّد، فمنها ما تولّد أشباهها، ومنها ما تولّد غير أشباهها[٤٤٧]. أمّا المولّدة لأشباهها، فبمنزلة الخشونة والملاسّة، فإنّ الملاسّة[٤٤٨] تولّد[٤٤٩] ملاسة[٤٥٠] مثلها، والخشونة تولّد بطبعها خشونة مثلها، وكثيرًا مّا تولّد الخشونة بطريق العرض ملوسة[٤٥١] بمنزلة ما يفعل المبرّد، وأمّا المولّدة لغير أشباهها، فبمنزلة الكثافة والتّخلخل، فإنّ الكثافة يتولّد عنها السواد والثّقل والصّلابة، والتّخلخل يتولّد عنه البياض والخفّة واللين[٤٥٢]، وأمّا الكيفيّات المتولّدة عن الكيفيّات الملموسة، فمنها مبصورة ومنها مسموعة ومنها مشمومة ومنها مذوقة[٤٥٣]، والكيفيّات المبصورة منها البياض، وهو يتولّد عن تخلخل أجزاء الجسم، ومنها السّواد، وهو يتولّد عن تكاثف أجزاء الجسم، وسائر الألوان التي فيما بين هذينِ اللّونينِ تتولّد عن الحالات التي فيما بين هاتينِ الحالتينِ، وأمّا الكيفيّات المسموعة، فهي تابعة لصلابة الأجسام ولينها وعظم مقاديرها وصغرها، وأمّا الكيفيّات المشمومة والكيفيّات المذوقة، فهي تابعة للمزاج بمنزلة ما يتّبع الحارّ الحدّة والحرافة والمرارة والملوحة، ويتّبع البارد القبض والعفوصة والحموضة، ويتّبع الاعتدال بينهما الحلاوة والدّسومة[٤٥٤].

---

٤٤٥ DSY: أخرى أتمّ؛ M: أخرى | ٤٤٦ A¹؛ AS: – ولا تتولّد هي عن غيرها | ٤٤٧ DRY: أشباهه | ٤٤٨ FDY: الملوسة | ٤٤٩ FS: + منها | ٤٥٠ DFSUY: ملوسة | ٤٥١ A: ملموسة؛ A¹M: ملاسة | ٤٥٢ MS: واللّين والخفّة | ٤٥٣ AFM: مذاقة | ٤٥٤ F: والرّسومة؛ Y: والدّسوقة

(34) We also say that some of the tactile qualities are specified ambiguously by using the name of the genus that encompasses them.[71] Thus, they are called tactile—qualities apprehended only by the sense of touch and not known by any other sense. Others are not specified by the name "tactile." The effect of some of the qualities specified by the name "tactile" penetrates the entire substance, both outer and inner. The effect of others extends only to the exterior of the substance. Some have no effect on either the exterior or the interior of the substance. Some of the qualities that are not specified by the name "tactile" are visible, those being the qualities that affect only the sense of vision. Some are audible—those that affect only the sense of hearing. Some are olfactory—those that affect the two front ventricles of the brain, all or most of their effect being there. Some are gustatory—those whose effect is only, or mostly, in the tongue and the upper part of the throat.

<Their disagreement about [titles of books on] the elements>[72]

(35) The authors of treatises about the elements give different names and titles to their works. Some of them gave their books about the elements the title *On Nature*, as Parmenides and Melissus did. Some of them gave their books about the elements the title *On Substance*, as Chrysippus did. Some gave their books about the elements the title *On the Nature of Man*, as Hippocrates did. Some of them gave their books about the elements the title *On the Elements*, as Galen did. Aristotle

71. L 9.16–17; K 1:485; G^{a} 107–8: This paragraph comments on the end of Galen's chapter, in which he seeks to prove that all the other qualities of bodies derive from the four primary qualities of heat, coldness, moisture, and dryness. Having shown in the previous lines that the tactile qualities derive from these four primary qualities, and not the reverse, Galen here shows that even though the visible, audible, olfactory, and gustatory qualities are in a sense tactile, they do not themselves alter the substance of the body and thus cannot be primary qualities.

72. L 9.26–30; K 1:487–88; G^{a} 111–13: Galen explains why he entitled his book *On the Elements According to Hippocrates* when Hippocrates's book was known as *On the Nature of Man*. Earlier authors tended to use the title *On Nature* or various other titles when writing about the elements, while later authors consistently entitled such books *On the Elements*. Galen is following the modern custom. In addition to the authors referred to in the epitome, Galen mentions Asclepiades, Alcmaeon, Gorgias, and Prodicus.

(٣٤) ونقول أيضاً إنّ الكيفيّات الملموسة منها كيفيّات تخصّ على طريق اشتراك[٤٥٥] الأسماء باسم الجنس المشتمل[٤٥٦] عليها، فتدعى الملموسة، وهي الكيفيّات التي لا تدرك إلاّ بحاسّة اللّمس ولا يعرف بغيرها[٤٥٧]، ومنها ما لا تخصّ باسم الملموسة، والتي تخصّ باسم الملموسة منها ما تنفذ فعله في جميع الجوهر ظاهره وباطنه، ومنها ما لا يتجاوز فعله ظاهر الجوهر، ومنها ما لا يفعل لا في ظاهر الجوهر ولا في باطنه، وأمّا التي لا تخصّ باسم الملموسة، فمنها المبصورة، وهي التي تفعل في حاسّة البصر وحدها، ومنها المسموعة، وهي التي تفعل في حاسّة السّمع وحدها، ومنها المشمومة، وهي التي تفعل في البطنيْنِ المقدّميْنِ من بطون الدّماغ وخاصّة فعلها أو أكثر[٤٥٨] فعلها فيهما[٤٥٩]، ومنها المذوقة[٤٦٠]، وهي التي[٤٦١] فعلها في اللّسان وأعلى الحنك فقط أو أكثره[٤٦٢] هناك.

‹اختلافهم في عنوانات كتبهم في العناصر[٤٦٣]›

(٣٥) قد اختلف أصحاب الكتب في العناصر في عنوانات كتبهم وترجماتها[٤٦٤]، فمنهم من جعل ترجمة كتابه في العناصر «الكتاب في الطّبيعة[٤٦٥]» بمنزلة ما فعل بارمانيدس[٤٦٦] ومالسيس، ومنهم من جعل ترجمة كتابه فيها «الكتاب في الجوهر» بمنزلة ما فعل خروسبس، ومنهم من جعل ترجمة كتابه فيها «الكتاب في طبيعة الإنسان» بمنزلة ما فعل إبّقراط[٤٦٧]، ومنهم من جعل ترجمة كتابه «فيها الكتاب في العناصر» بمنزلة ما فعل

---

٤٥٥ $D^{1}Y^{1}$: اشتقاق | ٤٥٦ A: المستعمل؛ $A^{1}F$: المشترك | ٤٥٧ FS: تغيّرها | ٤٥٨ ADFY: وأكثر | ٤٥٩ DFMSY: فيها | ٤٦٠ AFM: المذاقة | ٤٦١ ADMY: + تفعل | ٤٦٢ A: وأكثر فعلها؛ DSY: وأكثره؛ F: وأكثرها | ٤٦٣ R^{2}: اختلافهم في العناصر | ٤٦٤ AF: ترجمتها | ٤٦٥ R: طبيعة العناصر | ٤٦٦ A: فارميندس؛ D: بارميليدس؛ R: بارميناذس؛ U: باربيندس؛ Y: بارمينيدس | ٤٦٧ ADSUY: بقراط

mentioned the problem of the elements in two places, one of which he entitled *On the Heavens*,[73] and the other, *On Generation and Corruption.*

<Their disagreement about the temperament>[74]

(36) The Ancients differed about the problem of the temperament. Among them, they had three general opinions: first, Asclepiades's opinion; second, the Stoics' opinion; and, third, Aristotle's opinion. Asclepiades claimed that the temperament arises from the adherence of the indivisible parts to each other. His view is false, because the adherence of the parts to each other is not a temperament but, rather, only contiguity and combination. That is because they are not joined in a true composition but, rather, are joined only by a concatenation that is apparent to the senses. The Stoics maintain that bodies interpenetrate each other. Finally, Aristotle says that bodies are divided into small parts that are then joined to each other, whereupon their qualities entirely mix in such a way that each of the qualities acts upon every other, and each of them receives the effects of the others. Let us disregard the first view, for it does not affirm that there is a temperament, but only that there exists a combination by means of contiguity. That reduces the three opinions to two: one of them the opinion of those who claim that the substances of corporeal bodies interpenetrate each other, and the other

73. Some MSS add: "*and the Universe.*"

74. L 9.20–21; K 1:486; G[a] 108–9: A continuation of the criticism of the Atomists' theory that substantial change—here, temperament—can result from the combination or separation of unchanged atoms or parts. Galen is grumbling that the Pneumatist Athenaeus bungled his argument against the Methodist Asclepiades, who held a doctrine that was essentially Atomist. Temperament (*mizāj;* κρᾶσις) is specifically the type of mixture from which a new substance results—what we would call a chemical compound—and not simply a mixture of unchanged parts, like powders or grains stirred together; see also pp. 142–43, nn. 30–31, above. Galen also complains that Athenaeus did not refute the objectionable parts of the physical theories of Aristotle and the Stoic Chrysippus—though, as a Pneumatist, Athenaeus would have been aligned with the Stoics. The Stoics, and specifically Chrysippus, held a strictly materialist theory in which πνεῦμα, a material spirit, interpenetrated other bodies, giving them their organic unity. This required them to hold that, in a true mixture or temperament, the parts are completely blended, such that every part, no matter how small, is perfectly

جالينوس، فأمّا أرسطوطاليس[٤٦٨] فإنّه ذكر أمر العناصر في موضعين، أحدهما جعل ترجمته[٤٦٩] «كتاب السّماء[٤٧٠]»، الآخر جعل ترجمته[٤٧١] «كتاب الكون والفساد».

‹اختلافهم في المزاج[٤٧٢]›

(٣٦) قد اختلف القدماء في أمر المزاج وجملة آرائهم فيه ثلثة، أحدها رأي أسقليبياذس، والآخر رأي الرّواقيّين[٤٧٣]، والثّالث رأي أرسطوطاليس[٤٧٤]، فأمّا أسقليبياذس[٤٧٥]، فيزعم[٤٧٦] أنّ المزاج يحدث عن لزوم الأجزاء التي لا تتجزّأ[٤٧٧] بعضها لبعض، ورأي هذا[٤٧٨] منفسخ لأنّ لزوم الأجزاء[٤٧٩] بعض[٤٨٠] لبعض ليس هو مزاج[٤٨١]، بل إنّما هو مجاورة وتركيب إذ كانت هذه المضامّة ليست مضامّة مخالطة بالحقيقة، بل إنّما هي مضامّة اقتران عند الحسّ فقط. وأمّا الرّواقيّون، فيزعمون أنّ الأجسام يدخل بعضها في بعض، وأمّا أرسطوطاليس[٤٨٢] فيقول إنّ الأجسام تتجزّأ أجزاء صغار، ثمّ يضامّ بعضها بعضاً، وكيفيّاتها تمازج بكلّيّتها عند ما تفعل كلّ واحدة[٤٨٣] من الكيفيّات في الأخرى[٤٨٤] ويقبل كلّ واحدة[٤٨٥] منهما[٤٨٦] فعل الأخرى، فيحصل من هذه الثّلثة الآراء إذا عزل عنها الرّأي الأوّل الذي لا[٤٨٧] يوجب مزاجاً، بل تركيباً[٤٨٨] على طريق المجاورة رأيان[٤٨٩]، أحدهما رأي من يزعم أن جواهر[٤٩٠] الأجسام يخالط[٤٩١]

---

٤٦٨ ARY: أرسطاطاليس | ٤٦٩ A¹R: ترجمة | ٤٧٠ DMSY: + والعالم | ٤٧١ DFMSY: – جعل ترجمته | ٤٧٢ R²: اخت . . . | ٤٧٣ F: أصحاب الرّواق؛ R: والرّواقين | ٤٧٤ F¹R: أرسطاطاليس | ٤٧٥ D: أسقلسادس؛ Y: اسليبياذس | ٤٧٦ AM: فزعم | ٤٧٧ F: ينقطع | ٤٧٨ A: وهذا الرّأي؛ F: + رأي | ٤٧٩ AF: – الأجزاء؛ DMSY: + التي لا تتجزّئ | ٤٨٠ DFMSUY: بعضها | ٤٨١ MS: مزاجاً | ٤٨٢ AFR: أرسطاطاليس | ٤٨٣ MS: واحد | ٤٨٤ FU: + منها فعلاً؛ S: + منها فعل | ٤٨٥ AS: واحد | ٤٨٦ ADFMY: منها | ٤٨٧ DFSUY: ليس | ٤٨٨ A: مزاج تركيب؛ F: مزاج بل تركيب | ٤٨٩ FY: ورأيان | ٤٩٠ DFY: جوهر | ٤٩١ DSY: يداخل؛ M: يدخل

the opinion of those who claim that bodies do not interpenetrate each other. Those who hold the first view claim that the corporeal substances interpenetrate each other. This is a principle whose basis is absurd and incoherent and from which, moreover, other absurdities follow. One of them is that the entire universe must be able to enter into a single grain of millet, which would contain it completely. That is because, if it is possible for one body to interpenetrate another, it is possible for a grain of millet to interpenetrate another grain of millet. If it is possible for one grain of millet to interpenetrate another, it must be possible for yet another to do so. This process could continue until the entire universe was divided into grains of millet and all of them had interpenetrated that first one.

(37) Those who hold the second view say that the bodies themselves do not interpenetrate each other, but they are joined to each other when they are divided into small parts.[75] Their qualities affect each other and receive the effects of each other until they come to resemble each other and the whole becomes homoeomerous. However, the whole does not become one with respect to those single parts from which the temperament arises. Instead, it becomes something intermediate between the two. It is still potentially those two single parts from which it actually arose, but is not either one of the two.

---

mixed with the other components while still preserving its own nature. Critics quoted Chrysippus as saying that a drop of wine could temper the whole sea. The theory also implied that two bodies could occupy the same space, giving rise to the argument against the theory given in the epitome. See von Arnim, *Stoicorum veterum fragmenta*, 2.463–81; Long and Sedley, *Hellenistic Philosophers*, frag. 48A–F; Themistius, *In Aristotelis physica paraphrasis*, 104.9–19; Alexander of Aphrodisias, *De mixtione*, 213–38; Sambursky, *Physics of the Stoics*, 11–17. Despite his criticism of the Stoics, a Stoic-influenced Pneumatic doctrine is central to Galen's physiology.

75. L 9.33–39; K 1:489–91; G[a] 113–16: Aristotle believed that temperament occurred through the interaction of qualities; cf. his *On Generation and Corruption* 1.10, where he explains that, although things broken into small parts and mixed together do not merge into each other, their qualities can affect each other, with the larger quantities having a greater effect than the smaller. The whole mixture thus becomes homoeomerous. Galen remarks that Hippocrates does not settle the question of the metaphysical nature of temperament, contenting himself with stating the fact of the phenomenon. Galen, the practical physician, states his tentative allegiance to Aristotle's position but then reframes it in the context of the preparation of drugs: drugs crushed finely share their medicinal qualities more thoroughly.

بعضها بعضاً، والآخر رأي من يقول إنّ الأجسام لا يدخل بعضها في بعض، وأهل القول الأوّل يزعمون أنّ الجواهر الجسميّة يدخل بعضها في بعض، وهذا أصل مبناه[٤٩٢] شنع لا يفهم[٤٩٣]، ويتّبعه مع ذلك[٤٩٤] شناعات أُخر، أحدها أنّه يجب من هذا القول أن يكون العالم كلّه يدخل في حبّة جاورس يحصره بأجمعه، وذاك[٤٩٥] أنّه إن كان[٤٩٦] يجوز[٤٩٧] أن يدخل جسم في جسم، فقد يجوز[٤٩٨] أن تدخل جاورسة في جاورسة، وإن جاز[٤٩٩] أن تدخل جاورسة في جاورسة فقد يجوز أن تدخل معها أخرى ولا يزال هذا[٥٠٠] حتّى يقسم العالم كلّه جاورساً[٥٠١]، ويدخل جميعها في تلك الواحدة.

(٣٧) وأمّا أهل القول الثّاني، فيقولون إنّ الأجسام أنفسها لا يدخل بعضها في بعض، لكنّها يضامّ بعضها بعضاً عندما ينقسم أجزاء صغار[٥٠٢]. فأمّا كيفيّاتها، فيفعل بعضها في بعض ويقبل بعضها فعل بعضٍ حتّى تصير شبيهة بعضها[٥٠٣] ببعض[٥٠٤]، ويصير الكلّ متشابه[٥٠٥] الأجزاء من غير أن يكون هذا الكلّ واحداً[٥٠٦] من تلك الأجزاء المفردة التي عنها حدث المزاج، لكنّ شيئاً وسطاً[٥٠٧] فيما بينهما[٥٠٨]، وهو بالقوّة ذانك[٥٠٩] الجزآنِ المفردانِ اللذانِ عنهما حدث وبالفعل[٥١٠] ليس بواحد منهما.

---

٤٩٢ DY: منفاه ومبنى؛ M: + واهن؛ RSU: + مبنى | ٤٩٣ MS: + ويدخل | ٤٩٤ ASY: - مع ذلك؛ R^{2}: + شناعات | ٤٩٥ AMS: وذلك | ٤٩٦ AS: - كان | ٤٩٧ AS: جاز | ٤٩٨ AR: فقد يجوز > جاز | ٤٩٩ F: كان يجوز؛ M: فإن جاوز | ٥٠٠ DFSY: + هكذا | ٥٠١ ADFY: جاورس | ٥٠٢ $D^{1}RY$: صغاراً | ٥٠٣ شبيهة بعضها > F: شبية؛ R: شبيهة بعضها > شبيه بعض؛ U: شبيهة أيضاً | ٥٠٤ DSY: لبعض | ٥٠٥ DMSY: متشابهة | ٥٠٦ AMSY: واحد | ٥٠٧ $FMSY^{1}$: وسط | ٥٠٨ A: فيها بينهما؛ A^{1}: فيما بينها؛ DRUY: بينها | ٥٠٩ A: ذينك؛ F: ذلك | ٥١٠ F: حدثا وبالقياس؛ R: بالفعل

[Two bodies]

(38) When two natural bodies are in contact, having active qualities and passive qualities, those qualities being opposites, if they are equal in potency or are close to being equal in potency or in magnitude, each affects the other, and each receives the effect of the other. In that case, neither one is transformed into the species of the other, nor is it totally transmuted into the other. Something intermediate is thus generated. This kind of alteration and transformation is called "temperament." If the two bodies are not equal, but, rather, one of them is more potent and greater in magnitude than the other, while the other is weaker and smaller than the first, then the body that is the weaker of the two and has a smaller magnitude is transformed, corrupted, and changed into the species of the more potent, and the one that is more potent is said to have increased. However, it is not said to have combined or formed a temperament, for food is not said to have combined with the flesh, nor firewood to have formed a temperament with the fire.

‹الجسمان[٥١١]›

(٣٨) متى التقيا[٥١٢] جسمان طبيعيّان لهما كيفيّات فاعلة وكيفيّات منفعلة، وتلك الكيفيّات أضداد، فإنّهما إن كانا متكافئيْن في القوّة أو قريبيْنِ من ذلك[٥١٣] في القوّة أو في المقدار، فعل كلّ واحد منهما في الآخر وقبل كلّ واحد منهما فعل الآخر، فلمْ ينقلب أحدهما إلى نوع الآخر، ولمْ يستحيل[٥١٤] إليه غاية الاستحالة، ويولّد بينهما شيء وسط، وهذا الصّنف من التّغيّر[٥١٥] والاستحالة يقال له مزاج، وإن كان[٥١٦] الجسمانِ ليسا بمتكافئيْنِ[٥١٧]، لكنّ أحدهما أقوى من الآخر، وأكثر منه مقدارًا والآخر أضعف من هذا وأقلّ[٥١٨] مقدارًا منه، فالجسم الذي هو منها[٥١٩] أضعف وأقلّ[٥٢٠] مقدارًا يستحيل ويفسد وينتقل إلى نوع الأقوى[٥٢١]، والذي[٥٢٢] هو أقوى[٥٢٣] يقال[٥٢٤] له[٥٢٥] قد تزيّد، فأمّا أنّه قد خالط أو قد مازج، فلا، وذاك[٥٢٦] أنّه ليس يقال إنّ الغذاء يخالط اللّحم، ولا أنّ الحطب يخالط النّار[٥٢٧].

٥١١ R^2: الجسميْن | ٥١٢ A: ألقوا؛ A^1: التقيا | ٥١٣ DFUY: تلك | ٥١٤ DMY: يستحل | ٥١٥ ADMSY: التّغيير | ٥١٦ DMS: كانا | ٥١٧ F: متكافئين؛ M: بمتكافئانِ؛ S: بالمتكافئين | ٥١٨ RU: أو أقلّ | ٥١٩ DFMY: منهما | ٥٢٠ R: أو أقلّ | ٥٢١ A: أقوى؛ R: – الأقوى | ٥٢٢ R: الذي | ٥٢٣ D: أقلّ قوى؛ F: أقلّ [F^2: + فعلًا] وأقوى [F^2: + فعلًا]؛ Y: – أقوى | ٥٢٤ R: فعلًا ويقال | ٥٢٥ RU: أنّه؛ Y: – له | ٥٢٦ ADY: وذلك | ٥٢٧ SY: + لكن يستحيل إليه

[Chapter 10][76]

<What is said to be potential>

(39) Some things that are said to be potential are remote, and some are proximate. An example of the remote is when it is said that a child is potentially knowledgeable about grammar, or that water is potentially fire. An example of what is proximately potential is when it is said that a sleeping grammarian is a grammarian, or when it is said that pitch and sulfur are potentially fire. Some of the elements of the body are general, prior in nature, and remote from sensation. These are the ones that we mentioned previously.[77] Others are specific, prior in sensation, and proximate; these are the tissues. Still others, the humors, are intermediate between these two classes. Because of that, we need to mention the humors first of all and only then undertake to mention the proximate elements.

---

76. L 10.5–7; K 1:493; G^{a} 118–19: Galen has just mentioned the primary qualities, the humors, and the tissues—all of which can be considered elements of the body. The tissues are the proximate elements of the body.

77. A gloss in one MS adds: "That is, fire, air, water, and earth."

[الفصل العاشر]

‹ما يقال بالقوّة[٥٢٨]›

(٣٩) الأشياء[٥٢٩] التي[٥٣٠] يقال إنّها[٥٣١] بالقوّة منها[٥٣٢] ما هو بعيد ومنها[٥٣٣] ما هو قريب. أمّا البعيد، فبمنزلة ما يقال إنّ الصّبيّ بالقوّة عالم بالنّحو، وإنّ الماء بالقوّة نار[٥٣٤]، وأمّا القريب، فبمنزلة ما يقال في النّحويّ النّائم أنّه نحويّ، وفي الزّفت[٥٣٥] والكبريت إنهما[٥٣٦] بالقوّة نار[٥٣٧]. عناصر البدن منها عاميّة متقدّمة في الطّبع بعيدة عن الحسّ، وهي التي ذكرناها[٥٣٨] فيما تقدّم[٥٣٩]، ومنها خاصّيّة متقدّمة عند الحسّ قريبة، وهي الأعضاء المتشابهة الأجزاء، ومنها وسط فيما بين هاتين الطّبقتين[٥٤٠]، وهي الأخلاط، ومن أجل ذلك ينبغي لنا أن نذكر أوّلاً هذه، ثمّ نأخذ في ذكر العناصر القريبة.

---

٥٢٨ R^2 | ٥٢٩ DFUY: –الأشياء | ٥٣٠ D^1؛ DFU: الذي | ٥٣١ DFU: إنّه | ٥٣٢ FU: –منها؛ DFUY: +شيء من الأشياء منه | ٥٣٣ DFUY: ومنه | ٥٣٤ AF: بارد؛ A^1: نار؛ RU: نار بالقوّة | ٥٣٥ A: الزّيت؛ A^1: خ الزّفت | ٥٣٦ DFSUY: إنّه | ٥٣٧ FMS: ناراً | ٥٣٨ FS: ذكرنا | ٥٣٩ S^h: أعني النّار والهواء والماء والأرض | ٥٤٠ F: الطّبيعتين

[Chapter 11][78]

<Their disagreement about the humors>

(40) Difference of opinion has also occurred concerning the humors, for one group said that the structure and composition of the living body was from a single humor, while another group said that it was from several humors. Some of those who said that it was from only one humor claimed that it was from blood only; some of them said that it was from yellow bile; some of them said that it was from black bile; and some of them said that it was from phlegm. Among those who said that the body was made up of several humors were Hippocrates and his disciples, for they said that the makeup and composition of the body were from blood, phlegm, and the two biles. If you examine these five theories, you will find that three of them are false and implausible. Those are the opinions of those who claimed that the body was composed of yellow bile, those who claimed that it was composed of black bile, and those who claimed that it was composed of phlegm. That is because life depends on heat and moisture, and in none of these three humors are both combined, for yellow bile is dry, phlegm is cold, and black bile is cold and dry. The opinion of those who held that the structure and composition of the body were from blood alone is plausible but is not true. The opinion of those who thought that the body was composed of the four humors is plausible and, in addition, is true.

---

78. L 11.1–13; K 1:494–97; G^a 119–22; Hippocrates, *On the Nature of Man* 2. A return to the criticism of those who held that the body was generated from only one of the humors, blood being the obvious candidate and the only humor with real defenders; see pp. 156–58, paragraphs 20–23, above. Galen understands this dispute to refer specifically to the generation of the fetus in the womb.

[الفصل الحادي عشر]

‹اختلافهم في الأخلاط[٥٤١]›

(٤٠) وقد وقع[٥٤٢] في أمر الأخلاط أيضاً اختلاف في الرّأي، وذلك[٥٤٣] أنّ قوم[٥٤٤] قالوا إنّ بنية البدن وقوامه من خلطٍ واحدٍ، وقوم[٥٤٥] قالوا إنّه من أخلاط كثيرة، فأمّا الذين قالوا إنّه من خلط واحد، فمنهم من زعم[٥٤٦] أنّه من الدّم وحده، ومنهم من قال إنّه من المرّة الصّفراء، ومنهم من قال إنّه من المرّة السّوداء، ومنهم من قال إنّه من البلغم، فأمّا[٥٤٧] الذين قالوا إنّه من[٥٤٨] أخلاط كثيرة، فمنهم[٥٤٩] إبقراط وأصحابه، فإنّ هؤلاء قالوا إنّ بنية البدن وقوامه من الدّم والبلغم والمرّتيْنِ[٥٥٠]، وهذه الخمسة الآراء إذا امتحنت وجدت ثلثة منها، وهي[٥٥١] رأي من يزعم[٥٥٢] أنّ البدن مركّب من الصّفراء، ومن يزعم أنّه مركّب من السّوداء، ومن يزعم أنّه مركّب من البلغم كاذبة لا مقنع فيها[٥٥٣]، وذلك[٥٥٤] أنّ الحياة إنّما تكون بالحرارة والرّطوبة، وليس من هذه الثّلثة الأخلاط واحد يجتمع[٥٥٥] فيه هذانِ[٥٥٦] لأنّ الصّفراء[٥٥٧] يابسة والبلغم بارد[٥٥٨] والسّوداء[٥٥٩] باردة يابسة، وأمّا رأي من يزعم أنّ بنية البدن وقوامه من الدّم وحده، فهو رأي مقنع إلاّ أنّه ليس بحقّ، وأمّا رأي من يزعم أنّ البدن مركّب من الأربعة الأخلاط، فهو رأي مقنع، وهو مع هذا[٥٦٠] حقّ.

---

٥٤١ R^2: اختلفو | ٥٤٢ FMS: + أيضاً | ٥٤٣ FR: وذاك؛ DY: فذلك | ٥٤٤ U: + من الأوائل | ٥٤٥ DFMY: وقوماً | ٥٤٦ AS: يزعم | ٥٤٧ DFU: وأمّا؛ Y: أمّا | ٥٤٨ M: + الأخلاط أعني من؛ S: + الأخلاط أعني | ٥٤٩ FY: فهم | ٥٥٠ ASU: والمرّة الصّفراء والمرّة السّوداء؛ DFY: والصّفراء والسّوداء؛ M: والسّوداء الصّفراء | ٥٥١ SY: وهو | ٥٥٢ FR: زعم | ٥٥٣ DFSUY: عندها؛ R: + عندنا | ٥٥٤ MS: وذاك | ٥٥٥ RU: يجمع | ٥٥٦ A: هاتان؛ F: هذا؛ R: هذيْن الأمريْن | ٥٥٧ DMSUY: + حارّة | ٥٥٨ R: - بارد؛ DMRSY: + رطب | ٥٥٩ AU: والمرّة السّوداء | ٥٦٠ FU: ذلك

(41) Those who claimed that the structure and composition of the body were from blood alone asserted their opinion by offering as convincing proof an argument from physics and the evidence of the senses.[79] They first offered a physical argument in proof of this. They said that the substance of the animal was by motion, and imparting motion was something specific to heat and conformable with it, while ease of motion was something specific to moisture. Since this was the case, the substance of the animal was dependent on heat and moisture. Therefore, the animal must come from the matter in which heat and moisture are predominant. Since that was so, neither yellow bile nor black bile nor phlegm was suitable to be that from which the substance of the animal came. Rather, it could come only from blood, for blood was hot and moist and the qualities that predominated in it corresponded to frequent motion. Nonetheless, it also contained coldness and dryness, by which it could thus have stability and solidity and not have the highest degree of liquidity, which would not have been stable at all. Since that was the case, the body came to be from blood alone.

(42) As further proof, they then said that if two corporeal bodies came into contact and it happened that the potency of one was equal to the potency of the other, the one could transform the other[80] and a temperament would be generated from the two.[81] If one of them dominated the other, the dominant one would transform the one dominated and convert it into its own species, so that the latter would become an increase in the first. When the body becomes larger by means of a nutriment, the body is stronger than it was. Were that not the case, the body would have been transformed into the species of the nutriment, rather than the nutriment being transformed into the species of the

79. L 14.1; K 1:506–7; G[a] 136–37: Galen says that the opinion of those who held that the generation of the fetus is from blood alone is reasonable. However, he maintains that Hippocrates understood nature better and that his view—that the fetus is generated from all four humors—is actually the correct view. The arguments cited in the epitome in favor of the generation of the fetus from blood are not given in Galen's *On the Elements*.

80. One early MS reads: "will not transform the other." Several other MSS add: "and will be transformed from it."

81. Cf. Aristotle, *On Generation and Corruption* 1.10, and pp. 171–72, paragraph 36 and n. 74, and p. 173, paragraph 38, above.

(٤١) الذين[٥٦١] يزعمون أنّ بنية البدن وقوامه من الدّم وحده يثبتون ذلك ويأتون فيه بالحجّة المقنعة من القياس على مجرى الكلام في الطّبائع ومن الحسّ. أمّا من القياس الطّبيعي، فأوّل ما احتجّوا به هذا، قالوا إنّ جوهر الحيوان إنّما هو بالحركة، والتّحريك هو شيء خاصّ بالحرارة ملائم لها وسهولة الحركة خاصّ بالرّطوبة، وإذا كان الأمر على هذا، فجوهر الحيوان إنّما هو بالحرارة والرّطوبة، فالمادّة إذن[٥٦٢] التي[٥٦٣] الغالب عليها الحرارة والرّطوبة، هي[٥٦٤] التي ينبغي أن يكون منها الحيوان، وإذا كان ذلك كذلك، فلا المرّة الصّفراء ولا المرّة[٥٦٥] السّوداء ولا البلغم يصلح أن يكون منه جوهر الحيوان، لكنّ الدّم وحده لأنه حارّ رطب، والأغلب عليه الكيفيّة الموافقة للحركة[٥٦٦] الكثيرة، وفيه مع هذا برودة ويبوسة ليكون له بذلك ثبات وجمود، ولا يكون في غاية الانحلال سيّالاً لا يثبت[٥٦٧]، وإذا[٥٦٨] كان ذلك كذلك فكون[٥٦٩] البدن إذن إنّما هو من الدّم.

(٤٢) ثمّ احتجّوا بعد ذلك، فقالوا إنّ كلّ جسمين يلتقيانِ، فهما إن كانت قوّة كلّ واحد منهما مساوية لقوّة الآخر[٥٧٠] يحيل الواحد منهما الآخر[٥٧١] ويتولّد عنهما مزاج، وإن كان أحدهما قاهراً[٥٧٢] للآخر أحال[٥٧٣] القاهر المقهور[٥٧٤] وأقلبه[٥٧٥] إلى نوعه[٥٧٦]، فصار زيادةً فيه، والغذاء لمّا كان يتزيّد به البدن صار البدن أقوى منه، ولولا ذلك لكان البدن يستحيل إلى نوع الغذاء، ولم يكن الغذاء يستحيل إلى نوع

٥٦١ DMSY: والذين | ٥٦٢ FU: إذا؛ M: – إذن | ٥٦٣ DFY: – التي | ٥٦٤ AFS: فهي | ٥٦٥ DFUY: – المرّة | ٥٦٦ F: للحياة؛ M: للحرارة | ٥٦٧ S: يلبث | ٥٦٨ DF: إذ | ٥٦٩ DFMSUY: فيكون | ٥٧٠ R: + لم | ٥٧١ ADMSY: للآخر ويستحيل منه | ٥٧٢ AUY: قاهر | ٥٧٣ AS: حال | ٥٧٤ A: لمقهور؛ F: للمقهور | ٥٧٥ DMSUY: وقلبه | ٥٧٦ A: + في الأخرى إلى طبيعته

body. Since all of this is the case, the composition of the body was from neither yellow bile nor black bile nor phlegm, but from blood only. This is because yellow bile is many times hotter and drier than the body. It is therefore more powerful than the body because of the intensity of its heat and because it is dry, the body being moist and the dryness being more powerful than the moisture of the body. Black bile is much colder and drier than the body, for which reason the body would be unable to transform and convert it to itself. Indeed, it would be more likely that the black bile would transform and convert the body. Phlegm is also much colder and more moist than the body. However, blood is similar in its temperament to the body, though it is less hot and more moist than the body. In both respects, it is weaker than the body. This is how they went about proving that the body was from blood by using an argument from physics.

(43) They argued from sensation using three things: first, generation; second, nutriment; and, third, that which is evacuated from and retained in the body. With respect to generation, they argued that we could see that it was blood that was deposited in the womb and from which the embryo came and that semen was of the nature of blood. Neither black bile nor yellow bile nor phlegm is deposited in the womb. The argument they gave concerning nutriment was that we could plainly see that the body was nourished from blood alone, since blood was the only one of the humors that the organs drew to themselves to take nourishment from. The organs forcibly repelled the other humors from themselves as they would repel a foreign body to which they were averse. That is why we see that even though the gallbladder, unlike

البدن، وإذا[٥٧٧] كان الأمر على هذا، فالبدن ليس قوامه من مرّة صفراء ولا من مرّة[٥٧٨] سوداء ولا من بلغم، لكنّ[٥٧٩] من الدّم فقط. وذلك لأنّ[٥٨٠] المرّة الصّفراء أشدّ حرارةً ويبوسةً من البدن بأضعاف كثيرة، فهي لذلك أقوى من البدن لشدّة حرارتها، ولأنّها يابسة، والبدن رطب واليابس أقوى من الرّطب. والمرّة السّوداء أبرد من البدن كثيرًا وأيبس منه، ولذلك ليس يستطيع البدن أن[٥٨١] يحيلها ويقلبها[٥٨٢]، بل هي أحرى أن يحيله ويقلبه. والبلغم هو أيضًا أبرد من البدن وأرطب منه كثيرًا. وأمّا الدّم فإنّه شبيه في مزاجه بالبدن، وهو مع هذا أقلّ حرارةً منه وأكثر رطوبةً منه، فهو للأمرين جميعًا أضعف من البدن. فهذا ما احتجّوا به على طريق القياس من الطّبع[٥٨٣].

(٤٣) وأمّا من[٥٨٤] الحسّ فاحتجّوا في ذلك بثلاثة أشياء، أحدها الكون، والآخر الغذاء، والثّالث ما يستفرغ من البدن ويحتبس[٥٨٥] فيه. أمّا[٥٨٦] من الكون، فاحتجّوا بأن قالوا إنّا نرى الشّيء[٥٨٧] الذي يقع في الأرحام، ويكون منه الجنين إنّما هو الدّم والمنى الذي من طبيعة الدّم، فليس[٥٨٨] يقع في الأرحام عند كون الجنين[٥٨٩]، ولا واحد[٥٩٠] من المرّتين ولا البلغم. وأمّا من الغذاء فاحتجّوا بأن قالوا إنّا نجد عيانًا أنّ البدن إنّما يغتذي من الدّم فقط. وذلك[٥٩١] أنّ الأعضاء إنّما تجذب إليها من الأخلاط ليغتذي[٥٩٢] به[٥٩٣] الدّم وحده، وأمّا سائر الأخلاط، فهي تدفعها عن أنفسها وتقذف بها كما تقذف بالشّيء الغريب[٥٩٤] المنافر[٥٩٥]. ولذلك قد نجد المرارة فضلًا عن غيرها

---

٥٧٧ A: فإذا؛ FY: وإذ | ٥٧٨ D: –مرّة؛ FUY: –من مرّة | ٥٧٩ FM: ولكن | ٥٨٠ AF: أنّ | ٥٨١ FU: –أن | ٥٨٢ DMSY: + إليه | ٥٨٣ DFMSUY: الطّبائع | ٥٨٤ M: + طريق | ٥٨٥ DY: وما يحتبس؛ FU: ويحبس | ٥٨٦ A: فأمّا؛ F: وأمّا | ٥٨٧ A¹: المنى | ٥٨٨ ADMSUY: وليس | ٥٨٩ R: الكون للجنين | ٥٩٠ ADFSY: واحدة | ٥٩١ DSY: وذاك | ٥٩٢ FSU: ما يغتذي | ٥٩٣ AFU: + من | ٥٩٤ R: –الغريب | ٥٩٥ A¹F¹R: + لها

other organs, attracts yellow bile to itself, it is not nourished by it. This is indicated by, among other things, the fact that there are blood vessels distributed through the body of the gallbladder by which it is nourished. Likewise, the spleen attracts black bile to itself but is not nourished by it. Instead, it is nourished by the blood that is mixed with it. The proof of that is that if this blood is purified and separated out from the black bile, it will reject the black bile and expel it from itself to the mouth of the stomach, as it does anything else that is not beneficial to it. Since this was the case, it could be known that the rest of the humors were superfluities consequent on the creation and generation of blood in the same way that making wine results in scum, which is analogous to yellow bile, and dregs, which are analogous to black bile. This is the argument they gave for it based on nutriment. Finally, they had an argument based on the excretion of what was excreted from the body and the retention of what was retained in it. They explained that the retention of the blood was necessary by nature and its excretion was unnatural. However, it was necessary by nature to excrete each of the others—the two biles and phlegm—and that to retain them was unnatural and harmful. Jaundice resulted from retaining yellow bile, and cancer and leprosy from retaining black bile. If phlegm was retained, its retention harmed the stomach and intestines.

(44) Those who said that the structure of the body originated from the four humors explained their view by three things: first, the diversity of organs; second, the diversity of the blood; and, third, the excretion of

من الأعضاء وإن كانت تجتذب[٥٩٦] إليها المرّة الصّفراء، فليس تغتذي بها، وممّا يدلّ على ذلك أنّ عروقاً يجري فيها الدّم، يتفرّق في جرم المرارة وتغذوها، وكذلك الطّحال أيضاً، فإنّه[٥٩٧] تجتذب إليه المرّة السّوداء ولكنّه[٥٩٨] لا يغتذي منها، بل إنّما يغتذي من الدّم المخالط لها، والدّليل على ذلك أنّه إذ أصفى هذا الدّم وميّزه وفصّله من المرّة السّوداء قذف بالمرّة السّوداء[٥٩٩] ودفعها عن نفسه إلى فم المعدّة كما يدفع الشّيء، الذي[٦٠٠] لا ينتفع به ، وإذا[٦٠١] كان الأمر على هذا فقد علم أنّ سائر الأخلاط إنّما هي[٦٠٢] فضول لازمة لتولّد[٦٠٣] الدّم وكونه بمنزلة ما يلزم في كون الشّراب من تولّد الزّبد الذي هو نظير للمرّة[٦٠٤] الصّفراء والدّردي الذي هو نظير للمرّة[٦٠٥] السّوداء، وهذا ما احتجّوا به من الغذاء، وأمّا من استفراغ ما يستفرغ من البدن واحتباس ما يحتبس فيه، فاحتجّوا بأن قالوا إنّ الدّم احتباسه شيء واجب في الطّبع[٦٠٦] واستفراغه شيء خارج عن الطّبع، وأمّا[٦٠٧] المرّتانِ[٦٠٨] والبلغم فاستفراغ كلّ واحد منهما واجب في الطّبع نافع[٦٠٩] واحتباسه خارج عن الطّبع ضارّ، فالمرّة الصّفراء إذا احتبست حدث عنها اليرقان، والسّوداء إذا احتبست حدث عنها السّرطان والجذام، والبلغم[٦١٠] إذا احتبس أضرّ احتباسه بالمعدّة والأمعاء.

(٤٤) فالذين[٦١١] قالوا إنّ بنية البدن من الأربعة الأخلاط بيّنوا[٦١٢] ذلك من ثلثة أشياء، أحدها اختلاف الأعضاء والآخر اختلاف الدّم والثّالث استفراغ ما

---

٥٩٦ DMY: تجذب | ٥٩٧ A: وإن كان؛ DU: فإن كان؛ R: – فإنّه؛ Y: فإن | ٥٩٨ F: لكنّه؛ M: إلاّ أنّه | ٥٩٩ M: بالمرّة السّوداء> بها؛ Y: – السّوداء | ٦٠٠ DY: ما؛ F: – الذي؛ U: ممّا؛ S: الذي ما | ٦٠١ ADFUY: وإن | ٦٠٢ F: + من؛ R: هو | ٦٠٣ R: لتوليد | ٦٠٤ DFMSUY: المرّة | ٦٠٥ DFMSUY: المرّة | ٦٠٦ R: الطّبيعة | ٦٠٧ ADMSUY: فأمّا | ٦٠٨ AF: المرّتيْن | ٦٠٩ AU: – نافع | ٦١٠ DMSY: وأمّا البلغم | ٦١١ DFSUY: والذين؛ R: الذين | ٦١٢ A: يثبتوا؛ R: أثبتوا

what is excreted from the body.[82] With respect to the diversity of organs, they said that each of the organs was nourished by a humor whose temperament was approximately similar to its temperament. Some organs were cold and moist, such as the brain; some were cold and dry, such as bone; some were hot and dry, such as the lungs; and some were hot and moist, such as flesh. Thus, it was clearly the case that flesh would have to be nourished by a humor that was hot and moist—a description that applies only to pure blood. Bone could only be nourished by a humor that inclines toward cold and dryness—an attribute existent in the humor belonging to the genus of black bile. The brain could only be nourished by a cold, moist humor—an attribute existent in phlegm. The lungs could only be nourished by a hot, dry humor—an attribute existent in yellow bile. With respect to the diversity of the blood, they argued that milk appears superficially to us to be a single thing, yet it is compounded from diverse substances. Part of it is water, part cheese, and part butter. In the same way, there is something coarse in blood, analogous to dregs; something dense, analogous to black bile; something else subtle and reddish, analogous to a red bile; and something white, analogous to phlegm. For this reason, blood has various states of color and composition corresponding to the difference of ages, the difference of times of the year, the difference of the animal's temperament, and the difference of the temperament of the body. With respect to the excretion of what is excreted from the body,[83] they argued that, when a person takes a purgative drug to excess, we see that the

82. L 11.3–15; K 1:494–97; G[a] 120–23: The epitome gives an expanded restatement of the argument from *On the Elements*. Galen mentions an alternative theory that the fetus was generated by blood and that the various tissues were generated by separating out the harder, moister, hotter, and cooler parts of the blood. However, he thinks it is more natural to suppose that blood contains the other humors from the beginning, as is indicated by the diversity of blood encountered by the physician.

83. L 11.16–19; K 1:497–98; G[a] 123–25; Hippocrates, *On the Nature of Man* 6. The rest of *On the Elements* deals with purgation, commenting on *On the Nature of Man* 5–7.

يستفرغ من البدن، أمّا من[٦١٣] اختلاف الأعضاء، فإنّهم قالوا[٦١٤] إن كلّ[٦١٥] واحد من الأعضاء إنّما يغتذي من خلطٍ مزاجُه شبيهٌ بمزاجه على التقريب، وكانت الأعضاء بعضها بارد رطب[٦١٦] بمنزلة الدماغ، وبعضها بارد يابس[٦١٧] بمنزلة العظم، وبعضها حارّ يابس[٦١٨] بمنزلة الرّئة[٦١٩]، وبعضها حارّ رطب[٦٢٠] بمنزلة اللّحم، فالأمر فيها بيّنٌ أنّ اللّحم إنّما يغتذي من خلط حارّ رطب، وهذه الصّفة إنّما هي للدّم[٦٢١] الخالص، والعظم إنّما يغتذي من خلط مائل إلى البرودة واليبوسة، وهذه الصّفة[٦٢٢] موجودة في الخلط الذي من جنس السّوداء، والدّماغ إنّما يغتذي من خلط بارد رطب، وهذه الصّفة موجودة في البلغم، والرّئة إنّما تغتذي من خلط حارّ يابس، وهذه الصّفة موجودة في المرّة الصّفراء، وأمّا من اختلاف الدّم، فإنّهم قالوا كما أنّ اللّبن في ظاهر أمره إنّما نراه شيئاً واحداً، وهو مركّب من جواهر مختلفة، فبعضه ماء وبعضه جبن وبعضه زبد[٦٢٣]، كذلك[٦٢٤] الدّم أيضاً فيه شيء غليظ من جنس[٦٢٥] الدّرديّ، والكثيف[٦٢٦] نظيره[٦٢٧] المرّة السّوداء[٦٢٨]، وشيء آخر رقيق يضرب إلى الحمرة من جنس المرّة الحمراء وشيء أبيض من جنس البلغم، ولذلك صار الدّم مختلف الحالات في لونه وفي قوامه بحسب اختلاف الأسنان وبحسب[٦٢٩] اختلاف أوقات السّنة وحسب اختلاف مزاج الحيوان وحسب اختلاف مزاج البدن، وأمّا من[٦٣٠] استفراغ ما يستفرغ من البدن، فإنّهم قالوا إنّا نرى الدّواء المسهل إذا أخذه إنسان فأفرط بصاحبه الإسهال حتّى

---

٦١٣ DSY: – من | ٦١٤ DFMY: + إنّه | ٦١٥ AM: كان؛ DFY: كان كلّ | ٦١٦ DY: بارداً رطباً؛ RU: باردة رطبة | ٦١٧ DY: بارداً يابساً؛ U: باردة يابسة | ٦١٨ DY: حارّاً يابساً؛ U: حارّة يابسة | ٦١٩ S: القلب؛ U: العلّة | ٦٢٠ Y: حارّاً رطباً؛ U: حارّ وطب < رطبة | ٦٢١ FMSU: الدّم | ٦٢٢ F: صفة؛ M: وهي الصّفة؛ U: – صفة | ٦٢٣ S: دسمي؛ Y: وبعضه زبد > ودسمي | ٦٢٤ MS: وكذلك؛ R: فكذلك | ٦٢٥ R: + المرار السّوداويّ من جنس | ٦٢٦ A: اللّطيف [؟] الكثيف؛ DY: اللّطيف؛ FMU: اللّيف | ٦٢٧ ADMUY: نظير | ٦٢٨ R: – والكثيف. . . السّوداء | ٦٢٩ DRSUY: وحسب | ٦٣٠ MS: – من

purgative acts on the person who has taken it until he has excreted all of that humor in his body to which the particular drug is specific. After that, it expels another humor that that drug is not intended to purge, and he dies.[84] Since that is the case, it is clear that the body needs all four humors to remain in health.

(45) The color of the blood[85] is known from the color of the outside of the body and by its appearance during bloodletting.

<*The rest of the drugs*>[86]

(46) If a purgative drug is administered and its purgative effect is excessive, the first thing to be expelled by the purgation will be the humor for which that purgative is specific. After that, the finest and most responsive of the remaining humors will be expelled, followed by the grosser and more difficult humors. Finally, blood, the humor most closely associated with nature, will be expelled. For example, if we were to give a drug that purges yellow bile, the first thing that it would expel would be yellow bile. After that, it would expel phlegm, then black bile, and finally it would expel blood. If we were to administer a drug that purges phlegm, the first thing that would be expelled would be phlegm, then yellow bile, then black bile, and finally blood. If we were to administer a drug that purges black bile, the first thing it would expel would be black bile; then after that it would expel yellow bile, then phlegm, and then finally it would expel blood. Each of the purgative drugs first attracts the humor that it specifically purges from the cavities of the

84. A gloss in one MS reads: "That drug will not cause the excretion of the other humor until all of the humor for which that drug is specific has been excreted."

85. L 11.15–16; K 1:497; G^a 122–23.

86. L 13.12–15; K 1:504; G^a 133; Hippocrates, *On the Nature of Man* 6. This account of the effects of excessive purgation comes from Hippocrates.

يستفرغ ما في بدنه من الخلط الذي ذلك الدّواء مخصوص بإسهاله ويخرج بعده[٦٣١] خلط آخر ليس من شأن ذلك الدّواء إسهاله مات[٦٣٢]، وإذا كان الأمر على هذا فقد تبيّن أنّ البدن يحتاج في البقاء على السّلامة إلى[٦٣٣] الأخلاط الأربعة.

(٤٥) لون الدّم يعرف من لون البشرة من خارج ومن رؤيته إذا استفرغ بفصد العرق.

‹باقي الأدوية[٦٣٤]›

(٤٦) إذا سقي الدّواء المسهل، فأفرط عمله في الإسهال، كان[٦٣٥] أوّل شيء يخرج بالإسهال الخلط الذي ذلك الدّواء مخصوص بإسهاله، ثمّ يخرج بعده أرقّ الأخلاط الباقية وأسهلها إجابةً، ثمّ يتبع ذلك أغلظ الأخلاط وأشدّها عسراً، وفي آخر الأمر يخرج بعدها كلّها أخصّ الأخلاط بالطبيعة، وهو الدّم. مثال ذلك أنّا إذا[٦٣٦] سقينا دواءً يسهل المرّة الصّفراء كان أوّل شيء يخرجه مرّة صّفراء، ثمّ يخرج بعدها بلغم، ثمّ مرّة سوداء، وفي آخر الأمر يخرج الدّم[٦٣٧]، وإن سقينا دواءً يسهل البلغم كان أوّل شيء يخرج البلغم، ثمّ يخرج بعده[٦٣٨] مرّة صفراء، ثمّ مرّة سوداء، ثمّ في[٦٣٩] آخر الأمر يخرج الدّم، وإن سقينا دواءً يسهل المرّة السّوداء كان أوّل شيء يخرجه[٦٤٠] مرّة سوداء، ثمّ يخرج بعدها[٦٤١] صفراء ثمّ بلغم[٦٤٢]، وفي آخر الأمر يخرج الدّم[٦٤٣]، وكلّ واحد من الأدوية المسهلة يجتذب في أوّل الأمر الخلط الذي هو

---

٦٣١ R: بعده > بعد ذلك ما في بدنه من | ٦٣٢ A: + ذلك الإنسان؛ R: + الإنسان؛ R^2: + ممّا به الموت؛ S^1: وليس يستفرغ هذا الدّواء لهذا الخلط الآخر إلاّ بعد أن قد استفرغ شأن الخلط الذي هو مخصوص باستفراغه. | ٦٣٣ DR: + جملة | ٦٣٤ R^2 | ٦٣٥ S: وكان؛ DY: فكان | ٦٣٦ ADFUY: إن | ٦٣٧ DMSY: دماً؛ F: دم؛ U: + الذي هو أخصّ الأخلاط بالطّبيعة ولنا | ٦٣٨ R: – بعده؛ U: – يخرج بعده | ٦٣٩ DMSUY: وفي | ٦٤٠ DMSY:: يخرج | ٦٤١ ADMSY: + مرّة | ٦٤٢ DMSY: بلغماً | ٦٤٣ A^1FR؛ AU: – وإن سقينا. . . الدم؛ DMSY: دماً

hollow organs. After it has cleansed all of that humor located there, it will harshly attract the humor specific to it from the substance of the original organs. It is the violent harshness of this attraction that extracts the other humors that it is not specific to from the organs, along with the humor that it is specific to. However, it will first attract and extract from the organs the other humor that is easiest to lead out, is least resistant and quickest to respond to it, and is least natural there. After that, it will expel what is less responsive and more natural there, and in the end it will expel what is most natural, which is blood.

<The division of the elements>[87]

(47) There are two kinds of elements. Some have qualities that remain in their natural state, and others have qualities that diverge from their natural state. The elements that are in their natural state are those that are existent in nature—such as fire, by which there is life, and elemental water, on which life is also based. The elements whose qualities diverge from the natural state are those in which the qualities peculiar to them have become excessive and thereby destroy life—for example, ice, in which the coldness has become excessive, and burning fire, in which heat has become excessive. The same is true of the humors, for some of them are elemental and natural. These humors

87. This paragraph consists of various observations about the elements and humors, most not closely connected to Galen's text.

مخصوص بإسهاله من تجويفات الأعضاء المجوّفة، ثمّ إنّه بعد ذلك إذا استنظف ما هناك من ذلك الخلط اجتذب ما في جوهر الأعضاء الأصليّة من ذلك الخلط الذي هو مخصوص[٦٤٤] به اجتذابًا عنيفًا، ولشدّة اجتذابه وعنفه ينتزع[٦٤٥] منها مع الخلط الذي هو مخصوص به الأخلاط التي ليس هو مخصوص بها إلاّ أنّ أوّل شيء يجتذبه وينتزعه منها أسهلها[٦٤٦] انقيادًا[٦٤٧] وأكثرها مؤآتاةً[٦٤٨] وأسرعها إجابةً وأقلّها خصوصيّةً[٦٤٩] بالطّبيعة، وبعده أعسرها إجابةً وأبعدها عن القرب من الطّبيعة، وفي آخر الأمر أخصّها بالطّبيعة، وهو الدّم.

‹تقسيم العناصر[٦٥٠]›

(٤٧) العناصر صنفانِ، فمنها عناصر كيفيّاتها باقية على الحال الطّبيعيّة، ومنها عناصر كيفيّاتها خارجة عن الحال الطّبيعيّة، أمّا[٦٥١] التي كيفيّاتها[٦٥٢] على حال[٦٥٣] الطّبيعيّة[٦٥٤]، فهي العناصر التي هي الموجودة في الطّبع بمنزلة النّار التي تكون بها[٦٥٥] الحياة والماء[٦٥٦] العنصريّ[٦٥٧] الذي هو أيضاً ممّا تكون به الحياة، وأمّا التي كيفيّاتها خارجة عن الحال الطّبيعيّة، فهي العناصر التي قد أفرطت فيها الكيفيّات التي هي مخصوصة بها، فهي لذلك مفسدة للحياة بمنزلة الجمد الذي قد أفرطت فيه البرودة ولهيب النّار الذي قد أفرطت فيه الحرارة، وكذلك[٦٥٨] الأمر في الأخلاط، فمنها عنصريّة طبيعيّة،

---

٦٤٤ DMY: الذي هو مخصوص > المخصوص | ٦٤٥ A: يجتذب؛ DY: ينزع | ٦٤٦ S: أساسها | ٦٤٧ M: انقياد؛ DRY: قيادًا؛ D^1Y^1: جذبًا | ٦٤٨ M: مواتاها؛ D^1Y^1: مساعدة | ٦٤٩ DF: خصوصة | ٦٥٠ R^2 | ٦٥١ A: فأمّا؛ DMY: + العناصر | ٦٥٢ DMY: + باقية | ٦٥٣ AY: − الحال | ٦٥٤ F^1؛ F: − أمّا. . . الحال؛ S: − ومنها عناصر. . . الطّبيعيّة | ٦٥٥ S: تكوّن منها | ٦٥٦ A^1DFSY: وأمّا | ٦٥٧ DSY: العنصر | ٦٥٨ A: مثال ذلك؛ A^2: وكذلك؛ D: ولذلك

are moderate, sweet to the taste, and nourishing to bodies, like those of them that are mixed in the blood. That is because a nutriment is sweet insofar as all of its parts are sweet to the taste. Those humors that are mixed in the blood we call yellow bile, black bile, and phlegm. Others of them are unnatural. These are the excesses of those qualities conformable to nature. Whatever is like this does not nourish but, rather, is repelled and expelled from the body as something to which the body is averse. When it is retained, it harms and corrupts the body. Whatever is like this[88] is never sweet. Yellow bile of this sort is bitter, black bile is sour and astringent, and phlegm is sour or salty.

(48) If blood or semen is deposited in the womb, nature separates it.[89] That portion of it that is coarse, cold, dry, and inclining toward black bile becomes matter for the generation of the bones. That portion of it that is cold, moist, and phlegmatic becomes matter for the generation of the brain. That portion of it that is hotter and more moist becomes matter for the generation of flesh. That portion of it that is hottest and driest becomes matter for the generation of the lungs.

88. Some MSS read: "Whatever is of this genus."

89. L 11.6–7; K 1:495; G[a] 120–21: Galen is speculating about the means by which the embryo is formed.

وهي معتدلة حلوة المذاق تغذو الأبدان بمنزلة ما هو مخالط منها[٦٥٩] للدّم[٦٦٠]، وذلك لأنّ كلّ غذاء فهو حلوٌ من قِبَل أنّ جميع الأعضاء حلوة الطّعم. وهذه الأخلاط المخالطة للدّم نحن نسمّيها مرّةً صفراء وسوداء[٦٦١] وبلغم[٦٦٢]، ومنها خارجة عن الطّبع، وهي إفراطات تلك الكيفيّات الملائمة للطّبع، وما كان كذلك فليس يغذو، بل يدفع ويخرج عن البدن بمنزلة الشّيء المنافر، ومتى احتبس أضرّ بالبدن وأفسده. وليس من هذا[٦٦٣] شيء له حلاوة، بل المرّة الصّفراء من هذا الجنس مرّة والسّوداء حامضة عفصة، والبلغم حامض أو مالح.

(٤٨) وإذا[٦٦٤] وقع الدّم في الأرحام أو المنى[٦٦٥] ميّزته الطّبيعة فصار ما هو منه غليظ بارد يابس مائل[٦٦٦] إلى السّوداء مادّةً لكون العظام وما هو منه بارد رطب بلغميّ مادّةً لكون الدّماغ وما هو منه أشدّ حرارةً و رطوبةً مادّةً لكون اللّحم وما هو منه أشدّ حرارةً ويبساً مادّةً لكون الرّئة.

---

٦٥٩ D: – منها؛ R: لها؛ SY: فيها | ٦٦٠ RU: الدّم | ٦٦١ AU: ومرّة سوداء؛ F: أوسوداء | ٦٦٢ DRY: وبلغماً | ٦٦٣ DMSY: + الجنس | ٦٦٤ MR: إذا | ٦٦٥ DUY: أو المنى < والمنى | ٦٦٦ A: غليظاً بارداً يابس مائل؛ DY: غليظ بارداً يابساً مائلا

[Chapters 12 and 13][90]

<Their disagreement about drugs>

(49) Opinions[91] differed about purgative drugs. In general, there were two opinions. One was the opinion of Hippocrates, who believed that every purgative drug attracted the humor to which it was specific. The other was the view of Asclepiades, who thought that each purgative drug generated the humor that it purged.[92] This second opinion is open to refutation in two ways. First, we find that it is difficult to treat someone phlegmatic with drugs that purge yellow bile, and they may actually harm him, whereas he can easily be treated with drugs that purge phlegm, and these are beneficial to him. On the other hand, it is difficult to treat the one who is dominated by yellow bile with drugs that purge phlegm, and these may actually harm him, whereas it is easy to treat him with drugs that purge yellow bile, and these are beneficial to him. Second, we find that, after a purgative drug has cleansed the humor to which it is specific, it then expels another humor to which it is not specific. This implies one of two things: first, the potency of the

---

90. L 12.1–8; K 1:499–500; G^{a} 125–28: Galen attributes to Asclepiades the theory that a purgative drug does not draw out the humor but rather converts what is in the body into something else, thus neutralizing that humor. By this account, all evacuations would be the same.

91. Some MSS read: "The Ancients."

92. A gloss in one MS reads: "This man [Asclepiades] believed that the purgative drug for yellow bile converts whatever humor is existent in the body into yellow bile and then purges it. Since scammony is hot and dry, it converts the humor existing in the body, thus making the body hot and dry. This results if what is excreted is yellow bile. If this were the case, then, after a suitable drug caused something to be purged, we would find relief, because two humors would never be in our bodies to cause us to suffer pain." Cf. Galen, *On the Natural Faculties* 1.13, where the subject is discussed at greater length.

[الفصلان الثّاني عشر والثّالث عشر]

‹اختلافهم في الدّواء[٦٦٧]›

(٤٩) قد اختلف[٦٦٨] الآراء[٦٦٩] في الدّواء المسهل، ومحصولها[٦٧٠] رأيانِ، أحدهما رأي إبقراط[٦٧١] الذي يعتقد أنّ كلّ واحد من الأدوية المسهلة إنّما يجتذب الخلط الذي هو مخصوص به، والآخر رأي أسقليبياذس[٦٧٢] الذي يظنّ أنّ كلّ واحد من الأدوية المسهلة إنّما يولّد الخلط الذي يسهله[٦٧٣]، وهذا الرأي ينفسخ من وجهينِ، أحدهما أنّا نجد من كان الغالب عليه البلغم يعسر إسهاله بالأدوية التي تسهل المرّة[٦٧٤] الصّفراء وضرّت به[٦٧٥] ويسهل الأمر في إسهاله بالأدوية التي تُسهِل البلغم وينتفع بها، ومن كان الغالب عليه المرّة[٦٧٦] الصفراء عسُر[٦٧٧] إسهاله بالأدوية التي تسهل البلغم وتضرّ به[٦٧٨]، ويسهل الأمر في إسهاله بالأدوية المسهلة للصّفراء[٦٧٩] وتنفعه[٦٨٠]، والوجه الآخر أنّا نجد الدّواء المسهل إذا استنظف الخلط الذي هو مخصوص به أخرج بعده خلطًا[٦٨١] آخر ممّا ليس هو مخصوص به، ولا بدّ في ذلك من أحد أمرينِ، إمّا أن تكون

---

٦٦٧ R^2 | ٦٦٨ R: قد اختلف > واختلاف | ٦٦٩ DMSY: الأوائل | ٦٧٠ FS: وحصولها؛ U: فحصولها | ٦٧١ DMSUY: بقراط | ٦٧٢ S^h: هذا يعتقد أنّ الدّواء المسهل للصّفراء يقلّب هذا الخلط الموجود في البدن، فيجعله مرّة صفراء ثمّ يستفرغه، وذاك أنّه لمّا كان السّقمونيا حارًّا يابسًا قلّب الخلط الموجود في البدن، فجعله حارًّا يابسًا، فيعرض من هذا أن كان ما استفرغ مرّة صفراء؛ U: أسيقليبيلاس | ٦٧٣ S^h: ولو كان الأمر على هذا كنّا لما كنّا بعد استفراغ بالدّواء الموافق نجد خفا [؟] لأنّه لم يكن في أبداننا خلطانِ أبدًا يؤلمنا. | ٦٧٤ DFMSY: – المرّة | ٦٧٥ DMY: أو تضرّ به؛: وضرّته؛ R: وضرّت به؛ S: وتضرّ به | ٦٧٦ DFMSY: – المرّة | ٦٧٧ RU: يعسر؛ DMS: + عليه | ٦٧٨ تضرّ به > A: وضرّته؛ D: ضرّ به | ٦٧٩ ADY: الصّفراء؛ R: للمرّة الصّفراء | ٦٨٠ D: ينفعه؛ F: ونفعته | ٦٨١ ASU: خلط

drug might decline, weakening until it disappears—in which case it would very likely not be able to expel anything, since it would be unable, in this circumstance, to generate anything, its potency being lost. The other possibility is that its potency continues. If it continues, the purgative must continue to expel that which it had been expelling since the beginning.[93] You cannot say that it was hot in the beginning and so generated yellow bile and, after that, changed back and so began to generate phlegm. If it did, its temperament would be entirely transformed, and it would cease to be purgative, since we do not find a cold purgative drug. However, it is utterly unacceptable for you to say that the purgative drug acquires differing temperaments in the body, each of which is opposite to its counterpart in this degree—in such a way that, at one time, it becomes hot and dry while it is purging yellow bile and, at another time, becomes cold and wet when it is purging phlegm, then becomes cold and dry when it is purging black bile, eventually becoming once more hot and moist when it is purging blood.[94]

93. A gloss in one MS reads: "One must know that the drug that purges a humor does not cause it all to be excreted, for the humors are continually in the body and merely increase and decrease there. The drug attracts due to the correspondence between it and the humor."

94. A gloss in **S**: "This refutation is directed against the opinion of Asclepiades that the patient finds relief in excretion in general. Were this true, when we treated someone with a drug that purges yellow bile, it would be harmful, not helpful, since a harmful humor remains as it was."

قوّة الدّواء قد[٦٨٢] خارت وضعُفت وبطلت فلا ينبغي له[٦٨٣] حينئذٍ أن يخرج شيئاً[٦٨٤] إذ كان لا يقدر عند هذه الحال أن يولّد وقوّته قد بطلت، وإمّا أن تكون قوّته باقية، وإن[٦٨٥] كانت[٦٨٦] باقية فينبغي[٦٨٧] أن يخرج ما لم يزل يخرجه منذ أوّل الأمر، فإنّه ليس يجوز لك[٦٨٨] أن تقول إنّه كان أوّلاً حارّاً، فكان يولّد الصّفراء، وبعد ذلك يردّ فصار يولّد البلغم لأنّه إن كان مزاجه قد استحال هذه الاستحالة كلّها، فقد بطل من أن يكون مسهلاً لأنّا ليس نجد دواءً مسهلاً بارداً، مع أنّ هذا شنع[٦٨٩] منكر جدّاً أن نقول إنّ الدّواء المسهل يكتسب في البدن أمزاجاً[٦٩٠] كلّ واحد منهما[٦٩١] من الخلاف[٦٩٢] لصاحبه في هذا الحدّ، فيصير مرّةً حارّاً يابساً عند إسهاله للصّفراء ومرّةً بارداً رطباً عند إسهاله للبلغم[٦٩٣]، ثمّ يصير بارداً يابساً عندما يسهل المرّة[٦٩٤] السّوداء، ويرجع في آخر الأمر فيصير[٦٩٥] حارّاً رطباً عندما يسهل الدّم[٦٩٦].

٦٨٢ DMSY: – قد | ٦٨٣ DSY: – له | ٦٨٤ DSY: شيء | ٦٨٥ F: فإن؛ S: فإذا | ٦٨٦ R: + قوّته | ٦٨٧ Sʰ: ينبغي أن يعلم أنّ الدّواء المسهل للخلط لا يستفرغه بأسره، فإنّ الأخلاط لا تخلو [؟] منها بدن دائماً، وإنّما يزيد وينقص فيه وجذب الدّواء بالمناسبة التي بينه وبين الأخلاط. | ٦٨٨ A¹ ؛ AR: ذلك؛ S: يجوز لك > ذلك | ٦٨٩ F: ممتع؛ M: أشنع | ٦٩٠ ADFM: أمزاج | ٦٩١ ADFSY: منها | ٦٩٢ A¹R: من الخلاف > مخالف | ٦٩٣ AM: البلغم | ٦٩٤ DFMSY: – المرّة | ٦٩٥ DMY: إلى أن يصير؛ S: يصير | ٦٩٦ Sʰ: هذا الرّدّ من جملة قول أسقليبياذس إنّ الاستفراغ على الإطلاق يجد الإنسان له خفّة ، ولوكان الأمر على هذا لما كنّا إذا أعطينا المستسقين دواء يسهل المرّة الصّفراء يستضرّ ولا ينتفع به لأنّ الخلط المؤذي يبقى على حاله.

[Chapter 14][95]

<The location of the humors>

(50) The humors in the body are in two locations. Some of them are contained in the cavities of the blood vessels. These are the first to be attracted by purgative drugs, which happens easily and quickly. Others are in the very substances of the primary organs. The purgative drugs attract these humors once they have cleansed those that are in the vessels. The strength and severity of the attraction draws not only the humor for which the purgative drug is specific, but also another humor to which the drug is not specific; and this too is purged.[96]

---

95. L 14.5–6; K 1:508; G[a] 139–40: Galen mentions that excessive purgation draws the humors out of the substances of the organs, thus destroying the body.

96. A gloss in one MS reads: "No one should say that there is only one humor; rather, the organs differ with respect to the difference of the effect of heat on it. Thus, in the part of the humor in which the heat acts more strongly, ripening and hardening, bone results. Flesh results from what is less than that. This is why bodies differ. We say that the difference of the effect of heat in a single thing is not due to the effect of diverse substances. Those who think that there is only one humor think that the cause of our differing complexions is that we are always in an unnatural state. Galen refuted their view and asked, in shock, 'Do you think that there is any time when it is in a natural state? Why is it not in a natural state all the time? That which is manifest in us is only blood.' It is due to these difficulties that Galen said that the theory of humors is convincing but not known inductively."

[الفصل الرّابع عشر]

‹موضع الأخلاط[٦٩٧]›

(٥٠) الأخلاط من البدن في موضعيْن، فبعضها في تجويفات العروق محصورة، وهذه هي أوّل شيء يجتذبه الأدوية المسهلة بسهولةٍ وسرعةٍ، وبعضها في نفس جواهر الأعضاء الأصليّة، وإذا[٦٩٨] اجتذبت الأدوية المسهلة هذه الأخلاط عندما يستنظف ما في العروق منها جذبت[٦٩٩] بشدّة اجتذابها وعنفها مع الخلط الذي[٧٠٠] الدّواء المسهل مخصوص بإسهاله[٧٠١] خلطاً[٧٠٢] آخر، ليس[٧٠٣] ذلك الدواء مخصوص به، فأسهله[٧٠٤].

---

٦٩٧ R^2 | ٦٩٨ DMSY: فإذا؛ F: إذا | ٦٩٩ ADFMY: جرت؛ U: واجبذبت | ٧٠٠ AM: + ذلك | ٧٠١ DFMSY: به | ٧٠٢ ADSY: خلط | ٧٠٣ A: ليست؛ S: وليس | ٧٠٤ AFU: وأسهلته؛ D: فأسهلته؛ R: به فأسهله > بإسهاله؛ || S^h: ليس لأحد أن يقول إنّ الخلط واحد، وإنّما يختلف الأعضاء منه لأجل اختلاف فعل الحرارة عنه، فما كان قد فعلت فيه الحرارة أكثر وأنضجته وصلبته يكون منها العظم، وما كان دون ذلك يكون منها اللّحم، ولهذا يختلف الأبدان، ونحن نقول إنّ اختلاف فعل الحرارة في الشّيء الواحد لا بعمل جواهر مختلفة، وأصحاب الخلط الواحد كانوا يعطون العلّة في اختلاف ألواننا كوننا أبداً على حال غير طبيعيّة، وجالينوس يردّ قولهم ويقول متعجّباً، أترى ليس يوجد زمان يكون فيه على حال طبيعيّة، ولِمَ لا يكون في جميع هذا الزّمان على حال طبيعيّة، ويكون الظّاهر فينا الدّم حسب، ولهذه الشّكوك قال جالينوس إنّ البيان على الأخلاط اقتناعيّ وغير الاستقراء.

The end of the Alexandrian epitome of Galen's book

On the Elements

According to the Opinion of Hippocrates

in the form of commentary and summary,

translated by

Ḥunayn ibn Isḥāq[97]

97. The colophons differ considerably, the translation being a composite of the significant elements of several.

Added in **M**: Written by Sallām ibn Ṣāliḥ ibn Khiḍr ibn Ibrāhīm, known as the teacher of Shafar^cām, on the fifth of Tishrīn II, in the year of our Father Adam 6749 [5 November 1240].

تمت جوامع^٧٠٥ كتاب^٧٠٦ جالينوس في العناصر^٧٠٧

ترجمة^٧٠٨ حنين بن إسحق^٧٠٩

---

٧٠٥ AM: تمّت جوامع > تمّ | ٧٠٦ F: الإسكندوانيين لكتاب | ٧٠٧ A: + على رأي إبقراط ممّا عنى بجمعه الإسكندرانين؛ F: + علي الشّرح والتّلخيص؛ في الهامش بخطّ آخر: بلغ مقابلة؛ R: الإسطقسات على رأي إبقراط للإسكندرانيّين | ٧٠٨ M: نقل | ٧٠٩ A: + والحمد لله كثيراً؛ DY: + رحمه الله؛ F: – ترجمة. . . إسحق؛ M: + جوامع الإسكندرانيّين؛ S: + والحمد لله حقّ حمده || M: + وكتب سلام بن صالح بن خضر بن إبرهيم المعروف بمعلّم شفرعام في خامس يوم من تشرين الآخرسنة ستّ ألف وسبع مائة وتسعة وأربعون لأبونا آدم عليه رحم الله من دعا [كذا] لكاتبه بالمغفرة آمين ـهـ || R: + والحمد لله وصلّى على النّبيّ محمّد وعلى آله وسلّم.

قوبل به الأمّ وصحّ

والحمد لله وحمده.

R³: عورض به نسخة أخرى وصحّ حسب الطّاقة والحمد لله. || U: + الخيريّ ولواهب العقل ومعطي الخير وكاشف الهمّ دائماً بلا نهاية.

Appendix 1: Greek and Islamicate Physicians

Few of the ancient and medieval physicians and philosophers referred to in this book are household names. As in the underlying works of Galen, the epitomes mention various ancient physicians and philosophers by name; Galen, indeed, is our most important source of information for the doctrines of Hellenistic physicians. The individuals mentioned in connection with the composition of the epitomes are even more obscure. The following are the ancient physicians mentioned in the text of the epitome of *On the Sects*, including a few names given only in certain manuscripts or mentioned elsewhere in the text. I have also included those mentioned in historical sources in connection with the composition of the epitomes and a few Islamicate physicians (most of whom are actually Christians) who appear in the historical documentation of the epitomes or in the manuscripts. In most cases, I have given references only to standard reference sources, notably Pauly-Wissowa and its recent updates *(RE, BNP)*, *The Complete Dictionary of Scientific Biography (CDSB)*, *Dictionnaire des philosophes antiques (DPA)*, *The Encyclopaedia of Islam (EI)*, and *Geschichte des arabischen Schriftums (GAS)*. Readers needing access to primary sources can easily trace them through these references.

Acron of Agrigentum (fifth century BCE). Physician, contemporary and fellow-townsman of Empedocles. Later Empiricists traced the origin of their school to him. He was known to Islamicate physicians through quotations as an authority on dietetics and as the first in the succession of physicians between Parmenides and Plato the Physician. *BNP* 1:113; *DPA* 1:50–51; *GAS* 3:22; *RE* 1:1199.

Agnellus of Ravenna (sixth century CE). Obscure medical teacher and writer to whom a Latin commentary on Galen's *On the Medical Sects* is attributed, as well as commentaries on *The Pulse for Teuthras* and *Therapeutics for Glaucon*. *BNP* 1:345; see p. xl and n. 28 above.

Akīlāʾus. A variant spelling in one source for Archelaus. See above, p. xl, and below, s.v. "Archelaus."

Alcmaeon of Croton (fl. early fifth century BCE). Mentioned by Galen in *On the Elements* as the author of a work entitled *On Nature*. He had a theory of opposites similar to that of the Pythagoreans and was interested in natural phenomena, medicine, and the functioning of the senses, though it is not clear whether he was a physician himself. Except for references in Aetius's doxography, he was unknown to the Islamic world. *BNP* 1:454–55; *CDSB* 1:103–4; DK 24; *DPA* 1:116–17; KR 232–35; *RE* 2:1556.

ʿAlī ibn Riḍwān (998–1061). A distinguished but self-taught and argumentative physician in Fatimid Egypt; his famous dispute with Ibn Buṭlān concerned the merits of learning from books without a teacher. He followed Galen closely and had an impressive knowledge of ancient medicine and science. Several of his works, notably a commentary on *The Small Art*, were translated into Latin, where his name was given as "Haly Abenrudian." *CDSB* 11:444–45; *EI*, s.v. "Ibn Riḍwān."

Anaxagoras of Clazomenae (500–428 BCE). Presocratic philosopher and one of the Ionian physicists. He lived and taught in Athens from 461 to 431, until he was tried for impiety and exiled. He taught a theory that accepted Parmenides's denial of material change by claiming that the material universe consists of an infinite number of tiny particles of each of the simple kinds of things—"homoeomerous seeds." He was known to the Islamic world through references in doxographical sources. *BNP* 1:656–57; *CDSB* 1:149–50; DK 59; *DPA* 1:183–87, supp. 751–59; KR 362–94; *RE* 2:2076–77, supp. 1:78, supp. 12:28–30.

Anaximenes of Miletus (sixth century BCE). Presocratic philosopher and Ionian physicist who held that the primal element was air, identified with soul, which condensed in stages to wind, clouds, water, earth, and stone. He was said to have been a student of Anaximander or Parmenides. He was known to the Islamic world through doxographies and is mentioned as a teacher of Pythagoras. *BNP* 1:661–62; DK 13; *DPA* 1:192–93, supp. 761–65; KR 143–62; *RE* 2:2086–98, supp. 1:78, supp. 12:69–71.

Angeleuas (before 600). An otherwise unknown physician quoted on the anatomy of the bladder by Stephanus in his commentary on Galen's *Therapeutics for Glaucon.* He was perhaps the Anqīlāʾus of the Arabic sources or, less plausibly, Agnellus of Ravenna. See p. xl above.

Anqīlāʾus (ca. sixth century). Mentioned by all the sources as one of the compilers of the Alexandrian epitomes, and by three as the leader of the group. His name is also given as Nīqālāʾus (Nicholas). Ibn al-Qifṭī gives a biography of him that contains little that could not have been inferred from the epitomes. See pp. xxxix–xl and s.v. "Angeleuas" above.

Apollonius. Three Empiricist physicians. Though the epitome of *On the Sects* refers to "Apollonius," the pseudo-Galenic *Introduction to Medicine* refers to "the two Apollonioi," by which is certainly meant Apollonius of Antioch, known as "the Empiricist," and his son Apollonius Byblas, "the bookworm," both of the second century BCE. The former carried on a dispute with the Herophilean physician Zeno on "characters." *DPA* 1:282–84; *RE* 3:149, no. 101. There was also a first-century-BCE Empiricist, Apollonius of Citium, whose book on joints survives and whose Empiricism is attested by his contempt for Herophilian anatomy. *BNP* 1:881–82; *RE* 3:149, no. 102.

Archelaus (ca. sixth century). One of the Alexandrian epitomists, whose name is given as "Arkīlāʾus" or "Akīlāʾus." He cannot be identified with certainty but may be the author of an extant Greek commentary on Galen's *On the Sects* and/or a work on urine quoted in Islamic sources. Ullman, *Medizin*, 83.

Aristotle (384–322 BCE). Greek philosopher, student of Plato, and major influence on the later medical and scientific tradition, particularly through his theory of the four primal qualities and four material elements.

Arkīlāʾus (ca. sixth century). One of the Alexandrian epitomists. Despite suggestions that this name is a duplication of Anqīlāʾus or is derived from the name of the Italian city of Aquilea, this clearly represents "Archelaus." See "Archelaus" and pp. xxxix–xl above.

Asclepiades of Bithynia (first century BCE). Advocate of a theory of medicine based on the flow of corpuscles through pores. Though his theory was later adapted by the Methodists, he himself is more properly classified as a Rationalist. He was famous for his mild treatments using such means as diet, massage, light exercise, and bathing and was the first Greek physician to achieve major success in Rome. He was known to the Islamicate scholars through citations in doxographies and Galen. *BNP* 5:13–16; *CDSB* 4:382–86; *DPA* 1:624–25; *GAS* 3:55; *RE* 4:1633–50; Vallance, *Lost Theory of Asclepiades.*

Athenaeus of Attaleia (fl. first century). The founder of the Pneumatic school of medicine, which applied Stoic ideas about materialism and *pneuma* to medicine. His works are lost, but he is cited often by Galen and Oribasius. He was little known in the Islamic world. *BNP* 2:244–45; *CDSB* 1:324–25; *DPA* 1:643–44; *GAS* 3:56–57; *RE* 4:2034–36.

Chrysippus of Soli (ca. 280–205 BCE). Third leader of the Stoic school whose extensive and systematic writings—all lost apart from fragments—standardized Stoic thought and were probably responsible for the success of the school into the Roman imperial period. The Stoic and Chrysippean doctrine of the *pneuma*, an all-pervading spirit composed of fire and air, was his most important contribution to medical thought. He is frequently criticized by Galen. *BNP* 3:288–93; *CDSB* 20:122–23; *DPA* 2:329–65; *RE* 6:2502–509, supp. 12:148–55.

Democritus of Abdera (ca. 460–370 BCE). With his teacher Leucippus, the founder of ancient atomism. His student Nausiphanes was the teacher of Epicurus, whose atomism carried on Democritus's ideas. He held that the universe was composed only of indivisible atoms and void. Differences among physical objects are to be explained by the sizes, shapes, and arrangements of atoms. The Islamic world knew him as an Atomistic philosopher through the doxographies and as an alchemist and physician, with several books on each subject being attributed to him. *BNP* 4:267–69; *CDSB* 4:30–35; DK 68; *DPA* 2:649–716, supp. 765–73; *GAS* 3:23; KR 400–426; *RE* 9:135–40, supp. 12:191–223.

Diocles of Carystus (fl. ca. 300). A Rationalist physician and possibly a student or associate of Aristotle. He wrote widely and influentially on medicine, anatomy, and hygiene, integrating philosophical ideas into medicine. In Arabic sources he appears in lists of ancient physicians and doxographies. *BNP* 4:424–26; *CDSB* 4:105–7, 20:301–4; *DPA* 2:772–77; *GAS* 3:51. Ullman, *Medizin*, 69.

Diodorus Cronus (fl. ca. 300 BCE). Megarian philosopher, best known for devising the "Master Argument," dealing with the problem of future contingents. De Lacy's Greek edition of Galen's *On the Elements*, following Sālim's Arabic edition, mentions him with Leucippus as an Atomist, but the reading should certainly be Democritus. *BNP* 4:439–40; *DPA* 4:779–81; *RE* 9:705–707. See p. 140, n. 24, above.

Diogenes of Apollonia (fl. ca. 440–430 BCE). A very late Ionian physicist who, like Anaximenes, held that the primal element was air. The author of one or four books on nature, he is probably the Diogenes cited by Galen as the author of a systematic treatise on the causes and remedies of diseases. Arabic sources mention a *Kitāb al-aghdhiyah* (book of nutriments), as well as various philosophical positions derived from the doxographies. He is not to be confused with Diogenes of Sinope, the Cynic, who was also known to the Islamic world. *BNP* 4:448–49; DK 64; *DPA* 2:801–802; *GAS* 3:47–48; *RE* 9:765–76, supp. 12:233–36.

Dogmatic school. See Rationalist school.

Empedocles of Agrigentum (ca. 490–430 BCE). Sicilian philosopher who was the first known exponent of the four material elements of earth, air, fire, and water, which he combined with two active principles: love and strife. He was well known in the Islamic tradition as one of the five pillars *(asāṭīn)* of philosophy and appears both in biographical dictionaries and doxographical sources. *BNP* 4:943–47; *CDSB* 4:367–69, 20:395–98; DK 31; *DPA* 3:66–88; KR 320–61; *RE* 10:2507–12, supp. 12:141–48.

Empiricist school. A medical sect founded around 260 BCE by Philinus of Cos, a student of Herophilus. The Empiricists stressed the role of medical experience in treating disease and denigrated the role of scientific understanding of the causes of disease. Galen disapproved of their theories, though he was often more respectful of them in practice. Nutton, *Ancient Medicine,* 146–50; *RE* 10:2516–24. See pp. 23–24 above.

Epicurus of Samos (342–270 BCE). Though best known for his ethical hedonism, he is relevant here as an advocate of atomism and a medical theory based on it. Though his name appears in the doxographies, the Islamic world knew little of him. *BNP* 4:1071–84; *CDSB* 4:381–82; *DPA* 3:154–81; *RE* 11:133–55, supp. 11:579–652.

Erasistratus of Ceos (ca. 304–ca. 240 BCE). A physician in Cos and Alexandria, one of the most important physicians of antiquity. Among the few premodern physicians to have done human dissection, he also performed vivisection, though it is not clear whether on animals or condemned criminals. Galen praised him for his anatomical research but criticized many of his physiological and therapeutic conclusions. *BNP* 5:13–16; *CDSB* 4:382–86; *GAS* 3:53–54; *RE* 11:333–50, supp. 3:440.

Galen of Pergamon (129–ca. 216). The most important physician of antiquity. The son of a prosperous architect, Galen studied medicine in Pergamon, Smyrna, and Alexandria. He then practiced in Pergamon, Rome, and elsewhere, eventually becoming a court physician. In addition to his extensive writings on medicine, he also wrote on logic and philosophy; his philosophical sophistication is evident throughout his medical writings. He was an advocate of Hippocratean medicine at a time when Empiricists and Methodists were in the ascendant. Despite his belligerent attitudes towards rival schools, he seems also to have drawn on them extensively. His reputation grew steadily after his death until, in late antiquity, his works largely supplanted those of other medical schools—except, as was the case with Hippocrates, when they were recommended by Galen himself. He was very well known in the Islamic world, with long entries in the biographical dictionaries of physicians and philosophers. A large number of his works (and some spurious works) were translated into Arabic; many survive, including some now lost in Greek. *BNP* 5:654–61, supp. *Dict.* 275–80; *CDSB* 5:227–33, 21:91–96; *DPA* 3:440–466; *GAS* 3:68–150; *RE* 13:578–91.

Gesius (or Gessius) of Petra (d. ca. 520). Philosopher, physician, author of now-lost commentaries on the Alexandrian medical canon, and one of the Alexandrian epitomists. *BNP* 5:824–25; *DPA* 3:477–78; *GAS* 3:160–61; *RE* 13:1324. See p. xxxix above.

Gorgias of Leontini (fl. second half of fifth century BCE). A sophist and teacher of rhetoric in Athens, now best known as the eponym of a dialogue of Plato. Galen mentions his treatise entitled *On Being* or *On Nature*, in which he sought to prove that nothing existed and, that if something does exist, it is unknowable and, therefore, incommunicable. *BNP* 5:933–35; DK 82; *DPA* 3:486–91; *RE* 14:1598–619, supp. 4:710.

Heraclides of Erythrae (fl. late first century). A follower of Herophilus known mainly through citations in Galen. He composed at least seven books on Herophilean medicine and was one of the first commentators on Hippocrates's *Epidemics*. He is mentioned unambiguously in *The Small Art* and Agnellus's lectures on *The Medical Sects*.

The "Eraclitus" mentioned in John of Alexandria's list of prominent Empiricists might also be him but is more likely to be the distinguished early first-century Empiricist Heraclides of Tarentum. *BNP* 6:173–74; Galen, *The Small Art*, ed. Boudon-Millot, 393–94; *RE* 15:493–96.

Heraclitus of Ephesus (fl. ca. 500 BCE). A philosopher whose central theme was the role of logos, an all-encompassing reason that orders the universe. Though best known now for his paradoxical dicta, the doxographers were interested in him as an advocate of the view that fire, which he held was the physical counterpart of logos, was the primal element. He was not well known to the Islamic world, though he appears in doxographical materials. *BNP* 6:176–78; *CDSB* 6:289–91; DK 22; *DPA* 3:573–627; KR 182–215; *RE* 15:504–8, supp. 10:246–326.

Herophilus of Chalcedon (ca. 330–260 BCE). A student of Praxagoras who spent most of his career in Alexandria, where he perfected his knowledge of anatomy through human dissection. Apart from anatomy, he is known for a new classification of the subject matter of medicine, later adopted by Galen in *The Small Art*, that divides it into matters related to health, matters related to disease, and matters related to neither—the last including therapeutics, surgery, and diet. He was known to the Islamic world, if at all, through citations in Galen and the doxographies. *BNP* 6:274–76; *CDSB* 6:316–19; *GAS* 3:52–53; *RE* 15:1104–10, supp. 8:179; von Staden, *Herophilus*.

Hippasus of Metapontum (fifth century BCE). A Pythagorean philosopher who appears in the doxographies as an advocate of the view that fire is the primal element, though his actual interests, so far as they are known, seem to have been mainly mathematical. Ancient tradition claims that he was expelled from the Pythagorean school for revealing its secrets. He was unknown to the Islamic world apart from the doxographies' reference to fire. *BNP* 6:339–40; DK 18; *DPA* 3:753–55; *RE* 16:1687–88.

Hippocrates of Cos (ca. 460–377 BCE). The semimythical "Father of Medicine" and, by Galen's account, the founder of the Rationalist School of medicine. About seventy works attributed to him survive, many or most of which are clearly not authentic. Similar doubts attach to his ancient biographies. He is mentioned as a great physician by both Plato and Aristotle, but little more than that can be known with certainty. Much of the traditional view of Hippocrates is based on Galen's interpretations of his works. However, it is clear that Galen is projecting his own views back on Hippocrates, as can be seen by a comparison of Hippocrates's *The Nature of Man* with its supposed commentary, *On the Elements*. *BNP* 6:354–63; *CDSB* 6:418–31, 21:321–26; *DPA* 3:771–90; *GAS* 3:23–47; *RE* 16:1801–52, supp. 3:1154, supp. 6:1290–1345, supp. 12:486–96.

Hippon of Samos (fl. mid-fifth century BCE). A Pythagorean with medical interests, he is best known for holding that the primal element is water. He was led to this view by physiological considerations, thus making his notion of primal water quite different from Thales's. Aristotle dismisses him as unworthy of mention. He seems unknown in the Islamic world. *BNP* 6:372; DK 38; *DPA* 3:799–801; *RE* 16:1889.

Ḥunayn ibn Isḥāq al-ᶜIbādī (808–873). Probable translator of the Alexandrian epitomes and the most important medieval translator of Greek texts into Arabic. As an Iraqi Christian, he knew both Arabic and Syriac and is said to have learned Greek during two years of intensive study. His importance rests on the large number of Greek texts that he translated—directly into Arabic and Syriac, and into Arabic via Syriac—and the sophistication of his translation techniques in comparison to his predecessors. He was particularly interested in medicine, and his auto-bibliography of his translations of Galen survives. He was assisted in his translation work by his son Isḥāq and his nephew Ḥubaysh ibn Ḥasan. Bergsträsser, *Neue Materialen*; Ḥunayn ibn Isḥāq, "Risālah"; *CDSB* 7:24–26, 15:230–49; *EI*, s.v. "Ḥunayn b. Isḥāq"; *GAS* 3:247–56; Meyerhof, "New Light."

Ibn Abī Uṣaybiᶜah (d. 1270). Medical biographer and bibliographer known for his *ᶜUyūn al-anbāʾ fī ṭabaqāt al-aṭibbāʾ*, the largest and most important medieval biographical dictionary of Greek, eastern Christian, and Muslim physicians. *EI*, s.v. "Ibn Abī Uṣaybiᶜa."

Ibn Buṭlān, al-Mukhtār ibn al-Ḥasan (d. 1066). Christian physician and theologian of Baghdad. A leading student of Abū'l-Faraj ibn al-Ṭayyib, he was a learned, well-traveled, but somewhat difficult man. He is best known now for his controversy with the Egyptian physician ᶜAlī ibn Riḍwān and his *Taqwīm al-ṣiḥḥah*, a manual of medicine in the form of tables that was translated into Latin. *EI*, s.v. "Ibn Buṭlān"; *CDSB* 2:619–20; Schacht and Meyerhof, *Medico-Philosophical Controversy*.

Ibn Juljul (944–after 994). Arab physician of Cordova and author of an early Arabic history of medicine, *Ṭabaqāt al-aṭibbāʾ wa-al-ḥukamāʾ*. *EI*, s.v. "Ibn Djuldjul"; *GAS* 3:309–10.

Ibn al-Nadīm (c. 995 or 998). A bookseller in Baghdad who wrote the *Kitāb al-fihrist*, a catalog of all books in Arabic known to the author. The book is the most important source of information on works translated from Greek into Arabic. *EI*, s.v. "Ibn al-Nadīm."

Ibn al-Qifṭī (1172–1248). Egyptian scholar in Aleppo and author of a biographical dictionary of philosophers and physicians, the *Kitāb ikhbār al-ʿulamāʾ fī akhbār al-ḥukamāʾ*, usually known as *Tārīkh al-ḥukamāʾ* (History of the sages). *EI*, s.v. "Ibn al-Ḳifṭī."

Ibn Rushd / Averroës (1126–98). Philosopher and legal scholar best known for his commentaries on Aristotle. He also wrote a medical textbook, *al-Kullīyāt*, and summaries (*talkīṣ*) of Galen's works, a number of which survive, though they were better known in Latin and Hebrew in Europe than in the Islamic world. *CDSB* 12:1–9; *EI*, s.v. "Ibn Rushd."

Ibn Sīnā / Avicenna (980–1037). Iranian philosopher and physician. His *al-Qānūn fī al-ṭibb* (Canon of medicine) synthesized Galenic medicine and then largely supplanted Galen's own works in the Islamic world. *CDSB* 15:494–501; *EI*, s.v. "Ibn Sīnā"; *Encyclopaedia Iranica*, s.v. "Avicenna, Medicine and Biology."

Ibn al-Ṭayyib, Abū'l-Faraj (d. 1043). Christian philosopher and physician in Baghdad. He wrote commentaries on or abridgments of a number of Hippocrates's and Galen's works, including most of those in the Alexandrian curriculum. Among others, Ibn Buṭlān and al-Yabrūdī were his students. *EI*, s.v. "Ibn al-Ṭayyib," *GAS* 3: passim.

Ibn al-Tilmīdh, Abū'l-Ḥasan Hibat Allāh ibn Ṣāʿid, Amīn al-Dawlah (1073–1165). A learned Syrian Christian, the most famous physician in Baghdad in his time. He was perhaps a priest and knew Arabic, Persian, Syriac, and Greek. He was particularly interested in medical literature and is known to have commented on Greek and Arabic medical classics. His own works were mainly on pharmacology. *CDSB* 13:415–16; *EI*, s.v. "Ibn al-Ṭilmīdh"; IAU 1:259–76; Ibn al-Qifṭī, 340–42; Ullman, *Medizin*, 163–64, 306–7.

Ion of Chios (ca. 480–423 BCE). Poet and philosopher of Pythagorean bent. In addition to plays, he wrote a prose philosophical work entitled *The*

Triad, in which he argued that there were three elements and that the virtues of each thing were threefold. *BNP* 6:907–8; DK 36; *DPA* 3864–66; *RE* 18:1861–68; see pp. 138–39, no. 20, above.

Isḥāq ibn Ḥunayn (d. 910). The son of the famous translator, Ḥunayn ibn Isḥāq. Like his father, he knew Arabic, Syriac, and Greek; and some thought his Arabic style was better than his father's. He mainly translated philosophy, mathematics, and astronomy, though he did translate a few works of Galen and wrote some medical works of his own, including a short chronological history of medicine. *CDSB* 7:24–26, 15:236–37; *EI* 2, s.v. "Isḥāḳ b. Ḥunayn"; *GAS* 3:267–68.

John of Alexandria (poss. fl. first half of seventh century). Late Alexandrian medical writer mentioned as one of the authors of the Alexandrian epitomes. Islamic sources refer to him as Yaḥyā al-Naḥwī, thus confusing him with the more famous John Philoponus. Commentaries in the Alexandrian style on works of Hippocrates and Galen attributed to him survive in Greek, Latin, and Arabic. *BNP* 6.897; *RE* 18:1800; see p. xli above.

John the Grammarian (fl. mid-seventh century). Bishop of Alexandria at the time of the Muslim conquest of Egypt according to the famous (but certainly false) account of the destruction of the great library. He is known in Arabic as Yaḥyā al-Naḥwī and is thus conflated with the famous sixth-century philosopher and the medical writer John of Alexandria. It is not clear that there was such a person. *EI*, s.v. "Yaḥyā al-Naḥwī." See pp. xxxvii–xxxviii, xl–xli above.

John Philoponus (ca. 490–ca. 570). A Christian natural philosopher and commentator on Aristotle, one of the possibly three individuals conflated in Arabic sources under the name Yaḥyā al-Naḥwī. He was very well known in the Islamic world, both through biographies and doxographies and through translations of his works, some of which survive in Arabic. *BNP*, s.v. "Philoponus"; *CDSB* 7:134–39, 22:51–53; *EI*, s.v. "Yaḥyā al-Naḥwī"; *GAS* 3:157–60; *RE* 18:1764–95.

Leucippus (5th century BCE). A poorly documented Atomist philosopher and the teacher of Democritus. He was said to have written a book entitled *The Great System of the Universe*. Even in ancient times, he was known entirely through the writings of Democritus and the comments of early critics like Aristotle. He was known to the Islamic world through doxographies. *BNP* 7:447–48; DK 67; *DPA* 4:97–98; KR 400–26; *RE* 13:2264–77.

Marinus (poss. sixth century). One of the Alexandrian epitomists. He is known only from two Arabic lists of the epitomists. There is no other information on him in Arabic sources, and he cannot be convincingly identified with anyone known from Greek sources. See pp. xli–xlii above.

Melissus of Samos (fl. ca. 440 BCE). A monist who held that the universe was one, unchanging, and infinite. He took Parmenides's doctrine that only Being is and drew the conclusion that the universe must be infinite in space and time, since limit would imply nonbeing. He was disliked by Aristotle and those influenced by him. He is probably the only philosopher ever to win a naval battle, having commanded the Samian fleet that defeated the Athenians in 441. *BNP* 8:635–36; DK 30; *DPA* 4:391–93; KR 298–306; *RE* 19:530–32.

Menemachus of Aphrodisias (first century). Methodist physician of whom little is known beyond a few citations in Galen and Oribasius. *RE* 15:838.

Menodotus of Nicomedia (fl. ca. 125). A prominent Empiricist physician and skeptical philosopher. MSS **A** and **M** of the epitome of *On the Sects* list him as a Methodist, though in fact he was an active polemicist against the Methodists. Perhaps his name was added from another source and misplaced. *BNP* 8:695; *DPA* 4:476–82; *GAS* 3:56; *RE* 29:901–16.

Methodists (ca. first to fifth [?] century). A school of medicine founded by Thessalus of Tralles and said to have been based on the earlier theories of Asclepiades and Themison. The Methodists analyzed diseases in terms of "communities" and common features of diseases and believed that diseases reflected either restricted or excessive flows within the body. They were notable for their antitheoretical bent and their comparatively gentle treatments. Galen's savage attacks on them, reflected in the epitomes, has affected most later evaluations of them, but recently scholars have begun to reevaluate them more favorably. *BNP* 8:801–2; Nutton, *Ancient Medicine*, 187–201; *RE* supp. 6:358–73.

Mnaseas (fl. late first century). An obscure Methodist physician known to us through a few citations in Galen and several other late medical writers and to the Muslims through citations of a *Kitāb al-qawābil*, "Book of Midwives [?]" *GAS* 3:56; *RE* 30:2252–53.

Mnesitheus of Athens (fl. late fourth century BCE). Mentioned in MSS **A** and **M** as a Methodist, but actually a Hippocratic physician who predated the Methodists by several centuries and who should thus be listed as a Rationalist. He was best known for writings on dietetics and known to the Arabs through citations with his name usually garbled, sometimes in the form Minīthānūs al-Qadīm, "the elder," to distinguish him from Mithīnānūs al-Thānī, "the second," by whom is meant Mnaseas. *BNP* 9:102; *GAS* 3:51–52; *RE* 30:2281–84.

Oribasius or *Oreibasius of Pergamon* (ca. 320–400). A pagan iatrosophist and the personal physician of the Emperor Julian the Apostate. He is of particular importance because his surviving work contains extensive extracts from the works of earlier physicians. Five works were attributed to him in Islamic sources, and quotations from some of them survive in the works of Arabic authors—who, however, knew nothing else about him. *BNP* 10:203–5; *CDSB* 10:230–31; *DPA* 4:800–4; *GAS* 3:152–54; *RE* 7:797–812.

Palladius of Alexandria (sixth century). An otherwise unknown iatrosophist known to have written commentaries on Hippocrates and Galen. The Islamic world knew him as the author of a commentary on Hippocrates's *Aphorisms* and as one of the compilers of the Alexandrian epitomes. *BNP* 10:393; *GAS* 3:161–62; *RE* 36.2:211–14. See p. xlii above.

Parmenides of Elea (fl. ca. 475 BCE). Presocratic philosopher who held an extreme form of monism in which reality was one, unchanging, and finite, a view known to the Islamic world through Aristotle and the doxographies. Islamic sources also mention him as a physician. *BNP* 10:537–40; *CDSB* 10:324–25; DK 20; KR 263–85; *RE* 36.3:1553–59.

Paul of Aegina (fl. 640). A physician in Alexandria known mainly for his seven-book manual of medicine. This work was based on Oribasius's much larger medical encyclopedia and, in Ḥunayn's translation, was widely cited by Islamic authors as his *Kunnāsh*. The Arabic tradition knows five other titles, one of which survives in Hebrew translation. *BNP* 10:635–36; *CDSB* 10:417–19; *GAS* 3:168–70; *RE* 36.3:2386–98.

Philinus of Cos (mid-third century BCE). He broke with his teacher Herophilus to found the Empiricist school, asserting that the causes of disease could not be known and rejecting the use of the pulse in diagnosis. He was unknown to Islamicate scholars except as a name in a list of Empiricist physicians. *BNP* 11:22; *CDSB* 10:581; *RE* 38:2193–94.

Philotimus or *Phylotimus of Cos* (late fourth century BCE). A student of Praxagoras and thus a Rationalist in the Galenic classification of medical schools. He was known to Islamicate scholars through citations of his work on dietetics. *GAS* 3:52; *RE* 39:1030–32.

Plato (ca. 427–348 BCE). Eminent Greek philosopher, much respected by Galen. He was known in the Islamic world as a medical authority through a summary by Galen of the medical doctrines of the *Timaeus.* Islamic authors also knew of a "Plato the Physician," said to have been a teacher of Galen, whose book on cauterization seems to be extant in Arabic. *GAS* 3:48–49.

Pneumatic School (first to second century). Medical school founded by Athenaeus in opposition to the Methodists. Its teachings were a combination of Hippocratic medicine and the doctrine of *pneuma* of the Stoics, apparently as transmitted by Posidonius. Two other major medical writers belonged to this school: Archigenes of Apamea and Claudius Agathinus. *BNP* 11:433–34; Nutton, *Ancient Medicine*, 202–6.

Praxagoras of Cos (late fourth century BCE). Possibly the teacher of Herophilus. He wrote widely, developing a system with eleven humors, but was probably most influential as an anatomist. *BNP* 11:782–83; *CDSB* 11:127–28; *RE* 44:1735–43; Steckerl, *Fragments of Praxagoras.*

Prodicus (ca. 450–400 BCE). Sophist and teacher of oratory who appears several times in Plato's dialogues and is mentioned by Galen in *On the Elements* as the author of a work entitled *On Nature. BNP* 11:930–31; DK 84; *RE* 45:84–89.

Pythagoras (fl. ca. 500 BCE). Semilegendary founder of the Pythagorean school. He and his school were said to hold that the elements were numbers—in particular, the numbers one through ten—and that there were thus ten elements. He was well known to the Islamic world as one of the five "pillars" of philosophy. There were also two pseudepigraphic Arabic medical works attributed to him. *BNP* 13:276–87; *GAS* 3:2022; *RE* 47:171–209, supp. 10:843–64.

Rationalist school. Also known as the Dogmatic school. The school of medicine to which Galen assigns those physicians, beginning with Hippocrates, who believed it was necessary to understand the inner state of the body in order to maintain health and treat illness. It was not an organized school but rather a retrospective categorization of physicians sharing a common medical epistemology. *BP* 4:612–13; *RE* supp. 10:179–80. See pp. xlvi, 10, 30–34 above.

Rāzī, Abū Bakr al- / Rhazes (ca. 854–925 or 935). Iranian physician with unconventional philosophical views. His most important medical work was *al-Ḥāwī fī al-ṭibb*, which contains citations and quotations from many sources now lost, many of them Greek, as well as al-Rāzī's own observations. He also wrote a critique of Galen's views entitled *al-Shukūk ʿalā Jālīnūs* (Doubts concerning Galen). *CDSB* 11:323–26; *EI*, s.v. "al-Rāzī, Abū Bakr"; *Encyclopaedia Iranica*, s.v. "*Ḥāwī, al-*."

Ruhāwī, Isḥāq ibn ʿAlī al- (fl. second half of ninth century). Author of at least five medical works, of which only a work on medical ethics survives. *GAS* 3:263–64; al-Ruhāwī, *Medical Ethics.*

Serapion of Alexandria (late second century BCE). The second major Empiricist, later known best as a pharmacologist. His works are lost apart from citations in Galen and other late authors. *RE* 2.4:1667–68.

Sextus Empiricus (fl. late second century). The well-known skeptical philosopher. He was also a physician, but his works on medicine are lost. In his *Outline of Pyrrhonism* 1:236–41, however, he praises the Methodists as being closer to Skepticism than the Empiricists of his own day. *BNP* 13:370–72; *CDSB* 12:340–41; *RE* 2.4:1667–68.

Shimshon ben Shlomo (fl. 1322). Translator of the Alexandrian epitomes into Hebrew. Nothing else is known of him. Steinschneider, *Hebraeischen*, 654.

Socrates (ca. 470–399 BCE). Athenian philosopher and teacher of Plato. He appears in the epitome as an advocate of the view that there are three elements. See p. 140 and n. 26 above.

Soranus of Ephesus (early second century). A major Methodist physician, best known now for his surviving work on gynecology, though it is clear from surviving citations in the works of Galen and others that he was rivaled only by Galen as a medical authority in later Roman times. He was known to Islamicate scholars through citations, lists of ancient doctors, and a book on enemas, evidently extracted from a larger Greek work on therapeutics. *BNP* 13:653–55; *CDSB* 12:538–42; *DPA* 4:476–82; *GAS* 3:61; *RE* 2.5:1113–30; Ullman, *Medizin*, 76–78.

Stephanus of Athens (ca. 600). Author of two surviving commentaries on works of Hippocrates and another on a work of Galen. He is possibly to be identified with one or more others of the same name: a philosopher active in Alexandria around 580 who wrote on Aristotle's logic, an Athenian who taught philosophy in Constantinople around

606, and an alchemist known from much later sources. *BNP* 13:824–25; *RE* 2.6:2404–405. See p. xlii above.

Stoics (fourth century BCE–third century CE). Hellenistic philosophical school whose importance for medicine was its materialistic physics, its conception of *pneuma* (spirit), and its theories of scientific inference. Though the Stoics were most naturally associated with the Pneumatic school of medicine and were often criticized by Galen, it is clear that they had a great influence on various Rationalist theories of medicine. *BNP* 13:852–57; Hankinson, "Stoicism and Medicine."

Thales of Miletus (fl. first half of sixth century BCE). Traditionally the first true philosopher, though little is known about him with certainty. Though various scientific, astronomical, and mathematical discoveries are attributed to him, the doxographers and later Muslim scholars knew him mostly for his theory that water was the primal element. *BNP* 14:360–62; DK 1; KR 74–98; *GAS* 4:45; *RE* 2.9:1210–12, supp. 10:930–47.

Themison of Laodicea (first century BCE). Physician whose theories were a transition between those of his teacher Asclepiades and the later Methodist school. The traditional genealogies begin the Methodist school with him. *BNP* 14:426; Nutton, *Ancient Medicine*, 189–91; Pigeaud, "L'introduction," 566–87; *RE* 2.12:1632–38.

Theodosius (sixth century?). One of the Alexandrian epitomists, according to two of the sources. He has not been otherwise identified.

Thersites. A Greek soldier at Troy, malformed in body and character, who became a symbol of ugliness in Greek literature. *BNP* 14:556; *RE* 2.10:2455–71. See p. lxviii and p. 67, n. 52.

Thessalus of Tralles (first century). Probably the true founder of Methodism. He taught that all disease could be explained, and thus easily diagnosed and treated, by some combination of states of constriction and looseness of the bodily pores. He was more or less unknown to the Islamic world. *BNP* 14:578–79; Pigeaud, "L'introduction," 587–99; *RE* supp. 11:168–82.

Xenophanes of Colophon (ca. 570–467 BCE). One of the earliest Greek philosophers. He is best remembered for his interesting critique of the anthropomorphism of traditional Greek religion. The doxographers, probably incorrectly, say that he held that the primal element was earth; more likely, he thought that the primal elements were earth and water. The Islamic world seems to have been largely unaware of him. *BNP* 15:819–22; DK 21; KR 163–81; *RE* 2.18:1542–62.

Yabrūdī (or Bīrūdī), Abū'l-Faraj Jirjis ibn Yūḥannā ibn Sahl al- (eleventh century). A Syrian Orthodox Christian physician who worked in Damascus. He was a student of Abū'l-Faraj Ibn al-Ṭayyib and was known for his copies of and commentaries on medical works, especially those of Galen. IAU 2:140–43.

Yaḥyā al-Naḥwī. See John of Alexandria, John the Grammarian, John Philoponus, and pp. xl–xli above.

Appendix 2: The Three Schools of Medicine

Greek names, especially those of more obscure figures, are often corrupted in Arabic translations from Greek, the Arabic texts citing those translations, and the Hebrew and Latin texts dependent upon Arabic sources. The list of prominent Greek adherents of the Empiricist, Rationalist (or Dogmatist), and Methodist schools of medicine as found in the introduction to the epitome of *The Medical Sects* has not escaped this fate; I have not bothered to record all the variations of dotting and the like found in my six manuscripts. These lists obviously represent a Late Antique trope preserved with more or less fidelity in various Greek, Latin, and Arabic texts.

The list of members of the three schools as found in our text is essentially identical to that of two other works: Yaḥyā al-Naḥwī's *talkhīṣ* of *The Medical Sects* and a set of Latin lectures on *The Medical Sects* by one Agnellus of Ravenna (see table 4). Closely related is the list of names from the pseudo-Galenic *Introduction to Medicine (Eisagōgē Iātros)*, which I have translated below. Another commentary on *The Medical Sects* by one John of Alexandria gives a somewhat different list, as does the list of Empiricists in Ibn Abī Uṣaybiᶜah, most of whose names are indecipherable. The one major textual difficulty with regard to these lists involves the list of Methodists, where MSS **A** and **M**, which were both copied around 1240 from an exemplar originating in the circle of Ibn al-Tilmīdh, add two names, Menodotus and Mnesitheus, and give the names in a slightly different order. This variant list can safely be disregarded since the version I give is supported by all the other MSS and by the versions of Yaḥyā and Agnellus. Moreover, Menedotus was an Empiricist who actively opposed the Methodists, and Mnesitheus a Hippocratic of the fourth century BCE, well before the founding of the Methodist school.

Table 4: Members of the three medical sects

The numbers refer to the order in which the names are given in each list.

Alexandrian Epitome of *The Medical Sects*, para. 4	**ps.-Galen: *Introduction to Medicine***	**Ibn Abī Uṣaybiʿa**	**Yaḥyā al-Naḥwī, *Talkhīṣ al-Firaq***	**John of Alexandria**	**Agnellus of Ravenna, *Lectures on* De Sectis**
		Empiricists / Ἐμπειρικοί / أصحاب التجارب			
1. Acron of Agrigentum أقرن الأقراغنيطي	1. Ἄκρων Ἀκραγαντῖνος	1. أقرن الأقراغنطي	1. أقرن الأقراغيطي		1. Acron Cacrantinus
2. Philinus of Cos فيلنوس القزاقي	2. Φιλῖνος Κῷος	4. فيلنبس	2. فيلنوس القوقاي		2. Philon de Cho
3. Serapion of Alexandria سارافيون الإسكندراني	3. Σεραπίων Ἀλεξανδρεύς		3. سرافيون الإسكندري	1. Serapion	3. Seraphion de Alexandria
4. Sextus Empiricus سخطس	7. Σέξτος		4. سخطوس	10. Sextus Afer	4. Sextus
5. Apollonius [of Antioch] أبولونيوس	4–5. Ἀπολλώνιοι δύο, πατήρ τε καὶ υἱός, Ἀντιοχεῖς		5. أبولونيس	2–3. Apollonius senior et Apollonius junior	5. Apollonius

Alexandrian Epitome of *The Medical Sects*, para. 4	**ps.-Galen: *Introduction to Medicine***	**Ibn Abī Uṣaybiʿa**	**Yaḥyā al-Naḥwī, *Talkhīṣ al-Firaq***	**John of Alexandria**	**Agnellus of Ravenna, *Lectures on* De Sectis**
	6. Menodotus Μηνόδοτος	2. بنتخلس 3. أنقلس 5. غافرطيمس 6. الحسدروس 7. ملسيس All unidentifiable		6. Eraclitus 7. Nicomachus 8. Glaucias 9. Menedotus	
		Rationalists / Λογικοί / أصحاب القياس			
1. Hippocrates إبقراط	1.Ἱπποκράτης Κῶος		1. بقراط	1. Ypocras	1. Ypocrates de Cho
2. Diocles of Carystus ديوقليس	2. Διοκλῆς ὁ Καρύστιος		2. ديوقلس	3. Diocles	2. Diocles
3. Praxagoras of Cos فركساغورس	3. Πραξαγόρας Κῶος		3. فركساغورس	2. Praxagoras	3. Praxegoras
4. Philotimus of Cos فولوطيموس		2. فولوطيمس	4. فولوطيمس		4. Philotemus

Alexandrian Epitome of *The Medical Sects*, para. 4	**ps.-Galen: *Introduction to Medicine***	**Ibn Abī Uṣaybiʿa**	**Yaḥyā al-Naḥwī, *Talkhīṣ al-Firaq***	**John of Alexandria**	**Agnellus of Ravenna, *Lectures on* De Sectis**
5. Erasistratus of Chios and Alexandria أراسسطراطيس	5. Ἐρασίστρατος Χῖος		5. أرسسطراطس	4. Erasistratus	5. Erasistratus
6. Asclepiades of Bithynia أسقليباذس	7. Ἀσκληπιάδης Βιθυνός		6. أسقليباذس	8. Asclipiades	6. Asclepiades
	4. Herophilus Ἡρόφιλος Χαλκηδόνιος 6. Mnesitheus Μνησίθεος Ἀθηναῖος	1. أنكساغورس 3. ماخاخس 4. سقولوس 5. سوفورس All unidentifiable		5. Crisippus 6. Erofilus 7. Leufastus 9. Galienus	7. Gallienus
Methodists / Μεθοδικοί / أصحاب الحيل					
1. Themison of Laodicea ثاميسن الأوذيقي	1. Θεμίσων ὁ Λαυδικεύς τῆς Συρίας		1. ثاميس الأوذيقي	1. Themison	1. Fimision de Laodicia

Alexandrian Epitome of *The Medical Sects*, para. 4	**ps.-Galen: *Introduction to Medicine***	**Ibn Abī Uṣaybiʿa**	**Yaḥyā al-Naḥwī, *Talkhīṣ al-Firaq***	**John of Alexandria**	**Agnellus of Ravenna, *Lectures on* De Sectis**
2. Thessalus of Tralles ثاسلس	2. Θεσσαλὸς ὁ Τραλλιανός		2. ثاسلس	2. Thesalus	2. Tessalus de Roma
3. Menemachus مناماخوس	8. (Μενάμαχος ὁ Ἀφροδισεύς)	1. ماناخص [?]	3. مناماخس	8. Menemachus	3. Mimomachus
4. Mnaseas مناساس	3. Μνασέας	2. ماساوس	4. مناسياوس	4. Manaseus	
5. Soranus سورانوس	9. (Σωρανὸς ὁ Ἐφέσιος)		5. سورانوس	7. Soranus	4. Soranus
AM: Themison, Thessalus, Menodotus, Menemachus, Mnesitheus, Mnaseas	4. Dionysius Διονύσιος 5. Proclus Πρόκλος 6. Antipatrus Ἀντίπατρος; forming a subsect 7. Olympiacus Ὀλυμπιακὸς ὁ Μιλήσιος 8. Menemachus Μενάμαχος 9. Σωρανός	3. غوريانس 4. غرغوريس 5. قونيس All unidentifiable		3. Dionysius 5. Philon 6. Olimpicus 9. Avidianus	

Translation of Pseudo-Galen. *Introductio seu medicus*, K 14:683–84, on the medical sects

4. [Who led the three sects?]

The Rationalist sect was led by Hippocrates of Cos, who was also its founder and the one who first established the Rationalist sect, then after him Diocles of Carystus, Praxagoras of Cos, Herophilus of Chalcedon, Erasistratus of Chios, Mnesitheus of Athens, and Asclepiades of Cian in Bithynia, which is also called Prusias.

The Empiricist sect was led by Philinus of Cos, who breaking first with the Rationalist sect sought occasions to dispute with Herophilus, whose student he had been. Wishing to lead their own sect that would nonetheless be older than the Rationalist sect, they claimed that it had been founded by Acron of Agrigentum. After Philinus there was Serapion of Alexandria, then the two Apolloniuses, father and son, of Antioch. After them was Menodotus and Sextus, who brought it to perfection.

Methodism was founded by Themison of Syrian Laodicea, who had acquired from Asclepiades the Rationalist what he needed to devise the Methodist sect. It was then perfected by Thessalus of Tralles. After them were Mnaseas, Dionysius, Proclus, and Antipatrus. Forming their own faction within it were Olympiacus of Milesia, Menemachus of Aphrodisias, and Soranus of Ephesus. Moreover, some, like Leonidas of Alexandria, combined the sects, while others were eclectics, such as Archigenes of Syrian Apamea.

Appendix 3: The Structure and Terminology of the Eye in the Epitome of *The Small Art*

Galen, as one of the glosses correctly remarks, does not give an account of the anatomy of the eye in the passage corresponding to para. 40 of the epitome of *The Small Art*. The epitomist's attempt to do so is not a masterpiece either of anatomy or clarity, but with one major problem it is accurate so far as it goes. He explains that the eye is composed of two layers, *ṭabaqāt*, of solid material—"cloaks," to use the ancient term—and three fluids, *ruṭūbāt*, or humors (not to be confused with *akhlāṭ*, the four humors). He does not mention the optic nerve or, properly speaking, the retina, which he has confused with the sclera.

The eye is divided into two parts separated by the lens, which the epitomist properly treats separately. The larger back portion of the eye, the eyeball, is a spherical sack of transparent jelly called vitreous humor, "a fluid resembling liquid glass." He says that the sack within which it is contained has two layers, the inner "resembling a placenta," the choroids; and the outer "resembling a net," the retina. In fact, it has three layers: the outermost being the sclera, the white of the eye, a tough white sack of material; the middle being the choroid, which contains blood vessels; and the innermost being the retina, a continuation of the optic nerve. It may be that the epitomist uses the term "net-like" for the sclera because of its attached muscles, but more likely this is either a case of error in manuscript transmission or the blunder of a fellow iatrosophist lacking practical knowledge of anatomy. A gloss in several of our manuscripts points out that his usage of "retina" differs from the usage of Ḥunayn, who does indeed have it right (see Ḥunayn ibn Isḥāq, *Ten Treatises on the Eye*, Arabic pp. 74–75, English pp. 4–5). Ḥunayn notes that there is disagreement among anatomists over the number of layers comprising the eye, based not so much on disagreement about the anatomical structure as on issues of which elements should be combined and which should be seen as distinct (Arabic pp. 80–81, English pp. 10–11).

The following is a list of the terms used for the parts of the eye, arranged in the order we know them from modern anatomy with the terms used and parts identified by the epitomist along with Ḥunayn's more complex and accurate account.

The modern term is given first, followed by Galen's Greek term, as given in Meyerhof's edition of the ten treatises. E: term used in *Epitome*. Ḥ: term used in Ḥunayn, *Ten Treatises.*

Parts of the eyeball from the outside in

- *Sclera*, σκληρὸς χιτών, lit. "hard tunic," the white of the eyeball: E: *al-ṭabaqah al-shabīhah bi-al-shabakah*, lit. "the layer resembling a net," the term normally used for the retina, probably an error on the part of the epitomist; Ḥ: *al-ṭabaqah al-ṣalbah.*
- *Choroid*, χοριοειδὴς χιτών, a layer carrying blood vessels; E: *al-ṭabaqah al-shabīhah bi-al-mashīmah*, "the layer resembling the placenta"; Ḥ: *al-ṭabaqah al-mashīmah.*
- *Retina*, ἀμφιβληστροειδὴς χιτών, the light-sensitive inner coating of the eyeball; E: omitted; Ḥ: *al-ṭabaqah al-shabakīyah*. The epitomist either did not know this feature or confused it with the sclera. Its function was not known to Ḥunayn, who did however understand it as an extension of the optic nerve.
- *Vitreous humor*, ὑαλοειδὲς ὑγρόν, the transparent jelly contained in the eyeball; E: *al-ruṭūbah al-shabīhah bi-al-zujāj al-dhāʾib*, "the fluid resembling liquid glass"; Ḥ: *al-ruṭūbah al-zujājīyah.*

The anterior of the eye, from the front to the back

- *Conjunctiva*, ἐπιπεφυκὼς χιτών, a clear membrane covering the white of the eye; E: *ifīfāfūqūs*; Ḥ: *al-ṭabaqah al-multaḥamah*, lit. "the connecting layer."
- *Cornea*, κερατοειδὴς χιτών, a transparent portion of the eyeball allowing light to reach the lens; E: *al-ṭabaqah al-shabīhah bi-al-qarn*, "the layer resembling horn," because it is thin, white, and translucent; Ḥ: *al-ṭabaqah al-qarnīyah.*

- *Aqueous (or albuminoid) humor*, ὠοειδὲς ὑγρόν, the fluid filling the anterior chamber of the eye between the cornea on the outside and the lens and iris on the inside; E: *al-ruṭūbah al-shabīhah bi-bayāḍ al-bayḍ*, "the fluid resembling egg white; Ḥ: *al-ruṭūbah al-bayḍīyah*.
- *Uvea*, στραφυλοειδὴς or ῥαγοειδὴς χιτών, the iris, which is the colored part of the eye and acts as an aperture; and the ciliary body, which is the hidden part of the iris; E: *al-ṭabaqah al-shabīha bi-al-ᶜinaba*, "the layer resembling a grape," due to its dark color; Ḥ: *al-ṭabaqah al-ᶜinabīyah*.
- *Zonula*, ἀραχνοειδὴς χιτών, a network of fibers controlling the shape and focus of the lens; E: omitted; Ḥ: *al-ṭabaqah al-ᶜankabūtīyah*, "the cobweb layer."
- *Lens (or crystalline humor)*, κρυσταλλοειδὲς ὑγρόν, the transparent body focusing light on the retina; E, Ḥ: *al-ruṭūbah al-jalīdīyah*, "the ice-like layer."

Arabic-Greek-English Glossary

The following glossary records the English equivalents that I have used in the translations of these three texts. In general, I have recorded the English renderings of all Arabic words with a technical or semitechnical sense. Where possible, I have added what I believe to be the original Greek word of which the Arabic is a translation. Since I do not have the original Greek text—only the Galenic text that the epitomes are commenting on—I cannot always know the underlying Greek word. In some cases, it is very clear: *ᶜunṣur* represents the Greek στοιχεῖον. In many other cases, however, the text of the epitome is not closely linked to Galen's text—most commonly, adding examples to clarify Galen's statements. *Kharbaq aswad* (black hellebore) is mentioned as an example but is not in Galen's text, and while there are several Greek words for hellebore or its use found in Galen's surviving works, I do not, in fact, know which one was used in the Greek version of the epitome. Moreover, there are cases where someone—presumably the translator but possibly the epitomist—has replaced remedies given by Galen with remedies current in his own time. Therefore, with only a few exceptions, I have given the Greek only when the word occurs in one of the three texts of Galen in a context that makes it reasonably plausible that it is the word translated in the epitome. Finally, the Greek word is not necessarily the same part of speech as its Arabic equivalent—a Greek noun corresponding to an Arabic verb, for example. Thus, the Arabic and English terms here are not necessarily direct translations of the Greek.

آ

needle	βελόνη	إبرة
stagnant	—	آجاميّ
second	ἕτερος	آخر
posterior	ὀπίσθιος	مؤخّر
culture	—	أدب
device	ὄργανον	أداة
ear	οὖς	أُذُن
earth	γῆ	أرض
element	στοιχεῖον	إسطقس
finger	δάκτυλος	إصبع
source, principle	ἀρχή	أصل
licorice	—	أصل السّوس
to remove surgically, extirpate	ἀφαίρεσις	استأصل
viper	ἔχιδνα	أفعى
conjunctiva	ἐπιπεφυκώς	أفيافوقوس
epilogism	ἐπιλογισμός	إفيلوجسموس
to eat	ἐσθίειν	أكل
sweet melilote	—	إكليل الملك
coronal	στεφάνη, στεφανιαῖος	إكليليّ
to be combined	τὸ δρᾶν ἄλληλα	تأليف، ائتلف
composed	κεκρᾶσθαι	مؤلّف
to experience; to be hurt, suffer	πάσχειν, πάθος, ἀλγεῖν	ألم
state; entity	πρᾶγμα	أمر
relating to health	ὑγιεινός	أمرصحيّ
relating to disease	νοσώδης	أمرمرضيّ
original, fundamental	—	أمهات
analogism	ἀναλογισμός	أنالوجسموس

generating females	θηλυγόνος	توليد الأناث
testicles	ὄρχεις	أنثيان
man, person	ἀνήρ, ἄνθρωπος	إنسان
human (adj.)	ἀνθρώπινος	إنسيّ
future	μέλλων	ما يستأنف
defect, disorder	σφάλμα, κακία	آفة
occurrence of a disorder	ἁμαρτάνειν	حدث من الآفة
edema	οἴδημα	أوذيما
instrument, tool	ὄργανον, ὀργανικός	آلة
instrumental; organic; functional (of organs)	ὀργανικός	آليّ
primary	πρότερος	أوّل، أوّليّ
now	νῦν	الآن
which thing it is	—	أيّ شيء هو

ب

chamomile	—	بابونج
sprinkled	—	مبثوث
hoarseness	βραγχώδης	بحّة
hoarse	βραγχώδης	أبحّ
crisis	κρίσις	بحران
vapor, smoke	ἀτμός	بخار
to be dispersed	—	تبدّد
principle, basis; beginning	—	مبدأ
body, living body	σῶμα	بدن
healing	θεραπευτική	برء
cold	ψυχρός	بارد
coldness	τὸ ψυχρόν, ψυχρότης, ψυχρότερος, ἕξις ψυχροτέρα	برد، برودة
easily cooled	εὔψυκτος	إسراع البرودة
difficult to cool	δύσψυκτος	عسر قبول البرودة

feces; cf. ثفل, which some MSS use in its place	διαχώρημα	براز
demonstration	ἀπόδειξις	برهان
linseed	λινόσπερμον	بزر الكتّان
ispaghula	—	بزر قطونا
expansion	ἐξάπλωσις	بسط
spreading apart	—	انبساط
simple	ἁπλοῦς	بسيط
autopsy	αὐτοψία	مباشرة
vision	ὄψις	بصرٌ
to see	—	بصرَ، أبصر
seen	ὁρατός	التي تُبصَر
optic	—	باصر
optic spirit	—	روح باصرة
visible	ὁρατός, πρὸς τὴν ὄψιν	مبصور
to have sexual intercourse	ἀφροδίσια	باضعَ
beans	κύαμος	باقلاء
incision	—	بطّ
retarding, slowness	βραδυτής	إبطاء
implacability	δύσπαυστος	إبطاء الغضب
horizontal	—	مبطوح
abdomen, belly; ventricle	γαστήρ	بطن
inner	—	باطن
flux	ῥύσις	انبعاث
fluent disease	νόσημα ῥοῶδες	علّة انبعاثيّة
to deviate	ἐξίστασθαι	بعدَ
remote; far from	—	بعيد
bug	—	بقّ
purslane	ἀνδράχνη	بقلاء حمقاء
to remain	λοιπός	بقي

permanence	ἀμετάβλητος	بقاء
country	χώρα, οἴκησις	بلد
phlegm	φλέγμα	بلغم
phlegmatic	φλεγματικός	بلغميّ
daughters of the elements	—	بنات الأركان
makeup, structure	κατασκευή	بنية
method (part of a science or art)	ἀρχή	باب
pylorus	πυλωρός	بوّاب
urine	οὗρον	بول
whiteness; pallor	λευκός, τὸ λευκόν	بياض
white	λευκός	أبيض
resembling the white of an egg	ὠοειδές	الشبيهة ببياض البيض
to explain	—	بيّنَ
obvious	ἐναργής, γνώρισμα	بيّنٌ
not obvious	ἀμυδρός	غير بيّن
to be obvious, to be clearly found	—	تبيّن

ت

following; variety; resulting	ἀκολουθῶν	تابعٌ
aspect	—	شيء تابع
to follow	ἕπεσθαι	اتّبع
title (of book)	ἐπιγράφειν	ترجمة
to overlook	—	ترك
theriac	θηριακή	ترياق
spittle	σίαλος	تفل
tasteless	—	تفه
to complete	συμπληροῦν	تمّ
complete	ἄμεμπτος	تامّ

end	τέλος	تمام
unimpaired functions	τὰς ἐνεργείας ἡ τελειότης	تمام الأفعال
completeness	τελειότης	تماميّة
mulberry	—	توت
yearning	—	توقان

ث

stability	μόνιμος	ثبات
without stability	ἀσύστατος	لا ثبات لها
snake	ἔχιδνα	ثعبان
feces	διαχώρημα	ثفل
aperture; hole	στόμιον	ثقبة
deep (of voice)	βαρύς	ثقيل
triangle	τρίγωνος	مثلّث
secondary	—	ثانٍ

ج

millet	—	جاورس
cheese	τυρός	جبن
extreme timidity	ἄτολμος	إفراط الجبن
absorption, draw out, attract	ἕλκειν, ἐφέλκειν	جذب
leprosy	—	جذام
daring	ἑτοῖμος	جرأة
experience	ἐμπειρία	تجربة
Empiricists	ἐμπειρικοί	أصحاب التّجارب
empirical sect	τηρητική	الفرقة المجرّبة
duct	πόρος	مجرى
part	μόριον	جزء
particular	κατὰ μέρου	جزئيّ
indivisible	ἀδιαίρετος	لا يتجزّأ

atom	ἄτομος	جزء لا يتجزأ
body, corporeal body	σῶμα	جسم
firm	—	جاسئ
belch	ἐρυγή	جشاء
sour belch	ὀξυρεγμιώδης	جشاء حامض
curly	οὖλος	جعد
desiccate, dry	ξηραίνειν	جفّف
eyelid	—	جفن
higher rank	ἀξίωμα	جلالة القدر
therapy	θεραπεία	اجتلاب الصّحّة
crystalline	κρυσταλλοειδές	جليديّة
lens (of eye)	κρυσταλλοειδὲς ὑγρόν	الرطوبة الجليدية
to cleanse	ῥυπτικός	جلى
ice	κρύσταλλος	جمد
solidified; solidity	στερεός	جمود
to concentrate; to bring together	συνάγειν	جمَعَ
epitome	σύνοψις	جوامع
sexual intercourse	συνουσία	جماع
frequent sexual intercourse	ἀφροδισιαστικός	كثرة الجماع
to be joined	σύγκρισις	اجتمع
combination, concentration	σύγκρασις	اجتماع
beauty	κάλλος	جمال
community (of diseases)	κοινότης	جملة
pleurisy	—	ذات الجنب
to avoid	ἀποστροφή, ἀποχωρῆσαι	تجنّب، اجتنب
genus	γένος	جنس
generic	—	جنسيّ
unknowable	ἄγνωστος	مجهول
good quality, ease	—	جودة

acuity	—	جودة نظر
contiguity	συντίθεσθαι	مجاورة
cavity; concavity	κοιλότης	تجويف
substance	οὐσία	جوهر
essential	οὐσιώδης, σύμφυτος	جوهريّ

ح

grain	—	حبّة
love	φιλία	محبّة
to prevent	ἐπέχειν	حبس
retention, to retain	ἐπέχειν, ἰσχόμενος	احتباس
retained	ἰσχόμενος	محتبس
to argue	ἐπιλέγειν	احتجّ
to refute	—	احتجّ على
orbit (of eye)	—	حجاج
stone	λίθος	حجارة
definition	ὅρος, οὐσιώδης	حدّ
acute; high-pitched	ὀξύς	حادّ
pungent	δριμύς	حادّ
sharpness	ἐκδριμέων	حدّة
violent character	θυμικός	حدّة الأخلاق
steel	σμίλη	حديد
thing defined	ὃς ὁρίζεται πρᾶγμα	شيء محدود
proper	—	محدود
hump	—	حدبة
convex	κυρσός	محدّب
to change, influence, effect, come to be; to result, arise from; changeable	γίγνεσθαι, πάσχειν, προσγίγνεσθαι, προσπίπτειν, τυγχάνειν	حدَث
to bring about	ἐργάζεσθαι, ποιητικός	أحدث
change, effect	—	حدُث

able to be affected, changeable, to accept effects	πάσχοντες, παθητικός	قابل الأحداث
unchanging	ἀμετάβλητος, ἀπαθής	غير قابل الأحداث
occurence	—	حدوث
to drop down	—	انحدر
skill	—	حذق
heat	τὸ θερμόν, θερμότης	حرّ، حرارة
hot	θερμός, θερμότερος	حارّ
pungency	—	حرافة
pungent	—	حريف
to inflame; consume	—	أحرق
caustic (of drugs)	καυστικός	محرق
motion, exercise	κινεῖσθαι, κίνησις, γυμνάσιον	حركة
setting in motion, exercising	ἐνέργεια	تحريك
requiring exercise	μεγάλη ἐνέργεια	تحريك بالعمل
motor	πρακτικός	محرّك
moving	κινούμενος	متحرّك
unmoving	ἀκίνητος	غير متحرّك
grief	λύπη	حزن
sensation	αἴσθησις, αἰσθητικός	حسّ
sense organ	αἴσθησις, αἰσθητικός	حاسّة
to sense	αἰσθάνεσθαι, αἰσθητικός	أحسّ
not having sensation	ἀναίσθητος	لا يحسّ
sensory	αἰσθητικός	حسّاس
sensible; perceptibly	αἰσθητός	محسوس
envy	φθόνος	حسد
comeliness	κάλλος	حُسن
well formed	εὔρυθμος	حسن الشكل
appetite	χαίρειν	حسن القبول
gloss	—	حاشية

retention of the breath	κατάληψις	حصر النفس
study	—	تحصيل
stone	λίθος	حصًى
present	νῦν	حاضر
preservation, prophylaxis; to remember	φυλάττειν, μνήμη	حفظ
prophylaxis	προφυλακτικός	تقدّم بالحفظ
hygiene	φυλακτικός	حفظ الصحّة
preserving	φυλακτικός	حافظ
sect that relies on memory	μνημονευτικός	الفرقة الحافظة والمتذكّرة
truly	ἀληθής, ὄντως	حقّ
true	ἀληθής	حقّ، حقيقيّ
true	κατὰ τὴν ἀλήθειαν	بالحقيقة
grasp its reality	ἀληθής βέβαιος	وقف على حقيقته
enema	κλύσμα	حقنة
costiveness	στέγνωσις	احتقان
costive disease	νόσημα στεγνόν	علّة احتقانيّة
to scrape	ξύειν	حكّ
to exercise judgment upon	κρίσις	حكَم
resolution; judgment	κρίσις	حكْمٌ
irresolvable	ἀνεπίκριτος	لا يقع عليه الحكم
firm; established	—	مستحكم
to loosen (bowels)	ὑπάγειν	حلّ
analysis	ἀνάλυσις	تحليل
dialysis of the definition	ὅρου διάλυσις	تحليل الحدّ
resolution, dissolution	διαίρεσις	تحلّل
resolvent	—	محلّل
formless	ἀραιός	منحلّ
fenugreek	—	حلبة
sweet	γλυκύς	حلوٌ

sweetness	τὸ γλυκύ	حلاوة
bathing	λουτρόν	حمام
red	ὑπόπυρρος	أحمر
light (of hair color)	ξανθός	أحمر ناصع
erysipelas; redness; ruddiness	ἐρυσίπελας, ἐρυθρός	حمرة
red color; flushing	—	حمرة اللّون
chickpea	ἐρέβινθος	حمص
sour	ὀξυρεγμιώδης	حامض
sourness	—	حموضة
ability	φέρειν	احتمال
fever	πυρετός	حمّى
wheat	πυρός	حنطة
palate	ὑπερῴα	حنك
surrounding	περιέχων	محيط
state	διάθεσις, ἀρετὴ τε καὶ κακία	حال، حالة
to transform	συναλλοιοῦν	أحال، استحال
transformation; transmutation	μεταβολή	استحالة
rapid change	εὐμετάβολος	سرعة الاستحالة
absurd	ἀδύνατος	محال
transmuted	μεταβολή	مستحيل
to contain	περιέχειν	احتوى
life	βίος	حياة
animal, animate	ζῷον	حيوان
inanimate	—	غير حيوان
vicious animal	θηρίον	حيوان مفسد
vital	—	حيوانيّ
vital faculty	—	قوّة حيوانيّة
method	μέθοδος	حيلة
Methodists	Μεθοδικοί	أصحاب الحيل

خ

black hellebore	—	خربق أسود
to expel	ἐκκρίνεσθαι	أخرج
removal	κομιδή	إخراج
to acquire (knowledge)	εὕρεσις	استخرج
outside; protuberant (of eyes)	ἐκτός; προπετής	خارج
unnatural	παρὰ φύσιν	خارج الطبيعة
external	ἔξοθεν	من خارج
deviation	—	خروج
autumn	—	خريف
tear	ῥῆξις	انخراق
wood	—	خشب
rough (of voice)	τραχύτης	خشن
roughness	τραχύτης, τετράχυνθαι	خشونة
peculiar to	οἰκεῖος	خاصّة ل
specifically	κατ᾽ εἶδεος	في خاصيّة
property	ἴδιος	خصوصيّة
specific	αὐτός, τοιοῦτος	مخصوص
lightness	τὸ κοῦφον	خفّة
hidden	ἄδηλος	خفيّ
wine vinegar	ὄξος	خلّ الخمر
nail	ὄνυξ	مخلب
to be rarefied	ἀραιοῦσθαι	تخلخَل
rarefaction	διακρινεῖν	تخلخُل
spongy, rarified	ἀραιός	متخلخل
pure	ἄμικτος	خالص
humor	χυμός	خلط
to combine	ἀναμιγνέναι	خالَط
composition, combination	συγκεῖσθαι, κεραννύναι	مخالطة

dislocation	ἐξάρθρημα	خلع
dislocation	—	انخلاع
back	ὄπισθεν	خلف
diarrhea	—	خلفة
to counteract	ἀντιπράττειν	خالَفَ
opposite	ἕτερος, ἐναντία	خلاف، متخالف
contrary	ἐναντίος	مخالف
backwardness	—	تخلّف
to disagree; become different	διιστάναι	اختلف
disagreement	διαφορία	اختلاف
different; disagreeing	χωρίζειν, διαφέρειν	مختلف
disposition	ἦθος	خلق
structure, construction	—	خلق، خلقة
good structure	εὐρυθμία	خلقة حسنة
void	τὸ κενόν	خلاء
spiced, honeyed wine	—	خنديقون
scrofula	—	خنازير
to decline	—	خار
subordinate to	ὑπηρετεῖν	خول وخدم
imagination	φαντασία	تخيّل

د

regimen	διαίτημα	تدبير
governing	ἡγεμονικός	مدبّرة
inside; sunken (of eyes)	ἐντός, ἐν βάθει	داخل
beginner	εἰςαγόμενος	داخل في علم
to interpenetrate	—	دخل بعضه في بعض
diuretic	—	مدرّ للبول
dregs	τρύξ	درديّ
suture	ῥαφή	درز

to apprehend, comprehend	γνωρίζειν, ταῖς αἰθήσεσιν ὑποπίπτειν	أدرك
apprehension, comprehension; to perceive	κατάληψις	إدراك
to fail to comprehend	ἀκαταληψία	لم يدرك
to apprehend	τήρησις	استدرك
fatty	—	دسم
fattiness	—	دسومة
explusion; to repel	ὠθεῖν	دفع
sudden	ἀθρόος	دفعةً
expulsive	—	دافع
resistant to touch	ἀντίτυπος	مدافعة اللمس
hectic (of fever)	—	دِقّ
fine	—	دقيق
barley meal	—	دقيق الشّعير
to indicate	ἔνδειξις	دلّ
indicating; diagnostic	δηλῶν, δηλωτικός; διαγνωστικός	دالّ
deduction	ἔνδειξις	استدلال
indication, symptom; denotation	σημεῖον, γνώρισμα	دلالة
massage	τρῖψις	دلك
running with tears	ῥευματιζόμενος	كثيرة الدّموع
brain	ἐγκέφαλος	دماغ
blood	αἷμα	دم
rose oil	ῥόδινος	دهن الورد
worms	—	دود
round	στρογγύλος	استدارة
ileum	—	استدارات
round	στρογγύλος	مستدير، مدوّر
constantly; chronic	τὸ διὰ παντός	دائم

drug	βοήθημα, φάρμακον	دواء
treatment (not necessarily by drugs)	θεραπεία, θεραπευτικός	مداواة

ذ

to remember	μιμνήσκειν	ذَكَرَ
memory	μνήμη	ذِكر
procreation of males	ἀρρενογόνος	توليد الذكور
mnemonic	ἀναμνηστικός	مذكّر
sect that relies on memory	μνημονευτικός	الفرقة الحافظة والمتذكّرة
without technical method	ἄτεχνος	ليس لها مذهب صناعيّ
to taste	γευέιν	ذاق
tasted	γευστός	التي تذاق
tasted; taste	γεῦσις	مذاق
gustatory, tasted	γευστός	مذوق
tastable	γευστός	مذاقة

ر

head; principle	ἀρχή, κεφαλή	رأس
principal	ἀρχή	رئيس
lung	πνεύμων	رئة
opinion	δόξα, γνώμη	رأي
mulberry	—	رُبّ
ligament; binding	σύνδεσμος, ἐπίδεσις	رباط
quartan (of fever)	—	ربع
square	τετράγωνος	مربّع
spring, of spring	ἔαρ, ἐαρινός	ربيع
order	μέθοδος	ترتيب
in order	κατὰ μέθοδον	على ترتيب
level	—	مرتبة

to return	παλινδρόμησις	رجع
feces	κόπρος	رجيع
leg, foot	πούς	رجل
womb	μήτηρ	رحم
to relax	χάλασις	رخي
soft	χαλαρός, ἁπλοῦς	رخو
inducing relaxation	χαλᾶν	إرخاء
flabbiness	τὸ λεπτόν	رخاوة
to restore	ἀνασῷζειν	ردّ
to return to natural position	ἀνορθωτικός	ردّ إلى موضع طبيعيّ
defective, bad, injurious	πλημμελής, μοχθηρός, φαῦλος	رديء
badly formed	—	رديء الشّكل
to block	στέλλειν	ردع
dew	—	رذاذ
to send away	παραπέμπειν	أرسل
flowing, fluency	ῥοώδης	استرسال
description	ὅρος ἐννοηματικός	رسم
descriptive	ἐννοητικός	رسميّ
ooze	ῥευματίζεσθαι	رشح
observation	τήρησις	رصد
bruise	—	رضّ
moist	ὑγρός, ὑγρότερος	رطب
to moisten	ὑγραίνειν	رطّب
moisture; fluid (part of the eye)	τὸ ὑγρόν, ὑγρότης	رطوبة
lens (of eye)	κρυσταλλοειδὲς ὑγρόν	الرطوبة الجليدية
aqueous humor	ᾠοειδὲς ὑγρόν	الرطوبة الشبيهة ببياض البيض
vitreous humor	ὑαλοειδὲς ὑγρόν	الرطوبة الشبيهة بالزجاج الذائب
nosebleed	αἵματος ῥύσις ἐν ῥινῶν	رعاف

helpful	—	مرفد
to be removed	—	ارتفع
to become finer	λεπτυνόμενος	رقّ
fineness	—	رقّة
finer	—	أرقّ
runny; fine	λεπτομερής, διαυγής	رقيق
soft parts of the belly	ὑποχόνδριον	مراقّ
neck	αὐχήν	رقبة
knee	γόνυ	ركبة
synthesis (as method of instruction)	σύνθεσις	تركيب
collection, combination	μικτός, συναγωγή	تركيب
structure (of organs)	κατασκευή	تركيب
compound, composite	ἐπιπεπλέγμενος, συζύγιος, συγκείμενος, σύνθετος	مركّب
to trust	—	ركن
puberty	—	مراهقة
voluntary	αὐτοσκεδιός	إراديّ
spirit	πνεῦμα	روح
optic spirit	—	روح باصرة
wind	—	ريح
odor, scent	—	رائحة
history	ἱστορία	رواية
exercise	αἰώρα	رياضة

ز

shaggy	—	أزبّ
scum; butter	ἰλύς	زبد
vitreous	ὑαλοειδής	الشبيهة بالزجاج الذائب
omen	—	زجر
blueness	γλαυκός	زرقة

blue	γλαυκός	زرقاء
thin (of hair)	ἄτριχος	أزعر
medlar	μέσπιλον	زعرور
to claim	ἀποφαίνειν	زعم
pitch	—	زفت
to diverge	—	زال
inseparable	ἀχώριστος	لا يزايل
to increase	αὐξάνεσθαι	تزيّد
excess; increase	περιττεύειν, αὔξησις	زيادة

س

cause, means	αἰτία, αἴτιος	سبب
necessary cause	ἐξ ἀναγκῆς	سبب اضطراري
antecedent cause	αἰτία κατακριτικόν, προκατακριτικόν	سبب بادئ
preceding cause	αἰτία προηγούμενον	سبب سابق
cause effecting or bringing about	αἰτία ποιητική	سبب فاعل
cohesive cause	αἰτία συνεκτικόν	سبب ماسك
prior cause	—	سبب متقادم
transformative cause	ἐπανορθωτικός	سبب ناقل
cohesive cause	αἰτία συνεκτικόν	سبب واصل
lank, straight (of hair)	ἁπλοῦς	سبط
beast of prey	θηρίον	سبع
brutishness	ἄγριος	سبعيّة
prognosis	πρόγνωσις	سابق النظر
before	τάχιστα	مسابقة
thinness	τὸ μανόν	سخافة
to warm	θερμαίνειν	سخّن
to block	ἐπέχειν	سدّ
to be blocked	—	انسدّ

blockage	ἔμφαξις	سدّة
cancer	—	سرطان
rapidity	εὐκίνητος	سرعة
quick-wittedness	ἀγχίνοια	سرعة الفهم
accepting knowledge quickly	εὐμάθεια	سرعة قبول العلم
quickly	ταχύς	سريع
quick to be imprinted	εὐτύπωτος	سريع الانطباع
surface	—	سطح
cough, coughing	βήξ	سعال
skewerlike	ὀβελιαῖος	سفّوديّ
quince	μῆλον	سفرجل
long and narrow of head	φοξότερος	مسفّط الرأس
scirrhus	σκίρρος	سقيروس
sick	νοσώδης	سقيم
sickly	νοσώδης, οὐδέτερος	مسقام
douche	—	إسقاء
to rest; to quench (thirst)	καταπαύειν	سكن
rest	ἡσυχία	سكون
oxymel	—	سكنجبين
consumption	—	سلّ
negation	—	سلب
loose	ἀραιός	سلس
inducing flow	χάλασις	تسليس
method	τάξις	مسلك
to follow instructions	—	استسلم
in health	ὑγιαίνων	على السّلامة
venom, poison	φάρμακον	سمّ
hot winds	—	سمائم
pores	πόροι	مسامّ
to hear	ἀκούειν	سمعَ

hearing	ἀκοή	سمعٌ
heard	ἀκουστός	التي تسمع
auditory	ἀκουστός	مسموع
fat	πιμελώδης	سمن
age	ἡλικία	سنّ
insomnia	ἀγρυπνία	سهر
purge, purgation	ἐκκαθαίρειν, ἐκκάθαρσις	إسهال
excessive purgation	ὑπερκάθαρσις	إفراط في الإسهال
purgative	καθαίροντα	مسهل
anger quickly placated	εὐκατάπαυστος	سهولة سكون [الغضب]
sagittal	ὀβελιαῖος	سهميّ
black	μέλας	أسود
melancholic	μελαγχολῶν	سوداويّ
deliberation, deliberative	ἡγεμονικός	سياسة، سياسيّ
calf, leg	σκέλος, κνήμη	ساق
bad temperament	δυσκρασία	سوء المزاج
equally	ἐξ ἴσου	على المساواة
unequally	τῷ πλέον	على غير المساواة
symmetry	συμμετρία	استواء
asymmetrical	ἀσύμμετρος	خروج عن الاستواء
balanced; symmetrical	σύμμετρος	مستوٍ
uniform	—	متساوٍ
sword	ξίφος	سيف
flowing	ῥυτός	سيّال

ش

seam	ῥαφή	شأن
alum	στυπτηρία	شبّ
sclera, white of the eye (probably an error for الطبقة الصلبة)	σκληρὸς χιτών	الطبقة الشبيهة بالشبكة

like		شِبه
imitation	μιμητικός	تشبيه
imitative, similar	μιμητικός, ὅμοιος	مشبّه
homogeneous, homoeomerous	ὁμοιομερής	متشابهة الأجزاء
tissues	ὁμοιομερεῖς	الأعضاء المتشابهة الأجزاء
systemic diseases	—	الأمراض المتشابهة الأجزاء
winter	χειμών	شتاء
tree	δένδρον	شجرة
fat	πιμελή	شحم
abundant fat	πιμελώδης	كثرة الشّحم
individual	—	شخص
strength	βία	شدّة
to drink	πίνειν	شرب
wine, drink	οἶνος, ποτόν	شراب
drink	πινόμενος	مشرب
commentary	—	شرح
anatomy	ἀνατομή	تشريح
upper abdomen	ὑποχόνδριον	شراسيف
scarification	ἀμυχή	شُرْط
best	ἄριστος	شرف
association, commonality; contiguity	ἀνάλογος, συντετραίνειν	مشاركة
to be in agreement, share a common view	ὁμολογεῖν	اشترك
common	κοινός	مشترك
arterial	ἀρτηριώδης	شرياني
hair	θρίξ	شعر
hairless	ψιλὰ τριχῶν	قلّة الشّعر
hairiness	δασύτης	كثرة الشّعر

thick, coarse hair	λασιώτατος	كثرة الشّعر وكثافته
barley	κριθή	شعير
lip	—	شفة
cure	θεραπεύειν	شفاء
blond	ξανθός	أشقر
shape	σχῆμα	شكل
well formed	εὔρυθμος	حسن الشكل
badly formed	—	رديء الشّكل
to smell	ὀσφραίνεσθαι	شمّ
smell	ὄσφρησις	شمّ
smelled	ὀσφρητός	التي تشمّ
olfactory	ὀσφρητός	مشموم
sun	—	شمس
northerly	ὑπὸ ταῖς ἄρκτοις	شماليّ
convulsions	σπασμός	تشنّج
absurd	ἄτοπος	شنع
absurdity	ἀτοπία	شناعة
observation	τήρησις	مشاهدة
evidences	—	شواهد
desire; appetite	ὀρέγειν	شهوة
old man, elderly	ὁ ἐν ἐσχάτῳ γήρᾳ	شيخ
eye salve	—	شيافة
placenta	—	مشيمة
choroid	χοριοειδὴς χιτών	الطبقة الشبيهة بالمشيمة

ص

to flow	ἀποχεῖν	انصبّ
heap	σωρός	صبّة
child	παῖς	صبيّ
healthy, healthful, relating to health	ὑγιεινός	صحّيّ

healthy	ὑγιεινός	صحيح
creating health (of signs)	ὑγιεινός	مصحّح
ichor	ἰχώρ	صديد
trunk; chest	θώραξ, στέρνον	صدر
smallness	μικρότης	صغر
small; soft (of voice)	μικρός, σμικρότης, ὀλίγος	صغير
yellow	ξανθός	صفراء
clarity	ὑδατώδης	صفاء
spine	—	صلب
hard; indurated (of swellings)	σκληρός	صلب
hardness; hard swelling, induration; see also الورم الصلب، أوذيما	οἴδημα, σκληρότης, σκληρότερος, σκῖρος, τὸ σκληρόν	صلابة
cure	ἰᾶσθαι	إصلاح
surgery	χειρουργία	إصلاح باليد
gum	—	صمغ
art; profession	ἐπιτήδευμα, τέχνη	صناعة
kind	—	صنف
reddish	—	أصهب
voice; sound	φωνή	صوت
form	εἶδος	صورة
jejunum	νῆστις	صائم
summer	θέρος	صيف

ض

opposition	ἀντίθεσις	تضادّ
opposite	—	مضادّ
to harm	βλάπτειν	ضرّ، أضرّ
harmful	βλάβη	ضارّ
harm	ἀνία	ضررٌ
necessarily	ἐξ ἀνάγκης	ضرورةً

absolutely necessary	ἀναγκαῖος	لا بدّ له ضرورةً
necessary	ἐξ ἀνάγκης	اضطراريّ
by compulsion	ἐξ ἀνάγκης	باضطرار
injury, harm done; impairment	βλάβη	مضرّة
approach, way	τρόπος	ضرب
in two senses	διχῶς	على ضربين
weakness	ἀσθένεια	ضعف
weak	ἄρρωστος, ἀσθενής	ضعيف
rib	πλευρά	ضلع
being joined	—	مضامّة
poultice	κατάπλασμα	ضماد
luminosity	λαμπρότης	ضياء
luminous	λαμπρός τε καὶ αὐγοειδής	مضيئ
relation	—	إضافة
narrowness	στενότης	ضِيقٌ
narrow	στενός	ضيّقٌ
shortness of breath	—	ضيق النفس

ط

medicine	ἰατρικός	طبّ
physician	ἰατρός	طبيب، متطبّب
nature	φύσις	طبع، طبيعة
unnatural, contrary to nature	παρὰ φύσιν	خارج الطّبيعة، خارج عن الطّبيعة
natural; physical	φυσικός, κατὰ φύσιν	طبيعيّ
imprinted with difficulty	δυστύπωτος	سريع الانطباع
quick to be imprinted	εὐτύπωτος	عسر الانطباع
membrane	—	طبقة
sclera, white of the eye (probably an error for الطبقة الصلبة)	σκληρὸς χιτών	الطّبقة الشبيهة بالشبكة
iris	στραφυλοειδὴς or ῥαγοειδὴς χιτών	الطّبقة الشبيهة بالعنبة

cornea	κερατοειδὴς χιτών	الطّبقة الشبيهة بالقرن
choroid	χοριοειδὴς χιτῶν	الطّبقة الشبيهة بالمشيمة
zonula	ἀραχνοειδὴς χιτών	الطّبقة العنكبوتية
spleen	σπλήν	طحال
extreme	ὅρος	طرف
method	μέθοδος	الطّريق الصّناعيّ
food; flavor	θρέψις	طعم
food	ἔδεσμα	طعام
to condemn	κατηγορεῖν	طعن
to seek	ἐρευνᾶν	طلب
question	—	مطلب
to be investigated	—	مطلوب
to release	ὑπάγειν	طلق
diarrhea	διάρροια	استطلاق
absolutely	τὸ ἁπλῶς	مطلق
menstrual blood	—	دم الطّمث
oblong	—	مطاول
clay	πηλός	طين

ظ

nails	ὄνυξ	أظفار
manifest; outer, exterior	φαινόμενος	ظاهر
manifest to sense	φαινόμενος, φαίνεσθαι	ظاهر للحسّ
not manifest	ἄδηλος	غير ظاهر

ع

pellitory (anacyclus pyrethrum)	—	عاقرقرحا
plumpness	—	عبول
insolent	τυραννικός	عاتٍ
electuary	—	معجون
number	ἀριθμός, ποσότης	عدد

predisposed	ῥᾳδίως	مستعدّ
moderation	συμμετρία	اعتدال
immoderate	ἄμετρος	جاوز الاعتدال
moderate	σύμμετρος	معادل، معتدل
moderate temperament	εὔκρατος	معتدل المزاج
nonbeing	—	عدم
symptom; accident	συμβεβηκότος, σύμπτωμα	عرض
duration	—	عرض من الزّمان
accidentally	τύχῃ	بالعرض، على طريق العرض
accidental	τυχικός	عرضيّ
disturbance; accident	πάθος	عارض
broad	—	عريض
interference	—	معارضة
to know, to diagnose	γιγνώσκειν	عرف
to diagnose	διοριστέον	تعرّف
understanding, knowledge, to know	γνῶσις, ἀξιώματα	معرفة
sweat	ἱδρώς	عَرَق
blood vessel, vein	φλέψ	عِرق
arteries	ἀρτηρίαι	عروق ضوارب
veins	φλέβες	عروق غير ضوارب
venous	φλεβώδης	عروقيّ
resolution	—	عزيمة
honey-water	—	ماء العسل
nerve	νεῦρον	عصب
genera of nerves	τὸ νευρῶδες	أجناس العصب
having to do with nerves	νευρώδης	عصبيّ
arm	βραχίων	عضد
muscle	μῦς	عضل
organ	μέρος, μόριον, ὄργανον, τόπος	عضو

functional organ	ὀργανικός	عضو آليّ
internal organ	—	عضو باطن
organ, compound organ	ὄργανον	عضو مركّب
tissue; lit. homoeomerous organ	ὁμοιομερής	عضو متشابه الأجزاء
destruction	—	عطب
thirst	διψώδης	عطش
to be enlarged	μείζων	عظُم
large size; magnificence (of pulse)	μέγας	عِظَم
bone	ὀστέον	عَظم
great quantity; loud	μέγας	عظيم
astringent	στρυφνός	عفص
putrefaction	σήπεσθαι, σηπεδών	عفونة
putrid	σηπεδονώδης	عفونيّ
outcome	—	عاقبة
coagulate	—	انعقاد
reason, mind	λόγος, νοῦς	عقل
intelligible	—	بالعقل، معقول
conversion	—	عكس
disease; see also عارض	πάθος	علّة
diseased	πεπονθώς	عليل
surgery	χειρουργία	علاج باليد
theory, knowledge, science	ἐπιστήμη	عِلم
theoretical	—	علميّ
instruction	διδασκαλία	تعليم
difficulty in receiving instruction	δυσμαθία	عسر قبول التّعليم
sign	σημεῖον, γνώρισμα	علامة
student	—	متعلّم
general	καθόλου	عامّيّ

practice	—	عمل
requiring exercise	μεγάλη ἐνέργεια	تحريك بالعمل
practical	—	عمليّ
blindness	—	عمىً
garden nightshade	—	عنب الثّعلب
iris	σταφυλοειδὴς or ῥαγοειδὴς χιτών	الطبقة الشبيهة بالعنبة
element	στοιχεῖον	عنصر
elemental	στοιχιώδης	عنصريّ
harshness	—	عنف
harsh, strenuous	μόλις	عنيف
neck	αὐχήν	عنق
strong neck	κρατεραύχην	قويّ العنق
zonula	ἀραχνοειδὴς χιτών	الطبقة العنكبوتيّة
name (of book)	ἐπιγράφειν	عنوان
meaning; topic	σημαινόμενος	معنى
to be twisted	βλαισός, ῥαιβός, λοξός	عُوّج، أُعوج
habit; customary	ἔθος	عادة
caecum	—	أعور
pubic	γεννητικός	حول العانة
to have an aversion to	—	عاف
aversion	—	عيافة
eye	ὀφθαλμός	عين
clearly	σκοπᾶσθαι τῷ λόγῳ, σαφῶς ὁρῶν	عياناً

غ

tertian (of fever)	—	غبّ
early morning	—	غدوات
to nourish	τρέφειν	غذا
nutriments; diet	τροφή	غذاء، أغذية
nutrition	τρέφειν	تغذية

to be nourished	τρέφεσθαι	اغتذى
foreign	—	غريب
to prick	τιτρώσκειν	غرز
innate	σύμφυτος	غريزيّ
goal	σκοπός, τέλος	غرض
to quench	—	غرق
glutinous	γλίσχρος	غرويّ، شيء يغرّى
membrane	—	غشاء
pia mater	—	الغشاء الرّقيق
dura mater	—	الغشاء الصّلب
film	—	غشاوة
fainting	—	غشي
irascibility, anger	ὀξυθυμία, ὀργή, θυμός	غضب
placidity	εἰς ὀργήν, οὐχ ἑτοῖμος	قلّة الغضب
cartilage	χονδρός	غضروف
to overcome; to predominate	νικᾶν, ἐπικρατεῖν	غلب
strife	νεῖκος	غلبة
predominant	ἄκρως	أغلب
fallacy	σόφισμα	غلط
to become coarse	παχυμερής, παχύτερος	غلَظ
coarseness	παχύς	غِلظ
to coarsen	—	غلّظ
coarse, thick	παχύς	غليظ
grief, anxiety	πόνος	غمّ
feces	—	غائط
to alter, change	ἀλλοιοῦν	غيّر
to be altered, changed	ἀλλοιοῦσθαι, ἀλλοίωσις	تغيّر
alteration	ἀλλοίωσις	تغيُّر
easily transformed	εὐαλλοίωτος	سريع التغيّر
end, extreme	ἄκρος	غاية

ف

opening up	διάπτυξις	تفتيح الحدّ
opening	—	انفتاح
torpor	—	فتور
rupture	ἔντερόν ἐν ὀσχέῳ	فتق
intestinal rupture	ῥῆξις	فتق الأمعاء
coils (of intestines)	—	فتل
hemorrhage	αἱμορραγία	انفجار الدّم
thigh	μηρός	فخذ
joy	—	فرح
individual, singular	πρῶτος	مفرد
excess	ἐκτροπή, πλείων	إفراط
excessive purgation	ὑπερκάθαρσις	إفراط في الإسهال
evacuation, to remove	ἐκκρινεῖν, κένωσις	استفراغ
evacuant	ἐκκρινόμενος	مستفرغ
sexual climax	ἐμπιπλάμενος	فراغ
difference	διακριτικός, διαφέρειν	فرق
sect	αἵρεσις	فرقة
to disperse	διϊστάναι	فرّق
inseparable	ἀχώριστος	غير مفارقة لـ
separation	διάκρισις, λύσις	تفرّق
dissolution of continuity, dierisis	λύσις τῆς συνεχείας	تفرّق الاتّصال
fear	φόβος	فزع، تفزّع
hydrophobia	τὸ ὕδορ δείδειν	تفزّع من الماء
break	σχίσμα	فسخ
to be refuted	ἀνατρέπεσθαι	انفسخ
to destroy	διαφθείρασθαι	أفسد، انفسد
corruption, distortion	σφάλμα	فساد

explanation	ἐξήγησις	تفسير
the underlying text commented on	—	فصّ
bloodletting	φλεβοτομεῖν	فصد
joint	—	مفصل
excretion, superfluity	αὐξάνειν, διαχώρησις, περίττωμα	فضْل
excellent, optimal	ἄριστος	أفضل
excellent	ἄριστος	فاضل
excellence	ἀρετή	فضيلة
space	κενός	فضاء
to affect	δρᾶν	فعَل
effect, action, function	ἐνέργεια, ἔργον, ποιητικός	فِعل
unimpaired functions	τὰς ἐνεργείας ἡ τελειότης	تمام الأفعال
actual, in fact	ἐνεργείᾳ	بالفعل
effecting, active	ποιητικός, δραστικός	فاعل
to be affected, passion	πάσχειν	انفعل
passive	παθητικός	منفعلة
vertebra	σφόνδυλος	فقار
thought	—	فكر
to think	—	تفكّرَ
philosophical	φιλοσοφία	فلسفيّ
phlegmona	φλεγμονή	فلغمونيّ
pepper	—	فلفل
mouth	στόμα	فم
to destroy	—	أفنى
incoherent	—	لا يُفهم
quick-wittedness	ἀγχίνοια	سرعة الفهم
to acquire	κτῆσις	إفادة

ق

laws, canon, standard	κανών	قانون
to contract	—	قبَض
constriction; causing constipation	—	قبْض
astringent; restraining, constricting	στύφειν	قابض
receptive	ἐπιδεκτικός	قابل
appetite	χαίρειν	حسن القبول
skull	—	قحف
rank	ἀξίωμα	قدر
ability	—	قدرة
size, magnitude; pl. dimensions	μέγεθος, πηλικότης, πόσος	مقدار
prophylaxis	προφυλακτικός	تقدّم بالحفظ
front	ἔμπροσθεν	قدام
Ancients	παλαιοί	قدماء
anterior	πρόσθιος	مقدّم
premise	πρότασις	مقدّمة
to repel forcibly	—	قذفَ
speck	—	قذىً
menses	καταμήνια	أقراء
proximate	προσεχής	قريب، أقرب
ulcer; lesion (does not refer to internal abnormalities)	ἕλκος	قرحة
growling in the stomach	—	قراقر
horn	κέρας	قرْن
cornea	κερατοειδὴς χιτών	الطبقة الشبيهة بالقرن
concatenation	—	اقتران
forcible	βίαιος	قسر
class	—	قسم

division	διαίρεσις	قسمة
crust	—	قشرة
squamous	κροτάφιος	قشريّ
flake off	—	انقشر
windpipe	ἀρτηρία, τράχυς	قصبة الرّئة
moderation	—	قصدٌ
extreme	—	أقصى
leanness	ὀλιγόσαρκος	قضافة
lean	ὀλιγόσαρκος	قضيف
to cut	κόπτειν, τομή, τμητικός	قَطَعَ
to be cut	κόπτεσθαι	انقطع
concave	σιμός	مقعّر
less	βραχύς	أقلّ
heart	καρδία	قلب
to transform, convert	—	قلّب، أقلب
transformation	μεταπίπτειν	انقلاب
eradicate	ἕλκειν	قلع
block	ἐπέχειν	قمع
plausible	ὡς εἰκότα γιγνωσκότων	مقنع
to fall victim to	νικᾶσθαι	قُهِر، انقهر
dominant	—	قاهر
dominated	—	مقهور
concavity	κοιλότης	تقوير
tar	—	قار، قير
curvature, bowing	—	تقوّس
bowed	—	مقوّس
book	βιβλίον	مقالة
colon	κῶλον	قولون
posture	—	قامة
to oppose; to counteract	νικᾶν, ἀντιπράττειν	قاوم

straightness	—	استقامة
composition; continuance	γένεσις, γίγνεσθαι	قوام
erect	—	مستقيم
rectum	—	معاء مستقيم
faculty; strength, potency; potential	δύναμις, ῥώμη	قوّة
potentially	δυνάμει	بالقوّة
pharmacological potency	—	قوة دوائية
vital faculty	—	قوّة حيوانيّة
strong (of body)	εὔτονος	قويّ
strong (of foods that are hard to digest)	δυσαλλοίωτος	قويّ
vomit; emetic	ἔμετος	قيء
pus	πύον	قيح
tar	—	قير، قار
inference; syllogism	λόγος, σχῆμα συλλογιστικά	قياس
formal syllogism	σχῆμα συλλογιστικόν	قياس وضعيّ
Rationalists	λογικοί, δογματικοί	أصحاب القياس
on balance	μετρίως	مقايسة
intestinal hernia	κήλη, ἐντερόν ἐν ὀσχέῳ	قيلة الأمعاء

ك

liver	ἧπαρ	كبد
large size	μέγεθος	كبر
large	μέγας	كبير
sulfur	—	كبريت
sulfurous	—	كبريتي
treatise, referring to a complete work of Galen	σύγγραμμα	كتاب
linseed	—	كتّان
to become larger in quantity	πλείονος γίγνεσθαι	كثُر
abundance	πλῆθος, πολύς	كثرة

usually	ἐπὶ τὸ πολύ	على الأكثر، في أكثر الأمر
abundant; multiple	πλῆθος, πολύς	كثير
gum tragacanth	—	كثيراء
to condense, be compressed	συγκρινεῖν, χεῖσθαι	تكاثف
density	τὸ πυκνόν	كثافة
dense	πυκνός, παχύς	كثيف
collyrium	—	كحل
blackness (of eye)	μέλας	كحالة، كحولة، كحلاء
turbidity	οὐ καθαρός	كدورة
false	ψευδής	كذب، كاذب
cranes	—	كراكيّ
fracture	κάταγμα	كسر
sluggishness	—	كسل
barley water	—	كشك الشعير
repletion, surfeit	βαρύνειν	كظّة
without surfeit	ἄλυπος	من غير كظّة
equal	ἴσος	متكافئ
to sate	ἀρκεῖν	اكتفاء
universal	καθόλου	كلّيّ
rabid dog	κύων λυττῶν, λύττα	كلب كَلِب
discussion, views	—	كلام
quantity	ποσότης	كمّيّة
insufficient	ἀτελής	ليس تكمل
firmness	τὸ παχύ	اكتناز
soothsaying	—	تكهّن
soothsayer	—	متكهّن
to come to be, coming to be, generation, genesis	γίγνεσθαι, γένεσις	كون
cauterization	διὰ πυρός	كيّ
quality	ποιότης	كيفية
having qualities	ποιότητα δεδεγμένον	مكيّف

ل

why it is	—	لِما هو
suitability	—	ملاءمة
to be compatible; to be combined	—	التأم
to rest; to continue	—	لبث
milk	γάλα	لبن
flesh	σάρξ	لحم
symphysis	σύμφυσις	لحام
to be attached	—	التحم
knitting	σύμφυσις	التحام
abridgment	—	تلخيص
pleasure	ἡδονή	لذّة
sticky	γλίσχρος, ἄδηκτος	لزج
correlated, consequent	—	لازم
adherence	—	لزوم
tongue	γλῶττα	لسان
fractured (of skull)	—	لاطئ
to become subtle	λεπτυνόμενος	لطف
subtlety	λεπτομερής	لطافة
subtle	καθαρός	لطيف
contact, encounter	προσπίπτειν	لقاء، ملاقاة
to touch	ἅπτειν	لمس
contact	—	ملامسة
palpable, tactile	ἁπτός, πρὸς τὴν ἁφήν	ملموس
burning	καῦμα	التهاب
burning	—	لهيب
uvula	γαργαρεών	لهاة
lambdoid	λαμβδοειδής	لاميّ
color; complexion	χροιά	لون

neither	οὐδέτερος	ليس. . . ولا
tenderness; softness	τό μαλακόν, μαλακότης	لين
soft	μαλακός	ليّن

م

what it is	τὸ ὁποίων	ما هو
solidity	σκληρός	متانة
pattern; image	παράδειγμα	مثال
following the example of	παράδειγμα	امتثال، تمثّل
bladder	κύστις	مثانة
to examine	—	امتحن
marrow	μυελός	مُخّ
distention	ἀνεύρυνσις, διατείνειν, συντείνειν, τάσις	مدّ، تمدّد
matter	ὕλη, οὐσία	مادّة
bitter	—	مرّ
bile	χολή	مرّة
black bile	χολὴ μέλαινα	مرّة سوداء
yellow bile	χολὴ ξανθή	مرّة صفراء
bitterness; gall bladder	—	مرارة
bilious	χολῶν	مراريّ
to be diseased	νοσεῖν	مرَض
disease	νόσημα	مرض
of disease, diseased	νοσώδης, νοσερός	مرضيّ
creating disease (of causes)	νοσερός	ممرّض
diseased; patient	νοσώδης	مريض
sickly	νοσώδης	ممراض
to assimilate	—	استمراء
to mix	κεραννύναι	مزج
temperament	κρᾶσις	مزاج، امتزاج
mixture, tempering	κεραννύναι, κρᾶσις	ممازجة

to be mixed	κεράννυσθαι	امتزج
retention	στάλσις	إمساك
adhesion; costiveness	στεγνόν	استمساك
bowels	—	مصران
to be past	προγίγνεσθαι	مضى
stomach	γαστήρ	معدة
small intestines	λεπτὸν ἔντερον	أمعاء دقاق
large intestines	—	أمعاء غلاظ
rectum	—	معاء مستقيم
plausible	δυνατός	ممكن
salty	—	مالح
saltiness	—	ملوحة
to smooth	ἐκλεαίνειν	ملّس
smoothness	τὸ λεῖον, λειότης	ملاسة
smooth (of voice)	λειότης	أملس
plethora	πληθωρικὰ διάθεσις	امتلاء
to restrain	ἐπέχειν	منع
contraindicant	—	مانع
semen	σπέρμα, σπερματικός	منىً
having moist semen	ὑγρόσπερμος	رطوبة المنى
having thick semen	παχύσπερμος	غلظ المنى
having a small quantity of semen	ὀλιγόσπερμος	قلّة المنى
abundance of sperm	πολύσπερμος	كثرة المنى
water	ὕδωρ, ὑγρός	ماء
like water	ὑδατώδης	شبيهة بالماء
mortal	θνητός	مائت

ن

to grow from	φύειν ἀπό	نبت من
plant	φυτόν	نبات

pulse	σφυγμός	نبض
protruding	προπετής	ناتئ
protrusion	ἐξοχή	نتوء
to deduce, infer	ἀποδεικνύναι	أنتج
conclusion	ἀκόλουθος	نتيجة
fetid	—	منتنة
to hold (a belief)	φάναι, ἀξιοῦν	انتحل
kind; sense	—	نحو
grammarian	γραμματικός	نحويّ
side	—	ناحية
prickly	—	ناخس
spinal cord	νωτιαῖος	نخاع
prognostic	προγνωστικός	منذر
extraction	ἀντίσπασις	انتزاع
bleeding	—	نزف
emission	—	إنزال
relation	—	نسبة
forgetfulness	ἐπιλησμοσύνη	نسيان
that originate from it	ἀπ’ αὐτοὺ πεφυκώς	التي منشأه منه
emission	—	انتشار
inhaling	εἰσπνοή	استنشاق
erect	—	منتصب
light (of hair color)	ξανθός	أحمر ناصع
coction	πέττειν	نضج، إنضاج
sperm	σπέρμα	نطفة
rational, rationalist	λογικός	ناطق
irrational	ἄλογος	غير ناطق
science of logic	διαλεκτική, θεωρία	علم المنطق
of logic	λογικός	منطقيّ
douche	—	نطول

to look; to investigate	ἐρευνᾶν, εὑρίσκειν	نظرَ
investigation	εὕρεσις	نظرٌ
prognosis	πρόγνωσις	سابق النّظر
acuity	—	جودة النّظر
to debate	ἀμφισβητεῖν, διαβάλλειν	ناظرَ
to cleanse	ἐκκενοῦν	استنظف
arrangement	τάξις	نظام
recuperation, convalescence	ἀναληπτικός, ἀναθρεπτικός	إنعاش
inflation	διατείνασθαι	انتفاخ
to penetrate	συντετραίνειν	نفذَ
orifice	ἐκροή	منفذ
averse to	—	منافر
to breathe	—	تنفّس
respiration, breath	ἀναπνοή	تنفّس
soul	ψυχή	نَفْس
of itself	αὐτόματος	من قبل نفسه
breath	πνεῦμα	نَفَس
psychic	ψυχικός	نفسيّ، نفسانيّ
to help	ὀφέλλειν	نفع
benefit	χρεία	انتفاع
beneficial	χρήσιμος	نافع
unprofitable	ἄχρηστος	لا ينفع
uses	χρεία	منفعة
gout	—	نقرس
flaw, deficiency	ἔνδεια	نقص
lack, deficiency	λεπτός, λείπειν	نقصان
conversion (method of investigation)	—	نقْض
falls short	ἀπολειπόμενος	ناقض
point	—	نقطة

to transform; transfer	ἀλλοιωτικός, μεταβαίνειν, μετοχέτευσις	نقل
transition to that which is similar to it	τοῦ ὁμοίου μετάβασις	نقل إلى ما هو شبيه به
transition	μετάβασις	نقلة
to move	—	انتقل
transformative cause	ἐπανορθωτικός	سبب ناقل
to deny	μή συγχωρεῖν	أنكر
false	ψευδής	منكر
convalescent	ἀναληπτικός, ἀναθρεπτικός	ناقِه
herpes	ἕρπης	نَمْلة
growth	αὔξησις	نماء
bite	δάκνειν	نهشة
finite	πεπερασμένος	متناهٍ
infinite	ἄπειρος	غير متناهٍ
completion	τελειοῦσθαι, τέλος	منتهى
periodic	—	نائبة
fire	πῦρ	نار
fire-colored	ὑπόπυρρος	لون النّار
form, species	εἶδος	نوع
sleep	ὕπνος	نوم
dream	ὄνειρος	منام
tooth	—	ناب

ه

to tear	ἄμυγμα	هتك
brittleness	κραῦρον	هشاشة
to digest	πέττειν	هضم
worry	—	همّ
air; weather	ἀήρ	هواء
form	κατασκευή	هيئة

edema	σκίρρος, οἰδήμα	تهيّج
prime matter	ὕλη	هيولى

و

continuous (of pulse)	πυκνότης	تواتر
to find	εὑρίσκειν	وجَد
to exist	ὑπάρχειν	وُجد
pain	ἄλγημα	وجع
to feel pain	ἀλγεῖν	اتّجع
cheek	—	وجنة
one; unitary	ἕν	واحد
like an animal	—	وحشيّ
rose oil	ῥόδινος	دهن الورد
swelling	φλεγμονή	ورم
hematoma	φλεγμονή	ورم دمويّ
soft swelling, edema	οἴδημα	ورم رخو
indurated swelling; see also صلابة	σκίρρος	ورم صلب
swollen	ὄγκος, φλεγμαίνων	وارم
intermediate	μέσος, μεταξύ	وسطٌ، متوسّط
intermediate	μεταξύ	وسطى
to expand	—	وسع
to expand	—	اتّسع
breadth	εὐρύτης	سعة
melancholia	—	وسواس سودانيّ
description	—	وصف
attribute	—	صفة
to touch	ἅπτεσθαι	اتّصل
continuity	ἕνωσις, συνέχεια	اتّصال
to posit	τιθέναι	وضَع

orientation, position	θέσις	وضعٌ
formal syllogism	σχῆμα συλλογιστικόν	قياس وضعيّ
place	τόπος, χωρίον	موضع
subject, substratum	ὕλη, τὸ κοινῇ πᾶσιν	موضوع
sperm ducts	σπερματικὰ ἀγχεία	أوعية المني
according with, analogous	σύμφωνος	موافقة
to be in agreement	ὁμολογεῖν	اتّفق
incidence (technical term of the Empiricists for a natural or chance occurrence of a medical effect)	περίπτωσις	اتّفاق
most suitable	—	أوفق
moment	χρόνος	وقت
at the present moment	ἐν τῷ νῦν	في الوقت الحاضر
season	ὥρα	وقت السنة
at some particular time	πρόσκαιρος	وقت من الأوقات
to take care	προνεῖσθαι	وقّى
to generate, procreate		ولّد
procreation of females	θηλυγόνος	توليد الأناث
procreation of males	ἀρρενογόνος	توليد الذكور
fertile; generative	γόνιμος	مولّد

ي

dryness	ξηρότης, ἕξις ξηροτέρα, τὸ ξηρόν	يبس، يبوسة
dry	ξηρός, ξηρότερος	يابس
arm	χείρ	يد
jaundice	ἰκτεριᾶν	يرقان
slight	βραχύς	يسير
by a little	οὐ πολῷ	بيسير
wakefulness	ἐγρήγορσις, γρηγορεῖν	يقظة

Bibliography

Editions and translations of *On the Medical Sects for Beginners*, *The [Small] Art of Medicine*, and *On the Elements According to the Opinion of Hippocrates* used in the preparation of this edition and translation

Works of Galen are normally referenced by the page numbers of the edition of Karl Gottlob Kühn, the indispensable—if erratic—nineteenth-century Greek edition and Latin translation of most of Galen's surviving works in twenty-two volumes. As it happens, all three of the works epitomized in this volume appear in Kühn's edition but have more recent Greek editions and modern English translations. Ḥunayn ibn Isḥāq's Arabic translations of all three works have been edited by Muḥammad Salīm Sālim.

On the Medical Sects for Beginners. The standard Greek edition of Galen's *The Medical Sects* was edited by Georg Helmreich in Galen, *Scripta minora*, 3:1–32, supplanting Kühn's edition in Galen, *Opera omnia*, 1:64–105. A Latin translation accompanies Kühn's edition. Ḥunayn ibn Isḥāq's Arabic translation is in Galen, *Kitāb Jālīnūs fī firaq al-ṭibb*, edited by Muḥammad Salīm Sālim. Michael Frede's English translation is in Galen, *Three Treatises*, 3–20. References in notes to the text of Galen's *Medical Sects* are of the form: K 1.67, H 3, G^a 20, referring volume 1, page 67, of Kühn's Greek edition; page 3 of Helmreich's Greek edition; and page 20 of the Arabic edition. Michael Frede's English translation gives the Helmreich pagination in the margins, and Sālim gives both Helmreich's and Kühn's pagination in the notes. I use Helmreich's text when citing the Greek.

The [Small] Art of Medicine. Kühn's Greek edition in Galen, *Opera omnia*, 1:305–412, is supplanted by Véronique Boudon-Millot's Greek edition with French translation in Galen, *Galien*, vol. 3: *Exhortation à l'étude de la medicine; Art medical*, 274–391, which also has an admirable introduction and extensive notes. The edition of Ḥunayn's Arabic translation is Galen, *Jālīnūs: al-Ṣināᶜah al-ṣaghīrah*, edited by Muḥammad Salīm Sālim. The modern English translation is by P. N. Singer, in Galen, *Selected Works*, 345–96. References in the notes to Galen's original text read thus: B 1.5; K 1:307; G[a] 7, referring to chapter 1, sentence 5, of the Boudon-Millot edition; volume 1, page 307, of Kühn's edition; and page 7 of Sālim's Arabic edition. Singer's translation contains Kühn's pagination.

On the Elements According to the Opinion of Hippocrates. The Greek editions of Kühn in Galen, *Opera omnia*, 1:413–508, and Helmreich, *Galeni de elementis ex Hippocrate*, have been supplanted by Phillip De Lacy's edition and English translation, *On the Elements According to Hippocrates*, which includes both the Kühn and Helmreich paginations. The edition of Ḥunayn's Arabic translation is Galen, *Kitāb Jālīnūs fī al-isṭiqsāt ᶜalā raᵓy Ibbuqrāṭ*, edited by Muḥammad Salīm Sālim. References in the notes to Galen's original text read thus: L 1.1; K 1:413; G[a] 9, meaning paragraph 1, sentence 1, of De Lacy's edition; volume 1, page 413, of Kühn's edition; and page 9 of Sālim's Arabic edition.

Other works cited or consulted

Adler, Ada, ed. *Suidae lexicon*. 4 vols. Leipzig: Teubner, 1935. Reprint 1971.

Aetius [Pseudo-Plutarch]. *Aetius Arabus: Die Vorsokratiker in arabischer Überlieferung*. Edited by Hans Daiber. Veröffentlichungen der orientalischen Kommission. Wiesbaden: Franz Steiner, 1980.

[Agnellus of Ravenna?] *Lectures on Galen's De sectis*. Edited and translated by L. G. Westerink, Arethusa Monographs 8. Buffalo, NY: Department of Classics, SUNY Buffalo, 1981.

Al-Dubayan, Ahmad M., ed. *Galen: Über die Anatomie der Nerven: Originalschrift und alexandrinisches Kompendium in arabischer Überlieferung*. Islamkundliche Untersuchungen 228. Berlin: Klaus Schwarz Verlag, 2000.

Alexander of Aphrodisias. *De mixtione*. In *Alexandri Aphrodisiensis praeter commentaria scripta minora*, edited by Ivo Bruns, book 2, part 2. Berlin: Reimer, 1892.

———. *Jawāmiᶜ Kitāb al-nabḍ al-ṣaghīr*. Aligarh, India: Ibn Sina Academy of Medieval Medicine and Sciences, 2007.

ᶜAlī ibn Riḍwān. *Al-Kitāb al-nāfiᶜ fī kayfīyat taᶜlīm ṣināᶜat al-ṭibb*. Edited by K. al-Samarrāᵓī. Baghdad: Jāmiᶜat Baghdād, 1986.

———. See also Schacht, Joseph, and Max Meyerhof.

Andalusī, Abū'l-Qāsim Ibn Ṣāᶜid al-. *Kitāb ṭabaqāt al-umam*. Edited by Louis Cheikho. Beirut: Imprimerie catholique, 1912. Reprinted as Is. Phil. 1.

———. *Kitāb ṭabaqāt al-umam*. Edited by Ḥayāt Bū ᶜAlwān. Beirut: Dār al-ṭalīᶜa, 1985.

———. *Kitāb ṭabaqāt al-umam*. Edited by Ḥusayn Muʾnis. Dhakhāʾir al-ʿArab 74. Cairo: Dār al-maʿārif, 1998.

———. *Science in the Medieval World: "Book of the Categories of Nations."* Translated by Semaʿan I. Salem and Alok Kumar. History of Science Series 5. Austin, TX: University of Texas Press, 1991.

Arberry, Arthur J. *Fihris al-makhṭūṭāt al-ʿarabīya fī Maktabat Tshistir Baytī [Chester Beatty]*. Vol. 2. Amman: al-Majmaʿ al-malikī li-buḥūth al-ḥaḍārah al-islāmīyah, 1994.

Arikha, Noga. *Passions and Tempers: A History of the Humors*. New York: Ecco, 2007.

Aristotle. *The Complete Works of Aristotle: The Revised Oxford Translation*. Bollingen Series 71.2. 2 vols. Princeton: Princeton University Press, 1984.

———. *On the Heavens*. Translated by J. L. Stocks. In Aristotle. *Complete Works*. Vol. 1. 447–511.

———. *The Physics*. Loeb Classical Library 228. Cambridge, MA: Harvard University Press, 1993.

———. *Physics*. Translated by Hippocrates G. Apostle. Bloomington: Indiana University Press, 1969.

Ātish, Aḥmad. "Al-Makhṭūṭāt al-ʿarabīyah fī maktabāt al-Anāḍūl. 1. Makhṭūṭāt Maktabat Maghnīsā al-ʿumūmīyah." *Majallat Maʿhad al-makhṭūṭāt al-ʿarabīyah*, vol. 4.1 (1958/1377), 1–42. Reprinted in *Beiträge zur Erschliessung der arabischen Handschriften in Istanbul und Anatolien*, edited by Fuat Sezgin. Frankfurt am Main: IGAIW, 1986. 3:827–72.

Baffioni, G. "Inediti di Archelao da un codice bolognese." *Bollettino del Comitato per la preparazione dell'edizione nazionale dei classici greci e latini*, n.s., 3 (1955): 57–76.

———. "Scolii inediti di Palladio al *De sectis* di Galeno." *Bollettino del Comitato per la preparazione dell'edizione nazionale dei classici greci e latini*, n.s., 6 (1958): 61–78.

Bergsträsser, G. *Neue Materialen zu Ḥunain ibn Isḥāq's Galen-Bibliographie*. Leipzig: Deutsche Morgenländische Gesellschaft, 1932. Reprint Is. Med. 19.

Boudon, Véronique. "Les définitions tripartites de la médicine chez Galien." In Haase, *Wissenschaften (Medizin und Biologie)*, 2.37.2:1468–90.

———. "Les œuvres de Galien pour les débutants ('De sectis,' 'De pulsibus ad tirones,' 'Ad Glauconem de methodo medendi,' et 'Ars medica'): médecine et pédagogie au IIe s. ap. J.-C." In Haase, *Wissenschaften (Medizin und Biologie)*, 2.37.2.1421–67.

Brill's New Pauly: Encyclopaedia of the Ancient World. Edited by Hubert Cancik and Helmuth Schneider. Eng. edited by Christine F. Salazar. Leiden: E. J. Brill, 2002.

Brock, Sebastian. "The Syriac Background to Ḥunayn's Translation Techniques." *Aram* 3 (1991): 139–62. Reprinted in Sebastian Brock, *From Ephrem to Romanos: Interactions between Syriac and Greek in Late Antiquity*, xiv. Aldershot, Engl.: Ashgate Variorum, 1999.

Brockelmann, Carl. *Geschichte der arabischen Litteratur*. 2nd ed. 5 vols. Leiden: E. J. Brill, 1937–49.

Casiri, Michaelis. *Bibliotheca Arabico-Hispana Escurialensis*. 2 vols. Madrid: Antonius Perez de Soto, 1760–70. Reprinted Osnabrück, Austria: Biblio Verlag, 1969.

Complete Dictionary of Scientific Biography. Detroit: Charles Scribner's Sons, 2008.

Dānish-Pashūh, Muḥammad-Taqī. *Fihrist-i nuskha-hā-yi kitābkhānah-yi markazī-yi Dānishgāh-i Tihrān*. Vols. 14–15. Intishārāt-i Dānishgāh-i Tihrān 722, 1096. Tehran: Dānishgāh-i Tihrān, 1340/1962, 1345/1967.

Deichgräber, Karl. *Die griechische Empirikerschule*. 1930. Reprint, Berlin: Weidmann, 1965.

Dickson, Keith, M., ed. *Stephanus the Philosopher and Physician: Commentary on Galen's* Therapeutics to Glaucon. Studies in Ancient Medicine 19. Leiden: E. J. Brill, 1998.

Dietrich, Albert. *Medicinalia Arabica: Studien über arabische medizinische Handschriften in türkischen und syrischen Bibliotheken*. Göttingen, Ger.: Vandenhoeck und Ruprecht, 1966.

Derenbourg, Hatwig. *Les manuscrits arabes de l'Escurial*. Edited by H.-P.-J. Renaud. 2.2: *Médicine et histoire naturelle*. Paris: Paul Geuthner, 1941.

Diels, Hermann Alexander. *Die Fragmente der Vorsokratiker*. Revised by Walther Kranz. 6th ed. Berlin: Weidmann, 1952.

Duffy, John. "Byzantine Medicine in the Sixth and Seventh Centuries: Aspects of Teaching and Practice." *Dumbarton Oaks Papers* 38 (1984): 21–27.

Encyclopaedia of Islam. 2nd ed. Leiden: E. J. Brill, 1960–2008.

Fitzpatrick, Coeli. "Galen's Six Non-Naturals in the Medieval Arabic Tradition." PhD diss., Binghamton University, 2002.

Frede, Michael. "The Method of the So-Called Methodical School." In Michael Frede, *Essays in Ancient Philosophy*, 261–78. Minneapolis: University of Minnesota Press, 1987.

Freeman, Kathleen. *Ancilla to the Pre-Socratic Philosophers*. Cambridge: Harvard University Press, 1948.

———. *The Pre-Socratic Philosophers*. 2nd ed. Oxford: Basil Blackwell, 1959.

Galen, Claudius. *De usu partium libri XVII*. 2 vols. Leipzig: B. G. Teubner, 1907-1909. Reprint, Amsterdam: A. M. Hakkert, 1968.

———. *Galeni De elementis ex Hippocratis sententia*. Edited by G. Helmreich. Erlangen: Deichert, 1878.

———. *Galen on the Affected Parts*. Translated by Rudolph E. Siegel. Basel: S. Karger, 1976.

———. *Galen on the Usefulness of the Parts of the Body*. Translated by Margaret Tallmadge May. Cornell Publications in the History of Science. Ithaca, NY: Cornell University Press [1968].

———. *Galen's Art of Physick, wherein is laid down, 1. a description of bodies, healthful, unhealthful, and neutral 2. signs of good and bad constitutions 3. signs of the brain, heart, liver, testicles, temperature, lungs, stomach, &c., being too hot, cold, dry, moist, hot and dry, hot and moist, cold and dry, cold and moist 4. signs and causes of sickness, with many other excellent things, the particulars of which, the Table of contents will specifie: translated into English, and largely commented on : together with convenient medicines for all particular distempers of the parts, a description of the complexions, their condition, and what diet and excercise is fittest for them*. Translated by Nicholas Culpepper. London: J. Streater, 1671. Early English Books Online.

———. *Galien*. Vol. 1, *Introduction generale; Sur l'ordre de ses propres livres; Sur ses propres livres; Que l'excellent médicin est aussi philosophe*. Vol. 2, *Exhortation à l'étude de la médecine; Art medical*. Vol. 3, *Le medicine; Introduction*. Collection des Universités de France publiée sous le patronage de l'Association Guillaume Budé. Edited and translated by Véronique Boudon-Millot. Paris: Les Belles Lettres, 2000–2009.

———. "Jawāmi[c] Tashrīḥ al-[c]Aṣab." In Al-Dubyan, *Galen*, 96–117.

———. *Kitāb Jālīnūs fī firaq al-ṭibb*. Edited by Muḥammad Salīm Sālim. Muntakhabāt al-Iskandarānīyin 1. Cairo: Dār al-kutub, 1977.

———. *Kitāb Jālīnūs Fī al-istiqsāt [c]alā ra[ʾ]y Ibbuqrāṭ*. Edited by Muḥammad Salīm Sālim. Muntakhabāt al-Iskandarānīyin 5. Cairo: al-Hay[ʾ]ah al-[c]āmmah al-miṣrīyah li'l-kitāb, 1977.

———. *Method of Medicine*. Loeb Classical Library, 516–18. 3 vols. Cambridge: Harvard University Press, 2011.

———. *On Anatomical Procedures: De anatomicis administrationibus*. Translated by Charles Singer. Publications of the Wellcome Historical Medical Museum, n.s., 7. Oxford: Oxford University Press, 1956.

———. *On Antecedent Causes*. Edited and translated by R. J. Hankinson. Cambridge Classical Texts and Commentaries 35. Cambridge: Cambridge University Press, 1998.

———. *On Diseases and Symptoms*. Cambridge: Cambridge University Press, 2006.

———. *On the Elements According to Hippocrates*. Edited by Phillip De Lacy. Corpus medicorum graecorum 5.1.2. Berlin: Akademie Verlag, 1996.

———. *On the Parts of Medicine; On Cohesive Causes; On Regimen in Acute Diseases in Accordance with the Theories of Hippocrates*. Arabic edition and English translation by Malcolm Lyons; Latin edition by H. Schoene and K. Kalbfleisch; Latin reedition by J. Kollesch, D. Nickel, and G. Strohmaier. Corpus medicorum graecorum, Supplementum orientale II. Berlin: Akademie-Verlag, 1969.

———. *Opera omnia*. Edited by Karl Gottlob Kühn, Franz Wilhelm Assmann, and Konrad Schubring. 20 vols. in 22. 1821–33. Reprint, Hildesheim: Georg Olms, 1964–65.

———. *Protreptici*. Edited by Georgius Kaibel. 1894. Reprint Berlin: Weidmann, 1963.

———. *Scripta minora*. Edited by Johann Marquardt, Iwan von Müller, and Georg Helmreich. 3 vols. Leipzig: B. G. Teubner, 1884–93.

———. *Selected Works*. Translated by P. N. Singer. Oxford World's Classics. Oxford: Oxford University Press, 1997.

———. *Al-Ṣinā[c]ah al-ṣaghīrah*. Edited by Muḥammad Salīm Sālim. Muntakhabāt al-Iskandarānīyin 2. Cairo: al-Hay[ʾ]a al-miṣrīyah al-[c]āmmah li'l-kitāb, 1988.

———. *The Small Art*. In Galen, *Selected Works*, 345–96.

———. *Three Treatises on the Nature of Science:* On the Sects for Beginners; An Outline of Empiricism; *and* On Medical Experience. Translated by Michael Frede and Richard Walzer. Indianapolis, IN: Hackett, 1985.

Garcia-Ballester. "On the Origin of the 'Six Non-Natural Things' in Galen." In Kollesch and Nickel, *Galen*, 105–15.

Garofalo, Ivan. "I Sommari degli alessandrini." In Garofalo and Roselli, *Galenismo*, 203–27.

———. "La traduzione araba dei compendi alessandrini delle ope del canone di Galeno: Il compendio dell'*Ad Glauconem*." *Medicina nei secoli arte e scienza* 6 (1994): 329–48.

———. "Una nuova opera di Galeno: La *Synopsis* del *De methodo medendi* in versione araba." *Studi classici e orientali* 47, no. 1 (1999): 1–19.

Garofalo, Ivan, and Amneris Roselli, eds. *Galenismo e medicina tardoantica: Fonti greche, latine e arabe.* Annali dell'Instituto universitario oriental di Napoli 7. Naples: Dipartimento di studi del mondo classico e del Mediterraneo antico, sezione filologico-letteraria, 2003.

Goulet, Richard, ed. *Dictionnaire des philosophes antiques*. 4 vols. and 1 supplement to date. Paris: Éditions du Centre national de la recherché scientifique, 1989–.

Gundert, Beate. "Die *Tabula Vindobonenses* als Zeugnis alexandrinischer Lehrtätigkeit um 600 N. Chr." In *Text and Tradition: Studies in Ancient Medicine and Its Transmission Presented to Jutta Kollesch*. Edited by Klaus-Dietrich Fischer, Diethard Nickel, and Paul Potter. Studies in Ancient Medicine 18. Leiden: E. J. Brill, 1998.

Gutas, Dimitri. *Greek Thought, Arabic Culture: The Graeco-Arabic Translation Movement in Baghdad and Early ʿAbbāsid Society (2nd–4th/8th–10th Centuries)*. New York: Routledge, 1998.

Haase, Wolfgang, ed. *Wissenschaften (Medizin und Biologie)*. ANRW 2.37. Berlin: Walter de Gruyter, 1993–95.

Ḥāʾirī, ʿAbd al-Ḥusayn. *Fihrist-i Kitābkhānah-yi Majlis-i shūrā-yi millī*. Vol. 10. Tehran: Kitābkhānah-yi Majlis-i shūrā-yi millī, 1350/1972.

Hamarneh, Sami K. *Catalogue of Arabic Manuscripts on Medicine and Pharmacy at the British Library*. History of Arabic Medicine and Pharmacy 3. Cairo: n.p., 1975.

Hankinson, R. J. "Stoicism and Medicine." In *The Cambridge Companion to the Stoics*, edited by Brad Inwood, 295–309. Cambridge: Cambridge University Press, 2003.

———, ed. *The Cambridge Companion to Galen*. Cambridge: Cambridge University Press, 2008.

Hippocrates. *De natura hominis*. In *Hippocrates*, edited and translated by W. H. S. Jones. Loeb Classical Texts. Cambridge, MA: Harvard University Press, 1943. 4:1–41.

———. *De natura hominis*. In *Hippocratic Writings*, edited by G. E. R. Lloyd, translated by J. Chadwick and W. N. Mann. Harmondsworth, Engl.: Penguin, 1978.

Hitti, Philip K., Nabih Amin Faris, and Buṭrus ʿAbd-al-Malik. *Descriptive Catalog of the Garrett Collection of Arabic Manuscripts in the Princeton University Library*. Princeton Oriental Texts 5. Princeton: Princeton University Press, 1938.

Homer. *Iliad*. Translated by Robert Fagles. New York: Penguin, 1990.

Ḥunayn ibn Isḥāq al-ʿIbādī. *The Book of the Ten Treatises on the Eye Ascribed to Hunayn ibn Is-haq*. Edited and translated by Max Meyerhof. Cairo: Government Press, 1928. Reprinted as Is. Med. 22.

———. *Masāʾil fī al-ṭibb li-al-mutaʿallimīn*. Edited by Muḥammad–ʿAlī Abū-Rayyān, Mursī Muḥammad ʿArab, and Jalāl Muḥammad Mūsā. Cairo: Dār al-jāmiʿah al-miṣrīyah, 1978.

———. *Neue Materialen*. See Bergsträsser, G.

———. *Questions on Medicine for Scholars*. Translated by Galal M. Moussa. Cairo: al-Ahram Center for Scientific Translations, 1980.

———. "Risāla fī dhikr mā turjima min kutub Jālīnūs bi-ᶜilmihi wa-baᶜḍ mā lam yutarjam." In *Ḥunayn ibn Isḥāq über die syrischen und arabischen Galen-Übersetzungen*, edited by G. Bergsträsser. Leipzig: Brockhaus, 1925. Reprinted in Is. Med. 13, 185–306.

Hunger, Herbert, and Otto Kresten. *Katalog der griechischen Handscriften der Österreichischen Nationalbibliothek*. Museion: Veröffentlichungen der Österreichischen Nationalbibliothek 2. Vienna: Georg Prachner Verlag, 1969.

Ibn Abī Uṣaybiᶜa, Aḥmad ibn al-Qāsim. *ᶜUyūn al-anbāʾ fī ṭabaqāt al-aṭibbāʾ*. Edited by August Müller. 2 vols. Cairo: al-Wahbīyah, 1299/1882. Reprinted as Is. Med. 1–2.

Ibn Juljul, Abū Dāʾūd Sulaymān ibn Ḥassān al-Andalusī. *Ṭabaqāt al-aṭibbāʾ wa-al-ḥukamāʾ*. Edited by Fuʾād Sayyid. Cairo: Institut français d'archéologie orientale, 1955. Reprinted as Is. Phil. 57.

Ibn Jumayᶜ. *Treatise to Ṣalāḥ ad-Dīn on the Revival of the Art of Medicine*. Edited and translated by Hartmut Fähndrich. Abhandlungen für die Kunde des Morgenlandes 46, no. 3. Wiesbaden: F. Steiner, 1983.

Ibn al-Nadīm. *Al-Fihrist*. Edited by Riḍā Tajaddud. 2nd ed. Tehran: Marvi, n.d.

———. *Al-Fihrist*. Edited by Ayman Fuʾād Sayyid. London: Al-Furqan Islamic Heritage Foundation, 1430/2009.

———. *The Fihrist of al-Nadīm: A Tenth-Century Survey of Muslim Culture*. Translated by Bayard Dodge. 2 vols. Records of Civilization: Sources and Studies 83. New York, Columbia University Press, 1970.

Ibn Rushd, Abū'l-Walīd. *Kitāb al-kullīyāt: Les généralites d'Ibn Rochd*. Edited by M. Belkeziz ben Abdeljalil. Casablanca, Morocco: Najah al Jadida, 2000.

———. *Al-Kullīyāt fī al-ṭibb*. Edited by Muḥammad ᶜĀbid al-Jābirī. Silsilat al-turāth al-falsafī al-ᶜarabī. Muʾallifāt Ibn Rushd. Beirut: Markaz darāsāt al-waḥdah al-ᶜarabīyah, 1999.

———. *Medical Manuscripts of Averroes at el-Escorial*. Translated by Georges. C. Anawati and P. Ghalioungui. Cairo: al-Ahram Center for Scientific Translations, 1986.

———. *Rasāʾil Ibn Rushd al-ṭibbīya*. Edited by Georges C. Anawati and Saᶜīd Zāyed. Cairo: al-Hayʾah al-miṣrīyah al-ᶜāmmah li-al-kitāb, 1987.

Ibn Sīnā. *The Canon of Medicine (al-Qānūn fī al-ṭibb)*. Adapted by Laleh Bakhtiar from a translation by O. Cameron Gruner. Great Books of the Islamic World. Chicago: Kazi Publications, 1999.

———. *Al-Qānūn fī al-ṭibb*. 3 vols. Būlāq, 1294/1877. Reprint [Cairo]: Dār al-fikr, n.d.

———. *Al-Qānūn fī al-ṭibb*. 5 bks. in 6 vols.; vols. 1–2: New Delhi: Institute of the History of Medicine and Medical Research, 1402/1982–1408/1988; vol. 3.1–2 and vols. 4–5: New Delhi: Hamdard University, 1411/1989–1417/1996.

———. *Al-Qānūn fī al-ṭibb*. Edited by Muḥammad Amīn al-Ḍannāwī. 3 vols. Beirut: Dār al-kutub al-ᶜilmīyah, 1420/1999.

İhsanoğlu, Ekmeleddin, ed. *Fihris makhṭūṭāt al-ṭibb al-islāmī bi-al-lughāt al-ᶜarabīyah wa-al-turkīyah wa-al-fārisīyah fī maktabāt Turkīyah*. Silsilat darāsāt wa-maṣādir fī tārīkh al-ᶜulūm. Istanbul: Markaz al-abḥāth li-al-tārīkh wa-al-funūn wa-al-thaqāfah al-islāmīyah, 1984.

Irvine, Judith T., and Owsei Temkin. "Who was Akilāōs? A Problem in Medical Historiography." *Bulletin of the History of Medicine* 77, no. 1 (Spring 2003): 12–24.

Isḥāq ibn Ḥunayn. *Tārīkh al-aṭibbāʾ*. In Franz Rosenthal, ed. and trans., "Isḥāq b. Ḥunayn's *Tāʾrīkh al-aṭibbāʾ*." *Oriens* 7 (1954): 55–80.

Iskandar, Albert Z. "An Attempted Reconstruction of the Late Alexandrian Medical Curriculum." *Medical History* 20 (1976): 235–58.

Jaeger, Werner. *Diokles von Karystos: Die griechische Medizin und die Schule des Aristoteles.* Berlin: W. de Gruyter, 1938.

Jarcho, Saul. "Galen's Six Non-Naturals: A Bibliographic Note and Translations." *Bulletin of the History of Medicine* 44/4 (July–August 1970): 372–77.

Jihāmī, Jīrār. *Mawsūʿat muṣṭalaḥāt Ibn Rushd al-faylasūf.* Silsilat mawsūʿāt muṣṭalaḥāt aʿlām al-fikr al-ʿarabī wa-al-islāmī. Beirut: Maktabat Lubnān, 2000.

———. *Mawsūʿat muṣṭalaḥāt Ibn Sīnā (al-Shaykh al-Raʾīs).* Silsilat mawsūʿāt muṣṭalaḥāt aʿlām al-fikr al-ʿarabī wa-al-islāmī. Beirut: Maktabat Lubnān, 2004.

John of Alexandria. *Iohannis Alexandrini Commentaria in librum De sectis Galeni.* Edited by C. D. Pritchet. Leiden: E. J. Brill, 1982.

Kant, Immanuel. *Critique of Pure Reason.* Edited and translated by Paul Guyer and Allen W. Wood. Cambridge Edition of the Works of Immanuel Kant. Cambridge: Cambridge University Press, 1997.

Kantūrī, Taṣadduq Ḥusayn al-. *Fihrist-i kutub-i ʿarabī, fārisī wa-urdu makhzūnah-i Kutubkhānah-i āṣafīyah sarkarī ʿālī.* Hyderabad, Deccan: n.p., 1355/1936.

Karabulut, Ali Rıza. *İstanbul ve Anadolu kütüphanelerinde mevcut el yazması eserler ansiklopedisi (Muʿjam al-makhṭuṭāt al-mawjūdah fī maktabāt Istānbūl wa Ānāṭūlī).* 3 vols. Kayseri : Akebe Kitabevi, [2005].

Karabulut, Ali Rıza, and Ahmet Turan Karabulut. *Dünya kütüphanelerinde mevcut İslâm kültür tarihi ile ilgili eserler ansiklopedisi (Muʿjam al-tārīkh al-turāth al-islamī fī maktabāt al-ʿālām: al-Makhṭūṭāt wa-al-maṭbūʿāt).* 6 vols. Kayseri, Turkey: Akebe Kitabevi, [2006].

Karatay, Edhem. *Topkapı Sarayı Müzesi Arapça yazmalar Kataloğu.* 4 vols. Istanbul: Millî Eğitim Bakanlığı Yayınları, 1962–66.

King, Lester S. *Medical Thinking: A Historical Preface.* Princeton, NJ: Princeton University Press, 1982.

Kirk, G. S., and J. E. Raven. *The Presocratic Philosophers: A Critical History with a Selection of Texts.* Cambridge: Cambridge University Press, 1966.

Kollesch, Jutta, and Diethard Nickel. "Bibliographia Galeniana: Die Beiträge des 20. Jahrhunderts zur Galenforschung." In Haase, *Wissenschaften (Medizin und Biologie),* 2.37.2:1351–420, 2063–70.

Kollesch, Jutta, and Diethard Nickel, eds. *Galen und das hellenische Erbe: Verhandlungen des iv. Internationalen Galen-Symposiums . . . 1989.* Sudhoffs Archiv: Zeitschrift für Wissenschaftsgeschichte 32. Stuttgart: Franz Steiner, 1993.

Levey, Martin. See Ruḥāwī, Isḥāq ibn ʿAlī al-.

Lieber, Elinor. "Galen in Hebrew: The Transmission of Galen's Works in the Mediaeval Islamic World." In *Galen: Problems and Prospects,* edited by Vivian Nutton, 167–83. [London]: Wellcome Institute, 1981.

Littman, R. J. "Medicine in Alexandria." In Haase, *Wissenschaften (Medizin und Biologie)*, 2.37.3:2678–708.

Long, A. A., and D. N. Sedley. *The Hellenistic Philosophers*. 2 vols. Cambridge: Cambridge University Press, 1987.

Meyerhof, Max. "Johannes Grammatikos (Philoponos) von Alexandrien und die arabischen Medizin." *Mitteilungen des Deutschen Instituts für ägyptischen Altertumskunde in Kairo* 2 (1932):1–21. Reprinted in Is. Med. 96:149–69.

———. "New Light on Ḥunain Ibn Isḥâq and His Period." *Isis* 8 (1926): 685–724. Reprinted in Is. Med. 20:1–40. Reprinted in Max Meyerhof, *Studies in Medieval Arabic Medicine: Theory and Practice* I, London: Variorum Reprints, 1984.

———. "Von Alexandrien nach Bagdad: Ein Beitrag zur Geschichte des philosophischen und medizinischen Unterrichts bei den Arabern." *Sitzungsberichte der Preußischen Akademie der Wissenschaften: Philosophisch-Historische Klasse* (1930): 389–429. Reprinted in Is. Med. 20:223–63.

Muᶜjam al-muṣṭalaḥāt al-ṭibbīya wa-al-adwiyah al-mufradah al-mustaᶜmalah fī al-Qānūn fī al-ṭibb. New Delhi: Hamdard University, Qism al-dirāsāt al-islāmīyah, 1418/1998.

Nachmanson, E. "Ein neuplatonischer Galenkommentar auf Papyrus." *Götebergs Högskolas Årskrift* 31 (1925): 201–17.

Nutton, Vivian. *Ancient Medicine*. Sciences of Antiquity. London: Routledge, 2004.

———. "From Galen to Alexander: Aspects of Medicine and Medical Practice in Late Antiquity." *Dumbarton Oaks Papers* 38 (1984): 1–14.

———. "John of Alexandria Again: Greek Medical Philosophy in Latin Translation." *Classical Quarterly* 41, no. 2 (1991): 509–19.

———. Review of *Stephanus the Philosopher and Physician: Commentary on Galen's "Therapeutics to Glaucon,"* by Keith M. Dickson. *Classical Review*, n.s., 50, no. 1 (2000): 34–35.

Ottosson, Per-Gunnar. *Scholastic Medicine and Philosophy: A Study of Commentaries on Galen's* Tegni *(ca. 1300–1450)*. Naples: Bibliopolis, 1984.

Palmieri, Nicoletta. "Fonti Galenische (e non) nella lettura Alessandrina dell'*Ars medica*." In Garofalo and Roselli, *Galenismo*, 133–60.

Pauly, August Friedrich von, and Georg Wissowa. *Real-Encyclopädie der classischen Alterthumswissenschaft*. Stuttgart: J. B. Metzler, 1894–1963.

Pellegrin, Pierre. "Ancient Medicine and Its Contribution to the Philosophical Tradition." In *A Companion to Ancient Philosophy*, edited by Mary Louise Gill and Pierre Pellegrin, 664–85. Blackwell Companions to Philosophy 31. Oxford: Blackwell, 2006.

Pigeaud, Jackie. "L'introduction du Méthodisme à Rome." In Haase, *Wissenschaften (Medizin und Biologie)* 2.37.1:565–99.

Pormann, Peter E. "The *Alexandrian Summary (Jawāmiᶜ)* of Galen's *On the Sects for Beginners:* Commentary or Abridgement?" In *Philosophy, Science and Exegesis in Greek Arabic and Latin Commentaries*, edited by Peter Adamson, Han Baltussen, and M. W. F. Stone, 2:11–33. London: Institute of Classical Studies, University of London, 2004.

———. "Jean le Grammarien et le *De sectis* dans la littérature médicale d'Alexandrie." In Garofalo and Roselli, *Galenismo*, 233–63.

Pormann, Peter E., and Emilie Savage-Smith. *Medieval Islamic Medicine*. Washington, DC: Georgetown University Press, 2007.

Prioreschi, Plinio. *History of Medicine*. 6 vols. to date: vol. 2, *Greek Medicine*; vol. 3, *Roman Medicine*; vol. 4, *Byzantine and Islamic Medicine*. Omaha, NE: Horatius Press, 1996–.

Qifṭī, ᶜAlī ibn Yūsuf al-. *Tārīkh al-ḥukamāʾ*. Edited by Julius Lippert. Leipzig: Dieterich'sche Verlagsbuchhandlung (Theodor Weicher), 1903. Reprinted as Is. Phil. 2.

Rāhāward, Ḥasan. *Fihrist-i kutub-i khaṭṭī-yi Kitābkhānah-yi Dānishkadah-yi pizishkī*. Tehran: Dānishgāh-i Tihrān, 1333/1957.

Rather, L. J. "The 'Six Things Non-Natural': A Note on the Origins and Fate of a Doctrine and a Phrase." *Clio Medica* 3 (1968): 337–47.

Rāzī, Abū Bakr Muḥammad ibn Zakarīyā, al-. *Kitāb al-ḥāwī fī al-ṭibb*. Edited by Muḥammad ᶜAbd al-Muᶜīd Khān. 23 vols. in 25. Hyderabad, Decca: Dāʾirat al-maᶜārif al-ᶜuthmānīyah, 1381/1962.

Ritter, Helmut, and Richard Walzer. "Arabsiche Übersetzungen griechischer Ärtze in Stambuler Bibliotheken." *Sitzungsberichte der Preußischen Akademie der Wissenschaften: Philosophisch-Historische Klasse* (1934): 801–46. Reprinted in Is. Med. 21, pp. 1–48; and in Fuat Sezgin, ed., *Beiträge zur Erschliessung der arabischen Handschriften in Istanbul und Anatolien*. Veröffentlichengen des IGAIW B. Nachdrucke: Abteilung Hanschriftenkunde, 2:477–524. Frankfurt am Main: IGAIW, 1986.

Rosenthal, Franz. See Isḥāq ibn Ḥunayn.

Roueché, Mossman. "Did Medical Students Study Philosophy in Alexandria?" *Bulletin of the Institute of Classical Studies of the University of London* 43 (1999): 153–69.

Ruḥāwī, Isḥāq ibn ᶜAlī al-. *Adab al-ṭabīb*. Edited by Murayzin Saᶜīd Murayzin ᶜAsīrī. Riyadh, Saudi Arabia: Markaz al-Malik Fayṣal, 1992.

———. *The Conduct of the Physician: Adab al-ṭabīb*. Edited by Fuat Sezgin. Frankfurt am Main: IHAIS, 1985. Facsimile of MS Edirne, Selimiye 1658.

———. *Medical Ethics of Medieval Islam, with Special Reference to al-Ruḥāwī's 'Practical Ethics of the Physician.'* Translated by Martin Levey. Transactions of the American Philosophical Society, n.s. 57, no. 3 (1967).

Sambursky, Samuel. *The Physics of the Stoics*. Princeton: Princeton University Press, 1987.

Savage-Smith, Emilie. "Galen's Lost Ophthalmology and the *Summaria Alexandrinorum*." In *The Unknown Galen*, edited by Vivian Nutton, 121–38. London: Institute of Classical Studies, University of London, 2002.

Schacht, Joseph, and Max Meyerhof. *The Medico-Philosophical Controversy between Ibn Butlan of Baghdad and Ibn Ridwan of Cairo: A Contribution to the History of Greek Learning among the Arabs*. The Faculty of Arts Publication No. 13. Cairo: The Egyptian University, 1937.

Schoeler, Gregor. *Arabische Handschriften*. Vol. 17/B2. Stuttgart: Franz Steiner Verlag, 1990.

Sezgin, Fuat, ed. *The Alexandrian Compendium of Galen's Work: Jawamiᶜ al-Iskandaraniyyin*. 3 vols. PIHAIS Series C.68. Frankfurt am Main: IHAIS, 2001–2004.

———. *Geschichte des arabischen Schrifttums*. Leiden: E. J. Brill, 1967–.

———, ed. Islamic Medicine Reprint Series. 100 vols. Frankfurt am Main: PIHAIS, 1995–2004.

———, ed. Islamic Philosophy Reprint Series. 120 vols. Frankfurt am Main: PIHAIS, 1999–2000.

Simms, R. Clinton. "The Missing Bones of Thersites," *American Journal of Philology* 126 (2005): 33–40.

Smith, William. *Dictionary of Greek and Roman Antiquities*. Boston: C. Little and J. Brown, 1870.

Sorabji, Richard, ed. *Philoponus and the Rejection of Aristotelian Science*. 2nd ed. London: Institute of Classical Studies, 2010.

Soranus of Ephesus. *Soranos d'Éphèse: Maladies des femmes*. Edited and translated by Paul Burguière, Danielle Gourevitch, and Yves Malinas. Budé series. Paris: Les Belles Lettres, 1988.

———. *Soranus' Gynecology*. Translated by Owsei Temkin. Baltimore: Johns Hopkins University Press, 1991.

Steckerl, Fritz. *The Fragments of Praxagoras of Cos and His School*. Leiden: E. J. Brill, 1958.

Steinschneider, Moritz. *Die arabischen Übersetzungen aus dem Griechischen*. Graz, Austria: Akademische Druck- und Verlagsanstalt, 1960.

———. *Die hebraeischen Übersetzungen des Mittelalters und die Juden als Dolmetscher*. Graz, Austria: Akademische Druck- und Verlagsanstalt, 1956.

Strohmaier, Gotthard. "Der syrische und der arabische Galen." In Haase, *Wissenschaften (Medizin und Biologie)*, 2.37.2:1987–2017.

———. "'Von Alexandrien nach Bagdad'—eine fiktive Schultradition." In *Aristoteles Werk und Wirkung*, edited by Jürgen Wiesner, 380–89. Berlin: Walter de Gruyter, 1987.

Tecusan, Manuela, ed. and trans. *The Fragments of the Methodists. Methodism outside Soranus*. Vol. 1: *Text and Translation*. Studies in Ancient Medicine, 24/1. Leiden: E. J. Brill, 2004.

Temkin, Owsei. "Studies on Late Alexandrian Medicine I: Alexandrian Commentaries on Galen's *De Sectis ad introducendos*." *Bulletin of the Institute of the History of Medicine* 3 (1935): 405–30. Reprinted in *The Double Face of Janus and Other Essays in the History of Medicine*, by Owsei Temkin, 178–97. Baltimore: The Johns Hopkins University Press, 1977.

Themistius. *In Aristotelis physica paraphrasis*. In *Commentaria in Aristotelem Graeca*, edited by H. Schenkl, 5.2. Berlin: Reimer, 1899.

Ullman, Manfred. *Die Medizin im Islam*. Handbuch der Orientalistik 1, Nahe und mittlere Osten 6.1. Leiden: E. J. Brill, 1970.

Vallance, J. T. *The Lost Theory of Asclepiades of Bithynia*. Oxford: Clarendon, 1990.

———. "The Medical System of Asclepiades of Bythinia." In Haase, *Wissenschaften (Medizin und Biologie)*, 2.37.1.693-727.

Von Arnim, Hans Friedrich August. *Stoicorum veterum fragmenta*. 4 vols. Leipzig: Teubner, 1903–24.

Von Staden, Heinrich. *Herophilus: The Art of Medicine in Early Alexandria*. Cambridge: Cambridge University Press, 1989.

Walbridge, John. "The Alexandrian Epitomes of Galen and the Transmission of Greek Philosophy to the Muslims." In *Islamic Philosophy and Western Philosophies*. Tehran: Sadra Islamic Philosophy Research Institute, 2004 [actually 2008]). 2:217–30.

Walzer, Richard. "Arabische Aristoteles-Übersetzungen in Istanbul." *Gnomon* 9 (1934): 277–80.

———. "Codex Princetonianus Arabicus 1075." *Bulletin of the History of Medicine* 28 (1954): 550–52.

Westerink, L. G. "Philosophy and Medicine in Late Antiquity." *Janus* 51 (1964): 169–77.

Wolska-Conus, Wanda. "Les commentaires de Stéphanos d'Athènes au *Prognostikon* et aux *Aphorismes* d'Hippocrate: De Galien à la pratique scolaire Alexandrine." *Revue des études Byzantines* 50 (1992): 5–86.

———. "Stéphanos d'Athènes et Stéphanos d'Alexandrie: Essai d'identification et de biographie." *Revue des études Byzantines* 47 (1989): 5–89.

Yaḥyā al-Naḥwī. *Talkhīṣ Kitāb al-firaq li-Jālīnūs*. In Pormann, "Jean le Grammarien," 253–58.

Index

The index contains all significant proper names, all Greek words including all Greek words in the glossary, and significant English terms. While there is not an index of citations, classical and medieval Arabic authors cited in the notes are indexed by the author and sometimes also by title. Arabic words mentioned in the English text are indexed, but not Arabic words in the glossary.

English

ᶜAbd al-Wāḥid ibn Muḥammad al-Ṭabīb, lix, 131
ᶜAbd Allāh ibn al-Ḥusayn al-Mutaṭabbib, lix, 131
abdomen, lxix, 216, 233. *See also* belly, chest
Abenrudian, Haly. *See* ᶜAlī ibn Riḍwān
ability, 223, 244
abnormal, lxix. *See also* unnatural
al-Abrahāmī, Jār Allāh, lx
abridgment, xxxii–xxxiii, 248. See also *jumlah, talkhīṣ*
absolute, of time, 58–59
absolutely, 236–37
absorption, 79, 218
absurd, 223, 234
Abūʾl-Faraj ibn al-Ṭayyib. *See* Ibn al-Ṭayyib
Abūʾl-Khayr Sahl. *See* Ibn Ṭūmā
abundance, abundant, 246–47
Abyssinia, Ethiopians, 27, 43
accident, accidental, 51, 54–55, 238; experience, 14–17
Acre, lxi, lxiv, 48
Acron of Agrigentum, 10, 187, 204, 208
actions, 8, 25, 33, 133, 148, 243
active, 243. *See also* qualities
actual, 243
acuity, 220, 252
acute, 220
aᶜḍāʾ murakkabah, lxix, 239. *See also* organs
aᶜḍāʾ mutashābihat al-ajzāʾ, lxi, 239. *See also* tissues
adherence, 248
adhesion, 35, 250
adolescence, 23, 26
Aetius, pseudo-Plutarch, 138
to affect, affected, 144, 146, 220–21, 243; affected parts, 38
agent, lxix. *See also* cause
age of patient, 18, 23, 26, 40, 42, 179, 232; Methodists on, xlviii, 35, 38
Agnellus of Ravenna, 188–89; commentary on *The Medical Sects*, xxxix–xl, xlviii, 30, 192, 203
to be in agreement, 233, 255
Aḥmad Shaykhzāda, 48
air, 253; as element, 52, 134–35, 137, 146, 153, 156–58, 160, 168, 174; as non-natural cause, 18, 27, 46, 111, 113, 115–16; Presocratics on, 153, 164, 188, 191; temperament of, 18–19, 26, 43, 102, 115
akḥāl qābiḍah, 43
ākhar, xlv, 214
akhlāṭ, 88, 209, 224
akhlāṭ zaᶜm [?], 116
Akīlāʾus, xxxvii–xxxviii, xl. *See also* Archelaus, Arkīlīʾus
albuminoid humor, 211. *See* aqueous humor
alchemy, 190, 201
Alcmaeon of Croton, 188
Alexander of Aphrodisias, *De mixtione*, 172
Alexandria, library of, 196; medical education in, xx, xxiii, xliii, 191–92

Alexandrian Epitomes, authors, xxxviii–xliii, l; language, style, form, xxxii–xxxiv, xliv–xlv, xlix, liii; manuscripts, liii–lviii; stemma, lxiii–lxvi, translations, xliii–xlv, liii
Specific works: 1. *The Medical Sects, Firaq al-ṭibb*, xxvi, xxxiv, xlviii, lv–lviii, lxi, lxiii–lxv; 2. *The Small Art, al-Ṣināʿah al-ṣaghīrah*, xxvi, liv–lix, lxi, lxiv–lxv, 18, 49, 192; 3. *On the Pulse for Teuthras, Fī ʾl-nabḍ al-ṣaghīr ilā Ṭuthrūn*, xxvi, lv–lviii, lxi; 4. *Therapeutics for Glaucon, Ilā Ighlūqun*, lv–lviii; 5. *On the Elements, al-Isṭiqsāt ʿalā raʾy Ibbuqrāṭ, al-ʿAnāṣir*, liv–lviii, lxiv–lxvi; 6. *On the Temperament, Fī ʾl-mizāj*, xvi, xxiv, xxvi, lv–lviii, 74, 131; 7. *On the Natural Faculties, Fī ʾl-quwā al-ṭabīʿiyah*, lv–lviii; 8. *Anatomy for Beginners, Small Anatomy, Fī ʾl-tashrīḥ*, liv–lviii, lxi–lxiii; 9. *On Diseases and Symptoms, Fī ʾl-ʿilal waʾl-aʿrāḍ*, lv–lviii, lxi–lxiii; 10. *On Affected Parts, Taʿarruf ʿilal al-aʿḍāʾ al-bāṭinah*, lv–lviii, lxiii; 11. *Large Pulse, al-Nabḍ al-kabīr*, lv–lviii, lxii–lxiii; 12. *On Crises, Fī ʾl-buḥrān*, lv–lviii, lxii; 13. *Days of Crisis, Ayyām al-buḥrān*, liv–lv, lvii–lviii, lxiii, 4; 14. *The Kinds of Fever, Aṣnāf al-ḥummayāt*, lv, lvii–lviii; 15. *The Method of Healing, Ḥīlat al-burʾ*, xxviii, lii, lv–lviii, lxiv, lxviii; 16. *Regimen of the Healthy, Tadbīr al-aṣiḥḥāʾ*, xxiv, liv, lxiii.
See also Galen, Works studied in Alexandria; Ibn al-Tilmīdh
Alexandrians. *See* Alexandrian epitomes
ʿAlī ibn Riḍwān, xxii–xxiii, xxxvii, 68, 188, 194; commentary on *The Medical Sects*, xlviii; *Kitāb al-nāfiʿ*, xxiii–xxiv
alima, 141, 214
to alter, be altered, alteration, 161–62, 241
alum, 21, 232; fissile, Yemeni, 43
Amīn al-Dawla. *See* Ibn al-Tilmīdh
Ammonius of Alexandria, xxxix
analogism, 33, 214
analogy, analogous, xlvii, 10, 255
analysis, method of instruction, 50–52, 54–56, 134, 163, 222
anatomy, xxxi, 32–33, 46, 233; in Alexandria, xx–xxi, xxv, 189–93, 199
Anaxagoras of Clazomenae, 138–39, 141, 145–46, 164, 188
Anaximander, 188
Anaximenes of Miletus, 137–38, 153, 156, 188, 191
Ancients, 244; on elements, 146, 164
Andalusi. *See* Ṣāʿid
Angeleuas, xl, 189
anger, 18, 91–92, 112, 115, 232, 241
anḥāʾ, 107, 251
animal faculty. *See* vital faculty
animals, liii, 17, 33, 111, 136, 158, 223, 254
animate, 223
Anqīlāʾus, xxxv–xl, xliii, 1, 189. *See also* Arqīlāʾus
antecedent cause, xxix, 8, 25, 27–28, 32, 40–43, 230
anterior, 244
Antipatrus, 208
Anṭūn al-Ḥakīm al-Yāfūtī, lxi
anxiety, 112, 241
apertures, 71, 118, 218
Apollonius of Antioch, of Citium, Byblas, 10, 189, 204, 208
appetite, 101, 221, 234, 244
to apprehend, apprehension, 226
approach, 236
aqueous, albuminoid humor, 86–88, 211, 228
Aquileia, xl
Arabic, xix, xlv, 194–96
Archelaus, xxxvii–xl, xlii–xlii, xlviii, 188–89
Archigenes of Apamea, 199, 208
Arctic, 43
to argue, 220
to arise from, 220
Aristotle, xxx, xxxii, xxxix, lxii, 134, 189; commentaries on, 195–96, 200; on elements, liii, 133, 136, 140, 145, 148, 156, 170–72, 190; logical works, xxiii; on Presocratics, 138, 193, 196–98; *Categories, al-Maqūlāt*, xxiii, xxx, 148; *On Interpretation, al-ʿIbāra*, xxiii, xxx, xlii; *Prior Analytics, al-Qiyās*, xxiii, xxx; *Posterior Analytics, al-Burhān*, xxiii, xxx; *Physics, al-Ṭabīʿa*, xxx, 134, 137–38, 145; *On Generation and Corruption, al-Kawn waʾl-fasād*, xxx, 134, 171, 176; *On the Heavens, al-Samāʾ*, xxx, 134, 137, 171
Arkīlāʾus, xxxvii–xxxviii, xl, 189. *See also* Archelaus
arm, 14–15, 47, 52, 73, 77, 80, 238, 255
arrangement, 252
arrows, 18

art, 5, 235. *See also* profession
arterial, 233
arteries, 25, 40, 73, 77, 89, 100, 123, 238. *See also* blood vessels, veins
asāṭīn al-ḥikmah, 191
Asclepiades of Bithynia, lii, 10, 30, 51, 183–84, 189–90, 197, 201, 206, 208; on elements, 138–39, 141, 170–71
Asclepius, xl. *See also* Anqīlāʾus
aṣḥāb al-ḥiyal, 10, 223. *See* Methodists
aṣl, xxxvi, xliii, 214
aspect, 217
Asqalan, lix–lx
to assimilate, 249
association, 233
astringency, astringent, astringent drugs, 25, 43, 48, 127, 169, 182, 239, 244
astrology, xxiii
astronomy, 57
asymmetry, asymmetrical, 66–68, 73, 232
Athenaeus of Attaleia, lii–liii, 31, 55, 138, 140, 142–44, 146, 159, 162, 171, 190–91, 199, 219
atoms, atomism, xlvii, lii, 140, 142–46, 171, 190–91, 219
to be attached, 248
to attract, attraction, 127, 218
attractive faculty, 26
attribute, attribution, 51, 254
audible qualities. *See* sound
auditory, 232
autopsy, 16, 216
autumn, 18–19, 23, 157, 224
Averroës. *See* Ibn Rushd
averse to, aversion, 150, 240, 252
Avicenna. *See* Ibn Sīnā
Avidianus, 207
to avoid, 219
Aya Sofya Library, lxi, 48

babies. *See* child, children, childhood
back, 225
backwardness, 225
bad, 228
bad temperament, 232
badan, 136
badly formed, 228, 234
balance, balanced, l, 232, 246
barley, 143, 234; meal, 36, 226; water, 162, 247
Bārsīyūs, xli. *See* Marinus
basis, 215
basket, 143
bath attendants, 20
bathing, 12, 21, 129, 189, 223
baṭn, lxix, 216
beans, 143, 216
Bears, constellation, 43
beasts of prey, 18, 230. *See also* animals
beauty, 72, 219
Becoming, Platonic, 140
before, 230
beginners, l, 225
beginning, 215
belch, 101, 219
belly, lxix, 216, 229
benefit, beneficial, 252
bile, lii, 83, 249. *See also* black bile; red bile; yellow bile
bilious, 249. *See also* yellow bile
binding, 120, 122–23, 227
al-Bīrūdī. *See* al-Yabrūdī
bites, 27, 29, 41, 253
bitter, bitterness, 48, 169, 182, 249
black, blackness, 29, 86, 166–67, 169, 232, 247
black bile, 93, 176, 179, 249; as humor, 135, 156, 175, 177, 182; purgation, 126, 180, 184; and swelling, 24, 108
blacksmiths, 20, 116
bladder, 36, 104–5, 124, 189, 249. *See also* stone
bleeding, 251
blending, 145
blindness, 61, 106, 240
to block, to be blocked, blockage, 121, 228, 230–31, 245
blond, 234
blood, lii–liii, 15, 42, 93–94, 98, 100, 128, 182, 186, 226; diseases, 24, 26; diversity, 157, 178–80; evacuation, 45, 108, 181, 184; as humor, 134–35, 156, 175–77
blood vessels, 28, 94, 113, 136, 178, 185, 209; in eye, 85, 210; and liver, 67, 93, 105. *See also* arteries; veins
bloodletting, 12, 15–16, 26, 42, 45, 121, 127, 180, 243
blow, as cause, 25, 27, 41
blue, blueness, 86, 229–30
body, human, l, lii, 9, 54, 113, 136, 215, 219; degrees of health and disease, 57–63, 66–68, 75; and organs, 53, 90–91, 163; subject of medicine, 3, 11, 18, 132
body, corporeal, 53, 136, 139, 160, 219
bone, bones, xxvii, 81, 164, 239; of skull, 52, 80–81; temperament, 77, 99, 123,

136, 157, 179, 182, 186; treatment, 9, 122, 125
book, 245
bowed, bowing, 245
bowels, 250. *See also* intestines
brain, l, 25, 28–29, 52, 77–84, 226; anatomy of, 166, 170; temperament, 90–91, 101, 179, 182
bread, 21
breadth, 254
break, 125, 242
breath, to breathe, 28, 90–92, 102, 120, 150, 222, 236, 252
brittleness, 166–67, 253
broad, 238
bruise, bruising, 28, 108, 228
brutishness, 91, 230
bug, 216
burning, 28, 102, 248
butter, 179, 229

caecum, 45, 240
calf, 47, 232
cancer, 178, 231
canon, 244
to take care, 255
cartilage, 77, 241
case histories, xlix
causes, xxxi, l–li, lxix, 7–9, 11–12, 46, 116, 120, 132, 139–40, 198, 230; of health and disease, xxvii, xlix, 4–5, 27–29, 57–64, 73, 111–113, 125; Methodists on, 35, 38, 40; Rationalists on, 18, 27, 32, 43. *See also* antecedent cause; cohesive cause; efficient cause; external cause; necessary causes; non-naturals; preceding causes; prior cause; prophylaxis; transformative cause
caustic, 221
cauterization, 9, 122–23, 247
cavities, 127, 185, 220; defective, 115, 119–20
change, changeable, to be changed, 137, 144–45, 148, 220–21, 223, 241. *See also* motion
character traits, 92
cheek, 254
cheese, 218
chest, 90–92, 104, 179, 235
chicken, as food, 129
chickpeas, 143, 223
child, children, childhood, 83, 123, 157, 234; babies, 45, 83; regimen and treatment, 8, 23, 26, 42, 128
chills, 169
choroid, 89, 209–10, 234, 237
Christianity, xx–xxii
chronic, 59–60, 62–63, 66, 226
Chrysippus of Soli, 170–72, 190, 206. *See also* Stoicism, Stoics
ciliary body, 86, 89, 211
to claim, 230
clarity, 235
class, 244
Claudius Agathinus, 199
clay, 237
to cleanse, 219, 252
clearly, 240
cleidocranial dysplasia, 67
climate, xlviii, 18, 42–43
cloaks in eye, 88, 209
coagulate, 239
coarse, coarseness, to coarsen, become coarse, 109, 168, 241
cobweb layer. *See* zonula
coction, 79, 82, 110, 128, 251
cohesive cause, 8, 27, 32, 40, 230
coils, of intestines, 242
cold, coldness, 48, 108–9, 145, 162, 215; causes of, 20–21, 99, 115–17; extreme of, 28, 160, 181; as quality, 68, 71, 82, 142, 159, 166–70; temperament, 18, 22, 42, 90–94, 97, 101–3, 165, 176–77
collection, 229
collyria, 31, 43, 247
colon, 45, 245
color, 97, 128, 166–67, 169, 248; of eye, 86–87
combination, to combine, 214, 219, 224, 229, 248
comeliness, 221
commentaries, 1, 186, 233; ancient, xxi–xxii, xxxii–xxxv, 134
common, commonality, 233
communities of diseases, xlv, 35, 38, 44, 47, 197, 219
to be compatible, 248
compendia, xxii, xxxv. See also *kunnāsh*
completeness, 71–72, 218
completion, 253
complexion, 98–99, 106, 157, 186, 248
composed, 214
composite, 136, 145, 229
composition, 162, 224, 246
compound organs, lxix, 239. *See also* organs
compounds, 133–34, 142–43, 149, 164–65, 229
to comprehend, comprehension, failure to

comprehend, 226
compression, liii, 116, 169
by compulsion, 236
concatenation, 244
concave, concavity, 71, 125, 220, 245
to concentrate, concentration, 219
conclusion, in logic, 152, 251
to condemn, 237
condensation, to condense, 140, 153–55, 247
conjunctiva, 89, 210, 214
consequent, 248
Constantinople, 200
constipation, 15, 36, 244
constitution, lxix. See also *mizāj*; temperament
constriction, constricting, xlvii, 36, 169, 244
construction, of organs, 71, 225
consume, consumption, 221, 231
contact, 116, 142, 248
contiguity, 71, 142, 149, 171, 220, 233
continual. *See* chronic
continuance, to continue, 246, 248
continuity, 71, 73, 254. *See also* dieresis
continuous, of pulse, 90, 254
to contract, 244
contraindicant, 26, 42, 250
contrary, 225
contrary to nature, lxix, 236. *See also* unnatural
convalescence, convalescent, 8–9, 60–61, 128–29, 252–53
conversion, method of instruction, 50–52, 54–56, 162, 239, 252
to convert, 245
convex, 220
convulsions, 28–29, 74, 234
easily cooled, 215
cooling, of drugs, 15, 25
copyists. *See* scribes
cornea, 89, 210–11, 237, 244
coronal suture, 80–81, 214
corporeal. *See* body, corporeal
correlated, 248
correspondence, of symptoms, 10
corruption, 114–15, 242
costive disease, costiveness, 35–36, 38, 44, 47, 222, 250
cough, 28, 107, 121, 231
to counteract, 92–94, 225, 245
countries 19, 35, 40, 42–43, 217
craft. *See* art; profession
crane, bird, 247
crises, xxv, xxviii, 44
crust, 107, 245
crystalline humor, 88–89, 211, 219
cucumbers, 21
culture, 214
cure, 114–15, 234–35
curly, 71, 78, 97, 219
curriculum, medical, xxi, xxiv–xxxi, 8
curvature, 245
customary, 123, 132, 240
cut, 9, 16, 108, 245
Cynics, 51

Damascus, lix, 202
ḍarabān, 30
daring, 92, 218
daughters of the elements, 22–23, 27, 38, 217
to debate, 252
deduction, 12, 15, 109, 226, 251; Rationalist, 27, 32–33, 37–38
deep, of voice, 218
defecation, 14
defects, defective, 1, 73, 118–24, 215, 228
deficiency, 252
definition, 33, 51, 54–55, 57, 220
deformity, 122
deliberation, deliberative functions, 26, 78–79, 81, 83, 150, 232
Democritus of Abdera, 139–41, 143–45, 153, 164, 190–91, 196
demonstration, logical, xxiii, 54–55, 151, 155, 216
denotation, 226
dense, density, 48, 99, 166–67, 247
to deny, 253
dependents, 77
derivation of names, 13
description, descriptive definition, l, 33, 51, 54–55, 228, 254
desiccate, 219
desire, 150, 234; sexual, 95–96
to destroy, destruction, 239, 242–43
to deviate, deviation, 216, 224
device, 214
dew, 100, 228
diagnosis, xxxi, xlv, li, 2, 238; in Alexandrian curriculum, xxiv–xxv, xxviii; of organic diseases, 74, 103–8
diagnostic, of signs, 69–70, 226
dialysis, l, 50–56, 222
diarrhea, 15, 36, 45, 225, 237
dieresis, 41–42, 106–8, 125, 242
diet, 23, 115–16, 189, 193, 240
difference, 242
different, 225

differentia, xxviii, 57
to digest, digestion, 7, 22, 101, 253
dimensions, of organs, 71, 73, 118–20, 244
Diocles of Carystus, 10, 190, 205, 208
Diodorus Cronus, 140, 190–91
Diogenes of Apollonia, 137, 153, 164–65, 191
Diogenes of Sinope, 191
Dionysius, 207–8
disagree, disagreement, 34, 225
discussion, 247
disease, xxii, l–li, 16, 74–75, 107, 214, 239, 249; ancients on, 4, 14–16, 18, 23, 27, 30, 32, 35, 44, 140; causes and signs, xlix, 9, 11, 27–28, 57–63, 66, 69, 73, 110–11, 249; compound, 30–31, 35–36, 47, 124; organic, xxv, xxxi, 4, 41, 43, 105–8, 118–24; systemic, xxxi, 41, 233; treatment, 17, 114–17;
diseased, 57, 239, 249
dislocation, 123–24, 225
disorders, 104, 215
to disperse, be dispersed, 215, 242
disposition, 225
dissection, 33, 191–93
dissolution, 116, 222. *See also* dieresis
distention, 24, 26, 28, 46, 124, 249
distortion, 242
disturbance, 238
diuretic, 127–28, 225
to diverge, 230
division, method of instruction, 51–52, 54, 245
dog, 51; rabid, 23, 29, 41–42, 247
Dogmatic school. *See* Rationalist school
dominant, dominated, 245
douche, 43, 126, 231, 251
to draw out, 218
dream, 14, 16–17, 253
dregs, 178–79, 225
drink, 21, 101–2, 233; as cause, 9, 12, 18, 111, 113, 116–17
to drop down, 221
drugs, 5, 46, 116, 122, 135, 151–52, 161–62, 227; ancient views, liii, 14–15, 17, 32, 35, 38, 183–84; simple and compound, xxiii, xxxi, 30–31; therapy by, 9, 12, 24–25, 29, 41–42, 172. *See also* purgatives
drunkenness, 46
dry, dryness, 109, 219, 255; causes of, 20–21, 28, 115; as cause, 82, 103, 108, 122, 142; of eye, 85, 87–88; as quality, 68, 160, 165–70; and temperament, 18–19, 90–94, 98, 101, 176–77
ducts, 104–5, 115, 118, 218
dura mater, 88, 241
duration, 238

ears, 83, 166, 214
earth, 160, 168, 214; ancients on, 164, 188, 201; element, 52, 134–38, 146, 153–54, 156–58, 174
ease, 219
east, 19
eating. *See* diet
edema, 24, 108, 122, 215, 254
education, medical, xix–xxxi. *See also* curriculum; syllabus
effect, effecting, 220-21, 243
efficient cause, 17, 54
egg white, 211, 217
Egypt, xix, 43
Egyptian thorn, 43
eight heads, lxviii, 3, 132
elderly. *See* old
electuary, 31, 237
element, lii, 3, 8, 46, 133, 136, 186, 214, 240; in Alexandrian curriculum, xxv, xxxi, 162; ancients on, xlv, 137–38, 141–46, 149, 189, 191, 196, 199–202; and body, 52–54, 132, 157, 161; classifications, 134–36, 153, 159–60, 174, 181; and principles, 139–140; transmutation, liii, 154, 164
elemental, 240
elements, daughters of, 22–23, 27, 38, 217
embryo. *See* fetus
emetic, 246
emission of sperm, 95, 251
Empedocles of Agrigentum, 138, 140, 145–46, 164, 187, 191
Empiricist school, xx, xlv–xlvii, 8, 10–11, 64, 191–93, 208, 218; members, 187, 198, 200, 203–4; names, xlix, 12; and Rationalists, 18, 23–27, 30–34; views, 14, 16, 37–38, 40–41
end, goal, 3, 51, 133, 218, 241
enema, 121, 127–28, 222
to be enlarged, 239
entity, 57, 214
envy, 112, 221
Epicurus of Samos, 138, 190–91
epidemics, 19
epilogism, 33, 214
epistemology, medical, xlv, 12
epitome, xxii, xxxii, xxxv, xxxvii, xlix, 219

equal, equality, equally, 62–64, 232, 247
Eraclitus, 205. *See* Heraclitus of Ephesus
to eradicate, 245
Erasistratus of Ceos, 10, 30, 191-92, 206, 208
erect, 71, 246, 251
Erofilus, 206. *See* Herophilus
erysipelas, 14–15, 24, 31, 108, 223
essential, 220. *See also* definition
ethics, 92, 200
Ethiopians. *See* Abyssinia
evacuant, 242
evacuation, 22, 36, 44–46, 110, 162, 177, 242; as cause, 12, 18, 112–13, 115; as treatment, 24–26, 126–28
evidences, 27, 234
to examine, 249
example, 249
excellent, excellence, 243
excess, 230, 242
excretions, 103, 106–8, 178–79, 243
exercise, 27, 189, 221, 229, 240; as cause, 12, 18, 20, 115–16, 120; regimen, 9, 129
to exist, existence, 17, 148, 254
to expand, expansion, 216, 254
to expel, 224. *See* expulsion
experience, to experience, xlix, 30, 144, 214, 218; Empiricists, xlvii, 10–17, 32, 37, 191; Rationalists, 30–31
explanation, 243
expulsion, expulsive, expulsive faculty, 25–26, 79, 224, 226
exterior, 237
external, 224
external cause, 28, 78
to extirpate, 214
extraction, 251
extreme, 60, 237, 241, 245
eye, l, 43, 128, 166, 40; anatomy, 88–89, 209–11; diseases, 36, 47, 107; temperament, 85–88
eyeball, 89
eyelid, 120, 219

faculties, xxxi, 4, 8, 77, 133, 246; in treatment, 26, 35, 79, 127–28; types, 25, 132, 156. *See also* attractive faculty; natural faculty; psychic faculty; retentive faculty; transformative faculty; vital faculty
fainting, 29, 241
fallacy, 46, 241
false, 247, 253
fandīqūn. See *khandīqūn*
fascial, 45
faṣṣ, lxviii
fat, 97–100, 136, 226, 232–33; flavor, 48
fear, 18, 112, 116, 242
feces, 46, 113, 121, 216, 218, 228, 241; diagnosis by, xxxi, 25, 45, 107
females, procreation of, 95, 215, 255
fenugreek, 24, 36, 222
fertile, fertility, 95, 255
fetid, 251
fetus, 175–77, 179, 182
fever, xxiv, xxxi, 4, 15–16, 27–28, 46, 66, 223; in Alexandrian curriculum, xxv, xxviii, 1–2. *See also* hectic fever; periodic fever; quartan fever; septic fever; tertian fever
filaghmūnī, 108, 243
film, 107, 241
Fimision de Laodicia, 206. *See* Themison
fine, fineness, finer, 109, 168, 226, 229
finger, 67, 119–20, 214
fingernails, 55, 109
finite, finitude, 137, 253
fire, 18, 28, 42, 159, 162, 174, 253; ancients on, 138, 153–56, 164, 193; element, 52, 134–37, 146, 157, 160, 168, 174; and life, 158, 181
fire-colored, 253
firm, firmness, 166, 219, 247
firqah, 10, 242
fish, 116, 129
fisherman, 20
fissile alum. *See* alum
flabbiness, 166, 228
to flake off, 245
flavors, 48, 166–69, 237
flaw, 252
flesh, 77, 136, 164, 248; generation of, 100, 123, 182, 186; temperament of, 97–99, 179
flow, flowing, 36, 228, 231–32, 234
fluency, 228
fluent disease, 45, 216. *See also* flux
fluids, of eye, 88, 209–11, 228
flush, flushing, 25, 28, 223
flux, 216; in Methodism, xlvii, 35–36, 38, 44–45, 47. *See also* fluent disease
food, 7, 157, 161, 173, 237; as cause, 9, 11, 18, 21–22, 111, 113, 115–17, 129
foot, 51, 67, 228
forcible, 244
foreign, 241
forgetfulness, 82, 251

form, 17, 150, 154, 160, 235, 253; of body, 111, 113, 157; in organ, 73, 75, 114–15, 118; and prime matter, 139–40, 160
formless, 222
fracture, 125, 247
fractured, skull, 248
front, 244
fruit, 21, 116
fulling, 116
function, 18, 79, 86, 103, 132, 218, 243; impairment, 71–74, 106–10
functional organs. *See* organs
fundamental, 214
future, 215

Galen of Pergamon, xx, xxxvi, 192; Arabic sources, xxxiv, 200; on medical topics, 3, 11, 18, 57, 64, 124; on predecessors, xxii–xxiii, xlv–xlviii, 23, 30, 40, 133, 154, 159, 197
Works in general; Arabic translations, commentaries, glosses, lxii–lxviii, 195, 202; Greek commentaries, xxii–xxiii, xlii–xliii, 196, 198, 200; methods of instruction, 51, 55–56; order of books, xxi, xxxiii, 5, 129, 133
Works studied in Alexandria, xix, xxi, xxiv, xxxiii–xxxvii, xlv, 1. *See also* Alexandrian Epitomes
1. *The Medical Sects*, *Firaq al-ṭibb*, *De sectis ad eos qui introducuntur*, ix–x, xxi–xxii, xxiv, xxvi, xxxi, xxxix, xlv–xlix, liv, lix, lxviii, lxx, 1, 5; Arabic translation, xliv, liv, lxx, l; commentaries and glosses, xl–xlii, xlviii, lxvii, 42, 188
2. *The Small Art*, *al-Ṣināʿah al-ṣaghīrah*, *Ars medica*, *Ars parva*, *Microtechne*, *Tegni*, xxiv–xxvi, xlix–li, lix, lxiv, 5, 11, 21, 49, 73, 162, 192, 209; in Alexandrian curriculum, xxvi, 1; Arabic translation and reception, xxxi, xliv, liv, 188; method of, 53, 56, 105
3. *On the Pulse for Teuthras*, *Fī ʾl-nabḍ al-ṣaghīr ilā Ṭuthrūn*, *De pulsibus ad tirones*, xxiv, xxvi, 4; Arabic version, l, liv
4. *Therapeutics for Glaucon*, *[Fī mudāwāt al-amrāḍ] ilā Ighlīqūn*, *Ad Glauconem de methodo medendi*, xxiv, xxvi, 1, 4; Arabic reception, xxxi, l; commentaries, xlii, 189
5. *On the Elements According to the Opinion of Hippocrates*, *al-Isṭiqsāt ʿalā raʾy Ibbuqrāṭ*, *al-ʿAnāṣir*, *De elementis secondum Hippocratem*, xxiv, xxvi, li–liii, lxiii, lxvii, 131, 193; and *On the Humors*, lxvii, 3; in Alexandrian curriculum, xxvi, 2–4; Arabic translation and reception, xxxi, xliv, liv; title and method, 56, 133–36, 162, 170. See also *On the Humors* below
6. *On the Temperament*, *Fī ʾl-mizāj*, *De temperamentis*, xvi, xxiv, xxvi, xxxi, lii, 2–3, 56, 74, 131, 133
7. *On the Natural Faculties*, *Fī ʾl-quwā al-ṭabīʿīyah*, *De naturalibus facultatibus*, xvi, xxi, xxiv, xxvi, xxxi, 2, 4, 8, 56, 131, 133, 183
8. *Anatomy for Beginners*, *Small Anatomy*, *Fī ʾl-tashrīḥ liʾl-mutaʿallimīn*, xxiv, xxvi, xxxi, 131, 133; a. *Anatomy of the Bones*, *Tashrīḥ al-ʿiẓām*, *De ossibus ad tirones*, xxvii; b. *Anatomy of the Muscles*, *Tashrīḥ al-ʿaḍal*, *De musculorum dissectione*, xxvii; c. *Anatomy of the Nerves*, *Tashrīḥ al-ʿaṣab*, *De nervorum dissectione*, xxvii, liv; d. *Anatomy of the Veins*, *Tashrīḥ al-ʿurūq al-ghayr al-ḍawārib*, *De venarum arteriarumque dissectione*, xxvii; e. *Anatomy of the Arteries*, *Tashrīḥ al-ʿurūq al-ḍawārib*, *De venarum arteriarumque dissectione*, xxvii
9. *On Diseases and Symptoms*, *Fī ʾl-ʿilal waʾl-aʿrāḍ*, xxiv, xxvii, xxxi, 1, 2, 4, 56, 134; *Jumal al-ʿilal waʾl-aʿrāḍ*, *Jumal maʿānī al-ʿilal waʾl-aʿrāḍ*, lv, lxi; a. *On the Differentiae of Diseases*, *Fī aṣnāf al-amrāḍ*, *De morborum differentiis*, xxiv, xxvii; b. *On the Causes of Diseases*, *Fī asbāb al-amrāḍ*, *De causis morborum*, xxiv, xxvii; c. *On the Differentiae of Symptoms*, *Fī aṣnāf al-aʿrāḍ*, *De symptomatum differentiis*, xxiv, xxvii; d. *On the Causes of Symptoms*, *Fī asbāb al-aʿrāḍ*, *De symptomatum causis*, xxiv, xxvii, 1
10. *On Affected Parts*, *Diagnosis of Diseases of the Internal Organs*, *Fī taʿarruf ʿilal al-aʿḍāʾ al-bāṭinah*, *al-Mawāḍiʿ al-ālimah*, *De locis*

affectis, xxiv, xxviii, xxxi, 2, 4, 56, 107
11. *Large Pulse, al-Nabḍ al-kabīr, Fī ʾl-nabḍ, De pulsibus*, xxiv, xxvi, xxviii, xxxi, l, liv, lxi, 2, 4, 133; a. *The Kinds of Pulse, Aṣnāf al-nabḍ, De differentia pulsuum*, xxviii; b. *Diagnosis by the Pulse, Taʿarruf al-nabḍ, De dignoscendis pulsibus*, xxviii; c. *Causes of the Pulse, Asbāb al-nabḍ, De causis pulsuum*, xxviii; d. *Prognosis by the Pulse, Taqdimat al-maʿrifa min al-nabḍ, Sābiq al-ʿilm bi-mā yadillu ʿalayhi al-nabḍ, De praesagitione ex pulsibus*, xxviii
12. *On Crises, Fī ʾl-buḥrān, De crisibus*, xxiv, xxviii, 2, 4, 215
13. *Days of Crisis, Ayyām al-buḥrān, De diebus decretoriis*, xxviii, 2, 4
14. *The Kinds of Fever, Aṣnāf al-ḥummayāt, De typis febrium*, xxiv, xxviii, xxxi, 2, 4, 134
15. *The Method of Healing, Ḥīlat al-burʾ, De methodo medendi*, xxiv, xxvi, xxviii, xxxi, lii, lxviii, 4, 56, 134
16. *Regimen of the Healthy, Tadbīr al-aṣiḥḥāʾ, De sanitate tuenda*, xxi, xxiv, xxix, xxxi, 2

Other works of Galen:
On the Affirmation of Medicine, Fī ithbāt al-ṭibb, xxix, 56
On Anatomical Procedures, Kitāb al-tashrīḥ, De anatomicis administrationibus, xx, xxiv, xxvi, xxxi, lxi, 2, 33. See also *Anatomy for Beginners* above
On Antecedent Causes, Fīʾl-asbāb al-muttaṣilah fīʾl-maraḍ, De causis contentibus, xxix, 8
On the Constitution of the Art of Medicine, De constitutione artis medicae ad Patrophilum, xxix, 56. See also *The Affirmation of Medicine*
On the Doctrines of Hippocrates and Plato, Fī ārāʾ Buqrāṭ wa-Falāṭūn, De placitis Hippocratis et Platonis, xxix, 4, 133
History of Philosophy, 5, 138
On the Humors, Fī al-akhlāṭ, De humoribus, xxix, lxvii, 3, 133.
On Medical Experience, Fī ʾl-tajribah al-ṭibbīyah, xxix, xlvii
On the Medical Doctrines of the Timaeus, 199
On My Own Books, xlviii
On the Order of My Own Books, xlviii
An Outline of Empiricism, Fī jumal al-tajribah, De subfiguratio empirica, xxix, xlvii, 8, 33
On the Parts of Medicine, 8
On the Uses of the Parts, Manāfiʿ al-aʿḍāʾ, De usu partium, xxix, xli, 4, 133

Galen, pseudo-, *Introduction to Medicine*, 80, 203, 208
Galienus, 206. *See also* Galen
gallbladder, 177–78, 249
garden nightshade. *See* nightshade
general, 239
general matters, xxxi
to generate, generation, 156, 160, 177, 215, 247, 255
generation and corruption, 138
generative, 255
generic, 219
genesis, lii, 132, 247
genus, xlvi, 25–26, 34, 51, 54, 57, 170, 219
geometry, 5, 54, 57
Gesius, Gessius of Petra, xxxv–xxxix, xlii–xliii, 1
ghayr majrā ʾl-ṭabʿ, lxix
ghayr mutaḥarrik, 137, 221
ghāʾiratān, 42
glass, liquid, 210
Glaucias, 205
glosses and scholia, xxxiii, xlix, lxvii–lxviii, 221
glutinous, 241
goal. *See* end
God, 34, 139–40
goldsmith, 20
Gondeshapur, xx
Gorgias of Leontini, 170, 192
gout, 66, 252
governing faculty, function, 17, 26, 225. *See also* deliberative function
grain, 220
grammar, in medical curriculum, xxiii
grammarian, 174, 251
grape, 211
greasiness, 169
Greek, lxvi, 45, 135, 196, 203
grief, 18, 112, 145, 221, 241
gross, of flavors, 48

to grow from, growth, 250, 253
growling in stomach, 109, 157, 244
gum, 122, 235
gustatory, gustatory qualities, 227. *See* flavors
gynecology, 11, 200

habit 23, 35, 240
hair, hairiness, hairless, 67, 83, 230, 233–34; and brain, 71, 78, 90–95, 97–99
Hairesis, 7
ḥālāt al-abdān, 44
Haly Abenrudian. *See* ᶜAlī ibn Riḍwān
hands, 51
ḥarārah, 159, 221
hard, hardness, 97, 125, 142, 235; quality, 71, 166–69; symptom, 24, 82, 106, 109
harm, harmful, harm done, 235–36
ḥārr, 159, 221
harsh, harshness, 240
Ḥasan ibn ᶜAbd Allāh ibn al-Ḥusayn al-Mutaṭabbib, lix
hayʾah, 73, 253
head, 51–52, 101, 104, 227; form of, 67, 73, 78–80, 121, 231
healing, 215
health, 18, 132, 193, 214, 231, 234–35; body, sign, and cause, xlix, 11, 57– 63, 69, 110–11; degrees, 71, 73–75; end of medicine, l, 3, 11–12
healthful, healthy, 57, 234–35
heap, 143, 234
hearing, to hear, heard, 16, 78, 83, 166, 170, 231–32
heart, l, 25, 27–29, 77, 245; temperaments, 90–94
heat, 15, 157–58, 162, 175, 221; causes of, 20–21, 115–116, 145; effects of, 20, 22, 82, 103, 108, 176, 186; excessive, 18–19, 28, 43, 109, 160; quality, 68, 71, 159, 165–70; and temperament, 42, 90–94, 97, 101, 103, 177
heaviness, 169; and plethora, 25, 28
Hebrew, translations, liii, 200
hectic fever, 31, 226
hellebore, 126; black, 213, 224
to help, helpful, 229, 252
hematoma, 23–25, 30, 43, 108, 122, 254
hemorrhage, 45, 242
Heraclides of Erythrae, 55, 192, 193
Heraclides of Tarentum, 193
Heraclitus of Ephesus, 137–38, 153, 156, 164–65, 193, 205
herbs, succulent, 21
hernia, 36, 124, 242, 246. *See also* rupture
Herophilus of Chalcedon, 55, 191–93, 198–99, 206, 208; definition of medicine, 11, 57
herpes, 14–15, 24, 253
Hibat Allāh ibn [Hay?]kal al-Muṭabbib, lix
hidden, hidden entities, 33, 224
higher rank, 219
high-pitched, 220
hindering, 25
Hippasus of Metapontum, 137, 193
Hippocrates of Cos, xxix, li, 4, 33, 141, 165, 192–94, 205, 208; in Alexandrian curriculum, xxiv; commentaries on, xxxiv, xlii–xliii, lxviii, 195–96, 198, 200; on elements and principles, liii, 137–38, 140–41, 143–45, 151, 164; Galen on, xxii, 133–34; on humors, 156, 175; medical views, 42, 48, 176, 183; on temperament, 148, 160, 172; as Rationalist, xlvi, 10, 199
Works of, xxii, xxv, xxx; *Airs, Waters, Places*, *al-Ahwiyah waʾl-azminah waʾl-miyāh waʾl-buldān*, *De aëre aquis et locis*, xxiv, xxx; *Aphorisms*, *al-Fuṣūl*, xxiv, xxx, xlviii, 37, 198; *On the Nature of Man*, *Ṭabīᶜat al-insān*, *De natura hominis*, xxx, lii, 136–37, 142, 144, 151, 153, 155, 157, 160, 170, 175, 179–80, 193; *Oath*, xix; *Prognostics*, *Taqdimat al-maᶜrifah*, *Prognostica*, xxiv, xxx, 230, 252; *Regimen in Acute Diseases*, *Tadbīr al-amrāḍ al-ḥāddah*, *Regimen acutorum*, xxiv, xxx
Hippon of Samos, 137, 154, 193
history, Empiricist, lxv–lxvi, 16, 196, 229
ḥiyal, 10, 223
hoarse, hoarseness, 103, 215
to hold a belief, 251
hole, 218
Homer, *Iliad*, 67
homoeomerous, 233; bodies, 138, 146, 164; organs and parts, lxix, 51, 239. *See also* tissues
homogeneous, 233
honey, 21, 121, 143
honey-water, 121, 238
horizontal, 216
horn, 136, 210, 244
hot, 48, 159, 221. *See also* flavors; heat; wind
house, 149
Ḥubaysh ibn Ḥasan, xliv, 194
human, 215

human body. *See* body
humidity. *See* moisture
humors, xxv, xxix, xxxi, 8, 224; ancients on, 44, 156, 175–80, 186, 199; and body, liii, 83, 88, 134–35, 185; in *On the Elements*, lii–liii, 133–34; of eye, 209–11; medical aspects, 23–24, 121, 128, 181, 183; sources of, 3, 52–53, 132, 163, 174
hump, 220
Ḥunayn ibn Isḥāq al-ᶜIbādī, xxxviii–xxxix, xliii, xlv, lii, lxv, lxviii, 50, 57, 129, 131, 135, 186, 193, 196, 198, 210; on Alexandrians, xxi, xxxix; as translator of Alexandrian epitomes, xix, xxxiii, xxxv, xliii–xliv, 1, 48; *Questions on Medicine for Scholars*, 27; *Risāla fī dhikr mā turjima min kutub Jālīnūs*, xxi, xxxiii, xliv, xlvi, xlix, lii; *Ten Chapters of the Eye*, 89, 209
hunchback, 67
hunters, 20
to hurt, 214
hydrophobia, 29, 242. *See also* rabies
hygiene, 8–9, 190, 222

iatrosophists, xxi, xlii–xliii, li, 194, 198, 203, 209
Ibn Abī Uṣaybiᶜa, xxxvii, xxxix–xlii
Ibn al-Nadīm, *al-Fihrist*, xxxvi–xxxix, xliv, 195
Ibn al-Qiftīᵓ, *Tārīkh al-ḥukamā*, xxxv–xxxvi, xxxix, xli, xliii, 195
Ibn al-Shaykh al-Makkī al-Mutaṭabbib, lx
Ibn al-Ṭayyib, Abūᵓl-Faraj, xxv, 42, 194–95, 202; *Thimār Firaq al-ṭibb*, xlviii, liv
Ibn al-Tilmīdh, Abūᵓl-Ḥasan Hibat Allāh ibn Ṣāᶜid, Amīn al-Dawla, lix, 195, 203; glosses, xxxiii, lx–lxii, lxiv, lxvii–lxviii, 3, 8, 21, 73
Ibn Buṭlān, al-Mukhtār ibn al-Ḥasan, xxxvii–xxxix, 188, 194–95
Ibn Firās, xxxvi
Ibn Juljul, *Ṭabaqāt al-aṭibbāᵓ*, xxxvi, xxxviii, xliii–xliv, 195
Ibn Jumayᶜ, *Treatise to Ṣalāḥ ad-Dīn on the Revival of the Art of Medicine*, xxiii
Ibn Rushd, xxxi– xxxii, lxii, 195
Ibn Shimᶜūn, xxxvi
Ibn Sīnā, medical poetry, 7; *al-Qanun fī al-ṭibb*, xxxi, xxxiv, xlviii, 7, 14, 21, 195
Ibn Ṭūmā, Abūᵓl-Khayr Sahl ibn ᶜAbd Allāh, lxi–lxii, lxiv
Ibrāhīm ibn Sulaymān ibn Ḥakīm al-Ruhāwī, lx
ice, 21, 181, 219
ice-like layer. *See* lens
ichor, 125, 235
Ideas, Platonic, 140
ifīfāfūqūs, 210, 214
ikhtiṣār, xxxii, liv
ileum, 45, 226
images, 150, 249
imagination, 79, 81–83, 150, 162, 225
imitation, 14–17, 233
imitative, 15, 233
immoderation, immoderate, immoderate temperaments, 66– 67, 73, 84, 133, 238
impairment, 74–75, 106–8, 110, 236
implacability, 92, 216
impossibility, 148
imprinted, 231, 236
inanimate, 223
incidence, 14–16, 255
incision, 9, 216
incoherent, 243
increase, 230
indication, to indicate, indicating, 4–5, 8, 79, 103, 226
individual, 51, 54, 233, 242. *See also* particulars
indivisible, indivisibility, 142, 218. *See also* atoms
induction, 33
induration, indurated swelling, 108, 122, 235, 254
inequality of health, 62–64
inference, to infer, 10, 12–13, 32–34, 40, 43, 144–45, 201, 246, 251
infinite, infinity, 137–38, 253
to inflame, 221
inflammation, 109
inflation, 24, 28, 252
to influence, 220
inhaling, 251
injurious, injury, 228, 236
innate, 241
inner, 216
inseparable, 230, 242
inside, 225
insolence, insolent, 92, 237
insomnia, 115–16, 232
instruction, 239; methods of, li, 50–56, 162–63
instructions, to follow, 231
instrument, instrumental, 215
insufficient, 247

intelligible, 239
interference, 238
intermediate, 254; of flavors, 48
internal organ, 239
to interpenetrate, interpenetration, 171–72, 225
intestinal hernia, rupture, 242, 246
intestines, 45, 250; disorders, 104, 107, 120–21, 178, 242, 246
intuition in diagnosis, 124
investigation, to investigate, to be investigated, 237, 252
Ion of Chios, 138, 195–96
Ionian physicists, liii, 188, 191; on elements, 46, 138, 153–54, 156, 164
irascibility, 90–92, 241
iris, 86, 89, 211, 236, 240
irrational, 251
irresolvable, 222
Isḥāq ibn ᶜAlī al-Ruhāwī. *See* al-Ruhāwī
Isḥāq ibn Ḥunayn, xxxv–xxxix, xliv, 194, 196
ispaghula, 122, 216
Isṭafan, xxxviii. *See* Stephanus
isṭaqis, 135–36, 139–40, 214. *See also* element, στοιχεῖον
al-istifrāghāt al-ṭabīᶜīyah, 44

jāsī ṣalb, 108, 219
Jāsiyūs. *See* Gesius
jaundice, 29, 109, 178, 255
jawāmiᶜ, xxxii–xxxiii, xxxvi, xliii, 219
jejunum, 45, 235
Jirjis ibn Tādrus, lxi –lxii
jism, 136, 219
John of Alexandria, xl–xlii, liv, 193, 196; commentary on *The Medical Sects*, xli, xlviii, 203–7
John Philoponus, xxxvii–xli, liv, 139, 196
John the Grammarian, xxxvii–xxxviii, xl–xli, l, 196
joined, to be joined, 219, 236
joint, 243
joy, 18, 112, 115–16, 242
judgment, to exercise judgment upon, 150, 222
jumlah, pl. *jumal*, xxxiii, xxxvi, lxi, 219
Junayd ibn Kūnj ibn Junayd, lxii

khabar, 16
khandīqūn, 21, 225
khiyār, 21
Kibbutzei Galenos, liii
kind, 235, 251
knee, 36, 229
knitting, 125, 248
knowledge, 57, 231, 238–39
kunnāsh, xxii, 198

lack, 252
lambdoid suture, 80–81, 248
lancing, 43
lank, of hair, 71, 230
large, 246
large, to become larger, 239, 246
largeness, of sound, 168
Latin, liii
laws, 244
layers of eye, 88, 209–11
lean, leanness, 97–98, 245
legs, 52, 73, 77, 80, 228, 232; diseases of, 43, 67, 104, 121
lens, 86–89, 209–11, 219, 228
Leonidas of Alexandria, 208
leprosy, 178, 218
lesion, 28–29, 41, 109, 125, 244
less, 245
lettuce, 162
Leucippus, 138–39, 190–91, 196
Leufastus, 206
level, 227
liberal arts, xxiii– xxiv
licorice, 122, 214
life, 223
ligament, 77, 227
light, of hair color, 223, 251
lightness, 166–67, 169, 224
linseed, 24, 36, 216, 246
lips, 120, 122, 234
liquidity, 176
liver, l, 25, 77, 113, 128, 246; temperament, 83, 90–94; diseases of, 29, 67, 104–5, 121, 124
location of disease, 28
logic xxiii, xxv, 32–33, 46, 57, 132, 151, 192, 200, 251
logos, 193
loose, to loosen, of bowels, 222, 231
loud, 239
love, as principle, 138, 140, 191, 220
luminosity, luminous, 236
lungs, l, 40, 107, 109, 182, 227; temperament, 102–3, 179
lymphatic glands, 119

*mabda*ʾ, 139–40, 215
magnificence of pulse, 239
magnitude, 244
Magnus of Nisibis, xli
Maḥmūd I, Sultan, lx–lxi, 48
makeup, 217
males, procreation of, 95, 227, 255
man, 134, 215; definition of, 51, 54–55
managing, managing functions. *See* deliberation
Manaseus, 207. *See also* Mnaseas
manifest, manifest entities, manifest to sense, 33–34, 237
Manṣūr [?] ibn Muḥammad ibn al-Zakī al-ᶜAsqalānī al-Ṭabīb, lx
manuscripts, of Arabic Galen, xix; of Alexandrian epitomes, liv–lxiii; Aya Sofya 3588 (**S**), xxxvii–xxxviii, liv, lx, lxiv–lxviii; Aya Sofya 3609 (**A**), lv, lxi–lxii; Aya Sofya 3557, xlviii; Aya Sofya 3593, xxiv, 140; Aya Sofya 3701, xxiv; Azhar Ṭibb 79, lvi; Berlin Staats. Or. Oct. 1122, lvi; British Library Arund. Or. 17, xxv, xli; British Library Or. 9202 (**R**), liv, lvi, lix, lxv–lxvii; British Library Or. Add. 23407 (**D**) liv–lvi, lxii–lxv; Chester Beatty 4001, lvi; Escurial 797, 799, 848, xxv; Escurial 849, lvii; Fatih 3538 (**F**), liv–lv, lix–lx, lxiv–lxvi; Fatih 3539, lvii, lix; Florence Laurent. 235, xxiv; Haidarabad Āṣaf. Falsafa 371, lvii; Haidarabad Āṣaf. ṭibb 44, lvii, lxiii; Istanbul University Library A3559, lviii; Istanbul University Library A4712, xxiv; Istanbul University Library A6158 (**U**), liv, lvi, lxii–lxiii, lxvi; Köprülü Fazıl Ahmed Paşa 961, liv; Madrid Bib. Nac. 130, xxv; Manisa 1759 (**M**), lv, lx–lxii, lxiv–lxviii; Manisa 1772, xxv, xlviii; Paris 2860, xxiv; Princeton Garrett 1075, xxiv; Princeton Garrett 1G, lvii; Princeton NS 1532, lviii; Saray Ahmet III 2110, xxiv; Saray Ahmet III 2043, lvii; Tehran Majlis 521, 3974, xxiv; Tehran Majlis 3999, lvii; Tehran Majlis 6036, lviii, lxiii; Tehran Majlis 6037, liv; Tehran Majlis 6400, xxv; Tehran Sanā 3190, lviii; Tehran Univ. 4914, lviii, lxiii; Tehran Univ. 5217, lviii; Tehran Univ. Med. Fac. 167, lviii; Tehran Univ. Med. Fac 291, xxv; Wellcome Hist. Or. 62, lviii; Yeni Cami 1179 (**Y**), lvi, lxii, lxiv–lxvii
mariḍ, 74, 249
Marinus, xxxvi–xxxix, xli–xlii, 197
Marinus of Neapolis, xlii
marrow, 136, 249
massage, 189, 226
mathematicals, 140
mathematics, xxiii
matter, 20, 125, 159–60, 249; of experience, 15–16; prime, 139–40, 154, 254
meaning, 240
means, 230
meat, 21, 116
medicine, 236; definition, xxxi, 11, 53–54, 57, 64–65; divisions, xlix, 3, 7–9; physics, xxv, 138; subject matter, 3, 193; theory and practice, 5, 133–34
medlar, 14–15, 230
melancholia, melancholic, melancholic humor, 66, 232, 254. *See also* black bile
melilot, sweet, 36
Melissus of Samos, 137, 139, 153, 170, 197
membrane, 89, 236, 241
memory, 79, 81–83, 222, 227; Empiricists on, xlvii, 12–13, 27, 37–38
Menedotus, 205. *See* Menodotus
Menemachus of Aphrodisias, 10, 197, 207–8
Menodotus of Nicomedia, 10, 197, 203, 205, 208
menses, menstruation, menstrual blood, 46, 108, 237, 244
method of instruction, 50–56, 134, 217, 223, 227, 231, 237
Methodist school, xxiii, xlv, xlvii, 10, 139, 192, 197–201, 206, 208, 223; Galen on, xx, xlviii, 12, 40; members, 189, 203; and other schools, 37–38, 43–44; practice, 41, 45; theories, 35–40, 47, 171
mḥalḥal, Syriac, 47
middle age, 9, 23, 157
milk, 248
millet, 21, 172, 218
mimesis, 31
Mimomachus, 207. *See* Menemachus
mind, 12, 33–34, 139, 239
Minīthānūs al-Qadīm. *See* Mnesitheus
miqdār, 73, 244
misqām, 74, 231
Mīthinānūs al-Thānī. *See* Mnaseas
to mix, be mixed, 249–50
mixture, lxix, 3, 136, 249
mizāj, lxix, 148, 171, 249
Mnaseas, 10, 197–98, 207–8

mnemonic, 8, 69–70, 110, 227
Mnesitheus of Athens, 10, 198, 203, 206, 208
moderate, moderation, 18, 66, 71–73, 78, 97, 103, 112, 114, 133, 238, 245; of flavors, 48
moisture, moist, to moisten, 22, 95, 228; of body and organs, 42, 85, 87–88, 90–94, 100, 175–77; as cause, 82, 99, 103, 108–9, 122, 160; causes of, 20–21, 115; quality, 18–19, 68, 71, 165–70
moment, 255
monism, 137, 197–98
More, Thomas, *Utopia*, li
morning, 240
mortal, 250
motion, 17, 26, 74, 137, 221; non-natural cause, 111, 113; voluntary, 150, 176
motor, motor functions, 78, 221
mouth, 47, 243
to move, moving, 221, 253
mubāsharah, 16, 216
Muḥammad ibn Nāṣir al-Dīn ibn ᶜAlī ibn Muḥammad al-Bulyānī al-Shāfiᶜī al-Azharī, 131
mukhalkhal, 47, 224
mukhtaṣar, xxxii
mulberry, 43, 218, 227
multiple, 247
muscle, xxvii, 78, 238
mushāhadah, 16, 234
music, 5
mustard, 21, 116
mutaḥarrik, 137, 221
mutakhalkhal, 47, 224
mutilation, 122

al-Nadīm. *See* Ibn al-Nadīm
nail, 136, 224, 237
names, 13; Greek, lxvi
narrow, narrowness, 236
Nāṣir al-Dīn ibn ᶜAlī ibn Muḥammad al-Bulyānī al-Shāfiᶜī al-Azharī, 131
nāṭiʾatān, 42, 251
natural, 3, 9, 14, 132, 134, 236
natural faculty, functions, 4, 22, 25–26, 83, 103
nature, lxix, 1, 7, 34, 125, 133, 147, 180, 191, 236; and imitation, 14–17; Rationalists on, 18, 27, 32
nature, contrary to, lxix. *See* unnatural
Nausiphanes, 190
naval battles. *See* Mellisus of Samos
nawᶜ, 160, 253
necessarily, necessary, 235–36
necessary causes, 17–18, 115–16, 230. *See also* causes; non-naturals
neck, 79, 119, 229, 240
needle, 214
negation, 231
neither, 249. *See also* neutral
nerves, xxvii, 40, 51, 73, 77, 79–81, 123, 125, 136, 238. *See also* optic nerve; sensory nerve
neutral, neutral states, 11, 27, 193; body, sign, and cause, 57–64, 69, 110–11
Nicholas, 189
Nicomachus, 205
nightshade, 24, 31
Nīqulāʾus, Nīqālāʾus, xxxvi, xxxix, 189
nonbeing, 140, 148, 238
non-manifest entities, 33–34
non-naturals, six, 18n43, 21n47, 27n64, 111–12, 115n152–16
north, northerly, 19, 234
nose, 47, 83
nosebleed, 14–16, 228
to nourish, be nourished, 240–41. *See also* nutrition
now, 215. *See* present
number, 73, 237; as element, 140, 199; of organs, 68, 71, 115, 118–19, 122–23
nutrition, nutriments, 8–9, 128, 177, 182, 191, 240; from elements, 52–53, 157; from plants and animals, 158, 163, 176
Nutton, Vivian, xl

oblong, 237
observation, 103, 228, 234; Empiricists on, xlvi, 14, 27, 38
occupations, 18, 116
occurrence, 215, 221
odors. *See* scents
old, old age, old men, 60, 157, 234; regimen, 8–9, 128; treatment, 23, 26, 42
olfactory, 234. *See also* scents
Olimpicus, 207
olive oil, 43
Olympiacus of Milesia, 208
omen, 14, 16–17, 229
one, 254
On Nature, lii, 170
to ooze, 228
opening, 242
opinion, 83, 227
opposite, opposition, to oppose, 68, 225, 235, 245

optic, optic nerve, 89, 166, 209–10, 216, 229
optimal, 243
orbit of eye, 220
order, 227; in instruction, 51–53
Oreibasius. *See* Oribasius
organic, 215
organs, xxxi, lxix, 4, 8, 14–15, 17, 24, 64, 66, 99, 101, 238–39; anatomy, 33, 43, 46–47, 77, 100, 118, 178–79, 181; ancients on, liii, 35, 40, 44, 47, 133, 164; diagnosis, 76, 106–8, 124; disorders of, 23, 29, 61–62, 71–72, 74, 114–15, 122; functional, 215, 239; humors and tissues, 3, 52–53, 103–4, 123, 132, 163, 185. *See also* tissues
Oribasius of Pergamon, xxii, 190, 197–98
orientation, 255
orifice, 19, 118, 252
original, 214
to originate from, 251
orthography, lxvi
outcome, 239
outer, 237
outside, 224
to overcome, 241
oxymel, 121, 143, 231

pain, 108, 254; and elements, 141–42, 144–45, 151; symptoms and diagnosis by, 24, 28, 74, 93, 106, 109
palate, 83, 223
Palladius of Alexandria, xxxvii–xxxviii, xlii, xlviii, 198
pallor, 71, 217
palpable, palpable qualities, 71, 248
Parmenides of Elea, 137–39, 170, 187–88, 197–98
particulars, 64–65, 218
parts, 218; and elements, 136, 147
passion, 148, 243
passive, 243
passive primary qualities. *See* qualities
past, 250
pathology, 11
patient, 249
pattern, 249
Paul of Aegina, xxii, 198
peculiar to, 224
pellitory, 162, 237
to penetrate, 252
pepper, 21, 116, 162, 243
to perceive, 226
perceptibly, 221
Pergamon, xx, 192
periodic, periodic fever, 22, 253
peritoneum, 124
permanence, 217
Persius, xli. *See* Marinus
perspiration, 14, 36, 45, 113, 127–28
pharmacology, pharmacy technique, xxiv, 195, 200
Philinus of Cos, 10, 191, 198, 204, 208
Philon, 207
Philoponus. *See* John Philoponus; John of Alexandria; John the Grammarian, Yaḥyā al-Naḥwī
philosophical, philosophical terminology, lxx, 243
philosophy, xxv, xxxiv, xlvi, 192
Philotimus, Philotemus of Cos, 10, 199, 205
phlegm, lii, 24, 101–2, 108, 179, 217; in blood, 93–94, 182; humor, 135, 156–57, 175–77; purgation, 178, 180, 183–84
phlegmatic, 217
phlegmatic fever, 30–31
phlegmona, 24–25, 108, 243
physical, 236
physical therapy, 120
physician, xix, xxi–xxii, xxxv, xlvii, lxii, lxviii, 1, 3, 5, 15, 17, 23, 70, 131, 151, 236
physicists. *See* Ionian physicists
physics, xxx, 138, 176–77
physiology, xxv, xlv, 11
pia mater, 88, 241
pillars of philosophy, 191, 199
pitch, 174, 230
place, 23, 26–27, 38, 255
placenta, 89, 209–10, 234
placidity, 92, 241
plants, 53, 157, 163, 250
Plato, 31, 51, 133–34, 140, 189, 192–93, 199–200
Plato the Physician, 187, 199
plausible, 245, 250
pleasure, 145, 248
plethora, 25–28, 43, 250
pleurisy, 28, 219
plowman, 20
plumpness, 237
pneuma. *See* πνεῦμα
Pneumatic school, xx, 31, 171, 190, 199, 201
point, 252
poison. *See* venom, venomous animals
pores, 46, 231
Posidonius, 199
to posit, 254

position, 71, 73, 255; defect of, 68, 105, 108, 115, 118–20, 123–24, 127
possibility, 148
posterior, 214; of brain, 81–82
posture, 55, 245
potency, of drugs, 246
potential, potentiality, potentially, 174, 246
poultice, 12, 26, 36, 43, 127, 236
poultry, as food, 129
powder, 136, 142
powers, of drugs, 32
practical, practice, xx, xxiii, 3, 7–8, 58, 133, 240
Praxagoras of Cos, 10, 193, 199, 205, 208
preceding causes, 8, 25, 27, 32, 40, 43, 230
predisposed, 238
predominant, to predominate, 241
premise, 244
present, present moment or time, 58–59, 62–63, 66, 68, 76, 222, 255
preservation, preserving, 4, 111, 222
Presocratics. *See* Ionian physicists
to prevent, 220
to prick, 241
prickly, 251
primary, 215
primary qualities. *See* qualities
prime matter. *See* matter
principal, 227
principal organs. *See* organs, brain, liver, heart, testicles
principles, 132, 138–40, 155, 160, 214–15, 227
prior cause, 230
Proclus, xlii, 208
to procreate, 255
procreation, 95, 227, 255
Prodicus, 170, 199
profession, 20, 23, 26, 31, 235
prognostic, prognostication, 69–70, 110, 251
proper, 220
property, 224
prophylaxis, xxv, xxix, li, 8–9, 54, 126, 128, 222, 244
protruding, protrusion, 251
protuberance, protuberant, 67, 86, 224
proximate, 244
psychic, psychic faculty, 4, 22, 25–26, 133, 147, 252; psychic spirit, 82
Ptolemies, xx
puberty, 229
pubic, 240; hair, 95–96
pulse, xxv, xxviii, xxxi, 4, 25, 28, 90–92, 198, 251
pungency, pungent, 28, 169, 220–21
pure, 224
purgation, to purge, 12, 179–81, 232, 242
purgatives, 9, 179–80, 183–85, 232
purslane, 31, 216
pus, 125, 246
putrefaction, 27–29, 116, 126, 239
putrid diseases, fever, 31, 92, 94, 239
pylorus, 45, 217
Pythagoras, Pythagoreans, 140, 188, 193, 195, 199

qandīṭūn. See *khandīqūn*
al-Qifṭī. *See* Ibn al-Qifṭī
qithāʾ, 21
qiyās, 10, 144, 246
qualities, liii, 51, 149, 159–61, 172, 219, 247; as causes, 19, 22, 112; classes, 68, 71, 110, 148, 159, 166–170, 181
quantity, 22, 51, 73, 112, 128, 161, 239, 247
quartan fever, 126, 227
to quench, 231, 241
question, 132, 237
quickly, 231
quick-wittedness, 82, 231, 243
quince, 14–15, 36, 231

rabies, 23, 29, 41–42, 247
rain, 19, 21
rakhw manfūk, 108
rank, 244
rapidity, 223, 231
rarefaction, to be rarefied, liii, 47, 140, 154, 169, 224
to be removed, 229
rational, rationalist, 251
Rationalist school, xx, xlv–xlvi, xlix, 10–13, 28, 193, 199, 203, 205, 208, 246; and Empiricists, 23–27, 30–34; and Methodists, xlvii, 37–38, 43–44; methods of, 18, 33, 40
al-Rāzī, Abū Bakr, Rhazes, xxi, xli, 200
reality, to grasp, 222
reason, 136, 239
receptive, 244
rectum, 45, 246, 250
recuperation, 8–9, 128, 252
red, reddish, redness, 223, 235; symptom, 15, 24, 109
red bile, 179
reduction of dislocations, 9
to refute, 220, 242
regeneration, 122–23

regimen, 8–9, 11, 35, 128, 225
relation, 51, 236, 251
relaxation, to relax, 25, 36, 228
to release, 237
to remember, 222, 227
remote, 216
removal, to remove surgically, 214, 224
repletion, 247
repulsion, to repel, 25, 46, 127, 226, 244
resolution, 83, 222, 238
resolvent, 222
respiration, 252
rest, to rest, 9, 18, 20, 111, 113, 115, 120, 231, 248
to restore, 228
to restrain, restraining, 244, 250
to result, 220
retarding, 216
retention, retained, 22, 220, 250; of breath, 120, 222; effects, 18, 46–47, 79, 112, 115, 178; Methodists on, 36, 44
retentive faculty, 26
retina, 89, 209–11
to return, 228
Rhazes. *See* Rāzī
rib, 91, 236
rice, 21
riding, 129
riwāyah, 16
roots and heads, of elements, 138, 155
rope, 28
rose oil, 43, 226, 254
rough, of voice, 224
roughening, 122
roughness, 224; in organs, 71, 109, 115, 119, 122; as quality, 166–67, 169
round, 226
ruddiness, 71, 98, 223. *See also* red
al-Ruhāwī, Ibrāhīm ibn Sulaymān ibn Ḥakīm, lx
al-Ruhāwī, Isḥāq ibn ᶜAlī, *Adab al-Ṭabīb*, xxii, 200
runny, 229
rupture, 36, 120, 124, 242. *See also* hernia
rūsāṭan, *rūsātaq* fish, 116
ruṭūbāt, of eye, 88, 209–11, 228

sabab, lxix, 140, 230
sagittal suture, 81, 232
Sahl ibn ᶜAbd Allāh. *See* Ibn Ṭūmā
Ṣāᶜid al-Andalusī, *Kitāb ṭabaqāt al-umam*, xxxv–xxxvi, xxxviii
ṣalābah, 108, 235
Sallām ibn Ṣāliḥ ibn Khiḍr ibn Ibrāhīm, lx–lxi, 48, 186
salt water, 21
saltiness, salty, 48, 169, 182, 250
salve, 36, 234
al-Samarqandī, ᶜUthmān ibn ᶜAlī ibn Muḥammad, lix
sant tree, 43
to sate, 247
scammony, 126, 183
scarification, 233
scents, 109, 166, 169, 229
schole, xxi–xxii
scholia. *See* glosses and scholia
science, scientific method, xxv, xlv, 239
scirrhus, 24, 231
sclera, 89, 209–10, 232, 236
scorpion, 28
scraping, to scrape, 122, 222
scribes, xlv, xlviii, lix, lxi–lxiii, lxv–lxvi, 30, 48, 107, 137
scrofula, 119, 123, 225
scum, 178, 229
Scythians, 27
seafaring, 20, 116
seam, 232
seasons, xlviii, 18–19, 23, 26, 179, 255; Methodists on, 35, 38, 40
secondary, 218
secretions, 44
sects, medical, xlv, xlix, 9–10, 12–13, 242
seeds, 139, 142–43, 157
to seek, 237
seen, 216
semen, 95–96, 100, 122–23, 177, 182, 250–51. *See also* sperm
to send away, 228
not having sensation, 221
sensation, to sense, 221; and elements, 136–37, 141–43, 146–51, 153, 174, 177; medical uses, 12, 26, 34, 37–38, 40, 74, 78, 128
sense organ, 52, 221
sensible, sensible entities, sensory. *See* sensation
sensory nerve, 80
sentience, 141
separation, 28, 242
septic fever, 27
Serapion of Alexandria, 10, 200, 204, 208
servants of organs, 77
Sextus Empiricus, 10, 200, 204, 208
sexual intercourse, xxxv, 46, 95–96, 216,

219; climax, 95, 242
Sezgin, Fuat, xxii, xxxii, liv, lix
shabb yamānī. *See* alum
Shafarᶜām, Palestine, lx, 48
shaggy, of hair, 71, 229
shape, 71, 148–49, 234; defective, 115, 118, 120–21
sharḥ, xxxii
sharpness, 169, 220; in enemas, 121
shawkah miṣrīyah, 43
Sheikh, 3. *See also* Ibn al-Ṭayyib, Abūʾl-Faraj
Shimshon ben Shlomo, liii, 200
shingles, 14
shortness, of breath, 236
sick, 231
sickly, sickliness, 73–76, 118, 231, 249
side, 251
signs, xlvii, l–li, 8, 11, 35, 239; classes, 54, 57–63, 69–71, 75, 110; of diseases, 25, 28, 41, 46, 78–84; of temperament, 90, 92–99, 101–3
similar, 233
simple, 216
Simplicius of Cilicia, 138–39
singular, 242
size, 148–49, 244; of organs, 68, 72, 115
Skepticism, 200
skewerlike suture, 80–81, 231
skill, 221
skin, 14, 55, 98–99
skull, 89, 244
Slavs, 27, 42
sleep, 9, 253; as cause, 18, 21–22, 111, 113, 115
slight, 255
slowness, 216
sluggishness, 25, 28, 92, 247
smallness, 168, 235
smell, smelled, 78, 83, 166, 170, 234. *See also* scents
smoke, 139, 215
smooth, smoothness, to smooth, 250; as quality, 166–67, 169; in organs, 71, 109, 115, 119, 122
Smyrna, xx, 192
snake, 28, 218
snow, 21
Socrates, 51, 140, 200
soft, softness, 46, 228, 235, 249; diagnosis by, 97, 106, 109; quality, 71, 166–69
solidity, solidified, 166–67, 176, 219, 249
soothsaying, soothsayer, 14, 16–17, 247
Soranus of Ephesus, xlvii, 10–11, 200, 207–8
Sorenson Legacy Foundation, James and Beverley Sorenson, xvi
soul, 9, 34, 147, 150, 188, 252; accidents of, 18, 112–13, 115–16
sound, 166, 168–69, 235
sour, 48, 169, 182, 223
source, 214. *See also* principle
south, 19
space, 243
spear, 28, 41–42
species, xlvi, 11, 22, 48, 51, 54, 72, 91, 154, 156–57, 160, 253
specific, specifically, 224
speck, 244
sperm, 100, 251; ducts, 77, 255. *See also* semen
spinal cord, 52, 79–80, 91, 251
spine, 121, 235
spirit, 8, 229. *See also* πνεῦμα
spittle, 217
spleen, 40, 178, 237
sponges, spongy, 47, 224
spreading apart, 216
spring, season, 227; temperament, 18–19, 23, 26, 157
sprinkled, 215
squamous sutures, 80–81, 245
square, 148, 227
stability, 176, 218
stagnant, 214
standard, 244
state, 17–18, 57, 72, 161, 214, 223
steel, 220
Stephanus of Athens or Alexandria, xxxv–xxxviii, xl, xlii, 1, 189, 200–201
sticky, 248
stitching, 9
Stoicism, Stoics, 34, 57, 140, 159, 171–72, 190, 199, 201
stomach, l, 40, 128, 250; diseases of, 43, 101, 104–5, 109, 113, 121, 124, 178
stone, 220, 222; bladder, 36, 108, 121; thrown, 18, 28, 111
straight, of hair, 230
straightening, straightness, 120, 246
strength, 233, 246
strenuous, 240
stretching, 108, 124
strife, 138, 140, 191, 241
strong, 246
structure, 111, 127, 217; of organs, 71, 105, 115, 118, 225, 229
student, 239
to study, 222

subject, 255; of medicine, 3, 132
subordinate to, 225
substance, 17, 19, 21, 33–34, 48, 51, 148, 161, 168, 170, 220; in division, 53–55; alteration of, 161, 166, 171
substratum, 255
subtle, subtlety, 48, 168, 248
Suda, xxxix
to suffer, suffering, 141–42, 214
suitable, suitability, 248, 255
Süleymaniye Library, lxi
sulfur, sulfurous, 21, 174, 246
Summaria Alexandrinorum, xxxii. *See* Alexandrian epitomes
summaries, xxiii, 186. *See also* epitome
summer, 21, 26, 60, 68, 157, 235; temperament, 18–19, 23
sun, 234
sunken, of eyes, 225
superfluities, 74, 83, 113, 243; evacuation, 126–27
ṣūrah, 160, 235
surface, 231
surfeit, 247
surgery, li, 5, 9, 35, 120–23, 193, 235, 239
surrounding, 223
suture, 80–81, 225
sweat, 238. *See* perspiration
sweet, sweetness, 48, 169, 182, 222–23
swelling, 14–15, 41, 108, 121, 125, 235, 254; treatment of, 23–25, 36, 44, 47, 127. *See also* edema; erysipelas; hematoma; phlegmona
swollen, 254
sword, 18, 28, 41–42, 111, 232
syllabus, medical, xix–xxxi
syllogism, xxiii, xlvii, 10, 12, 151, 154, 246, 255
symmetry, symmetrical, 232; of organs, 66–68, 71
symphysis, 125, 248
symptoms, xxvii, xxxi, li, 4, 8–9, 27–29, 106, 125, 226, 238; medical sects on, xlvi, 23, 35
synthesis, 50–56, 163, 229
Syriac, xxxix, xliv, 47, 194–96

ṭabaqāt, 88, 209–11, 236
tactile, 166–67, 169–70, 248
taghdhiyah, 128, 240
tajribah, 144, 218
talkhīṣ, xxxii, 248
taqwiyah, 128
tar, 245–46; tar water, 21
taste, to taste, tasted, tastable, 78, 83, 166, 170, 227; tasteless, 48, 217. *See also* flavors
tear, to tear, 224, 253
tears, 36, 123, 125, 226
teeth, 123, 136
temperament, tempering, xxv, xxxi, lxix, l, 8, 18–19, 48, 95, 157, 249; ancients on, 35, 145, 147, 171–72; and elements, 137, 140, 142–43, 149, 151, 169, 173, 176; immoderate, 24, 108, 114, 117, 125–27, 133; moderate, 9, 17, 71–72, 82; nine kinds, 23, 84, 134; of organs, 66, 83–85, 90–99, 101–3; and treatment, 23, 179, 184
tenderness, 249
terminology, medical, lxix
tertian fever, 30–31, 126, 240
Tessalus de Roma, 207. *See* Thessalus of Tralles
testicles, l, 77, 95–96, 120, 215
texture, of eye, 87
Thales of Miletus, 137–38, 154, 156, 164–65, 193, 201
Thāwdhusiyūs, xxxviii. *See* Theodosius
Themison of Laodicea, xlvii, 10, 197, 201, 206–8
Themistius, 172
Theodosius, xxxvii–xxxix, xlii, 1, 201
theory, theoretical, xxiii, xxxi, xlvi, xlix, liii, 4, 45, 133, 239; divisions of, 3, 7
therapy, therapeutics, li, 1, 8–9, 35, 153, 193, 219; in curriculum, xxv, xxviii
theriac, 29, 41, 126, 217
Thersites, lxviii, 67, 201
Thessalus of Tralles, xlvii, 10, 38, 197, 201, 207–8
thick, 241
thigh, 14–15, 47, 121, 242
thimār, xxxii, liv
thin, thinness, 71, 166–67, 230
to think, 243
thirst, 101–2, 239
thought, 79, 81–83, 150, 243
throat, 107, 170
throbbing, 30
time, 110, 255
timidity, 91–92, 218
tissues, liii, lxix, 4, 52, 179, 233, 239; health and disease of, 66, 71–73, 106, 114–15, 117, 125, 140; and humors, 53, 132, 136, 157, 163, 174. *See also* homoeomerous
titles, of books, 133, 170, 217, 240

tongue, 40, 122, 166, 170, 248
tool, 215
tooth, 253
topic, 240
torpor, 25, 242
touch, 24, 78, 83, 166, 226, 248, 254
trade. *See* art; profession
tragacanth, gum, 122, 247
transfer, xlvii, 17, 253
transformation, to transform, 148, 154, 223, 241, 245, 253
transformative cause, 111, 230, 253
transformative faculty, 26
transition, 14–16, 253
translation, xliv–xlv, 194
transmutation, transmuted, liii, 153–55, 164, 223
transporting, 127
treatise, 246
treatment, xxxi, xlv, xlviii, 23–24, 70, 114–15, 126–27, 151–52, 161, 227; and causes, 12, 41–42; Empiricists on, xlvi, 15, 27, 38; Rationalists on, xlvi, 18, 23, 28
tree, 157, 233
triangle, 148, 218
trunk, 52, 77, 104, 235
tuberculosis, 119
turbidity, 247
to twist, 240

ulcer, 107, 122, 244
umūr kullīyah, xxxi
unchanging, 221
unconsciousness, 46
understanding, 238
unequally, 232
uniform, 232
unitary, 254
universal, 247
unknowable, 219
unmoving, 221
unnatural matter, states, things, 3–4, 9, 19, 27, 107–8, 110, 121, 132, 182, 186, 224, 236
unprofitable, 252
ᶜunṣur, 135–36, 240
urine, xlii, 113, 121, 189, 217; blockage, 36, 105; ducts, 104; unnatural, 45, 107–8, 125
urinoscopy, xxxi, 25
uses, 252
usual, usually, of health, 54, 59–63, 66, 247
ᶜUthmān ibn ᶜAlī. *See* al-Samarqandī
uvea, 86, 89, 211
uvula, 43, 248

vapor, 93, 115, 215
veins, xxvii, 12, 15, 40, 51, 73, 77, 89, 100, 123, 238. *See also* blood vessels; arteries
venom, venomous animals, 28–29, 41, 231
venous, 238
ventricles of brain, 166, 170, 216
vertebrae, 91, 243
vices, 81
to fall victim to, 245
vinegar, 21, 43, 143, 224
violent character, 92, 220
vipers, 28, 214
virtues, 81
visible, visible qualities, 71, 169, 216
vision, 16, 78, 83, 106, 166–67, 170, 216
vital faculty, 4, 25, 133, 223, 246
vitreous humor, 88, 209–10, 228–29
vivesection, 33, 191
voice, 102–3, 235
void, 225
volition, volitional incidence, 15–17, 25
voluntary, voluntary functions, 14, 83, 229
vomit, vomiting, 14, 45–46, 246

wakefulness, 9, 255; as cause, 18, 21–22, 111, 113, 115
walking, 129
waram, *wārim*, 108, 254
warming, to warm, 25, 46, 230
water, 12, 23, 51, 179, 181, 250; ancients on, 138, 153–60, 164, 188, 193, 201; coldness of, 14, 21, 46, 116; element, 52, 134–37, 146, 168, 174; and warmth, 43, 161–62
weak, weakness, 26, 236
weather, 23, 253
well formed, 221, 234
west, 19
what it is, 249
wheat, 143, 223
which thing it is, 215
white, whiteness, 166, 168–69, 217
white, of eye. *See* sclera
whole and parts, 51, 54, 218
why it is, 248
wind, 41, 157, 229
windpipe, 102, 107, 122, 245
wine, 14, 21, 23, 178, 225, 233; as treatment, 43, 116, 129
winter, 21, 26, 60, 157, 233; temperament, 18–19, 23
womb, 40, 122, 175, 177, 182, 228

wood, 139, 173, 224
wool, 47
worms, liii, 108, 119, 134, 226
worry, 18, 115, 253
wrestling, 116

Xenophanes of Colophon, 137–38, 154, 156, 201

al-Yabrūdī, Abū'l-Faraj Jirjis ibn Yūḥannā ibn Sahl, lxii, 195, 202
Yaḥyā al-Naḥwī, xxxvi–xxxviii, xl–xli; *ikhtiṣār* or *talkhīṣ* of *The Medical Sects*, xxv, xlviii, liv, 203. *See also* John of Alexandria; John Philoponus; John the Grammarian
yearning, 218
yellow, 235
yellow bile, 93, 175–80, 249; as fundamental humor, 135, 156; purgation, 126, 182–84; and swelling, 24, 108. *See also* bile
youth, 9, 42, 157
Ypocras, Ypocrates de Cho, 205. *See* Hippocrates

Zeno, Emperor, xliii
Zeno the Herophilean, 189
zonula, 211, 237, 240

Greek

ἄγνωστος, 219
ἄγριος, 230
ἀγρυπνία, 232
ἀγχίνοια, 231, 243
ἀδαίρετος, 218
ἄδηκτος, 248
ἄδηλος, 224, 237
ἀδύνατος, 223
ἀειπάθεια, 74
ἀήρ, 253
ἀθροῦς, 226
αἷμα, 226
αἵματος ῥύσις ἐν ῥινῶν, 228
αἱμορραγία, 242
αἵρεσις, 10, 242
αἰσθάνεσθαι, αἰσθητικός, αἰσθητός, 221
αἴσθησις, 221, 226
αἰτία, 140, 230
αἴτιος, 230
αἰώρα, 229
ἀκάνθης Αἰγυπτίας, 43
ἀκαταληψία, 226
ἀκίνητος, 137, 221
ἀκοή, 232
ἀκόλουθος, 251
ἀκολουθῶν, 217
ἀκούειν, 231
ἀκουστός, 232
ἄκρος, ἄκρως, 241
ἀλγεῖν, 141, 214, 254
ἄλγημα, 254
ἀληθής, 222
κατὰ τὴν ἀλήθειαν, 222
ἀλλοιοῦν, ἀλλοιοῦσθαι, ἀλλοίωσις, 241
ἀλλοιωτικός, 253
ἄλογος, 251
ἄλυπος, 247
ἁμαρτάνειν, 215
ἄμεμπτος, 217
ἀμετάβλητος, 217, 221
ἄμετρος, 238
ἄμικτος, 224
ἄμυγμα, 125, 253
ἀμυδρός, 217
ἀμυχή, 233
ἀμφιβληστροειδὴς χιτών, 210
ἀμφισβητεῖν, 252
ἀναγχαῖος, 236
ἐξ ἀνάγκης, 230, 235–36
ἀναθρεπτικός, 128, 252–53
ἀναίσθητος, 221
ἀναληπτικός, 128, 252–53
ἀναλογισμός, 33, 214
ἀνάλογος, 233
ἀνάλυσις, 51, 222
ἀναμιγνέναι, 224
ἀναμνηστικός, 227
ἀναπνοή, 252
ἀνασῴζειν, 228
ἀνατομή, 33, 233

ἀνατρέπεσθαι, 242
ἀνδράχνη, 216
ἀνεπίκριτος, 222
ἀνεύρυνσις, 249
ἀνήρ, 215
ἀνθρωπός, ἀνθρώπινος, 215
ἀνία, 235
ἀνορθοωτικός, 228
ἀντίθεσις, 235
ἀντιπράττειν, 225, 245
ἀντίσπασις, 251
ἀντίτυπος, 226
ἀξιοῦν, 251
ἀξίωμα, 219, 244
ἀξιώματα, 238
ἀπαθής, 221
ἄπειρος, 253
ἁπλοῦς, 216, 228, 230
τὸ ἁπλῶς, 237
ἀποδεικνύειν, 251
ἀπόδειξις, 216
ἀπολειπόμενος, 252
ἀποστροφή, 219
ἀποφαίνειν, 230
ἀποχεῖν, 234
ἀποχωρῆσαι, 219
ἅπτειν, 248
ἅπτεσθαι, 254
ἁπτός, 248
ἀραιός, 222, 224, 231
ἀραιοῦσθαι, 224
ἀραχνοειδὴς χιτών, 211, 237, 240
ἀρετή, 223, 243
ἀριθμός, 237
ἄριστος, 233, 243
ἀρκεῖν, 247
ὑπὸ ταῖς ἄρκτοις, 234
ἀρρενογόνος, 227, 255
ἄρροστος, 236
ἀρτηρία, 238, 245
ἀρτηριώδης, 233
ἀρχή, 138–40, 214, 217, 227
ἀσθένεια, ἀσθενής, 236
ἀσύμμετρος, 232
ἀσύστατος, 218
ἀτελής, 247
ἄτεχνος, 227
ἀτμός, 215
ἄτολμος, 218
ἄτομος, 219
ἀτοπία, 234
ἄτοπος, 234
ἄτριχος, 230
αὐγοειδής, 236
αὐξάνειν, 243
αὐξάνεσθαι, 230
αὔξησις, 253
αὐτόματος, 252
αὐτός, 224
αὐτοσκεδιός, 14, 229
αὐτοψία, 16, 216
αὐχήν, 229, 240
ἀφαίρεσις, 214
πρὸς τὴν ἀφήν, 248
ἀφροδίσια, 216
ἀφροδισιαστικός, 219
ἄχρηστος, 252
ἀχώριστος, 230, 242

ἐν βάθει, 225
βαρύνειν, 247
βαρύς, 218
βελόνη, 214
βήξ, 231
βία, 233
βίαιος, 244
βιβλίον, 245
βίος, 223
βλάβη, 235, 236
βλαισός, 240
βλάπτειν, 235
βοήθημα, 227
βραγχώδης, 215
βραδυτής, 216
βραχίων, 238
βραχύς, 245, 255

γάλα, 248
γαργαρεών, 248
γαστήρ, 216, 250
γένεσις, 246–47
γέννησις, 140
γεννητικός, 240
γένος, 219
γεῦσις, 227
γευστός, 227
γεύειν, 227
γῆ, 214
ἐν ἐσχάτῳ γήρα, 234
γίγνεσθαι, 246–47
γιγνωσκεῖν, 238
γλαυκός, 229, 230
γλίσχρος, 241, 248
γλυκύς, 222–23
γλῶττα, 248
γνώμη, 227

γνωρίζειν, 226
γνώρισμα, 217, 226, 239
γνῶσις, 238
γόνιμος, 255
γόνυ, 229
γραμματικός, 251
γρηγορεῖν, 255
γυμνάσιον, 221

δάκνειν, 253
δάκτυλος, 214
δασύτης, 233
δένδρον, 233
δηλῶν, 226
δηλωτικός, 226
διὰ πυρός, 247
διαβάλλειν, 252
διαγνωστικός, 226
διάθεσις, 44, 223
πληθωρικὰ διάθεσις, 250
διαίρεσις, 222, 245
διαίτημα, 225
διακρινεῖν, 224
διάκρισις, 242
διακριτικός, 242
διαλεκτική, 251
διαλεκτικὴ θεορία, 33
διάλυσις, 51, 222
διάπτυξις, 242
διάρροια, 237
διατείνασθαι, 252
διατείνειν, 249
διαυγής, 229
διαφέρειν, 225, 242
διαφθείρασθαι, 242
διαφορία, 225
διαχώρημα, 216, 218
διαχώρησις, 243
διδασκαλία, 50, 239
διϊστάναι, 225, 242
διοριστέον, 238
διφώδης, 239
διχῶς, 236
δογματικήν, 12
δογματικοί, 10, 246
δόξα, 227
δρᾶν, 243
τὸ δρᾶν ἄλληλα, 214
δραστικός, 243
δριμύς, 220
δύναμις, 246
δυνατός, 250
δυσαλλοίωτος, 246
δυσκρασία, 232
δυσμαθία, 239
δύσπαυστος, 216
δυστύπωτος, 236
δύσψυκτος, 215

ἔαρ, 227
ἐαρινός, 227
ἐγρήγορσις, 255
ἔδεσμα, 237
ἔθος, 240
εἶδος, 224, 235, 253
ὡς εἰκότα γιγνωσκότων, 245
ἐιςαγόμενος, 225
εἰσπνοή, 251
ἐκδριμέων, 220
ἐκκαθαίρειν, 232
ἐκκάθαρσις, 232
ἐκκενοῦν, 252
ἐκκρινεῖν, 242
ἐκκρίνεσθαι, 224
ἐκκρινόμενος, 242
ἐκλεαίνειν, 250
ἐκόπτειν, ἐκόπτεσθαι, 245
ἐκροή, 252
ἐκτός, 224
ἐκτροπή, 242
ἕλκειν, 218, 245
ἕλκος, 244
ἔμετος, 246
ἐμπειρία, 144, 218
ἐμπειρικοί, 10, 218
ἐμπιπλάμενος, 242
ἔμπροσθεν, 244
ἔμφαξις, 231
ἕν, 254
ἐναντία, 225
ἐναντίος, 225
ἐναργής, 217
ἔνδεια, 252
ἔνδειξις, 33, 226
ἐνέργεια, 218, 221, 240, 243
ἐνκέφαλος, 226
ἐννοηματικός, 55, 228
ἐννοητικός, 228
ἐντερόν ἐν ὀσχέῳ, 242, 246
ἐντός, 225
ἕνωσις, 254
ἐξάπλωσις, 216
ἐξάρθρημα, 225
ἐξήγεσις, 243
ἕξις, 215, 255
ἔξοθεν, 224

ἐξοχή, 251
ἐπανορθωτικός, 230, 253
ἕπεσθαι, 217
ἐπέχειν, 220, 230, 245, 250
ἐπιγράφειν, 217, 240
ἐπιδεκτικός, 244
ἐπίδεσις, 227
ἐπικρατεῖν, 241
ἐπιλέγειν, 220
ἐπιλησμοσύνη, 251
ἐπιλογισμός, 33, 214
ἐπιπεπλέγμενος, 229
ἐπιπεφυκώς, 89, 210, 214
ἐπιστήμη, 239
ἐπιτήδευμα, 235
ἐργάζεσθαι, 220
ἔργον, 243
ἐρέβινθος, 223
ἐρευνᾶν, 237, 252
ἕρπης, 14, 253
ἐρυγή, 219
ἐρυθρός, 223
ἐρυσαφέλη, 108
ἐρυσίπελας, 14, 223
ἐσθίειν, 214
ἕτερος, xlv, 214, 225
ἑτοῖμος, 218, 241
εὐαλλοίωτος, 241
εὐκατάπαυστος, 232
εὐκίνητος, 231
εὔκρατος, 238
εὐμετάβολος, 223
εὕρεσις, 224, 252
εὑρίσκειν, 252, 254
εὐρυθμία, 225
εὔρυθμος, 221, 234
εὐρύτης, 254
εὔτονος, 246
εὐτύπωτος, 231, 236
εὔψυκτος, 215
ἐφέλκειν, 218
ἔχιδνα, 214, 218

ζῷον, 223

ἡγεμονικός, 225, 232
ἡδονή, 248
ἦθος, 225
ἡλικία, 232
ἧπαρ, 246
ἡσυχία, 231

θεραπεία, 219, 227
θεραπευεῖν, 234
θεραπευτική, 215
θεραπευτικός, 227
θερμαίνειν, 230
θερμός, 159, 221
τὸ θερμόν, 159, 221
θερμότερος, 221
θερμότης, 221
θέρος, 235
θέσις, 255
θεωρία, 251
θηλυγόνος, 215, 255
θηριακή, 217
θηρίον, 223, 230
θνητός, 250
θρέψις, 237
θρίξ, 233
θυμικός, 220
θυμός, 241
θώραξ, 235

ἰᾶσθαι, 235
ἰατρικός, 236
ἰατρός, 236
ἴδιος, 224
ἱδρώς, 238
ἰκτεριᾶν, 255
ἰλύς, 229
ἴσος, 232, 247
ἱστορία, 16, 229
ἰσχόμενος, 220
ἰχώρ, 235

καθαίροντα, 232
καθαρός, 247–48
καθόλου, 239, 247
κακία, 215, 223
κάλλος, 219, 221
κανών, 244
καρδία, 245
κάταγμα, 247
κατακριτικόν, 230
κατάληψις, 34, 120, 222, 226
καταμήνια, 244
καταπαύειν, 231
καταπλάσμα, 236
καταρκτικόν, 8
κατασκευή, 217, 229, 253
κατηγοεῖν, 237
καῦμα, 248
καυστικός, 221
κεκρᾶσθαι, 214
τὸ κενόν, 225

κενός, 243
κένωσις, 242
κεραννύναι, 224, 249
κεράνυσθαι, 250
κέρας, 244
κερατοειδὴς χιτών, 210, 237, 244
κεφαλή, 227
κήλη, 246
κινεῖσθαι, 221
κίνησις, 221
κινούμενος, 137, 221
κλύσμα, 222
κνήμη, 232
κοιλότης, 220, 245
κοινός, 233, 255
κοινότης, 35, 219
κομιδή, 224
κόπρος, 228
τὸ κοῦφον, 224
κρᾶσις, 145, 171, 249
κρατεραύχην, 240
κραῦρον, 253
κριθή, 234
κρίσις, 215, 222
κροτάφιος, 245
κρυσταλλοειδὲς ὑγρόν, 211, 219, 228
κρύσταλλος, 219
κτῆσις, 243
κύαμος, 216
τὰς κυριῶς ἀρχάς, 138
κυρσός, 220
κύστις, 249
κύων λυττῶν, 247
κῶλον, 245

λαμβδοειδής, 248
λαμπρός, 236
λαμπρότης, 236
λασιώτατος, 234
τὸ λεῖον, 250
λειότης, 250
λείπειν, 252
λεπτομερής, 229, 248
λεπτὸν ἔντερον, 250
τὸ λεπτόν, 228
λεπτός, 252
λεπτυνόμενος, 229, 248
τὸ λευκόν, 217
λευκός, 217
λίθος, 220, 222
λινόσπερμον, 216
λογικοί, 10, 246
λογικός, 12, 251
λόγος, 144, 239–40, 246
λοιπός, 216
λοξός, 240
λουτρόν, 223
λύπη, 221
λύσις, 242
λύττα, 247

μαλακός, 249
μαλακότης, τό μαλακόν, 249
τὸ μανόν, 230
μεγάλη ἐνέργεια, 221, 240
μέγας, 227, 239, 246
μέγεθος, 244, 246
μεθοδικοί, 10, 223
μέθοδος, 223, 227, 237
μείζων, 239
μελαγχολῶν, 232
μέλας, 232, 247
μέλλων, 215
μέρος, 218, 238
μέσος, 254
μέσπιλον, 230
μεταβαίνειν, 253
μετάβασις, 14, 253
μεταβολή, 223
μεταξύ, 254
μεταπίπτειν, 245
μετοχέτευσις, 253
μετρίως, 246
μῆλον, 231
μήρηρ, 242
μήτρος, 228
μικρός, μικρότης, 235
μικτός, 229
μιμητικός, 14, 233
μιμνήσκειν, 227
μῖξις, 145
μνήμη, 222, 227
μνημονευτικός, 12, 222, 227
μόλις, 240
μόνιμος, 218
μόριον, 218, 238
μοχθηρός, 228
μυελός, 249
μῦς, 238

νεῖκος, 241
νεῦρον, 238
νευρώδης, τὸ νευρῶδες, 238
νῆστις, 235
νικᾶν, 241, 245
νικᾶσθαι, 245

νοσεῖν, 249
νοσερός, 249
νόσημα, 249
νοσώδης, 74, 214, 231, 249
νοῦς, 239
νῦν, 215, 222, 255
νωτιαῖος, 251

ξανθός, 223, 234, 235, 251
ξηραίνειν, 219
τὸ ξηρόν, 255
ξηρός, ξηρότης, ξηρότερος, 255
ξίφος, 232
ξύειν, 222

ὀβελιαῖος, 231, 232
ὄγκος, 254
οἴδημα, 108, 215, 235, 254
οἰκεῖος, 224
οἴκησις, 217
οἶνος, 233
ὀλίγος, 235
ὀλιγόσαρκος, 245
ὀλιγόσπερμος, 250
ὁμοιομερής, 233, 239
ὅμοιος, 14, 233
ὁμολογεῖν, 233, 255
ὄνειρος, 253
ὄντως, 222
ὄνυξ, 224, 237
ὀξυθυμία, 241
ὀξυρεγμιώδης, 219, 223
ὀξύς, 220
ὄξος, 224
ὄπισθεν, 225
τὸ ὁποίων, 249
ὁρατός, 216
ὀργανικός, 215, 239
ὄργανον, 214, 215, 238, 239
ὀργή, 241
ὀρέγειν, 234
ὅρος, 51, 220, 222, 228, 237
ὄρχεις, 215
ὀστέον, 239
ὀσφραίνεσθαι, ὄσφρησις, ὀσφρητής, 234
ἐντερόν ἐν ὀσχέῳ, 242, 246
οὐδέτερος, 74, 231, 249
οὖλος, 219
οὖρον, 217
οὖς, 214
οὐσία, 220, 249
οὐσιώδης, 55, 220
ὀφέλλειν, 252
ὀφθαλμός, 240
ὄψις, 216

παθητικός, 221, 243
πάθος, 214, 238, 239
παῖς, 234
παλαιοί, 244
παλινδρόμισις, 228
τὸ διὰ παντός, 226
παράδειγμα, 249
παραπέμπειν, 228
πάσχειν, 141, 214, 220, 243
πάσχοντες, 221
τὸ παχύ, 247
παχυμερής, 241
παχύς, 241, 247
παχύσπερμος, 250
παχύτερος, 241
πεπερασμένος, 253
πεπονθώς, 239
περιέχειν, περιέχων, 223
περίπτοσις, 14, 255
περιττεύειν, 230
περίττωμα, 243
πέττειν, 251, 253
ἀφ' αὐτοὺ πεφυκώς, 251
πηλικότης, 244
πηλός, 237
πιμελή, 233
πιμελώδης, 232, 233
πίνειν, 233
πινόμενος, 233
πληθωρικὰ διάθεσις, 250
πλείονος γίγνεσθαι, 246
πλείων, 242
πλευρά, 236
πληθωρικὰ διάθεσις, 250
πλῆθος, 246, 247
πλημμελής, 228
πνεῦμα, 159, 171, 190, 199, 201, 229, 252
πνεύμων, 227
ποιητικός, 220, 230, 243
ποιότης, 247
ποιότητα δεδεγμένον, 247
πολύς, 246–47, 255
πολύσπερμος, 250
πόνος, 241
πόρος, 218, 231
ποσός, 244
ποσότης, 237, 247
ποτόν, 233
πούς, 228
πρᾶγμα, 214, 220

πρακτικός, 221
προγίγνεσθαι, 250
προγνοστικός, 251
πρόγνωσις, 230, 252
προηγούμενος, 8, 230
προκατακριτικός, 230
προνεῖσθαι, 255
προπετής, 224, 251
προσγίγνεσθαι, 220
προσεχής, 244
πρόσθιος, 244
πρόσκαιρος, 255
προσπίπτειν, 220, 248
πρότασις, 244
πρότερος, 215
προφυλακτικός, 222, 244
πρῶτος, 242
πυκνός, τὸ πυκνόν, 247
πυκνότης, 254
πυλωρός, 217
πύον, 246
πῦρ, 253
πυρετός, 223
πυρός, 223

ῥαγοειδὴς χιτών, 211, 236, 240
ῥᾳδίως, 238
ῥαιβός, 240
ῥαφή, 225, 232
ῥευματίζεσθαι, 228
ῥευματιζόμενος, 226
ῥῆξις, 224, 242
ῥιζώματα, 138
ῥόδινος, 226, 254
ῥοώδης, 216, 228
ῥυπτικός, 219
ῥύσις, 216
ῥυτός, 232
ῥώμη, 246

σάρξ, 248
σαφῶς ὁρῶν, 240
σημαινόμενος, 240
σημεῖον, 226, 239
σηπεδονώδης, 239
σηπεδών, 239
σήπεσθαι, 239
σίαλος, 217
σιμός, 245
σκέλος, 232
σκῖρος, 235
σκίρρος, 231, 254
σκληρός, τὸ σκληρόν, 235, 249
σκληρὸς χιτῶν, 210, 232, 236
σκληρότερος, 235
σκληρότης, 235
σκοπεῖσθαι τῷ λόγῳ, 240
σκοπός, 241
σμικρότης, 235
σμίλη, 220
σόφισμα, 241
σπασθείς, 234
σπέρμα, 250–51
σπερματικὰ ἀγγεία, 255
σπερματικός, 250
σπλήν, 237
σπόνδυλος, 243
στάλασις, 250
στεγνόν, 222, 250
στέγνωσις, 222
στενός, στενότης, 236
στερεός, 219
στέρνον, 235
στεφάνη, στεφανιαῖος, 214
στοιχεῖον, lii, 135–36, 139–40, 163, 165, 213–14, 240
στοιχιώδης, 240
στόμα, 243
στόμιον, 218
στούφοντες, 43
στραφυλοειδὴς χιτών, 211, 236, 240
στρογγύλος, 226
στρυφνός, 239
στυπτηρία, 232
στύφειν, 244
σύγγραμμα, 246
συγκείμενος, 229
συγκεῖσθαι, 224
σύγκρασις, 219
συγκρινεῖν, 247
σύγκρισις, 219
μή συγχωρεῖν, 253
συζύγιος, 229
συλλογιστικός, 246, 255
συμβεβήκοτος, 238
συμμετρία, σύμμετρος, 232, 238
συμπληροῦν, 217
σύμπτωμα, 238
σύμφυσις, 248
σύμφυτος, 220, 241
σύμφωνος, 255
συνάγειν, 219
συναγογή, 229
συναλλοιοῦν, 223
σύνδεσμος, 227
συνεκτικός, 8, 230

συνέχεια, 242, 254
σύνθεσις, 51, 229
σύνθετος, 229
συνουσία, 219
σύνοψις, xxxiii, 219
συντείνειν, 249
συντετραίνειν, 233, 252
συντίθεσθαι, 220
σφάλμα, 215, 242
σφυγμός, 251
σχῆμα, 234, 246, 255
σχίσμα, 242
σῶμα, 44, 136, 215, 219
σωρός, 234

τάξις, 231, 252
τάσις, 249
τάχιστα, 230
τάχυς, 231
τελειότης, 218, 243
τελειοῦσθαι, 253
τέλος, 218, 241, 253
τετράγωνος, 227
τετράχυνθαι, 224
τέχνη, 235
τήρησις, 226, 228, 234
τηρητικήν, 12, 218
τιθέναι, 254
τιτρώσκειν, 241
τμητικός, 245
τοιοῦτος, 224
τομή, 245
τόπος, 238, 255
τράχυς, 245
τραχύτης, 224
τρέφειν, 240
τρέφεσθαι, 241
τρίγωνος, 218
τρῖψις, 226
τρόπος, 236
τροφή, 240
τρύξ, 225
τυγχάνειν, 220
τυραννικός, 237
τύρος, 218
τύχη, 238
τυχικός, 14, 238
τῷ πλέον, 232

ὑαλοειδὲς ὑγρόν, 210, 228
ὑαλοειδής, 229
ὑγιαίνων, 231
ὑγιεινός, 214, 234, 235
ὑγραίνειν, 228
ὑγρός, τὸ ὑγρόν, 228, 250
ὑγρόσπερμος, 250
ὑγρότερος, 228
ὑγρότης, 228
ὑδατώδης, 250
τὸ ὕδορ δείδειν, 242
ὕδωρ, 250
ὑδυτώδης, 235
ὕλη, 249, 254, 255
ὑπάγειν, 222, 237
ὑπάρχειν, 254
ὑπερκάθαρσις, 232, 242
ὑπηρετεῖν, 225
ὕπνος, 253
ὑπόξανθος, 83
ὑπόπυρρος, 223, 253
ὑποχόνδριον, 229, 233

φαίνεσθαι, φαινόμενος, 237
φάναι, 251
φαντασία, 225
φάρμακον, 227, 231
φαύλος, 228
φέρειν, 223
φθόνος, 221
φιλία, 220
φιλοσοφία, 243
φλέβες, 238
φλεβοτομεῖν, 243
φλεβώδης, 238
φλέγμα, 217
φλεγμαίνων, 254
φλεγματικός, 217
φλεγμονή, 108, 243, 254
φλέψ, 238
φόβος, 242
φοξότερος, 231
φορτικός, 138
φοῦν ἀπό, 250
φυλακτικός, 222
φυλάττειν, 222
φυσικός, 14, 236
κατὰ φύσιν, παρὰ φύσιν, 236
φύσις, 44, 224, 236
φυτόν, 250
φωνή, 235

χαίρειν, 221, 244
χαλᾶν, 228
χάλαρος, 228
χάλασις, 228, 231
χειμών, 233

χείρ, 255
χειρουργία, 9, 235, 239
χεῖσθαι, 247
χολή, χολῶν, 249
χονδρός, 241
χοριοειδὴς χιτῶν, 210, 234, 237
χρεία, 252
χρήσιμος, 252
χροιά, 248
χρόνος, 255
χυμός, 224
χώρα, 217
χωρίζειν, 225
χωρίον, 255

ψευδής, 247, 253
ψιλὰ τριχῶν, 233
ψυχή, 252
ψυχικός, 252
τὸ ψυχρόν, 215
ψυχρός, 215
ψυχρότερος, 215
ψυχρότης, 215

ὠθεῖν, 226
ὠοειδές, 217
ὠοειδὲς ὑγρόν, 211, 228
ὥρα, 255

About the Editor

JOHN WALBRIDGE is professor of Near Eastern languages and cultures at Indiana University Bloomington. He earned his PhD from Harvard University in 1983 and pursues research in Islamic philosophy and Islamic intellectual history with an emphasis on the cultural role of philosophy and science.

A Note on the Types

The English text of this book was set in BASKERVILLE, a typeface originally designed by John Baskerville (1706–1775), a British stonecutter, letter designer, typefounder, and printer. The Baskerville type is considered to be one of the first "transitional" faces—a deliberate move away from the "old style" of the Continental humanist printer. Its rounded letterforms presented a greater differentiation of thick and thin strokes, the serifs on the lowercase letters were more nearly horizontal, and the stress was nearer the vertical—all of which would later influence the "modern" style undertaken by Bodoni and Didot in the 1790s. Because of its high readability, particularly in long texts, the type was subsequently copied by all major typefoundries. (The original punches and matrices still survive today at Cambridge University Press.) This adaptation, designed by the Compugraphic Corporation in the 1960s, is a notable departure from other versions of the Baskerville typeface by its overall typographic evenness and lightness in color. To enhance its range, supplemental diacritics and ligatures were created in 1997 for exclusive use in this series.

The Arabic text was set in NASKH, designed by Thomas Milo (b. 1950), a pioneer of Arabic script research, typeface design, and smart font technology in the digital era. The Naskh calligraphic style arose in Baghdad during the tenth century and became very widespread and refined during the Ottoman period. It has been favored ever since for its clarity, elegance, and versatility. Milo designed and expanded this typeface during 1992–1995 at the request of Microsoft's Middle East Product Development Department and extended its typographic range even further in subsequent editions. Milo's designs pushed the existing typographic possibilities to their limits and led to the creation of a new generation of Arabic typefaces that allowed for a more authentic treatment of the script than had been possible since the advent of moveable type for Arabic.

BOOK DESIGN BY JONATHAN SALTZMAN

◆